W9-DFY-140

THE ELECTRONIC STRUCTURE
OF ATOMS

THE ELECTRONIC
STRUCTURE OF ATOMS

LEVENTE SZASZ
Department of Physics
Fordham University
Bronx, New York

A Wiley-Interscience Publication
JOHN WILEY & SONS, INC.
New York / Chichester / Brisbane / Toronto / Singapore

BIP-94

Copyright © 1992 by John Wiley & Sons, Inc.

Library of Congress Cataloging in Publication Data:

Szasz, Levente, 1931–
     The electronic structure of atoms / Levente Szasz.
          p. cm.

     "A Wiley-Interscience publication."
     Includes index.
     ISBN 0-471-54280-6
     1. Atomic structure.   2. Electronic structure.   3. Quantum chemistry.   I. Title.

     QC173.S975 1991
     539′.14—dc20   91-2551
                                                                            CIP

Printed in the United States of America

10 9 8 7 6 5 4 3 2 1

PREFACE

According to the mathematician Poincaré, scientific research is mostly a subconscious activity. My experience has been that the same is true for writing a book. It is for this reason that, having finished this volume on the electronic structure of atoms, I find it difficult to write down, in a systematic fashion, what the book is about. I will try to alleviate the difficulty by formulating a number of statements.

(A) *The goal and scope of the work*. The goal of this monograph, which is planned to consist of two volumes, is to present the quantum theory of the electronic structure of atoms and to explain what the electronic structure is like by presenting the main results of the theory. The monograph will consist of four parts. In this volume we present the first two parts. The first part deals with the Hartree–Fock model, and the second with the effective Hamiltonian theory for valence electrons. (The theory of electron correlation and the theory of density-functional formalism will be treated in a subsequent volume, along with some special topics that cut across the four main parts.)

In the background of the presentation is Dirac's formulation of quantum mechanics according to which for the solution of any quantum mechanical problem, we must select a complete set of commuting observables and construct eigenfunctions and eigenvalues for them. The central operator in every selection is the Hamiltonian; the other observables are various selections of angular momentum and spin operators. This monograph deals with the theory of the Hamiltonian; here the word "theory" must be understood in the broadest sense, including the construction of model Hamiltonians and the discussion of methods for the calculation of eigenfunctions and eigenvalues. The reason why we concentrate on the Hamiltonian is that, as we have discovered in the process of writing, most books published in the last 30 years on this topic are slanted in the direction of the theory of angular momentum and spin operators. On the other hand, most of the progress in the last 30 years has taken place in the area of Hamiltonian theory. Thus, there is a clear need for a treatise of this type.

(B) *For whom is the book written?* The book is written primarily for theoretical physicists interested in atomic structure theory. It is also meant for theoretical chemists specializing in quantum chemistry. It is potentially useful for students and scientists working in adjacent fields, that is, in molecular and solid-state physics, in molecular biology, quantum pharmacology, and so on. In fact, the book is potentially useful for any scientist whose work involves, at one time or another, a more than casual interest in atomic structure. We will establish "points of contact" with some of these fields in the book. However, the bulk of the work is written for theoretical physicists and chemists.

(C) *The treatment of numerical calculations.* It is a frequent occurrence to find a new publication the title of which sounds like theory but the work is, in fact, a compilation of numerical calculations. The reader will find nothing of the kind in this book. We are making an attempt to reestablish the primacy of theory. At the same time, the book contains a "first," which shows the importance of numerical calculations. This "first" is Chapter 5 and it is a systematic, critical evaluation of some of the numerical results obtained with the nonrelativistic and relativistic Hartree–Fock models. To the best of our knowledge, such an extensive evaluation has never been presented before in any book. Otherwise, calculations will be mentioned only as "representative calculations" serving to demonstrate the quality of the theory.

(D) *Novel features.* Apart from the discussions of Chapter 5, the main novelty is to treat relativistic atomic theory *on a par* with the nonrelativistic. Until quite recently, relativistic methods were viewed as the ultimate refinements to atomic structure theory and were treated in books as a kind of afterthought. In this work, we adopt the method of presenting the two approaches side by side.

The work contains a number of hitherto unpublished results that form integral parts of atomic structure theory. These are the detailed mathematical treatment of the "frozen core" and "extended frozen core" approximations (Secs. 6.3 and 6.4), the derivation showing the direct equivalence of Lagrangian multipliers and pseudopotentials (Sec. 7.2/C), the method for the elimination of many-electron projection operators in effective Hamiltonian theory (Sec. 8.4 and App. L), and the minor discovery that many of the Hamiltonian operators occurring in atomic structure theory are non-Hermitian (App. K).

(E) *The handling of the literature.* The difficult question in writing a book of this type is not what to include, but what to omit. We have found that if one wants to restrict oneself strictly to theory and some representative calculations, then approximately 100 references, apart from some important textbooks and handbooks, are sufficient to cover the topics that are handled in this volume. A larger number of references will probably be needed for the subsequent volume.

In presenting the work of any author, we have always tried to present the contents of the first paper of a series. We have found that, invariably, if an author writes a series of papers on any subject, the first paper is lucid and well structured, whereas the later additions have a tendency to get blurred.

The author's book on pseudopotentials must be mentioned (see References). Because we have written a book on pseudopotentials, those parts of this book that deal with this topic are abbreviated. Frequent references to the earlier book serve the purpose of making this book shorter, not as an advertisement for the earlier work.

(F) *Prerequisites.* The book starts on an elementary level, that is, Chapters 2 and 3 can be read by a college senior or a first-year graduate student having had some quantum physics. Over all, the prerequisite is two to three semesters of quantum mechanics, some electromagnetic theory, and some relativity theory besides the usual calculus courses. The atomic chapters in Eisberg and Resnick's *Quantum Physics* can be used as useful introductions to the book. The last chapters are leading the reader up to the point where research papers can be studied.

(G) *The "spirit" of the book.* It was Alexis de Tocqueville who wrote about the "spirit" of government as distinct from its constitutional form. It is meaningful to talk about the spirit of a book as distinct from its contents and structure. We have mentioned publications that are calculation-oriented. In this work, we have the opposite: it is strongly and exclusively theory-oriented. This means that our goal was to show that atomic structure theory is mathematically coherent and elegant and it is approaching a certain completeness. It means also that our goal was not only to present existing results, but also to stimulate the creation of new, bona fide scientific work.

LEVENTE SZASZ

CONTENTS

THE ELECTRONIC STRUCTURE OF ATOMS

CHAPTER 1

INTRODUCTION

1.1. THE ORGANIZATION OF ATOMIC STRUCTURE THEORY

For the framework of atomic structure theory, we adopt Dirac's description of how a theory based on quantum mechanics should be built up. The central task is to select a complete set of commuting observables, that is, Hermitian operators, the eigenvalues of which are the measurable quantities of the system. In the buildup of atomic theory, several different sets of operators are selected, each set representing a certain stage, or approximation, in the development of the theory. After each selection, methods for the calculation of the eigenfunctions and eigenvalues must be developed.

In atomic structure theory, the (not necessarily complete) set of operators always looks like this:

$$H, A, B, C, D, \dots . \qquad (1.1)$$

The central operator in every selection is the Hamiltonian operator, H. The operators $A, B, \dots$ are angular momentum and spin operators. Let us denote by L and L_z the orbital angular momentum and its z-component and by S and S_z the spin angular momentum and its z-component. Let J be the sum of L and S and J_z be its z-component. The symbols in the previous set stand for a particular selection of these operators. The selection is determined by the empirical data. We get different selections if we omit some of the data from consideration.

Strictly speaking, the selection of the angular momentum and spin operators is determined by the choice of H, because the other operators must commute with H and this requirement puts a condition on the selection. For

1

the choice of H, two main possibilities present themselves. H can be relativistic or nonrelativistic. These two possibilities mean different choices for the angular and spin operators. Which choice will match the empirical data?

It has been recognized, very early in the development of atomic structure theory, that much of the empirical data can be matched with LS coupling in which the good quantum numbers of the system are the quantum numbers associated with the operators L^2, S^2, L_z, and S_z. This implies the choice of a nonrelativistic Hamiltonian because that Hamiltonian commutes with these operators. If the fine structure of the atomic spectra is taken into account, then both the Hamiltonian and the corresponding set of angular and spin operators must be chosen differently. Again, if we want the doublet structure of the x-ray spectra to be explained we must introduce (jj) coupling either in the Dirac form or in the Breit–Pauli approximation. Either choice means different Hamiltonian operators.

What emerges from these considerations is that atomic structure theory can be divided into two main parts. One part deals with the central quantity, the Hamiltonian operator. The other part deals with the angular and spin operators. By taking a look at the literature, it is evident at once that the theory of angular and spin operators has been developed to a high degree of sophistication and completeness and there are numerous textbooks and monographs on the subject. Even the classic work of Slater, *Quantum Theory of Atomic Structure* (see References), is heavily slanted in the direction of the angular and spin operators. There is much less literature on the theory of the Hamiltonian and on the methods by which the eigenvalues of this operator (the observable energy levels of the atom) can be obtained.

It is for this reason that the object of this work is the theory of the Hamiltonian operator. The theory of the Hamiltonian will be developed in some detail; the theory of the angular and spin operators will be touched upon only to the extent that is necessary for the discussion of the Hamiltonian.

The organization of atomic structure theory is shown in Fig. 1.1. The starting point is the formalism of the quantum mechanics. The two main points of atomic theory are boxes 2 and 3. This book deals with box 3. In order to obtain results that can be compared with experiments, computational methods are needed: these are represented by box 4. Besides comparing the results with experiments, these results might also be used by scientists working in other fields, for example, in molecular and solid-state physics. These are indicated by boxes 5 and 6. As we said, this book deals with box 3; however, *points of contact* will be established with boxes 4, 5, and 6. This will mean discussions that have only marginal significance for atomic structure theory, but will be important as starting points for scientists interested in the topics indicated by boxes 4, 5, and 6.

We note that the scheme of Fig. 1.1 can be developed either nonrelativistically or relativistically. In this book we treat these two approaches side by

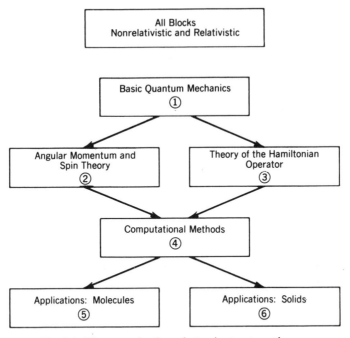

Fig. 1.1. The organization of atomic structure theory.

side, that is, in contrast to the earlier literature, we treat relativistic atomic theory *on a par* with the nonrelativistic.

The theory of the Hamiltonian operator, that is, the topic of this book, which we indicated by box 3, in Fig. 1.1, will be presented in three parts that form the first three parts of the two-volume work. As we see from the Table of Contents, the first part is the Hartree–Fock (HF) model. This is the center of the whole atomic structure theory and will be presented, in nonrelativistic form, in Chapters 2 and 3, and in relativistic form in Chapter 4. Chapter 5 contains a detailed, critical evaluation of some of the numerical results obtained with the HF model. Then, the second part deals with valence-electron theories, that is, with the (valence electron) effective Hamiltonian theory. This will be presented in nonrelativistic form in Chapters 6, 7, and 8, and in relativistic form in Chapter 9. The third part of the theory will be the theory of electron correlation and it will be presented in the second volume. The first three parts of the work form the bulk of atomic structure theory.

The fourth part of the book, to be presented in the second volume, will deal with density-functional formalism. It is not easy to fit density-functional formalism accurately into the scheme of Fig. 1.1. This scheme, as it is presented here, is a closed entity and it might be called wave-function formalism. With a certain degree of accuracy, density-functional formalism can be classified as a substitute for boxes 3 and 4 in the previous scheme. Thus, the discussion of this formalism, which can replace the Hamiltonian operator, should be a part of a book on atomic structure theory and as such it will be presented in the second volume.

In the discussions of this work, we will repeatedly use the word *model*. In the theory of atomic structure, as in any application of quantum mechanics, we are dealing with two basically different approaches. One of these we will call the ab initio approach or theory. The other will be called a *model*. It is important to clarify these concepts and we do so in the next section.

1.2. AB iNITIO VERSUS MODEL METHODS IN QUANTUM PHYSICS

The words *ab initio* are much misused. It is necessary to provide a clear definition for this concept. We will call a quantum mechanical calculation ab initio if the Hamiltonian operator, the eigenvalues and eigenfunctions of which we are seeking, is obtained by taking a classical energy expression and replacing the dynamical variables in this expression by the appropriate quantum mechanical operators. Typical such operators are the nonrelativistic Hamiltonian of Eq. (2.5) or the relativistic, Breit–Pauli Hamiltonian given by Eqs. (5.7)–(5.9). There are other such operators corresponding to other approximations.

Since the definition of a *model* is difficult, we solve the problem by simply calling everything that is not ab initio a model. Most models can be identified by a model Hamiltonian, H_M, which replaces an ab initio Hamiltonian, H. However, this definition excludes density-functional formalism because there is no Hamiltonian in that formalism. Thus, we will not attempt to give a definition more accurate than the previous one.

A moment of reflection will convince the reader that by this definition, the Hartree–Fock approach is not ab initio because in this approach, we are not seeking the eigenfunctions and eigenvalues of the nonrelativistic Hamiltonian operator, but only approximations to them. If we accept this definition, we are compelled to observe that at least 90% of atomic structure theory is based on models. This, of course, does not diminish the value of the theory. On the contrary, it enhances the value of models, especially the value of the HF model. In fact, from the four parts of this work, three are dealing with models and only one—electron correlation theory—deals with ab initio methods. The HF approach, the effective Hamiltonians discussed in Chapters 6–9, are all models; only the configuration interaction method and the method of correlated wave functions are ab initio procedures.

The distinction between these two kinds of approaches enables us to define and clarify two important concepts, which are the physical and mathematical approximations. Let us consider the ab initio Schroedinger equation

$$H\Psi = E\Psi. \tag{1.2}$$

Let us suppose that we replace this equation by the model equation

$$H_M\Phi = E_M\Phi. \tag{1.3}$$

Let us assume that we are able to solve both equations exactly. It is obvious that, generally,

$$E_M \neq E, \tag{1.4}$$

because by changing H into H_M, which usually means a simplification, we have also changed the eigenvalue. There is no such model that reproduces the ab initio equation exactly.[†] We call the approximation, inherent in the change from H to H_M, a physical approximation. The quality of this approximation may be judged, for example, by the quantity

$$\Delta E = E - E_M, \tag{1.5}$$

and, generally, we can say that

$$\Delta E \neq 0. \tag{1.6}$$

In contrast, we define a mathematical approximation as follows. Let us envisage a situation in which we construct approximate solutions of Eq. (1.2) by a variation method, for example, by applying the method of the superposition of configurations. Let E' be an energy obtained this way. Generally, for a many-electron system, we will have that

$$E' \neq E, \tag{1.7}$$

because of the slow convergence of the CI (configuration interaction) procedure. It is possible, however, that by improving our trial function, we can construct a variational solution for which

$$E' = E. \tag{1.8}$$

We call an approximation that deviates from the exact because of the inadequacy of the mathematical procedures a mathematical approximation.

[†]This statement must be qualified. What we mean is that it is not possible for a model to reproduce all energy levels and eigenfunctions of the exact Hamiltonian with arbitrary accuracy. It is often possible to reproduce a part of the energy spectrum at the expense of the eigenfunctions. A model Hamiltonian that would reproduce exactly the solutions of the ab initio Hamiltonian would be identical with it.

The error caused by such an approximation can be eliminated by improving the mathematical procedure. The error caused by a physical approximation cannot be eliminated because it is built into the model Hamiltonian, H_M.

In case of models, the prescription of Dirac for the buildup of a theory must be modified. Working with models means that one or more of the original ab initio operators are replaced by model operators. Very often, the operator that is replaced is the Hamiltonian. Then, the selection of commuting operators, which we displayed in Eq. (1.1), becomes

$$H_M, A, B, C, D, \ldots, \tag{1.9}$$

which is the same set as in Eq. (1.1) except H is replaced by H_M. All the rules laid down by Dirac for the construction of a quantum mechanical theory remain the same except that instead of using the ab initio operator, H, we use the model operator, H_M. Thus, for example, instead of demanding that the operators $A, B, C, \ldots$ commute with H, we demand that they commute with H_M. A typical example of this kind of procedure is how Slater built up the Hartree–Fock model for LS coupling. We present his method in detail in Sec. 3.1. We note also that while it is the Hamiltonian that is most often replaced by a model, it is by no means the only operator that can be modelized. In the method of configuration averages, the operators that are replaced by models are the angular momentum and spin operators. We present this method in detail in Sec. 3.3.

PART ONE

THE HARTREE – FOCK MODEL

CHAPTER 2

SINGLE-DETERMINANTAL
HARTREE-FOCK THEORY

2.6. KOOPMANS' THEOREM

The physical meaning of the orbital parameters in the HF equations; Koopmans' theorem.

2.7. DENSITY-MATRIX FORMULATION

Introduction of time-dependent HF equations; formulation of a formalism containing only the HF density operator and density matrix; demonstration of the technique in case of time-independent potentials.

2.1. THE IDEA OF INDEPENDENT PARTICLES: HARTREE'S WORK

Very shortly after the publication of the basic ideas of quantum mechanics, a general method for the determination of atomic structures was developed by Hartree.[1] This method, which we present in this section, can rightly be called the most important *model* of atomic physics.

Let us consider an atom of atomic number Z and containing N electrons. We know that the kth electron of this atom, in the presence of the $(N - 1)$ other electrons, has the Coulomb potential energy

$$V = \sum_{i=1}^{N} \frac{1}{r_{ki}} \qquad (k = 1, 2, \ldots, N), \qquad (2.1)$$

where r_{ki} is the distance between the ith and kth electrons. Thus, the potential energy depends on the position of the selected electron as well as on the positions of all other electrons.

In order to simplify the problem, Hartree suggested the following approximation. Let us assume that the potential in which the selected electron is moving depends only on the position of that electron, but not on the positions of the others. Of course, the other electrons will have an effect on the selected electron, but only in an *average fashion*. This assumption is called the independent-particle approximation.

From the concept of the independent-particle approximation, it follows that we can define a one-electron wave function ψ_k $(k = 1, \ldots, N)$ for each electron. Hartree assumed that the electric charge density ρ_k, associated with electron k, is given by

$$\rho_k = -|\psi_k|^2 \qquad (k = 1, 2, \ldots, N), \qquad (2.2)$$

where the sign reflects the negative charge of the electron.[†] Using this assumption, Hartree postulated that the potential in which the kth electron is moving is given by

$$V = -\frac{Z}{r} + \int \frac{\rho(\mathbf{r}') \, dv'}{|\mathbf{r} - \mathbf{r}'|}, \qquad (2.3)$$

[†]In most formulas, we will use atomic units. In Eq. (2.2), the electric charge is one atomic unit.

where ρ is the sum of the electron densities of all electrons minus the density of the kth electron. In this expression, the first term is the Coulomb potential of the nucleus and the second is the electrostatic potential of the $N-1$ other electrons. Accepting Hartree's assumption that the electric charge density of the individual electrons is given by Eq. (2.2), we see that the second term of the potential is simply the classical electrostatic potential of the $N-1$ electrons. In other words, our previous remark, that the other electrons will have an "average" effect on the selected electron, means that the selected electron is assumed to move in the classical electrostatic potential of the $N-1$ other electrons and that this potential is computed by assuming the validity of Eq. (2.2).

Finally, Hartree also postulated that for atoms with closed shells, the electron density should be spherically symmetric. With this assumption, the potential in which the electrons move also becomes spherically symmetric, that is, the approximation becomes a central field model.

On the basis of these ideas, Hartree wrote down the following Schroedinger equation for each electron in the atom:

$$\left\{ -\frac{1}{2}\Delta - \frac{Z}{r} + \int \frac{\rho(r')\,dv'}{|\mathbf{r} - \mathbf{r}'|} \right\}\psi_k = \varepsilon_k \psi_k \qquad (k = 1, 2, \ldots, N), \quad (2.4)$$

where ρ is defined as in Eq. (2.3), and Δ is the Laplacian operator.

The description that we have presented here contains all the basic physical ideas of the central-field, independent-particle model. These are the ideas that were postulated by Hartree in his first publications and their significance lies in the fact that they form the physical basis not only for the original Hartree method, but also for the Hartree–Fock model in its final formulation as well as for many other later developments in atomic structure theory.

Having surveyed the basic ideas of the model, we proceed now to the discussion of the mathematical formalism. In this section, the discussion is restricted to Hartree's original method, which is called, in the jargon of later developments, "self-consistent-field without exchange." The presentation of the full Hartree–Fock model begins in the next section.

We start with the nonrelativistic exact Hamiltonian of the system,

$$H = -\sum_{i=1}^{N} \frac{1}{2}\Delta_i - \sum_{i=1}^{N} \frac{Z}{r_i} + \frac{1}{2}\sum_{\substack{i,j=1 \\ (i \neq j)}}^{N} \frac{1}{r_{ij}}, \qquad (2.5)$$

and, according to the assumption of independent particles, we put the total wave function in the product form:

$$\Psi_T = \psi_1(1)\psi_2(2) \cdots \psi_N(N), \qquad (2.6)$$

where

$$\psi_k(k) = \psi_k(x_k y_k z_k), \tag{2.7}$$

that is, the k in the argument refers to the coordinates of the electron, and the subscript indicates the quantum state. By putting Ψ_T into the Schroedinger energy expression, we obtain the total energy of the atom,

$$
\begin{aligned}
E_T &= \int \Psi_T^* H \Psi_T \, dv \\
&= \sum_{i=1}^{N} \int \psi_i^*(i)\left(-\frac{1}{2}\Delta_i\right)\psi_i(i)\,dv_i + \sum_{i=1}^{N} \int \psi_i^*(i)\left(-\frac{Z}{r_i}\right)\psi_i(i)\,dv_i \\
&\quad + \frac{1}{2}\sum_{\substack{i,j=1 \\ (i\neq j)}}^{N} \int \psi_i^*(i)\psi_j^*(j)\frac{1}{r_{ij}}\psi_i(i)\psi_j(j)\,dv_i\,dv_j,
\end{aligned}
\tag{2.8}
$$

where dv means integration with respect to the coordinates of all electrons. Deriving E_T, we assumed that the one-electron functions are normalized, in which case Ψ_T is also normalized.

We now apply the variation principle to E_T. We vary the energy with respect to the one-electron wave functions and obtain N equations, one for each one-electron wave function. It is enough to vary the energy with respect to ψ_k^* ($k = 1, \ldots, N$) since the variation with respect to ψ_k yields the same results. The only subsidiary condition in the variation is that the one-electron functions are normalized. In this way, we obtain the Hartree equations:

$$\left\{-\frac{1}{2}\Delta - \frac{Z}{r} + \int \frac{\rho(\mathbf{r}')\,dv'}{|\mathbf{r}-\mathbf{r}'|} - \int \frac{|\psi_i(\mathbf{r}')|^2\,dv'}{|\mathbf{r}-\mathbf{r}'|}\right\}\psi_i = \varepsilon_i \psi_i \qquad (i = 1, 2, \ldots, N), \tag{2.9}$$

where now the ρ is the total electron density,

$$\rho = \sum_{j=1}^{N} |\psi_j|^2, \tag{2.10}$$

and ε_i is the one-electron energy.

Next, we establish the connection between the one-electron energies and the total energy. Multiplying the Hartree equations from the left by ψ_i^* and integrating, we obtain

$$\int \psi_i^* H_i \psi_i \, dv = \varepsilon_i \qquad (i = 1, 2, \ldots, N), \tag{2.11}$$

where H_i is the operator in the curly brackets of Eq. (2.9). Adding up the one-electron energies, we get

$$
\sum_{i=1}^{N} \varepsilon_i = \sum_{i=1}^{N} \int \psi_i^* H_i \psi_i \, dv
$$

$$
= \sum_{i=1}^{N} \int \psi_i^* \left(-\frac{1}{2} \Delta \right) \psi_i \, dv + \sum_{i=1}^{N} \int \psi_i^* \left(-\frac{Z}{r} \right) \psi_i \, dv
$$

$$
+ \sum_{i=1}^{N} \sum_{j=1}^{N} \int \frac{\psi_i^*(1)\psi_j^*(2)\psi_i(1)\psi_j(2)}{r_{12}} \, dv_1 \, dv_2
$$

$$
- \sum_{i=1}^{N} \int \frac{|\psi_i(1)|^2 |\psi_i(2)|^2}{r_{12}} \, dv_1 \, dv_2. \tag{2.12}
$$

The last term removes those terms from the electrostatic interaction in which $i = j$. Comparing Eq. (2.12) with Eq. (2.8), we see that they differ in a term that is ($\frac{1}{2}$ times) the electrostatic interaction. Thus, we obtain

$$
\sum_{i=1}^{N} \varepsilon_i = E_T + \frac{1}{2} \sum_{\substack{i,j=1 \\ (i \neq j)}}^{N} \varepsilon_{ij}, \tag{2.13}
$$

where

$$
\varepsilon_{ij} \equiv \int \frac{|\psi_i(1)|^2 |\psi_j(2)|^2}{r_{12}} \, dv_1 \, dv_2, \tag{2.14}
$$

and so

$$
E_T = \sum_{i=1}^{N} \varepsilon_i - \frac{1}{2} \sum_{\substack{i,j=1 \\ (i \neq j)}}^{N} \varepsilon_{ij}. \tag{2.15}
$$

The total energy of the atom is not equal to the sum of one-electron energies, but is given by Eq. (2.15).

The Hartree equations, as they are given by Eq. (2.9), are still equations in the three spatial coordinates of the electron. We reduce them now to one-dimensional equations. Originally, Hartree suggested that the charge density, which enters the electrostatic integral should be spherically symmetric for atoms with closed shells. Later, as the method was formulated for atoms with arbitrary electron configurations, it was postulated that the densities in Eq. (2.9) should be spherical averages. In order to explain this expression, first, let us assume that the one-electron wave functions are of a central field type, that is, let us put

$$\psi_i(x, y, z) = \frac{P_{n_i l_i}(r)}{r} Y_{l_i m_{l_i}}(\vartheta, \varphi), \tag{2.16}$$

where we have introduced spherical polar coordinates (r, ϑ, φ). The radial part of the wave function is

$$R_{nl} = \frac{P_{nl}}{r}, \tag{2.17}$$

where P_{nl} is normalized in the following way

$$\int_0^\infty P_{nl}^2(r) \, dr = 1, \tag{2.18}$$

and Y_{lm} is the normalized spherical harmonics.

Let us consider a point whose coordinates are (r, ϑ, φ). The electron density, arising from the kth electron that can be found in the volume dv, which is enclosing the space between r and $r + dr$, ϑ and $\vartheta + d\vartheta$, and φ and $\varphi + d\varphi$, is given by

$$|\psi_k(r, \vartheta, \varphi)|^2 \, dv = |\psi_k(r, \vartheta, \varphi)|^2 r^2 \, dr \sin \vartheta \, d\vartheta \, d\varphi. \tag{2.19}$$

As we see, this depends on r, ϑ, and φ. According to Hartree, we should replace this quantity by the spherical average, which is defined as follows. The electron density, arising from the kth electron, which can be found between the spheres with radii r and $r + dr$ regardless of the angular direction, is given by

$$\left(\int_\vartheta \int_\varphi |\psi_k(r, \vartheta, \varphi)|^2 \sin \vartheta \, d\vartheta \, d\varphi \right) r^2 \, dr$$

$$= \left(\int_{\vartheta=0}^\pi \int_{\varphi=0}^{2\pi} \frac{P_{n_k l_k}^2(r)}{r^2} |Y_{l_k m_{l_k}}(\vartheta\varphi)|^2 \sin \vartheta \, d\vartheta \, d\varphi \right) r^2 \, dr$$

$$= P_{n_k l_k}^2(r) \, dr \tag{2.20}$$

because of the normalization of the spherical harmonics.

According to Hartree's suggestion, we now replace the densities in Eq. (2.9) with their spherical averages, that is, we put

$$\rho = \sum_{j=1}^N P_{n_j l_j}^2, \tag{2.21}$$

and

$$|\psi_i|^2 = P_{n_i l_i}^2. \tag{2.22}$$

Next, we put the densities given by Eqs. (2.21) and (2.22) into the potentials of Eq. (2.9), write the Laplacian in terms of polar coordinates, and for the one-electron wave functions, we put the central-field expressions given by Eq. (2.16). The result is

$$\left\{ -\frac{1}{2}\frac{d^2}{dr^2} + \frac{l_i(l_i+1)}{2r^2} - \frac{Z}{r} + \sum_{\substack{j=1 \\ (j \neq i)}}^{N} \frac{1}{r}Y_0\big(n_jl_j, n_jl_j|r\big) \right\} P_{n_il_i}(r) = \varepsilon_{n_il_i} P_{n_il_i}(r)$$

$$(i = 1, 2, \ldots, N), \quad (2.23)$$

where

$$\frac{1}{r}Y_0(nl, nl|r) = \frac{1}{r}\int_0^r P_{nl}^2(r')\,dr' + \int_r^\infty \frac{P_{nl}^2(r')}{r'}\,dr'. \qquad (2.24)$$

Equation (2.23) is the Hartree equation in its final form. We note that the model is formulated here for an arbitrary atom, not just for atoms with closed shells.

The Pauli exclusion principle was satisfied by Hartree as follows. In writing down the total wave function Ψ_T, Eq. (2.6), Hartree satisfied the exclusion principle by assigning one wave function to each quantum-number quartet (n, l, m_l, m_s) that occurs in the atom. In view of the spherical averaging, this means that the radial parts of wave functions belonging to different m_l values will be identical; the same is true for the radial parts of wave functions belonging to different m_s values because the potential in the Hartree equation is independent of the spin. Thus, we obtain one Hartree equation like Eq. (2.23), for each nl pair that occurs in the electron configuration of the atom.

We note that the total potential in Eq. (2.23) is a central potential. The introduction of the spherical averages by Eqs. (2.21) and (2.22) makes the potential central; from this, it follows that the solutions of Eq. (2.9) will be central-field functions. This is consistent with the initial assumption of writing ψ_i in central-field form as we have done in Eq. (2.16).

For the total potential, Hartree introduced a special notation. Let

$$-\frac{Z_p(r)}{r} = -\frac{Z}{r} + \sum_{nl}\frac{1}{r}Y_0(nl, nl|r), \qquad (2.25)$$

where the summation is for all electrons. Using this notation, we get the Hartree equation in the form

$$\left\{ -\frac{1}{2}\frac{d^2}{dr^2} + \frac{l(l+1)}{2r^2} - \frac{Z_p(r)}{r} - \frac{1}{r}Y_0(nl, nl|r) \right\} P_{nl}(r) = \varepsilon_{nl} P_{nl}(r).$$

$$(2.26)$$

The potential $-Z_p/r$ is the Coulomb potential of the nucleus and the electrostatic potential of all electrons. As we see, each electron moves in the field consisting of $-Z_p/r$ corrected by the subtraction of the potential generated by itself.

We now turn to the problem of how the Hartree equation can be solved. As we see from Eq. (2.26), the Hartree equations are second-order, linear, integro-differential equations for the one-electron wave functions P_{nl} and the one-electron energies ε_{nl}. We have one such equation for each (nl) pair that occurs in the atom, that is, the number of equations will be considerably smaller than the number of electrons. However, the equations will be coupled; the potential occurring in the equation for P_{nl} will depend on all other one-electron functions (this can be seen clearly in Eq. (2.23)). Thus, the equations are linear only if we consider the total potential as a given function disregarding its dependence on the one-electron wave functions.

For the solution of these coupled integro-differential equations, Hartree designed the following procedure. First, a set of one-electron wave functions P_{nl} is chosen and the total potential that occurs in the equations is constructed. Then the equations are solved by numerical integration. In this step, the potentials are treated as fixed, prescribed functions, that is, the equations to be solved are ordinary second-order differential equations of the form

$$\left\{ -\frac{1}{2}\frac{d^2}{dr^2} + V_{\text{eff}}(r) \right\} P_{nl}(r) = \varepsilon_{nl} P_{nl}(r), \qquad (2.27)$$

where the effective potential is

$$V_{\text{eff}}(r) = \frac{l(l+1)}{2r^2} - \frac{Z_p(r)}{r} - \frac{1}{r} Y_0(nl, nl|r). \qquad (2.28)$$

By solving the equations, we have obtained a new set of one-electron functions and one-electron energies. The new set of one-electron functions will be different from the initial set that we used for the construction of the potentials. Let us denote the latter by $P_{nl}^{(1)}$ and the former by $P_{nl}^{(2)}$. Next, we use $P_{nl}^{(2)}$ for the construction of new potentials and solve the equations again, with these new potentials in them. In this way, we obtain a third set of wave functions $P_{nl}^{(3)}$ and a new set of one-electron energies. Thus, the sequence of steps in each cycle of iterations is as follows:

1. the initial potential field is constructed;
2. the equations are solved for one-electron functions;
3. the final potential is constructed with one-electron functions.

The correct solutions of the equations are obtained when the initial and first potential fields are identical. When this is accomplished, the field is said to be "self-consistent." For this reason, the model introduced by Hartree is called the method of "self-consistent fields."

The basic idea introduced by Hartree was to make the equations linear by treating the potential as a given function in each cycle of iteration. As to the actual calculations, they were considered very laborious when Hartree first developed the method. At that time (1928), only small desk calculators were available to scientists who wanted to do this kind of calculation. In view of this fact, it is fair to say that the calculation of a self-consistent field for a medium-sized atom was a formidable task. The lucidity of Hartree's original formulation, which we have tried to reproduce here, should not make us forget the great difficulties that Hartree had to overcome in such calculations. These difficulties were twofold: First, in each cycle of the iteration, a set of one-electron equations had to be solved by numerical integration; this was a laborious process in view of the small desk calculators that were then available. Second, the number of cycles was not necessarily small; the method as described here is not automatically convergent, that is, the potential obtained in the nth cycle is not necessarily appreciably better than the one obtained in the $(n - 1)$st cycle. However, it was shown by Hartree that convergence can eventually be achieved in the case of any atom.

The Discussion of a Representative Calculation: The Mercury Atom.

Hartree made self-consistent field calculations for a number of atoms and ions. We discuss here his results for the Mercury atom.[2] The atomic number of neutral Hg is $Z = 80$ and the electron configuration is

$$(1s)^2(2s)^2(2p)^6(3s)^2(3p)^6(3d)^{10}(4s)^2(4p)^6(4d)^{10}(4f)^{14}(5s)^2(5p)^6(5d)^{10}(6s)^2$$

We have 14 (nl) groups; thus, we have 14 Hartree equations. As we see, the number of equations is significantly less than the number of electrons; the figure is 14 *versus* 80. Hartree calculated the 14 one-electron wave functions and the 14 one-electron energies. We give the latter first (atomic units):

$$\varepsilon(1s) = -2776 \qquad \varepsilon(4p) = -19.45$$
$$\varepsilon(2s) = -462 \qquad \varepsilon(4d) = -12.89$$
$$\varepsilon(2p) = -446 \qquad \varepsilon(4f) = -4.19$$
$$\varepsilon(3s) = -108.4 \qquad \varepsilon(5s) = -3.46$$
$$\varepsilon(3p) = -100.3 \qquad \varepsilon(5p) = -2.30$$
$$\varepsilon(3d) = -85.25 \qquad \varepsilon(5d) = -0.46$$
$$\varepsilon(4s) = -23.03 \qquad \varepsilon(6s) = -0.235$$

From the one-electron wave functions, the total radial electron density can be computed by using the formula

$$D(r) = \sum_{(nl)} q(nl) P_{nl}^2(r), \qquad (2.29)$$

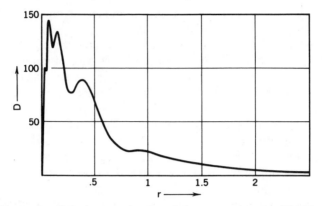

Fig. 2.1. The total radial electron density of the mercury atom (Hg, $Z = 80$). From Hartree's calculations without exchange.

where $q(nl)$ is the occupation member of the (nl) state and the summation is for all (nl) pairs. Figure 2.1 shows the total radial density.

The normalization of $D(r)$ is given by

$$\int_0^\infty D(r)\, dr = N, \tag{2.30}$$

which follows from Eq. (2.29) in view of the normalization of P_{nl},

$$\int_0^\infty P_{nl}^2(r)\, dr = 1. \tag{2.31}$$

In Fig. 2.2, we have the potential/ energy-level diagram. The curve in the figure represents

$$V = -\log\left(1 + \frac{Z_p}{r}\right), \tag{2.32}$$

that is, the ordinate scale on the left side of the diagram shows directly

$$-\left(1 + \frac{Z_p}{r}\right), \tag{2.33}$$

whereas the curve actually plotted gives V. The ordinate scale begins at -1 because for $r = \infty$, $Z_p/r \to 0$. The scale defined by Eq. (2.32) is linear for large r and logarithmic for small r. Indeed, for large r, where Z_p/r is small, we have

$$\log\left(1 + \frac{Z_p}{r}\right) \approx \frac{Z_p}{r}.$$

For small r, the 1 is negligible relative to Z_p/r, and we get

$$\log\left(1 + \frac{Z_p}{r}\right) \approx \log\frac{Z_p}{r}.$$

In the diagram, we also have the one-electron energies; more accurately, the quantity plotted is

$$-\log(1 + |\varepsilon_{nl}|),$$

which means that for small $|\varepsilon_{nl}|$, we have in the diagram $-|\varepsilon_{nl}|$, and for large $|\varepsilon_{nl}|$, we have the negative logarithm of this quantity. A look at the range of the ε_{nl}'s shows that this is the only way to present a potential/energy diagram that shows the optical as well as the x-ray spectrum. We call attention to the break in the abscissa at $r = 0.1$.

Now let us take a look at the diagram. The total radial electron density in Fig. 2.1 shows characteristic maxima and minima, demonstrating the *shell structure* of atoms. The curve shows that in the area in which a maximum occurs, there is a high concentration of electron density, that is, the electron density is concentrated in *shells*; a shell can be qualitatively defined as the volume between two spheres with radii that are slightly smaller and larger than the radius at which a maximum is located. This diagram can rightly be called the "picture" of the atom and it was in Hartree's calculations where such pictures first became available. The calculations with Hartree's method, as well as the calculations with the more accurate Hartree–Fock method, show that all atoms have electron densities of this general type with the characteristic maxima and minima. In the case of Hg, we can distinguish five shells (five maxima), indicating the position of the K, L, M, N, and O shells, that is, the groups of electrons with the principal quantum numbers $n = 1, 2, 3, 4,$ and 5.

In Fig. 2.2, we see the electrostatic potential inside the atom. The formula for this quantity is Eq. (2.25) and we note that this is not the actual potential in which the (nl) electron is moving because in order to obtain that, we would have to subtract from $-Z_p/r$ the potential generated by the (nl) election itself. This quantity is more like the potential that would be experienced by an infinitesimally small test charge inside the atom. Our diagram shows that this potential is an everywhere negative, attractive potential with an essentially Coulombic structure by which we mean a qualitative similarity to the Coulomb potential. For large r, the potential goes to zero, and for small r, the limit is $-Z/r$. The curve shows that despite the shielding

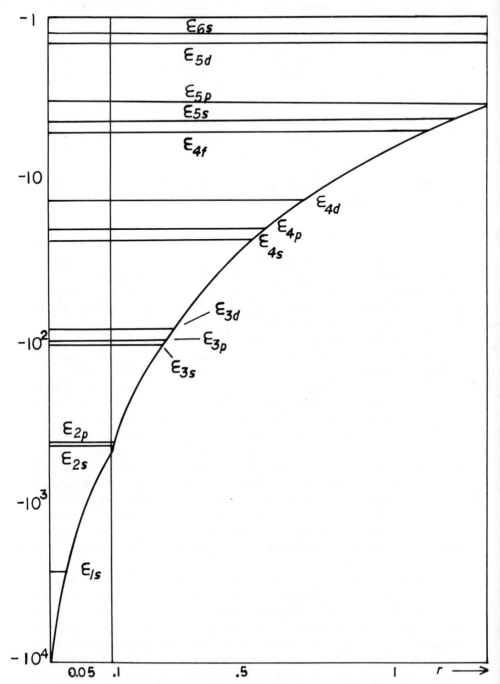

Fig. 2.2. The potential/energy diagram of the Hg atom. From Hartree's calculations without exchange.

provided by the second term of Eq. (2.25), which is the positive potential of the electrons, the total potential remains negative everywhere, demonstrating the dominant role of the nuclear potential.

In the diagram, we also have the one-electron energies. These encompass an enormous range of magnitudes, which is a consequence of the enormous range of the potential. The points of intersection between the one-electron energies and the potential are the classical turning points; the r value associated with these points increases as we go to higher energies. Of course, in quantum mechanics, the electron is not restricted to the area inside of the turning point, but still it can be said that the bulk of the one-electron density is always inside this point. As we go to higher energies, the associated one-electron functions expand more and more, corresponding to the increase in the r values of the turning points.

Another interesting feature is that the one-electron energies form clusters by which we mean that the energies of electrons with the same principal quantum number are grouped close to each other. Exceptions to this rule are the $\varepsilon(4f)$ and $\varepsilon(5d)$. If the energies are close to each other, the main maxima of the corresponding one-electron functions are also close to each other, which brings about the shell structure exhibited by the total radial density.

Summary of Hartree's Ideas. In his work, which can rightly be called the centerpiece of atomic structure theory, Hartree introduced a *model* whose basic ideas are as follows:

1. The electrons move independently from each other in an average potential. The total wave function is a product of one-electron functions.
2. The average potential is the nuclear potential plus the electrostatic potential of the electrons. This last is computed from the classical formula in which the total electron density is the sum of the one-electron densities.
3. The electric-charge density of an electron is equal to $-|\psi_k|^2$, where ψ_k is the one-electron wave function.
4. The resulting set of equations can be solved by iteration, where in each cycle of the iteration, the potential is considered as a given, fixed function.

We add a few words to idea 3. In his first papers, Hartree called the one-electron functions *orbitals*, emphasizing that this name is used in analogy to the Bohr theory. Strictly speaking, a wave function is not an orbital because the simultaneous knowledge of position and momentum is not possible in quantum mechanics. Nevertheless, this usage has become generally accepted especially by quantum chemists and will also be adopted in this book.

We observe also that Hartree talked about $-|\psi_k|^2$ as an electric-charge density and meant it quite literally. Again, strictly speaking, this quantity is a probability density and it does not mean that the electron is "smeared out" in space according to this density. In many publications, however, Hartree's picture has been accepted as being valid, and we adopt the same attitude in this book. In other words, we look at Fig. 2.1 as showing the "smeared out" electron density of the 80 electrons in the mercury atom whereas in reality the figure shows a probability distribution. We find this imagery very useful and it does not lead to inaccuracies as long as we remember the exact meaning of the densities.

2.2. SINGLE-DETERMINANTAL WAVE FUNCTIONS

Shortly after the introduction of Hartree's model, it was observed by Fock[3] that the total wave function suggested by Hartree does not have the symmetry required by quantum mechanics. According to the basic principles of quantum mechanics, the wave function of a fermion system must be antisymmetric in all particle coordinates. The Hartree wave function, as written in Eq. (2.6), does not satisfy the antisymmetry principle. This means that Ψ_T satisfies the Pauli exclusion principle only approximately; we have seen that Hartree has put only one electron into each state, but if the wave function is not antisymmetric, the compliance with the Pauli principle is only approximate.

In order to remedy this shortcoming of the theory, Fock suggested[3] that the total wave function should be written in a determinantal form, that is, into a form that satisfies the Pauli principle exactly. Fock has also shown that such a wave function leads to equations different from Hartree's. We present here Fock's theory.[3]

Let us consider an atom with atomic number Z and with N electrons. The Hamiltonian is

$$H = \sum_{i=1}^{N} \left\{ -\frac{1}{2} \Delta_i - \frac{Z}{r_i} \right\} + \frac{1}{2} \sum_{\substack{i,j=1 \\ (i \neq j)}} \frac{1}{r_{ij}}. \tag{2.34}$$

Let us select N one-electron orbitals, $\varphi_1, \varphi_2, \ldots, \varphi_N$, which are called spin-orbitals (Slater I, 271), and are defined as follows:

$$\varphi_k(q) = \psi_k(\mathbf{r})\eta_k(\sigma), \tag{2.35}$$

where q stands for the spatial coordinates $\mathbf{r}$ and for the spin-variable σ. The ψ_k is the spatial part of the spin orbital and η_k is the spin function. φ_k is normalized as follows:

$$\int |\varphi_k(q)|^2 \, dq = 1, \tag{2.36}$$

where the integration sign means integration with respect to the spatial coordinates and summation with respect to the spin variable.

Fock suggested that the total wave function should be written in the form

$$\Psi_T = A \det[\varphi_1(1)\varphi_2(2) \cdots \varphi_N(N)]$$

$$= A \begin{vmatrix} \varphi_1(1) & \varphi_1(2) & \cdots & \varphi_1(N) \\ \varphi_2(1) & \varphi_2(2) & \cdots & \varphi_2(N) \\ \vdots & \vdots & & \vdots \\ \varphi_N(1) & \varphi_N(2) & \cdots & \varphi_N(N) \end{vmatrix}, \qquad (2.37)$$

where A is a normalization constant to be determined from the requirement

$$\langle \Psi_T | \Psi_T \rangle = 1. \qquad (2.38)$$

Throughout this book, we will use the notation

$$\varphi_k(i) \equiv \varphi_k(q_i) = \varphi_k(x_i y_i z_i \sigma_i), \qquad (2.39)$$

that is, the argument of the function will contain the coordinates and the subscript the quantum state.

As we see from Eq. (2.37), the determinantal wave function Ψ_T satisfies the antisymmetry principle exactly for any set of one-electron orbitals because the interchange of any two particle coordinates q_i and q_j results in the wave function being multiplied by -1.

We note here that in this section, the one-electron orbitals will remain unspecified. This means that the wave function will not be subjected to any angular momentum and spin conditions. Here we concentrate exclusively on the discussion of the effects that the antisymmetry principle exerts on the theory.

We start building up the theory by stating two properties of the determinantal wave function. These are as follows:

1. Ψ_T does not change if the one-electron wave functions are orthogonalized to each other. That is, the generality of Ψ_T is not restricted by assuming that the one-electron orbitals are orthonormal.

2. If the one-electron orbitals are orthonormal, the determinantal wave function will be normalized with

$$A = (N!)^{-1/2}. \qquad (2.40)$$

These statements are proved in App. A. Thus, from here on, we assume that

$$\langle \varphi_i | \varphi_j \rangle = \delta_{ij} \qquad (i, j = 1, 2, \dots, N), \qquad (2.41)$$

where δ_{ij} is the Kronecker symbol. On putting the normalized total wave function into the Schroedinger energy expression and taking into account the orthonormality of the one-electron orbitals, we obtain the Hartree–Fock energy expression:

$$E_F = \langle \psi_T | H | \psi_T \rangle$$

$$= \sum_{i=1}^{N} \langle \varphi_i | t + g | \varphi_i \rangle$$

$$+ \frac{1}{2} \sum_{i,j=1}^{N} \left\{ \langle \varphi_i \varphi_j | \frac{1}{r_{12}} | \varphi_i \varphi_j \rangle - \langle \varphi_i \varphi_j | \frac{1}{r_{12}} | \varphi_j \varphi_i \rangle \right\}. \qquad (2.42)$$

The details of the derivation are given in App. A.

In this expression, t is the kinetic-energy operator, g is the nuclear potential, and the 2 electron integrals are defined by the general formula

$$\langle \varphi_\alpha \varphi_\beta | \frac{1}{r_{12}} | \varphi_\gamma \varphi_\delta \rangle \equiv \int \varphi_\alpha^*(1) \varphi_\beta^*(2) \frac{1}{r_{12}} \varphi_\gamma(1) \varphi_\delta(2) \, dq_1 \, dq_2. \qquad (2.43)$$

Thus, the two interaction integrals in the Hartree–Fock energy expression are

$$\langle \varphi_i \varphi_j | \frac{1}{r_{12}} | \varphi_i \varphi_j \rangle = \int \frac{1}{r_{12}} |\varphi_i(1)|^2 |\varphi_j(2)|^2 \, dq_1 \, dq_2, \qquad (2.44)$$

$$\langle \varphi_i \varphi_j | \frac{1}{r_{12}} | \varphi_j \varphi_i \rangle = \int \frac{1}{r_{12}} \left[\varphi_i^*(1) \varphi_j^*(2) \varphi_j(1) \varphi_i(2) \right] \, dq_1 \, dq_2. \qquad (2.45)$$

We see that Eq. (2.44) is simply the classical electrostatic interaction between the charge densities $-|\varphi_i|^2$ and $-|\varphi_j|^2$. Thus, the first two terms of the Hartree–Fock energy expression are identical with the Hartree energy expression, Eq. (2.8). The difference between the two expressions is the exchange-interaction integral given by Eq. (2.45). This term, which does not have a classical analog, is not present in Hartree's model.

We note that in the double summation of Eq. (2.42), the $i = j$ term is included, that is, the electrostatic interaction of an orbital with itself is included in the expression. This is permissible since the same expression appears also in the exchange term, which has a negative sign, thereby eliminating the self-energy from the electrostatic term.

We investigate now the following question. What is the best possible single-determinantal wave function? An equivalent question is: What are the best possible orbitals in a single-determinantal wave function? The answer is provided by the variation method. We derive the equations for the best one-electron orbitals by varying the Hartree–Fock energy expression with respect to the one-electron orbitals under the constraints of the subsidiary conditions. The subsidiary conditions are the orthonormality conditions given by Eq. (2.41). The details of this procedure are given in App. B.

According to the derivation given in App. B, we obtain the following Hartree–Fock (HF) equations for one-electron orbitals:

$$H_F\varphi_k = \varepsilon_k\varphi_k + \sum_{\substack{j=1 \\ (j\neq k)}}^{N} \bar{\lambda}_{kj}\varphi_j \qquad (k=1,2,\ldots,N), \qquad (2.46)$$

where H_F is the Hartree–Fock Hamiltonian operator,

$$H_F = t + g + U, \qquad (2.47)$$

with U being the total HF potential (operator),

$$U = \sum_{j=1}^{N} U_j, \qquad (2.48)$$

and

$$U_j(1)f(1) = \int \frac{1}{r_{12}}\left[|\varphi_j(2)|^2 f(1) - \varphi_j(1)\varphi_j^*(2)f(2)\right] dq_2. \qquad (2.49)$$

U_j is the HF potential (operator) generated by the orbital φ_j.

The $\bar{\lambda}_{kj}$ is the off-diagonal Lagrangian multiplier that is the result of the orthogonality conditions.

It is shown in App. B that the HF equations given by Eq. (2.46) preserve their form if the set of one-electron orbitals is subjected to a unitary transformation. Thus, we obtain infinitely many sets of HF orbitals, each set belonging to a particular Lagrangian matrix $\bar{\lambda}_{kj}$. The total energy and density belonging to these sets will be the same. We also show in App. B that Eq. (2.46) can be rewritten in the form

$$H_F'\varphi_k = \varepsilon_k\varphi_k, \qquad (2.46a)$$

where the operator H_F' is given by

$$H_F' = \Pi_k H_F, \qquad (2.46b)$$

and the projection operator Π_k is defined as

$$\Pi_k = 1 - \Omega_k, \qquad (2.46c)$$

with

$$\Omega_k = \sum_{\substack{j=1 \\ (j\neq k)}}^{N} |\varphi_j\rangle\langle\varphi_j|. \qquad (2.46d)$$

Thus, it is shown that the HF orbitals are not the eigenfunctions of H_F, but of H'_F. Because Π_k and H_F do not commute, the operator H'_F is not Hermitian. We obtained the peculiar result that, in general, *the effective HF Hamiltonian operator is not Hermitian*.

It is also shown in App. B that a unitary transformation can be used to bring the Lagrangian matrix to diagonal form. For that particular Lagrangian matrix, we get the HF equations in the form

$$H_F \varphi_k = \varepsilon_k \varphi_k \quad (k = 1, 2, \ldots, N). \tag{2.50}$$

In this case, the HF orbitals are the eigenfunctions of the Hermitian operator H_F. We note that the equations can always be brought to the form of Eq. (2.50) if the total wave function is a single determinant.

Let us establish the connection between the one-electron orbital energies and the total energy. Multiplying the equations from the left by φ_k^*, integrating, and summing them up, we get

$$\sum_{k=1}^{N} \varepsilon_k = \sum_{k=1}^{N} \langle \varphi_k | H_F | \varphi_k \rangle$$

$$= \sum_{k=1}^{N} \langle \varphi_k | t + g | \varphi_k \rangle + \sum_{k=1}^{N} \sum_{j=1}^{N} \langle \varphi_k | U_j | \varphi_k \rangle. \tag{2.51}$$

Let

$$\varepsilon_{kj} \equiv \langle \varphi_k | U_j | \varphi_k \rangle \equiv \langle \varphi_j | U_k | \varphi_j \rangle. \tag{2.52}$$

The property that the self-exchange interaction compensates for the self-electrostatic interaction is expressed by the relationship

$$\varepsilon_{kk} = 0. \tag{2.53}$$

Comparing Eq. (2.51) with the HF energy expression, we get

$$E_F = \sum_{k=1}^{N} \varepsilon_k - \frac{1}{2} \sum_{k,j=1}^{N} \varepsilon_{kj}. \tag{2.54}$$

Let us now discuss the structure of the HF equations, which we write according to Eqs. (2.50), (2.47), (2.48), and (2.49):

$$\left\{ -\frac{1}{2} \Delta_1 - \frac{Z}{r_1} + \sum_{j=1}^{N} \int \frac{1}{r_{12}} |\varphi_j(2)|^2 \, dq_2 \right\} \varphi_k(1)$$

$$- \sum_{j=1}^{N} \int \frac{\varphi_j(1)\varphi_j^*(2)\varphi_k(2)}{r_{12}} \, dq_2 = \varepsilon_k \varphi_k(1). \tag{2.55}$$

The original N-electron problem is transformed into a set of coupled, one-electron, integro-differential equations. A look at the Hartree equation, Eq. (2.9), shows that apart from the last term on the left side, the HF equation in Eq. (2.55) is identical with the Hartree equation. In order to formalize this difference, let us introduce the local potential V and the operator A:

$$V(1) \equiv \sum_{j=1}^{N} \int \frac{1}{r_{12}} |\varphi_j(2)|^2 \, dq_2, \tag{2.56}$$

$$A(1)f(1) \equiv \sum_{j=1}^{N} \int \frac{1}{r_{12}} \varphi_j(1) \varphi_j^*(2) f(2) \, dq_2. \tag{2.57}$$

The operator A, called the exchange operator, is a linear, integral operator with the kernel

$$K(1,2) = \sum_{j=1}^{N} \frac{1}{r_{12}} \varphi_j(1) \varphi_j^*(2). \tag{2.57a}$$

Using these notations, we get for the HF equations

$$\{t + g + V - A\} \varphi_k = \varepsilon_k \varphi_k \qquad (k = 1, 2, \ldots, N). \tag{2.58}$$

In order to shed some light on the meaning of the exchange operator, the best method is to consider an atom with N electrons among which p electrons have "up" spins and $N - p$ have "down" spins. Let

$$\varphi_i = \psi_i \alpha \qquad (i = 1, \ldots, p), \tag{2.59a}$$

$$\varphi_i = \psi_i \beta \qquad (i = p + 1, \ldots, N), \tag{2.59b}$$

where α and β are the "up" and "down" spin functions, respectively. Let us also introduce the spin-independent density matrices:

$$\rho_+(1,2) = \sum_{j=1}^{p} \psi_j(1) \psi_j^*(2), \tag{2.60a}$$

$$\rho_-(1,2) = \sum_{j=p+1}^{N} \psi_j(1) \psi_j^*(2), \tag{2.60b}$$

and the corresponding local densities:

$$\rho_+(1) = \rho_+(1,1), \tag{2.61a}$$

$$\rho_-(1) = \rho_-(1,1). \tag{2.61b}$$

We get, after spin integration,

$$V(1) = \int \frac{1}{r_{12}} \rho_+(2) \, dv_2 + \int \frac{1}{r_{12}} \rho_-(2) \, dv_2$$

$$= \int \frac{1}{r_{12}} \rho(2) \, dv_2, \tag{2.62}$$

where

$$\rho = \rho_+ + \rho_-. \tag{2.63}$$

From the structure of the exchange operator, we see that it will give different results in Eq. (2.55) depending on the spin direction of the orbital φ_k. Thus, for φ_k having "up" spin, we get

$$A\varphi_k = \sum_{j=1}^{p} \int \frac{1}{r_{12}} \varphi_j(1) \varphi_j^*(2) \varphi_k(2) \, dq_2 + \sum_{j=p+1}^{N} \int \frac{1}{r_{12}} \varphi_j(1) \varphi_j^*(2) \varphi_k(2) \, dq_2$$

$$= \sum_{j=1}^{p} \int \frac{1}{r_{12}} \psi_j(1) \psi_j^*(2) \psi_k(2) \, dv_2, \tag{2.64}$$

because in the second term, φ_j has "down" spin and, therefore, we get zero from the spin integration in dq_2. Likewise, for φ_k having "down" spin, we get

$$A\varphi_k = \sum_{j=p+1}^{N} \int \frac{1}{r_{12}} \psi_j(1) \psi_j^*(2) \psi_k(2) \, dv_2. \tag{2.65}$$

Introducing the exchange operators A_+ and A_- for the operators in the preceding two equations, we get from the HF equations, the following two equations for orbitals with "up" and "down" spins, respectively:

$$\left\{ -\frac{1}{2} \Delta - \frac{Z}{r} + V - A_+ \right\} \psi_k = \varepsilon_k \psi_k \qquad (k = 1, \ldots, p; \text{ spin "up"});$$

$$\tag{2.66a}$$

$$\left\{ -\frac{1}{2} \Delta - \frac{Z}{r} + V - A_- \right\} \psi_k = \varepsilon_k \psi_k \qquad (k = p + 1, \ldots, N; \text{ spin "down"}).$$

$$\tag{2.66b}$$

These two equations give us a clear idea about the meaning of the exchange operator. The first equation is for orbitals with "up" spin; the second is for orbitals with "down" spin. The equations are different: one has in it the operator A_+ and the other the operator A_-. If there would be no exchange interaction, the equations would be identical, that is, all electrons would be moving in the electrostatic potential V, which is the same for all orbitals. In reality, however, the HF Hamiltonian is different for orbitals with up and down spins. That means that, in general, the spatial parts of two orbitals with antiparallel spins might be different even if they belong to the same set of spatial quantum numbers. This effect is called spin polarization. For example, the spatial parts of two 1s orbitals might be different because the potential (operator) in the equation of the orbital with up spin is different from the potential in the equation of the orbital with down spin.

Looking over the derivation that led to Eqs. (2.66a) and (2.66b), we see that exchange interaction exists only between electrons with parallel spins. Thus, the exchange interaction introduces "correlation" between electrons with parallel spins. Such a correlation is not present in the Hartree model. There the potential does not depend on spin direction.

We want to look at the HF energy expression in the case of an atom in which we have p up spins and $N - p$ down spins. Using the relationships given in Eqs. (2.59), (2.60), (2.61), and (2.63), we obtain

$$E_F = E_\uparrow + E_\downarrow + E_{int}, \tag{2.67}$$

where

$$
\begin{aligned}
E_\uparrow = {}& \sum_{i=1}^{p} \langle \psi_i | t + g | \psi_i \rangle \\
& + \frac{1}{2} \int [\rho_+(1)\rho_+(2) - \rho_+(1,2)\rho_+(2,1)] \frac{1}{r_{12}} \, dv_1 \, dv_2, \tag{2.68a}
\end{aligned}
$$

$$
\begin{aligned}
E_\downarrow = {}& \sum_{i=p+1}^{N} \langle \psi_i | t + g | \psi_i \rangle \\
& + \frac{1}{2} \int [\rho_-(1)\rho_-(2) - \rho_-(1,2)\rho_-(2,1)] \frac{1}{r_{12}} \, dv_1 \, dv_2, \tag{2.68b}
\end{aligned}
$$

$$E_{int} = \int \frac{\rho_+(1)\rho_-(2)}{r_{12}} \, dv_1 \, dv_2. \tag{2.68c}$$

Equation (2.67) has a very plausible physical interpretation. The energy of the atom consists of three parts: the energy of the electrons with up spin, the energy of the electrons with down spin, and the interaction energy of the two groups of electrons. The energies $E_\uparrow$ and $E_\downarrow$ consist of the usual HF terms, kinetic energy, nuclear attraction, electrostatic interaction, and exchange interaction. The interaction energy between the two groups is a purely electrostatic interaction; there is no exchange term here. Thus, we may formulate the rule: In the HF approximation, the interaction between electrons with antiparallel spins is a purely electrostatic interaction. Exchange interaction exists only between electrons with parallel spins.

The derivation of the relationship given in Eq. (2.67) is presented at the end of App. A. A more detailed discussion of the exchange interaction follows in Sec. 2.4.

2.3. ATOMS WITH COMPLETE GROUPS

The discussions in the preceding section were fairly general in the sense that although we specified the number of electrons, we did assume nothing specific about the electron configuration. Now let us make the discussion more specific by assuming that we are talking about an atom with complete groups.

The electrons are said to belong to the same groups if their wave functions have the same principal and azimuthal quantum numbers. In contrast, we designate a shell as the collection of electrons with the same principal quantum number. Thus, atoms with complete groups do not necessarily possess closed shells.

According to angular momentum and spin theory, an atom with complete groups is in a ^1S state. Thus, we impose on the total wave function the following conditions. It must be a determinantal wave function and it must be in a ^1S state. Delbrück has shown[4] that the wave function satisfying these conditions will be a single-determinantal wave function with the one-electron orbitals being central-field functions:

$$\varphi_k(q) = \psi_k(\mathbf{r})\eta_k(\sigma) \tag{2.69a}$$

with

$$\psi_k(\mathbf{r}) = \frac{P_{n_k l_k}(r)}{r} Y_{l_k m_{lk}}(\vartheta, \varphi), \tag{2.69b}$$

where Y_{lm_l} is the normalized spherical harmonics and $\eta_k = \eta_{m_{sk}}$ is the spin function. Because a complete group, by definition, contains for each l, $2(2l + 1)$ electrons, the determinantal wave function will contain, for a given l, all the orbitals with m_l ranging from $-l$ to $+l$ and m_s being $+\frac{1}{2}$ or $-\frac{1}{2}$.

We develop now the Hartree–Fock model for complete groups. We present two formulations: Hartree's and Slater's. The final equations in these two formulations are, of course, identical, although their mathematical form is different. The reason for us to present both is that Hartree's formulation will be useful for the generalization of the model to atoms with incomplete groups, and Slater's formulations will be convenient for the theory of average configurations.

Hartree's Formulations [Hartree, 39]. In Hartree's formulation, we start by putting the central-field one-electron orbitals into the single-determinantal wave function. On putting the determinantal wave function into the Schroedinger energy expression, we obtain the HF energy expression in terms of radial integrals plus constants that are the results of angular and spin integrations. For the HF energy expression, we obtain a formula valid for all atoms with complete groups. By varying the HF energy expression with respect to the radial parts of the one-electron orbitals, we obtain the HF equations. These equations will be valid for any atom with complete groups.

The first step in this program is to obtain the integrals occurring in the HF energy expression, Eq. (2.42), in terms of central-field orbitals. This step is carried out in App. C and the results are

$$\langle \varphi_i | t + g | \varphi_i \rangle = I(n_i l_i), \tag{2.70a}$$

$$\langle \varphi_i \varphi_j | \frac{1}{r_{12}} | \varphi_i \varphi_j \rangle = \sum_{k=0}^{l_m} a_k(l_i m_{li}, l_j m_{lj}) F_k(n_i l_i, n_j l_j), \tag{2.70b}$$

$$\langle \varphi_i \varphi_j | \frac{1}{r_{12}} | \varphi_j \varphi_i \rangle = \delta(m_{si}, m_{sj}) \sum_{k=0}^{l_i+l_j} b_k(l_i m_{li}, l_j m_{lj}) G_k(n_i l_i, n_j l_j). \tag{2.70c}$$

Here we have the following radial integrals:

$$I(nl) = \int_0^\infty P_{nl}(r) \left[-\frac{1}{2} \frac{d^2}{dr^2} + \frac{l(l+1)}{2r^2} - \frac{Z}{r} \right] P_{nl}(r) \, dr, \tag{2.71a}$$

$$F_k(nl, n'l') = \int P_{nl}(1) P_{nl}(1) P_{n'l'}(2) P_{n'l'}(2) L_k(1, 2) \, dr_1 \, dr_2, \tag{2.71b}$$

$$G_k(nl, n'l') = \int P_{nl}(1) P_{n'l'}(1) P_{nl}(2) P_{n'l'}(2) L_k(1, 2) \, dr_1 \, dr_2, \tag{2.71c}$$

where

$$L_k(1, 2) \equiv L_k(r_1, r_2) = \frac{r_<^k}{r_>^{k+1}}. \tag{2.72}$$

Constants a_k and b_k are the results of angular integrations and are defined in App. C. In Eq. (2.70b) l_m means the lesser of $2l_i$ and $2l_j$.

Hartree evaluated the HF energy expression, Eq. (2.42), in terms of these integrals and obtained the result:

$$\begin{aligned} E_F = &\sum_{(nl)} q(nl) I(nl) + \sum_{(nl)} \frac{q(nl)}{2} [q(nl) - 1] F_0(nl, nl) \\ &+ \sum_{\substack{(nl)(n'l') \\ (n'l') \neq (nl)}} q(nl) q(n'l') F_0(nl, n'l') \\ &- \sum_{(nl)} \sum_{k>0} A_{lk} F_k(nl, nl) \\ &- \sum_{(nl)} \sum_{(n'l')k} \sum_{\substack{}} B_{ll'k} G_k(nl, n'l'). \end{aligned} \tag{2.73}$$

In this formula, $q(nl)$ is the occupation number of the group (nl); thus, for complete groups,

$$q(nl) = 2(2l + 1). \tag{2.74}$$

The summations indicated by $\Sigma_{(nl)}$ are for all groups and the double summations are for all pairs of groups; that means that these are not independent summations. $(n'l') \neq (nl)$ indicates that for a given (nl), either $n' \neq n$ or $l' \neq l$. The constants $A_{lk} \ B_{ll'k}$, which come from the summations over a_k and b_k, were tabulated by Hartree.

We will not need these constants because we will express them in what follows in terms of the c^k constants, which are defined by Eq. (C.19) in App. C. These constants were tabulated by Slater (Slater II, 281) and we also give an explicit formula for them in App. E. We note that the upper limit for A_{lk} is given by $k = 2l$ and for $B_{ll'k}$, the lower limit is zero and the upper limit is $k = l + l'$. Also, we note that

$$B_{ll'k} = B_{l'lk}. \tag{2.75}$$

The energy expression in Eq. (2.73) has a simple physical interpretation. The first term is the sum of the kinetic and nuclear potential energies of all orbitals. The second term is the intragroup electrostatic interaction summed over all groups. The third term is the electrostatic interaction between the (nl) and $(n'l')$ groups summed over all pairs of groups. The fourth term is the exchange interaction within the (nl) group summed over all groups, and the fifth term is the exchange interaction between the groups (nl) and $(n'l')$ summed over all pairs of groups.

By looking over this very plausible list, the question arises: Why are there no intergroup interaction terms with $F_k(nl, n'l')$? The reason for this is that the a_k coefficients have the property that for $k > 0$,

$$\sum_{m_{l\beta}} a_k(l_\alpha m_{l\alpha}, l_\beta m_{l\beta}) = 0, \tag{2.76}$$

for all $m_{l\alpha}$, if the summation is for all values of $m_{l\beta}$ in the complete $(n_\beta l_\beta)$ group. Thus, the only F_k integrals with $k > 0$ will be those with $(n'l') = (nl)$.

The HF energy can be written down at once, for any atom with complete groups, on the basis of Eq. (2.73).

The second step in our program is to derive the equations for the radial orbitals P_{nl} by varying the expression given by Eq. (2.73). In order to do this in one step, we make some mathematical preparations. First, let us write the F_k and G_k integrals in a more suitable form. Let

$$\frac{1}{r} Y_k(nl, n'l'|r) = \int_{r'=0}^{\infty} P_{nl}(r') P_{n'l'}(r') L_k(r, r') \, dr'. \tag{2.77a}$$

We see immediately that

$$\frac{1}{r}Y_k(n'l', nl|r) = \int_{r'=0}^{\infty} P_{n'l'}(r')P_{nl}(r')L_k(r, r')\, dr'$$

$$= \frac{1}{r}Y_k(nl, n'l'|r). \qquad (2.77b)$$

From the definition of F_k, it follows that

$$F_k(nl, n'l'|r) = \int P_{nl}(r)P_{nl}(r)\frac{1}{r}Y_k(n'l', n'l'|r)\, dr, \qquad (2.77c)$$

and also

$$F_k(n'l', nl) = \int P_{n'l'}(r)P_{n'l'}(r)\frac{1}{r}Y_k(nl, nl|r)\, dr$$

$$= F_k(nl, n'l'). \qquad (2.77d)$$

Thus, we have the relationship

$$\langle P_{nl}|\frac{1}{r}Y_k(n'l', n'l'|r)|P_{nl}\rangle = \langle P_{n'l'}|\frac{1}{r}Y_k(nl, nl|r)|P_{n'l'}\rangle. \qquad (2.77e)$$

Also,

$$G_k(nl, n'l') = \int P_{nl}(r)P_{n'l'}(r)\frac{1}{r}Y_k(nl, n'l'|r)\, dr$$

$$= G_k(n'l', nl). \qquad (2.77f)$$

Let δ mean the variation with respect to P_{nl}. Then we get, using the previous formulas,

$$\delta I(nl) = 2\langle \delta P_{nl}| -\frac{1}{2}\frac{d^2}{dr^2} + \frac{l(l+1)}{2r^2} - \frac{Z}{r}|P_{nl}\rangle, \qquad (2.78a)$$

$$\delta F_k(nl, n'l') = 2\langle \delta P_{nl}|\frac{1}{r}Y_k(n'l', n'l'|r)P_{nl}\rangle \qquad (n'l') \neq (nl), \quad (2.78b)$$

$$\delta F_k(nl, nl) = 4\langle \delta P_{nl}|\frac{1}{r}Y_k(nl, nl|r)|P_{nl}\rangle, \qquad (2.78c)$$

and

$$\delta G_k(nl, n'l') = 2\langle \delta P_{nl}| \frac{1}{r} Y_k(nl, n'l'|r)|P_{n'l'}\rangle \qquad (n'l') \neq (nl). \quad (2.78d)$$

The subsidiary conditions are the orthonormality of the radial orbitals,

$$\int P_{nl}(r) P_{n'l}(r)\, dr = \delta_{nn'}. \qquad (2.79)$$

The off-diagonal components of the Lagrangian multipliers can be eliminated by a unitary transformation; therefore, the only subsidiary condition that need to be considered is the normalization. Thus, if we put

$$E_0 = -\sum_{(nl)} \bar{\varepsilon}_{nl}\langle P_{nl}|P_{nl}\rangle, \qquad (2.80)$$

where $\bar{\varepsilon}_{nl}$ is a Lagrangian multiplier, we obtain our variation principle in the form

$$\delta(E_F + E_0) = 0, \qquad (2.81)$$

for all (n, l).

Now we substitute the HF energy expression, Eq. (2.73), and E_0, as given by Eq. (2.80), into the variation principle, Eq. (2.81). We carry out the variation with respect to P_{nl} and use the relationships given by Eqs. (2.78a) to (2.78d). In this way, we obtain the following equation:

$$2q(nl)\left[-\frac{1}{2}\frac{d^2}{dr^2} + \frac{l(l+1)}{2r^2} - \frac{Z}{r} \right]P_{nl} + \frac{q(nl)}{2}[q(nl) - 1]\frac{4}{r}Y_0(nl, nl|r)P_{nl}$$

$$+ \sum_{\substack{(n'l') \\ (n'l') \neq (nl)}} q(nl)q(n'l')\frac{2}{r}Y_0(n'l', n'l'|r)P_{nl}$$

$$- \sum_{k>0} A_{lk}\frac{4}{r}Y_k(nl, nl|r)P_{nl}$$

$$- \sum_{\substack{(n'l')\, k \\ (n'l') \neq (nl)}} B_{ll'k}\frac{2}{r}Y_k(nl, n'l'|r)P_{n'l'} - \bar{\varepsilon}_{nl}2P_{nl} = 0, \qquad (2.82)$$

for all (nl).

We want to put the equation into conventional form, that is, in a form where the second derivative is multiplied by $-1/2$. We divide by $2q(nl)$

and get

$$
\left\{ -\frac{1}{2}\frac{d^2}{dr^2} + \frac{(l+1)}{2r^2} - \frac{Z}{r} + [q(nl) - 1]\frac{1}{r}Y_0(nl, nl|r) \right.
$$

$$
+ \sum_{\substack{(n'l') \\ (n'l')\neq(nl)}} q(n'l')\frac{1}{r}Y_0(n'l', n'l'|r) - \sum_{k>0} \alpha_{lk}\frac{1}{r}Y_k(nl, nl|r) \left.\right\} P_{nl}(r)
$$

$$
- \sum_{(n'l')} \sum_{k} \beta_{ll'k}\frac{1}{r}Y_k(nl, n'l'|r)P_{n'l'}(r)
$$

$$
= \varepsilon_{nl}P_{nl}(r), \tag{2.83}
$$

where

$$
\alpha_{lk} = \frac{2A_{lk}}{q(nl)} \qquad (k > 0), \tag{2.84a}
$$

$$
\beta_{ll'k} = \frac{B_{ll'k}}{q(nl)}, \tag{2.84b}
$$

Incorporating the fourth term into the sixth, we obtain the Hartree–Fock equations, for an atom with complete groups, in Hartree's formulation:

$$
\left\{ -\frac{1}{2}\frac{d^2}{dr^2} + \frac{l(l+1)}{2r^2} - \frac{Z}{r} + \sum_{(n'l')} q(n'l')\frac{1}{r}Y_0(n'l', n'l'|r) \right.
$$

$$
- \frac{1}{r}Y_0(nl, nl|r) - \sum_{k>0} \alpha_{lk}\frac{1}{r}Y_k(nl, nl|r) \left.\right\} P_{nl}(r)
$$

$$
- \sum_{\substack{(n'l') \, k \\ (n'l')\neq(nl)}} \beta_{ll'k}\frac{1}{r}Y_k(nl, n'l'|r)P_{n'l'}(r) = \varepsilon_{nl}P_{nl}(r). \tag{2.85}
$$

Where now the first summation over $(n'l')$ includes (nl). The last term on the left side, which is arising from the exchange interaction, can be put into a simpler form by introducing the (nonlocal) exchange potential,

$$
X(nl|r)P_{nl}(r) = - \sum_{\substack{(n'l') \, k \\ (n'l')\neq(nl)}} \beta_{ll'k}\frac{1}{r}Y_k(nl, n'l'|r)P_{n'l'}(r). \tag{2.86}
$$

Using this notation, we can put the total HF potential, electrostatic plus exchange, in the form

$$U(nl|r) = \sum_{\substack{(n'l') \\ \text{(all pairs)}}} q(n'l') \frac{1}{r} Y_0(n'l', n'l'|r)$$

$$-\frac{1}{r} Y_0(nl, nl|r) - \sum_{k>0} \alpha_{lk} \frac{1}{r} Y_k(nl, nl|r) + X(nl|r), \quad (2.87)$$

and we note that the last term is an operator that, when applied to P_{nl}, produces the expression given by Eq. (2.86).

The structure of the HF potential is simple. The first term is the total spherically symmetric electrostatic potential of all electrons. The second term is the self-electrostatic potential of the orbital P_{nl}, which must be subtracted from the first term as physically meaningless. The third term is the exchange interaction of the orbital P_{nl} with other orbitals in the same group. The last (operator) term represents the exchange interaction between the orbital P_{nl} and the orbitals in the other groups.

Using Eq. (2.87), we can put the HF equations in the form

$$\left\{ -\frac{1}{2} \frac{d^2}{dr^2} + \frac{l(l+1)}{2r^2} - \frac{Z}{r} + U(nl|r) \right\} P_{nl}(r) = \varepsilon_{nl} P_{nl}(r), \quad (2.88)$$

for all (nl).

We note that the second and third terms of Eq. (2.87) can be combined by putting $\alpha_{lk} = 1$ for $k = 0$. We note also that the $\beta_{ll'k}$, which appears in the operator X, is not symmetric in l and l' due to the definition given by Eq. (2.84b).

Slater's Formulation [Slater II, 1]. We derive now the HF equations for complete groups in the form given by Slater. Actually, although the results are Slater's, the derivation is slightly different and was formulated by the author. The method here differs from Hartree's in that we start with the equations developed in Sec. 2.2. Instead of putting the central-field orbitals into the HF energy expression, integrating over the angles and spin and then varying the result with respect to the radial parts of the orbitals, we put the central-field orbitals directly into the HF equations, which were derived for single-determinantal wave functions in Sec. 2.2.

Let us clarify first the notation. We have an atom with complete groups; thus, we have $N/2$ up spins and $N/2$ down spins. We start with the HF equations, Eqs. (2.66a) and (2.66b), and we let $p = N/2$. For that case, the two equations are identical and we write down the equation valid for any orbital,

$$\left\{ -\frac{1}{2}\Delta - \frac{Z}{r} + V - A \right\}\psi_i = \varepsilon_i\psi_i \qquad (i = 1, 2, \ldots, N/2). \quad (2.89)$$

For the orbitals, we use central-field functions, that is, we put

$$\psi_i = \frac{P_{n_i l_i}(r)}{r} Y_{l_i m_{l_i}}(\vartheta, \varphi). \qquad (2.90)$$

Next, we investigate the structure of exchange operator A, which was defined by Eq. (2.64). We have

$$A(1)\psi_i(1) = \sum_{j=1}^{N/2} \int \frac{1}{r_{12}} \psi_j(1)\psi_j^*(2)\psi_i(2)\, dv_2. \qquad (2.91)$$

Using central-field orbitals for both ψ_i and ψ_j, we obtain, by a simple derivation that is reproduced in App. D, the result

$$A(1)\psi_i(1) = A(1)\psi_{n_i l_i m_{l_i}}(1)$$

$$= \sum_{(n_j l_j m_{l_j})} \int \frac{1}{r_{12}} \psi_{n_j l_j m_{l_j}}(1)\psi_{n_j l_j m_{l_j}}^*(2)\psi_{n_i l_i m_{l_i}}(2)\, dv_2$$

$$= \sum_{(n_j l_j)} \sum_k \sqrt{\frac{2l_j + 1}{2l_i + 1}}\, c^k(l_i 0, l_j 0)$$

$$\times \frac{1}{r_1} Y_k(n_j l_j n_i l_i | r_1) \frac{P_{n_j l_j}(r_1)}{r_1} Y_{l_i m_{l_i}}(\vartheta_1 \varphi_1). \qquad (2.92)$$

In this equation, the summation $\Sigma_{(n_j l_j m_{l_j})}$ is for all values of these quantum numbers occurring in the atom; constant c^k is defined by Eq. (C.19) in App. C, the potential function Y_k is defined by Eq. (2.77a), the summation $\Sigma_{(n_j l_j)}$ is for all complete groups and the range of k is $0 \le k \le l_i + l_j$. It is interesting that operator A does not change the angular part of ψ_i if the atom has complete groups; that is, the operator can be considered to act only on the radial part of ψ_i in the following way:

$$A(1)\psi_i(1) = A(1)\left[\frac{P_{n_i l_i}(r_1)}{r_1} Y_{l_i m_{l_i}}(\vartheta_1 \varphi_1) \right]$$

$$= f(r_1)Y_{l_i m_{l_i}}(\vartheta_1 \varphi_1), \qquad (2.93a)$$

where

$$f(r_1) = \sum_{(n_j l_j)} \sum_k \sqrt{\frac{2l_j + 1}{2l_i + 1}}\, c^k(l_i 0, l_j 0)$$

$$\times \frac{P_{n_j l_j}(r_1)}{r_1} \int_{r_2=0}^{\infty} P_{n_j l_j}(r_2) P_{n_i l_i}(r_2) L_k(r_1, r_2)\, dr_2. \quad (2.93b)$$

Thus, the effect of the operator is to change the radial part of the HF orbital into the function $f(r)$. This function depends on n_i and l_i.

Similarly, by using central-field functions, we get for the electrostatic potential, by a derivation that is given in App. D, the expression

$$V(1) = 2 \sum_{j=1}^{N/2} \int \frac{1}{r_{12}} |\psi_j(2)|^2\, dv_2$$

$$= \sum_{(n_j l_j m_{lj})} 2 \int \frac{1}{r_{12}} |\psi_{n_j l_j m_{lj}}(2)|^2\, dv_2$$

$$= \sum_{(n_j l_j)} 2(2l_j + 1) \frac{1}{r_1} Y_0(n_j l_j, n_j l_j | r_1), \quad (2.94)$$

where the summation over $(n_j l_j)$ is for all complete groups and Y_0 is again the potential function given by Eq. (2.77a).

Now let us put the expressions in Eqs. (2.90), (2.92), and (2.94) into the HF equation, Eq. (2.89). We obtain

$$\left\{ -\frac{1}{2}\Delta - \frac{Z}{r} + V - A \right\} \frac{P_{n_i l_i}}{r} Y_{l_i m_{li}} = \varepsilon_{n_i l_i m_{li}} \frac{P_{n_i l_i}}{r} Y_{l_i m_{li}}. \quad (2.95)$$

As we have seen in App. C, the Laplacian operates on the central-field orbital as follows [Eq. (C.10)]:

$$-\frac{1}{2}\Delta\psi_i = -\frac{1}{2r}\left(\frac{d^2 P_{n_i l_i}}{dr^2} \right) Y_{l_i m_{li}} + \frac{l_i(l_i + 1)}{2r^2} \frac{P_{n_i l_i}}{r} Y_{l_i m_{li}} \quad (2.96)$$

and using this line, we get

$$
\left[-\frac{1}{2r}\frac{d^2}{dr^2}P_{n_il_i} + \frac{l_i(l_i+1)}{2r^2}\frac{P_{n_il_i}}{r} - \frac{Z}{r}\frac{P_{n_il_i}}{r} \right.
$$

$$
+ \sum_{(n_jl_j)} 2(2l_j+1)\frac{1}{r}Y_0\!\left(n_jl_j,n_jl_j|r\right)\frac{P_{n_il_i}}{r}
$$

$$
\left. - \sum_{(n_jl_j)}\sum_{k}\sqrt{\frac{2l_j+1}{2l_i+1}}\,c^k(l_i0,l_j0)\frac{1}{r}Y_k\!\left(n_jl_jn_il_i|r\right)\frac{P_{n_jl_j}}{r} \right]Y_{l_im_{li}}
$$

$$
= \varepsilon_{n_il_im_{li}}\frac{P_{n_il_i}}{r}Y_{l_im_{li}}. \tag{2.97a}
$$

Multiplying the equation by $r/Y_{l_im_{li}}$, we obtain Slater's formulation of the HF equations for complete groups [Slater II, 21]:

$$
\left\{ -\frac{1}{2}\frac{d^2}{dr^2} + \frac{l_i(l_i+1)}{2r^2} - \frac{Z}{r} + \sum_{(n_jl_j)} 2(2l_j+1)\frac{1}{r}Y_0\!\left(n_jl_jn_jl_j|r\right) \right\}P_{n_il_i}
$$

$$
- \sum_{(n_jl_j)}\sum_{k}\sqrt{\frac{2l_j+1}{2l_i+1}}\,c^k(l_i0,l_j0)\frac{1}{r}Y_k\!\left(n_jl_jn_il_i|r\right)P_{n_jl_j}
$$

$$
= \varepsilon_{n_il_i}P_{n_il_i}. \tag{2.97b}
$$

We have dropped the magnetic quantum number from the orbital parameter because it is obvious that it will not depend on it.

Similarly, as we have done with Hartree's equation, we can write this equation in the form of Eq. (2.88). For the nonlocal HF potential U, we now get

$$
U(r)P_{n_il_i}(r) = \sum_{(n_jl_j)} 2(2l_j+1)\frac{1}{r}Y_0\!\left(n_jl_j,n_jl_j|r\right)P_{n_il_i}(r)
$$

$$
- \sum_{(n_jl_j)}\sum_{k}\sqrt{\frac{2l_j+1}{2l_i+1}}\,c^k(l_i0,l_j0)\frac{1}{r}Y_k\!\left(n_jl_j,n_il_i|r\right)P_{n_jl_j}. \tag{2.98}
$$

Let us compare this potential with Hartree's expression, Eq. (2.87). The outward forms of the two potentials are different, but we must assume that they are identical because of the underlying theory. The first terms are obviously the same in both potentials. With $\alpha_{lo} = 1$, we obtain, by equating the terms that contain $Y_k(nl, nl|r)$,

$$-\frac{1}{r}Y_0(nl, nl|r) - \sum_{k>0} \alpha_{lk}\frac{1}{r}Y_k(nl, nl|r)$$

$$= -\sum_{k=0}^{2l} \alpha_{lk}\frac{1}{r}Y_k(nl, nl|r)$$

$$= -\sum_{k=0}^{2l} c^k(l0, l0)\frac{1}{rY}(nl, nl|r), \qquad (2.99)$$

from which we obtain

$$\alpha_{lk} = \frac{2A_{lk}}{2(2l+1)} = c^k(l0, l0), \qquad (2.100a)$$

$$A_{lk} = (2l+1)c^k(l0, l0) \qquad (k > 0), \qquad (2.100b)$$

and

$$\alpha_{lo} = 1, \qquad A_{lo} = (2l+1), \qquad (k = 0), \qquad (2.100c)$$

because $c^0(l0, l0) = 1$. Likewise, by comparing the terms in which we have $Y_k(n'l', nl|r)$, we obtain

$$\beta_{ll'k} = \sqrt{\frac{2l'+1}{2l+1}}\, c^k(l0, l'0), \qquad (2.101a)$$

from which we get

$$B_{ll'k} = 2\sqrt{(2l+1)(2l'+1)}\, c^k(l0, l'0). \qquad (2.101b)$$

These equations give us Hartree's constants in terms of the single parameter c^k, which was tabulated by Slater [Slater II, 281]. (See also the explicit formula for c^k in App. E.) Using those tables or the explicit formula, we can easily construct the HF equations either from Hartree's formula, Eq. (2.85), or from Slater's, Eq. (2.97b).

As the final step in presenting Slater's results, we now derive the energy expression. Here, again, we start with the general formulas derived for single-determinantal wave functions in Sec. 2.2. For the purposes of the energy expression the formulas given by Eqs. (2.67), (2.68a), (2.68b), and (2.68c) will be convenient. From these formulas, we get for the electrostatic interaction of the atom:

$$E_{el} = \frac{1}{2} \int \frac{1}{r_{12}} [\rho_+(1)\rho_+(2) + \rho_-(1)\rho_-(2)$$

$$+ 2\rho_+(1)\rho_-(2)] \, dv_1 \, dv_2, \qquad (2.102)$$

where, according to Eqs. (2.60) and (2.61), we have

$$\rho_+ = \sum_{i=1}^{N/2} |\psi_i|^2 = \sum_{(n_i l_i m_{l_i})} |\psi_{n_i l_i m_{l_i}}|^2, \qquad (2.103)$$

and

$$\rho_+ \equiv \rho_-. \qquad (2.104)$$

Here the summation over $(n_i l_i m_{l_i})$ is for all values of these parameters occurring in the atom. Using these relationships, we get

$$E_{el} = 2 \sum_{(n_i l_i m_{l_i})} \sum_{(n_j l_j m_{l_j})} \int \frac{1}{r_{12}} |\psi_{n_i l_i m_{l_i}}(1)|^2$$

$$\times |\psi_{n_j l_j m_{l_j}}(2)|^2 \, dv_1 \, dv_2. \qquad (2.105)$$

Using Eq. (2.94), we get

$$E_{el} = \sum_{(n_i l_i m_{l_i})} \int |\psi_{n_i l_i m_{l_i}}(1)|^2$$

$$\times \sum_{n_j l_j} 2(2l_j + 1) \frac{1}{r_1} Y_0(n_j l_j n_j l_j | r_1) \, dv_1. \qquad (2.106)$$

Taking into account that the orbital $\psi_{n_i l_i m_{l_i}}$ is of central field form and that the spherical harmonics are normalized, we get

$$E_{el} = \sum_{(n_i l_i)} \sum_{(n_j l_j)} (2l_i + 1)2(2l_j + 1)$$

$$\times \left\langle \left| P_{n_i l_i}^2 \frac{1}{r} Y_0(n_j l_j n_j l_j r) \right| \right\rangle$$

$$= \frac{1}{2} \sum_{(nl)} \sum_{(n'l')} (4l + 2)(4l' + 2) F_0(nl, n'l'). \qquad (2.107)$$

Here, in the last line, we used Eq. (2.77c). We note that the summations over (nl) and $(n'l')$ are independent and they are for all complete groups.

We obtain the exchange energy from Eqs. (2.68a) and (2.68b):

$$E_{ex} = -\tfrac{1}{2}[\rho_+(1,2)\rho_+(2,1) + \rho_-(1,2)\rho_-(2,1)] \, dv_1 \, dv_2, \quad (2.108)$$

where the density matrices are given by Eqs. (2.60a) and (2.60b). For $p = N/2$, we get

$$\rho_+(1,2) = \sum_{j=1}^{N/2} \psi_j(1)\psi_j^*(2)$$

$$= \sum_{(n_j l_j m_{lj})} \psi_{n_j l_j m_{lj}}(1)\psi_{n_j l_j m_{lj}}^*(2), \quad (2.109)$$

and

$$\rho_-(1,2) \equiv \rho_+(1,2). \quad (2.110)$$

Here, again, the summation is for all values of (n_i, l_i, m_{li}) occurring in the atom. Using these relationships, we get

$$E_{ex} = -\sum_{(n_i l_i m_{li})} \sum_{(n_j l_j m_{lj})} \int \frac{1}{r_{12}} \psi_{n_i l_i m_{li}}^*(1)$$

$$\times \psi_{n_i l_i m_{li}}(2)\psi_{n_j l_j m_{lj}}(1)\psi_{n_j l_j m_{lj}}^*(2) \, dv_1 \, dv_2$$

$$= -\sum_{(n_i l_i m_{li})} \int \psi_{n_i l_i m_{li}}^*(1) \left\{ \sum_{(n_j l_j m_{lj})} \int \frac{1}{r_{12}} \psi_{n_j l_j m_{lj}}(1) \right.$$

$$\left. \times \psi_{n_j l_j m_{lj}}^*(2)\psi_{n_i l_i m_{li}}(2) \, dv_2 \right\} dv_1. \quad (2.111)$$

We have calculated the expression in the curly brackets and have given the result in Eq. (2.92). Taking into account that equation and the normalization of the spherical harmonics, we obtain

$$E_{ex} = -\sum_{(n_i l_i)} (2l_i + 1) \int \frac{P_{n_i l_i}(r)}{r}$$

$$\times \left\{ \sum_{(n_j l_j)} \sum_k \times \sqrt{\frac{2l_j + 1}{2l_i + 1}} \, c^k(l_i 0, l_j 0) \frac{1}{r} Y_k(n_j l_j n_i l_i | r) \frac{P_{n_j l_j}}{r} \right\} r^2 \, dr$$

$$= -\sum_{(nl)} \sum_{(n'l')} \sqrt{(2l + 1)(2l' + 1)} \sum_k c^k(l0, l'0) \times G_k(nl, n'l'), \quad (2.112)$$

where we have taken into account Eq. (2.77f). The summations over (nl) and $(n'l')$ are again independent and over all complete groups. We are now ready to write the total energy. It will be the sum of the electrostatic and exchange energies to which we must add the kinetic energies and the energies arising from nuclear attraction. These last two are the first terms in Eqs. (2.68a) and (2.68b). Using the results displayed in Eqs. (2.70a), (2.107), and (2.112), we get

$$
E_F = \sum_{(nl)} 2(2l+1) I(nl)
$$

$$
+ \frac{1}{2} \sum_{(nl)} \sum_{(n'l')} [4l+1][4l'+2] F_0(nl, n'l')
$$

$$
- \sum_{(nl)} \sum_{(n'l')} \sqrt{(2l+1)(2l'+1)} \sum_k c^k(l0, l'0) \times G_k(nl, n'l'). \quad (2.113)
$$

This is the HF energy expression in Slater's formulation. We can write this formula in such a form that will lend itself better to a plausible physical interpretation. Let us extract from the double summations those terms for which $(nl) = (n'l')$. From the electrostatic energy, we get

$$
\frac{1}{2} \sum_{(nl)} [4l+2]^2 F_0(nl, nl). \quad (2.114a)
$$

This is the electrostatic-interaction energy of the group (nl) summed over all groups (including the self-energy). Thus, we can write

$$
\frac{1}{2} \sum_{(nl)} \sum_{(n'l')} [4l+2][4l'+2] F_0(nl, n'l')
$$

$$
= \sum_{(nl)} \tfrac{1}{2}(4l+2)^2 F_0(nl, nl)
$$

$$
+ \tfrac{1}{2} \sum_{\substack{(nl)\,(n'l') \\ (n'l') \neq (nl)}} (4l+2)(4l'+2) F_0(nl, n'l'). \quad (2.114b)
$$

From the exchange energy, we get for $(n'l') = (nl)$

$$
- \sum_{(nl)} (2l+1) \sum_k c^k(l0, l0) G_k(nl, nl). \quad (2.115a)
$$

This is the exchange-interaction energy within the (nl) group summed over all groups. Thus, we can write

$$- \sum_{(nl)} \sum_{(n'l')} \sqrt{(2l + 1)(2l' + 1)} \sum_k c^k(l0, l'0) G_k(nl, n'l')$$

$$= - \sum_{(nl)} \frac{4l + 2}{2} \sum_k c^k(l0, l0) F_k(nl, nl)$$

$$- \sum_{\substack{(nl) (n'l') \\ (n'l') \neq (nl)}} \sqrt{(2l + 1)(2l' + 1)} \sum_k c^k(l0, l'0) G_k(nl, n'l'), \quad (2.115b)$$

where we have used the relationship

$$G_k(nl, nl) = F_k(nl, nl). \quad (2.115c)$$

Combining Eqs. (2.114b) and (2.115b), we obtain

$$E_F = \sum_{(nl)} 2(2l + 1) I(nl) + \sum_{(nl)} \frac{1}{2}[4l + 2]^2 F_0(nl, nl)$$

$$+ \frac{1}{2} \sum_{\substack{(nl) (n'l') \\ (n'l') \neq (nl)}} [4l + 2][4l' + 2] F_0(nl, n'l')$$

$$- \sum_{(nl)} \frac{4l + 2}{2} \sum_{k=0}^{2l} c^k(l0, l0) F_k(nl, nl)$$

$$- \sum_{\substack{(nl) (n'l') \\ (n'l') \neq (nl)}} \sqrt{(2l + 1)(2l' + 1)} \sum_{k=0}^{l+l'} c^k(l0, l'0) G_k(nl, n'l'), \quad (2.116)$$

where we note again that the double summations are independent.

The physical interpretation of this expression is simple because we can show easily now that it is identical with Hartree's energy expression, which we have given in Eq. (2.73). Let us compare the two expressions term by term. The first terms are obviously identical because $q(nl) = 2(2l + 1)$. The second term in Hartree's expression can be written as

$$\sum_{(nl)} \frac{q(nl)}{2}[q(nl) - 1] F_0(nl, nl)$$

$$= \sum_{(nl)} \frac{2(2l + 1)}{2}[2(2l + 1) - 1] F_0(nl, nl)$$

$$= \sum_{(nl)} \frac{1}{2}(4l + 2)^2 F_0(nl, nl) - \sum_{(nl)} (2l + 1) F_0(nl, nl). \quad (2.117)$$

The first term in this expression is identical with the second term of Slater's expression. The second term of Eq. (2.117) is also present in Slater's expression; it is the $k = 0$ term in the fourth sum. Indeed, we have for the $k = 0$ term

$$-\sum_{(nl)} \frac{4l + 2}{2} c^0(l0, l0) F_0(nl, nl)$$

$$= -\sum_{(nl)} \frac{4l + 2}{2} \frac{A_{l0}}{2l + 1} F_0(nl, nl)$$

$$= -\sum_{(nl)} (2l + 1) F_0(nl, nl), \tag{2.118}$$

which is the second term of Eq. (2.117). (We have used the relationship given by Eq. (2.100c).) The third sum in Hartree's expression is obviously identical with the third term in Slater's expression; the factor of $\frac{1}{2}$ difference is because in Slater's expression, the summations are independent, whereas in Hartree's expression, the summation is over all $(nl, n'l')$ pairs. For the fourth term in Hartree's expression, we get, using Eq. (2.100b),

$$-\sum_{(nl)} \sum_{k>0} A_{lk} F_k(nl, nl)$$

$$= -\sum_{(nl)} \sum_{k>0} (2l + 1) c^k(l0, l0) F_k(nl, nl), \tag{2.119}$$

which is identical with the $k = 0$ terms in Slater's fourth term. Finally, the last, the fifth, term in Hartree's expression becomes identical with Slater's last expression if we recall that because this is a double summation, we must multiply Hartree's expression by $\frac{1}{2}$ in order to get Slater's. Doing this for the last term, we obtain

$$-\sum_{(nl)} \sum_{(n'l')} \sum_k \frac{B_{ll'k}}{2} G_k(nl, n'l')$$

$$= -\sum_{(nl)} \sum_{(n'l')} \sum_k \sqrt{(2l + 1)(2l' + 1)}\, c^k(l0, l'0) G_k(nl, n'l'), \tag{2.120}$$

which is indeed Slater's last term. Thus, the identity of the two energy expressions is demonstrated.

From the identity of the two expressions, it follows that the plausible physical interpretation that we have attached to Hartree's expression is also valid for Slater's formula. It is clear, however, that although Slater's formula, as it is given in Eq. (2.113), does not have the simple physical interpretation that can be attached to Hartree's expression, it has certain advantages over

Hartree's formulation. One such advantage is conceptual simplicity: Looking at Eq. (2.113), we recognize at once that the three terms of this formula are coming directly from the three terms of the general HF energy expression given by Eq. (2.42). It is also clear that Slater's formula is much better suited for numerical calculations. On the other hand, Hartree's formula will be very useful when we come to the discussion of the HF model for atoms with incomplete groups.

2.4. EXCHANGE POTENTIALS

When we presented the HF theory for a single-determinantal wave function in Sec. 2.2, we pointed out that the main difference between the model of Hartree and the HF theory was the appearance of the exchange operator in the latter. We have analyzed the meaning of this operator and we have seen that it was the result of correlation between electrons with parallel spins. We have seen that due to the difference in the exchange operators, the spatial wave functions of orbitals with different spins are generally different when we consider an atom with p positive spins and $N - p$ negative spins.

A significant step can be made forward in the development of atomic theory by adopting an idea of Slater concerning the exchange operator.[6] We now show that on the basis of this idea, the exchange operator can be localized, that is, it can be formally written in the form of a local potential. Although the basic idea here is Slater's, the presentation is more a condensation of a procedure that evolved in atomic theory since the publication of Slater's work. (See especially the localization of pseudopotentials in Sec. 7.2/B.)

In order to appreciate the significance of this idea, let us first recall that in the HF equation, Eq. (2.50), in which we have the HF potential given by Eqs. (2.48) and (2.49), we have two terms: the electrostatic potential and the exchange operator. As we have seen in Fig. 2.2, the electrostatic potential can be plotted, that is, it is a function whose physical meaning can easily be demonstrated. This is not the case with the exchange operator. A nonlocal potential, even if its effect can be described, cannot be plotted; thus, its physical meaning cannot be demonstrated directly.

In his original paper, Slater pointed out that in the quantum theory of solids, the electrons are often considered to move in a joint potential. In atomic physics, the valence electron is often considered to move in a local (model) potential. Strictly speaking, neither of these pictures is compatible with the HF theory. We now show that there is a way of writing the nonlocal potential of the HF equations into the form of a local potential, which, in certain cases, can be plotted.

First, we consider the general case of an atom with p electrons with up spins and $N - p$ electrons with down spins. The HF formalism for that case was developed in Sec. 2.2, so we can write the equations immediately. We had separate HF equations for the orbitals with up and down spins; these were given in Eqs. (2.66a) and (2.66b). For positive spins, we had

$$\left\{-\frac{1}{2}\Delta - \frac{Z}{r} + V - A_+\right\}\psi_k = \varepsilon_k\psi_k \qquad (k = 1, 2, \ldots, p), \quad (2.121)$$

where, according to Eq. (2.62),

$$V(1) = \sum_{j=1}^{N} \int \frac{1}{r_{12}}|\psi_j(2)|^2 \, dv_2, \qquad (2.122)$$

and, according to Eq. (2.64),

$$A_+(1)\psi_k(1) = \sum_{j=1}^{p} \int \frac{1}{r_{12}}\psi_j(1)\psi_j^*(2)\psi_k(2) \, dv_2. \qquad (2.123)$$

Let V_k be the local exchange potential, which we introduce with the definition

$$V_k(1) \equiv \frac{A_+(1)\psi_k(1)}{\psi_k(1)}. \qquad (2.124)$$

Using this definition, we can write the HF equations in the form

$$\left\{-\frac{1}{2}\Delta - \frac{Z}{r} + V - V_k\right\}\psi_k = \varepsilon_k\psi_k \qquad (k = 1, 2, \ldots, p), \quad (2.125)$$

In these equations V_k is a *local* potential. Thus, the total potential, $V - V_k$, is now local and the exchange potential depends on ψ_k; thus, it will be different for each orbital even within the group of orbitals with parallel spins.

Next, we write down the HF equations for both groups. Let V_k^+ be the exchange potential of the orbital ψ_k among the orbitals with up spin and V_k^- the potential of the orbital ψ_k among the orbitals with down spin. Then from Eqs. (2.66a) and (2.66b), we get

$$\left\{-\frac{1}{2}\Delta - \frac{Z}{r} + V - V_k^+\right\}\psi_k = \varepsilon_k\psi_k \qquad (k = 1, \ldots p), \qquad (2.126a)$$

$$\left\{-\frac{1}{2}\Delta - \frac{Z}{r} + V - V_k^-\right\}\psi_k = \varepsilon_k\psi_k \qquad (k = p + 1, \ldots, N). \quad (2.126b)$$

Let us compare these equations with the originals, Eqs. (2.66a) and (2.66b). Mathematically, the equations have not been changed; localization keeps the equations intact. Formally, the equations are quite different. The exchange operators in the equations for up spins are different from the operators in the equations for down spins, although the operators in the equations for different orbitals with the same spin are the same. Thus, parallel-spin orbitals are the solutions of the same HF equation.

On the other hand, the exchange potentials are different for each orbital even if the orbitals have parallel spins. Therefore, although the mathematical contents of the two groups of equations are exactly the same, their external form is changed considerably.

What is the use of this transformation? In many cases, the physical interpretation of the equations is easier if the exchange operator is localized. This is especially true if the resultant local potential can be plotted in a diagram or can be represented by an analytic function. As we will see in the forthcoming parts of this book, Slater's idea proved to be significant in the development of atomic structure theory. One area in which this idea will prove to be important is the theory of valence electrons, especially the pseudopotential theory of valence electrons (Sec. 7.2).

We write down now the exchange potential for atoms with complete groups. For these atoms, we have $p = N/2$ electrons with up spins and $N - p = N/2$ electrons with down spins. Thus, in this case, the exchange operators A_+ and A_- are identical. For the joint exchange operator A, we have derived the expression given by Eq. (2.92):

$$A(1)\psi_i(1) = \sum_{(n_j l_j)} \sum_k \sqrt{\frac{2l_j + 1}{2l_i + 1}}\, c^k(l_i 0, l_j 0)$$

$$\times \frac{1}{r_1} Y_k\left(n_j l_j n_i l_i | r\right) \frac{P_{n_j l_j}(r_1)}{r_1} Y_{l_i m_{l_i}}(\vartheta_1 \varphi_1). \qquad (2.127)$$

Using this expression, we can write down the exchange potential immediately from the definition given by Eq. (2.124):

$$V_i(1) = \frac{A(1)\psi_i(1)}{\psi_i(1)}, \qquad (2.128)$$

and so

$$V_i(1) = \sum_{(n_j l_j)} \sum_k \sqrt{\frac{2l_j + 1}{2l_i + 1}}\, c^k(l_i 0, l_j 0)$$

$$\times \frac{1}{r_1} Y_k\left(n_j l_j, n_i l_i | r_1\right) \frac{P_{n_j l_j}(r_1)}{P_{n_i l_i}(r_1)}. \qquad (2.129)$$

As we see, this is a function of r_1 only; therefore, it can be plotted immediately if the radial orbitals and the Y_k functions are available. We note that the angular dependence disappeared from V_i because, as we see from Eq. (2.127), the angular part of $A\psi_i$ is the same as the angular part of ψ_i for atoms with complete groups.

Let us write down now the HF equations for complete groups with this new notation. The equations were given in Eq. (2.97b). Let $U(n_i l_i|r)$ be the total HF potential. Using Eq. (2.129), we obtain

$$U(n_i l_i|r) = \sum_{(n_j l_j)} 2(2l_j + 1)\frac{1}{r}Y_0(n_j l_j, n_j l_j|r)$$

$$- \sum_{(n_j l_j)} \sum_k \sqrt{\frac{2l_j + 1}{2l_i + 1}}\, c^k(l_i 0, l_j 0)$$

$$\times \frac{1}{r}Y_k(n_j l_j, n_i l_i|r)\frac{P_{n_j l_j}(r)}{P_{n_i l_i}(r)}, \qquad (2.130)$$

and the HF equations are

$$\left\{-\frac{1}{2}\frac{d^2}{dr^2} + \frac{l_i(l_i + 1)}{2r^2} - \frac{Z}{r} + U(n_i l_i|r)\right\}P_{n_i l_i}(r) = \varepsilon_{n_i l_i}P_{n_i l_i}(r). \quad (2.131)$$

Comparing Eq. (2.130) with Eq. (2.98), we can say that the original potential was nonlocal but was the same for all orbitals, whereas in our new notation, we have a local potential that is different for each $(n_i l_i)$. Nevertheless, the great advantage of the local potential over the nonlocal is that the former can be plotted and thus lends itself easily to physical interpretation.

For further elucidation of the meaning of the exchange potential, Slater presented some arguments that go back to the work of Wigner and Seitz.[7] Wigner and Seitz have shown that in a homogeneous electron gas, the density of electrons with parallel spins is not constant, but it is zero at the position of a selected electron, increases with the distance r from this electron, and goes asymptotically to a constant at $r \to \infty$. Thus, around each electron, there will be a hole in the density of electrons with spins parallel to the spin of the selected electron. This hole, called the "Fermi hole," diminishes the positive electrostatic-interaction energy of the parallel-spin electrons and this decrease in the electrostatic energy is called the exchange energy.

Slater transferred these arguments to finite systems and to the exchange potential. On the basis of Eqs. (2.123) and (2.124), for the exchange potential, we get

$$V_k(1) = \frac{A_+(1)\psi_k(1)}{\psi_k(1)}$$

$$= \sum_{j=1}^{p} \int \frac{1}{r_{12}} \frac{\psi_j(1)\psi_j^*(2)\psi_k(2)}{\psi_k(1)} \, dv_2. \tag{2.132}$$

Let us introduce the exchange-charge density by the formula

$$\rho_k(2) = \sum_{j=1}^{p} \frac{\psi_j(1)\psi_j^*(2)\psi_k(2)}{\psi_k(1)}. \tag{2.133}$$

This quantity is a function of $\mathbf{r}_2$ for every fixed position $\mathbf{r}_1$ of the selected electron whose wave function is ψ_k. Using this definition, we can consider V_k as the potential of the charge density ρ_k,

$$V_k(1) = \int \frac{1}{r_{12}} \rho_k(2) \, dv_2. \tag{2.134}$$

We can also define the corrected density of the electrons with up spins as

$$\rho_+^c(2) = \sum_{j=1}^{p} |\psi_j(2)|^2 - \rho_k(2), \tag{2.135}$$

and using this, we can write the total potential for up-spin electrons in the form

$$V - V_k = \int \frac{1}{r_{12}} \rho_+(2) \, dv_2 + \int \frac{1}{r_{12}} \rho_-(2) \, dv_2 - \int \frac{1}{r_{12}} \rho_k(2) \, dv_2$$

$$= \int \frac{1}{r_{12}} \rho_+^c(2) \, dv_2 + \int \frac{1}{r_{12}} \rho_-(2) \, dv_2. \tag{2.136}$$

The exchange-charge density has the following properties. Integrating over $\mathbf{r}_2$ with fixed $\mathbf{r}_1$, we get

$$\int \rho_k(2) \, dv_2 = \sum_{j=1}^{p} \int \frac{\psi_j(1)\psi_j^*(2)\psi_k(2) \, dv_2}{\psi_k(1)}$$

$$= \sum_{j=1}^{p} \frac{\psi_j(1)}{\psi_k(1)} \int \psi_j^*(2)\psi_k(2) \, dv_2 = 1, \tag{2.137}$$

because, obviously, k is among the first p indices. Also, for a fixed $\mathbf{r}_1$,

$$\lim_{(\mathbf{r}_2 \to \mathbf{r}_1)} \rho_k(2) = \rho_k(1) = \sum_{j=1}^{p} |\psi_j(1)|^2, \qquad (2.138a)$$

from which it follows that

$$\lim_{(\mathbf{r}_2 \to \mathbf{r}_1)} \rho_+^c(2) = \rho_+^c(1) = 0. \qquad (2.138b)$$

For a fixed $\mathbf{r}_1$, we have also

$$\lim_{\mathbf{r}_2 \to \infty} \rho_k(2) = 0. \qquad (2.139)$$

Thus, we see that the total charge density of electrons with positive spins, which is given by ρ_+^c, is equal to zero at the position of the selected electron and that the total amount of the exchange charge amounts to one electron charge. We can give Eq. (2.136) the following interpretation. Electron k is moving in an electrostatic potential in which the potential of the other electrons with positive spins is reduced by the exchange potential. This potential is generated by the exchange charge, which amounts to one electron. Thus, the electrostatic potential of the parallel-spin electrons will be reduced by 1 as the result of the exchange potential. The total charge of up-spin electrons is zero at the position of the selected electron. Thus, we can say again that there is a hole in the electron density around the selected electron that diminishes the electrostatic energy and this decrease is the exchange energy. Thus, the exchange interaction acts as if it would repel all parallel-spin electrons from the vicinity of the selected electron. The conclusions of Slater are identical with the conclusions of Wigner and Seitz in all aspects.

Finally, Slater suggested that the exchange potential, which is different for each orbital ψ_k, can be replaced in good approximation with an average potential that is the same for all orbitals. In the preceding discussion, we find two arguments for this assumption. First, the amount of the exchange-charge density is equal to one electron charge for each orbital. Second, the corrected charge density is zero at the position of each electron. Thus, each electron is surrounded by a Fermi hole amounting to one electron charge. Further arguments were given by Slater on the basis of the Thomas–Fermi model and he suggested, for the approximate average exchange potential, the formula

$$V_e = 6 \left(\frac{3}{8\pi} \right)^{1/3} \left\{ \sum_{j=1}^{N} |\psi_j(1)|^2 \right\}^{1/3}. \qquad (2.140)$$

The discussion of this formula belongs generically to the density-functional formalism and is planned for the second volume of this work.

As a qualitative application of the idea of the exchange hole, we show that Hund's rule can be explained by using the preceding discussion. Examining the atomic spectra, Hund stated the following empirical rule. Among multipletts with different L and S, that is, with different total angular and spin momenta, the multiplett with the lowest energy will be the one with the highest S. If there are several multipletts with the same S, the multiplett with the lowest energy will be the one with the highest L. We consider here only that part of the rule that associates the highest S with the lowest energy.

Let us consider an atom with N valence electrons, among which p have positive spins and $N - p$ have negative spins. We have seen that each electron has a Fermi hole around its position and that the presence of this hole lowers the total energy by diminishing the positive electrostatic-interaction energy of the electrons. The electrostatic-interaction energy of the positive-spin electrons will be diminished with respect to the other positive-spin electrons because there will be a Fermi hole in the density of these electrons around each selected electron. The electrostatic energy of these electrons with respect to the negative-spin electrons will not be influenced directly by the spin. Likewise, the electrostatic-interaction energy of the negative-spin electrons will be diminished with respect to the other negative-spin electrons, but not with respect to the positive-spin electrons.

Thus, the intragroup electrostatic interaction of both groups is reduced, but the spin direction does not directly affect the intergroup interaction.

Now let us consider the same system of N valence electrons, but now with all spins parallel. Then, clearly, every electron will be surrounded by a Fermi hole in the density of all other electrons. Thus, the electrostatic energy of every electron with respect to all other electrons will be diminished. The situation is this: In the first system, the electrostatic energy of the electrons is diminished with respect to electrons with parallel spins; in the second, the electrostatic energy is diminished with respect to all electrons. Clearly, the second system will have a lower energy than the first and this will be true regardless of the ratio $p/(N - p)$. Our result is, therefore, that the system in which all spins are parallel has the lowest energy. This corresponds to the first part of Hund's rule, which is shown here to be the direct result of the exchange interaction.

2.5. PERTURBATION THEORY

In this section, we formulate the perturbation theory in the framework of the HF model on the basis of the work of Allen.[8] Let us consider an atom with N electrons. Let us assume that for this atom, the HF equations have been solved and the HF orbitals and orbital parameters are available. Next, let us assume that the atom is subjected to an external potential that can be considered a perturbation relative to the internal HF potential in which the electrons are moving. Let us further assume that, in principle, we could obtain exact HF solutions also in the presence of this perturbing potential. Instead of obtaining these exact solutions, however, we want to obtain approximations to them by perturbation theory using the unperturbed HF solutions as starting points. Thus, the difference between perturbation theory in the HF model and perturbation theory in exact quantum mechanics is that in the former, the perturbed as well as the unperturbed solutions are HF solutions, whereas in the latter, we are trying to approach the solutions of the exact Schroedinger equation through perturbation theory.

The theory in its perturbed as well as in its unperturbed form rests on the single-determinantal approximation, therefore, the relevant formulas are those given in Sec. 2.2. Let φ_k be the kth orbital under perturbation and ε_k the corresponding orbital parameter. Let the one-electron perturbing potential be V. The HF equations under perturbation will be, according to Eq. (2.50),

$$H\varphi_k = \varepsilon_k \varphi_k \qquad (k = 1, 2, \ldots, N), \qquad (2.141)$$

where

$$H = H_F + \lambda V, \qquad (2.142)$$

with H_F being the HF Hamiltonian operator given by

$$H_F = t + g + U, \qquad (2.143)$$

where

$$U = \sum_{i=1}^{N} U_i, \qquad (2.144)$$

and

$$U_i(1)f(1) = \int \frac{1}{r_{12}} |\varphi_i(2)|^2 \, dq_2 \, f(1) - \int \frac{1}{r_{12}} \varphi_i(1)\varphi_i^*(2) f(2) \, dq_2. \qquad (2.145)$$

The parameter λ is a real number that is usually employed in perturbation theory to turn the perturbation on and off. We assume that φ_k and ε_k can be expanded in terms of λ:

$$\varphi_k = \varphi_k^{(0)} + \lambda \varphi_k^{(1)} + \lambda^2 \varphi_k^{(2)} + \cdots, \qquad (2.146)$$

$$\varepsilon_k = \varepsilon_k^{(0)} + \lambda \varepsilon_k^{(1)} + \lambda^2 \varepsilon_k^{(2)} + \cdots, \qquad (2.147)$$

Let us write Eq. (2.141) in detail:

$$H_1 \varphi_k(1) = \left(t_1 + g_1 + \lambda V_1 + \sum_{i=1}^{N} \int \frac{1}{r_{12}} |\varphi_i(2)|^2 \, dq_2 \right) \varphi_k(1)$$

$$- \sum_{i=1}^{N} \int \frac{1}{r_{12}} \varphi_i(1) \varphi_i^*(2) \varphi_k(2) \, dq_2$$

$$= \varepsilon_k \varphi_k(1). \tag{2.148}$$

We want to determine the total energy of the atom under perturbation up to and including the second-order perturbation energy. This will necessitate the comparison of equal powers of λ in Eq. (2.148). Let us consider, first, the density matrix formed from the orbitals given by Eq. (2.146):

$$\varphi_k(1) \varphi_k^*(2) = \left(\varphi_k^{(0)}(1) + \lambda \varphi_k^{(1)}(1) + \lambda^2 \varphi_k^{(2)}(1) + \cdots \right)$$

$$\times \left(\varphi_k^{(0)*}(2) + \lambda \varphi_k^{(1)*}(2) + \lambda^2 \varphi_k^{(2)*}(2) + \cdots \right)$$

$$= \varphi_k^{(0)}(1) \varphi_k^{(0)*}(2) + \lambda \left[\varphi_k^{(1)}(1) \varphi_k^{(0)*}(2) + \varphi_k^{(0)}(1) \varphi_k^{(1)*}(2) \right]$$

$$+ O(\lambda^2) + \cdots \tag{2.149}$$

By putting this expression along with the formulas given by Eqs. (2.146) and (2.147) into Eq. (2 148), we obtain

$$(t_1 + g_1 + \lambda V_1)\left(\varphi_k^{(0)}(1) + \lambda \varphi_k^{(1)}(1) + \lambda^2 \varphi_k^{(2)}(1) + \cdots \right)$$

$$+ \sum_{i=1}^{N} \int \frac{1}{r_{12}} \Big\{ |\varphi_i^{(0)}(2)|^2$$

$$+ \lambda \left[\varphi_i^{(1)}(2) \varphi_i^{(0)*}(2) + \varphi_i^{(0)}(2) \varphi_i^{(1)*}(2) \right] + O(\lambda_2) \Big\} \, dq_2$$

$$\times \left(\varphi_k^{(0)}(1) + \lambda \varphi_k^{(1)}(1) + \lambda^2 \varphi_k^{(2)}(1) + \cdots \right)$$

$$- \sum_{i=1}^{N} \int \frac{1}{r_{12}} \Big\{ \left[\varphi_i^{(0)}(1) \varphi_i^{(0)*}(2) + \lambda \left(\varphi_i^{(1)}(1) \varphi_i^{(0)*}(2) \right. \right.$$

$$\left. + \varphi_i^{(0)}(1) \varphi_i^{(1)*}(2) \right) + O(\lambda^2) \Big] \left[\varphi_k^{(0)}(2) + \lambda \varphi_k^{(1)}(2) \right.$$

$$+ \lambda^2 \varphi_k^{(2)}(2) + \cdots \Big] \Big\} \, dq_2$$

$$= \left(\varepsilon_k^{(0)} + \lambda \varepsilon_k^{(1)} + \lambda^2 \varepsilon_k^{(2)} + \cdots \right)$$

$$\times \left(\varphi_k^{(0)}(1) + \lambda \varphi_k^{(1)}(1) + \lambda^2 \varphi_k^{(2)}(1) + \cdots \right). \tag{2.150}$$

Now we are able to compare equal powers of λ. We get from the zeroth power,

$$(t_1 + g_1)\varphi_k^{(0)}(1) + \sum_{i=1}^{N} \int \frac{1}{r_{12}} \left| \varphi_i^{(0)}(2) \right|^2 dq_2 \, \varphi_k^{(0)}(1)$$

$$- \sum_{i=1}^{N} \int \frac{1}{r_{12}} \varphi_i^{(0)}(1)\varphi_i^{(0)*}(2)\varphi_k^{(0)}(2) \, dq_2$$

$$= \varepsilon_k^{(0)}\varphi_k^{(0)}(1). \tag{2.151}$$

This equation can be written in the form

$$H_F\varphi_k^{(0)} = \varepsilon_k^{(0)}\varphi_k^{(0)}, \tag{2.152}$$

where the HF Hamiltonian operator is a functional of the unperturbed orbitals only.

Thus, our first result is that the unperturbed orbitals are solutions to the single-determinantal HF equations. Equating the coefficients of the first power of λ, we obtain

$$\left(H_F(1) + V_1\right)\varphi_k^{(1)}(1)$$

$$+ \sum_{i=1}^{N} \int \frac{1}{r_{12}} \left(\varphi_i^{(1)}(2)\varphi_i^{(0)*}(2) + \varphi_i^{(0)}(2)\varphi_i^{(1)*}(2) \right) dq_2 \, \varphi_k^{(0)}(1)$$

$$- \sum_{i=1}^{N} \int \frac{1}{r_{12}} \left(\varphi_i^{(1)}(1)\varphi_i^{(0)*}(2) + \varphi_i^{(0)}(1)\varphi_i^{(1)*}(2) \right)\varphi_k^{(0)}(2) \, dq_2$$

$$= \varepsilon_k^{(1)}\varphi_k^{(0)} + \varepsilon_k^{(0)}\varphi_k^{(1)}. \tag{2.153}$$

Let us denote the operator in the third and fourth terms by $X(1)$. Then the equation becomes

$$\left(H_F - \varepsilon_k^{(0)}\right)\varphi_k^{(1)} = \left(\varepsilon_k^{(1)} - X - V\right)\varphi_k^{(0)}. \tag{2.154}$$

Let us recall that in this equation, H_F is the functional of the unperturbed orbitals and the operator X is a functional of the unperturbed orbitals as well as of the first-order perturbed functions. Equations (2.152) and (2.154) define the orbitals and the orbital parameters up to the first order of perturbation theory.

We can derive a relationship for the first-order correction to the orbital parameter by multiplying Eq. (2.154) from the left by $\varphi_k^{(0)*}$ and integrating:

$$\langle \varphi_k^{(0)} | H_F | \varphi_k^{(1)} \rangle - \varepsilon_k^{(0)} \langle \varphi_k^{(0)} | \varphi_k^{(1)} \rangle$$

$$= \varepsilon_k^{(1)} \langle \varphi_k^{(0)} | \varphi_k^{(0)} \rangle - \langle \varphi_k^{(0)} | X + V | \varphi_k^{(0)} \rangle. \qquad (2.155)$$

Using the Hermitian property of H_F, we obtain

$$\varepsilon_k^{(1)} = \langle \varphi_k^{(0)} | X + V | \varphi_k^{(0)} \rangle. \qquad (2.156)$$

Thus, we see that by using this equation, Eq. (2.154) can be solved by a self-consistent procedure. We have assumed that the zero-order solutions are available. Using these, one can obtain a first iteration of $\varepsilon_k^{(1)}$ from Eq. (2.156); one way of doing it is to put the first-order corrections on the right side of this equation equal to zero. Using this first iteration of $\varepsilon_k^{(1)}$, one can get a solution of Eq. (2.154) for the first iteration of $\varphi_k^{(1)}$. Using the $\varepsilon_k^{(1)}$ and the $\varphi_k^{(1)}$ obtained this way as the first iterations, one can get the next iteration for $\varepsilon_k^{(1)}$ from Eq. (2.156) and the next iteration for $\varphi_k^{(1)}$ from Eq. (2.154). Repeating this procedure until self-consistency is achieved, one can get the first-order correction to the orbitals and orbital parameters.

Next, we look at the orthonormality conditions. The solutions of Eqs. (2.141) can be assumed to be orthonormal without loss of generality. Thus,

$$\langle \varphi_k | \varphi_j \rangle = \delta_{kj} \qquad (k, j = 1, 2, \ldots, N). \qquad (2.157)$$

Let us substitute the expression given by Eq. (2.146) into the orthonormality relationships:

$$\left\langle \left(\varphi_k^{(0)} + \lambda \varphi_k^{(1)} + \lambda^2 \varphi_k^{(2)} + \cdots \right) \right|$$

$$\times \left(\varphi_j^{(0)} + \lambda \varphi_j^{(1)} + \lambda^2 \varphi_j^{(2)} + \cdots \right) \right\rangle = \delta_{kj}. \qquad (2.158)$$

Equating the zeroth powers of λ, we obtain

$$\langle \varphi_k^{(0)} | \varphi_j^{(0)} \rangle = \delta_{kj} \qquad (k, j = 1, 2, \ldots, N). \qquad (2.159)$$

This equation expresses the orthonormality of the unperturbed HF orbitals. Equating the first powers of λ, we get

$$\langle \varphi_k^{(1)} | \varphi_j^{(0)} \rangle + \langle \varphi_k^{(0)} | \varphi_j^{(1)} \rangle = 0, \qquad (2.160)$$

from which we obtain, for $k = j$,

$$\langle \varphi_k^{(1)} | \varphi_k^{(0)} \rangle = 0. \qquad (2.161)$$

The first-order correction is orthogonal to the unperturbed function in the same way as in the "exact" perturbation theory. Allen noted that Eq. (2.160) can be obtained directly from the equation defining $\varphi_k^{(1)}$, that is, from Eq. (2.154).

We turn now to the total energy of the atom. This quantity is given as

$$E_T = \langle \Psi_T^* | H | \Psi_T \rangle = E_0 + \lambda E_1 + \lambda^2 E_2 + \cdots \qquad (2.162)$$

where H is the exact Hamiltonian operator, and Ψ_T is the single-determinantal function formed from the φ_k orbitals. Allen obtained, by using the equations defining the unperturbed orbitals, the equations for the first-order corrections and the orthonormality conditions:

$$E_1 = \sum_{k=1}^{N} \int | \varphi_k^{(0)} |^2 V \, dq, \qquad (2.163)$$

and

$$E_2 = \sum_{k=1}^{N} \int \varphi_k^{(0)*} V \varphi_k^{(1)} \, dq. \qquad (2.164)$$

Thus, the total energy of the atom, under the influence of the perturbation, up to the second order of perturbation theory, is given by

$$E = \sum_{k=1}^{N} \varepsilon_k^{(0)} - \frac{1}{2} \sum_{k,j=1}^{N} \varepsilon_{kj}^{(0)} + \sum_{k=1}^{N} \int | \varphi_k^{(0)} |^2 V \, dq + \sum_{k=1}^{N} \int \varphi_k^{(0)*} V \varphi_k^{(1)} \, dq, \qquad (2.165)$$

where, for the unperturbed total energy, we have substituted the formula given by Eq. (2.54). The quantity ε_{kj} is defined by Eq. (2.52) and the zero superscript means that in Eq. (2.52), we have to substitute the unperturbed orbitals, $\varphi_k^{(0)}$.

In summing up, we can say that the energy of the atom up to the second order of the perturbation theory is given by Eq. (2.165), in which we have the unperturbed orbitals, the orbital parameters, and the first-order corrections to the HF orbitals. The unperturbed orbitals and orbital parameters are the solutions of Eq. (2.152). The first-order corrections are defined by Eq. (2.154). The orthonormality conditions are given by Eqs. (2.157), (2.159), and (2.161). As we see from Eq. (2.165), the first-order correction to the orbital parameter is not needed for the total energy, but it is needed for Eq. (2.154), which determines the first-order correction $\varphi_k^{(1)}$.

It is clear from the previous discussion that the formulas for the higher approximations can be developed similarly to the equations for the zeroth and first approximations. It is also evident that instead of approaching the perturbed HF orbitals in successive approximations, one can try to solve Eq. (2.141) directly. For the problem of the electric dipole polarizabilities of atoms, that is, for a constant electric field as the perturbation, a direct method for the solution of Eqs. (2.141) was developed by Cohen and Roothaan.[9] In this method, the HF orbitals were expressed as linear combinations of analytic Slater-type functions. This work is a special case of the general Hartree–Fock–Roothaan procedure, which is discussed in Sec. 5.6/D.

2.6. KOOPMANS' THEOREM

In Sec. 2.2, we have seen that the total energy of an atom with N electrons, in the single-determinantal HF approximation, is given by Eq. (2.54):

$$E_F = \sum_{i=1}^{N} \varepsilon_i - \frac{1}{2} \sum_{i,j=1}^{N} \varepsilon_{ij}, \tag{2.166}$$

where ε_i, the HF energy parameter, was defined by the HF equations:

$$H_F \varphi_i = \varepsilon_i \varphi_i. \tag{2.167}$$

The HF Hamiltonian operator was given by the formula

$$H_F = t + g + \sum_{j=1}^{N} U_j, \tag{2.168}$$

with

$$U_j(1)f(1) = \int \frac{1}{r_{12}} |\varphi_j(2)|^2 \, dq_2 \, f(1) - \int \frac{1}{r_{12}} \varphi_j(1) \varphi_j^*(2) f(2) \, dq_2. \tag{2.169}$$

The quantity ε_{ij} is an interaction integral defined by

$$\varepsilon_{ij} = \varepsilon_{ji} = \langle \varphi_i | U_j | \varphi_i \rangle. \tag{2.170}$$

The most conspicuous feature of the energy equation, Eq. (2.166), is that the sum of the orbital parameters of the HF equations is not equal to the total energy. This is, of course, the result of the electrons not being noninteracting particles in the HF approximation. Our goal here is to show that the orbital parameters do have a definite physical meaning. This will be shown by presenting Koopmans' theorem.[10] The theorem states that under a certain approximation, the orbital parameter $-\varepsilon_k$ is equal to the energy necessary for the removal of an electron of the atom from the orbital φ_k.

Let the HF energy of the atom be E_N and let us assign the orbitals $(\varphi_1, \ldots, \varphi_k, \ldots, \varphi_N)$ to the N electrons. The wave function of the atom is

$$\psi_N = \frac{1}{\sqrt{N!}} \det[\varphi_1, \ldots, \varphi_k, \ldots, \varphi_N], \tag{2.171}$$

and the use of this HF wave function leads to the formulas given in Eqs. (2.166)–(2.170).

Let us now remove one electron from the atom and let us assign to the remaining $N - 1$ electrons the orbitals $\varphi_1', \ldots, \varphi_{k-1}' \varphi_{k+1}', \ldots, \varphi_N'$. These orbitals have the same angular and spin parts as the corresponding unprimed orbitals; the prime means that the radial parts will be different because now we must reformulate the HF model, that is, we must recompute the radial orbitals. Let us write down the equations for the $(N - 1)$-electron atom.

Let E_{N-1} be the HF energy. We again have the formula,

$$E_{N-1} = \sum_{\substack{i=1 \\ (i \neq k)}}^{N} \varepsilon_i' - \frac{1}{2} \sum_{\substack{i,j=1 \\ (i,j \neq k)}}^{N} \varepsilon_{ij}', \tag{2.172}$$

where the energy parameters are now defined by the HF equations for the $(N - 1)$ electron system,

$$H_F' \varphi_i' = \varepsilon_i' \varphi_i'. \tag{2.173}$$

In this equation, H_F' is given by

$$H_F' = t + g + \sum_{\substack{j=1 \\ (j \neq k)}}^{N} U_j', \tag{2.174}$$

where g is the same as in Eq. (2.168) because we have not changed the nuclear charge. The prime on U_j' means that the potential is now built from the primed HF orbitals.

Next let us consider the energy difference between the N-electron and $(N - 1)$-electron atoms:

$$\Delta E = E_{N-1} - E_N. \tag{2.175}$$

Using Eq. (2.166) for E_N and Eq. (2.172) for E_{N-1}, we obtain

$$\Delta E = \sum_{\substack{i=1 \\ (i \neq k)}}^{N} (\varepsilon_i' - \varepsilon_i) - \varepsilon_k - \frac{1}{2} \sum_{\substack{i,j=1 \\ (i,j \neq k)}}^{N} (\varepsilon_{ij}' - \varepsilon_{ij})$$

$$+ \frac{1}{2} \sum_{j=1}^{N} \varepsilon_{kj} + \frac{1}{2} \sum_{i=1}^{N} \varepsilon_{ik}. \tag{2.176}$$

From the HF equations, we obtain

$$\langle \varphi_i | H_F | \varphi_i \rangle = \langle \varphi_i | t + g | \varphi_i \rangle + \sum_{j=1}^{N} \langle \varphi_i | U_j | \varphi_i \rangle = \varepsilon_i, \tag{2.177}$$

and

$$\varepsilon_i' = \langle \varphi_i' | t + g | \varphi_i' \rangle + \sum_{\substack{j=1 \\ (j \neq k)}}^{N} \langle \varphi_i' | U_j' | \varphi_i' \rangle. \tag{2.178}$$

Using Eqs. (2.177) and (2.178), we get

$$\varepsilon_i' - \varepsilon_i = \langle \varphi_i' | t + g | \varphi_i' \rangle - \langle \varphi_i | t + g | \varphi_i \rangle$$

$$+ \sum_{j=1}^{N} (\varepsilon_{ij}' - \varepsilon_{ij}) - \varepsilon_{ik}. \tag{2.179}$$

Using the relationship $\varepsilon_{kj} = \varepsilon_{jk}$ and putting the expression given by Eq.

(2.179) into Eq. (2.176), we obtain

$$
\begin{aligned}
\Delta E &= \sum_{\substack{i=1 \\ (i \neq k)}}^{N} \left\{ \langle \varphi_i' | t + g | \varphi_i' \rangle - \langle \varphi_i | t + g | \varphi_i \rangle + \sum_{\substack{j=1 \\ (j \neq k)}}^{N} (\varepsilon_{ij}' - \varepsilon_{ij}) - \varepsilon_{ik} \right\} - \varepsilon_k \\
&\quad - \frac{1}{2} \sum_{\substack{i,j=1 \\ (i,j \neq k)}}^{N} (\varepsilon_{ij}' - \varepsilon_{ij}) + \sum_{j=1}^{N} \varepsilon_{jk} \\
&= \sum_{\substack{i=1 \\ (i \neq k)}}^{N} \left\{ \langle \varphi_i' | t + g | \varphi_i' \rangle - \langle \varphi_i | t + g | \varphi_i \rangle + \sum_{\substack{j=1 \\ (j \neq k)}}^{N} (\varepsilon_{ij}' - \varepsilon_{ij}) \right\} \\
&\quad - \varepsilon_k - \frac{1}{2} \sum_{\substack{i,j=1 \\ (i,j \neq k)}}^{N} (\varepsilon_{ij}' - \varepsilon_{ij}) \\
&= \sum_{\substack{i=1 \\ (i \neq k)}}^{N} \left\{ \langle \varphi_i' | t + g | \varphi_i' \rangle - \langle \varphi_i | t + g | \varphi_i \rangle \right\} \\
&\quad + \frac{1}{2} \sum_{\substack{i,j=1 \\ (i,j \neq k)}}^{N} (\varepsilon_{ij}' - \varepsilon_{ij}) - \varepsilon_k.
\end{aligned}
\tag{2.180}
$$

We are now ready to formulate Koopmans' theorem. Let us assume that the removal of the orbital φ_k has no effect on the remaining $N - 1$ orbitals $\varphi_1, \ldots, \varphi_{k-1} \varphi_{k+1}, \ldots, \varphi_N$. That is, let us assume that

$$
\varphi_i' \equiv \varphi_i \quad (i = 1, \ldots, k - 1, k + 1, \ldots, N).
\tag{2.181}
$$

Then, obviously, we have

$$
\langle \varphi_i' | t + g | \varphi_i' \rangle = \langle \varphi_i | t + g | \varphi_i \rangle,
\tag{2.182}
$$

and

$$
\varepsilon_{ij}' = \varepsilon_{ij} \quad (i, j = 1, \ldots, k - 1, k + 1, \ldots, N).
\tag{2.183}
$$

Using these relationships in Eq. (2.180), we get

$$
\Delta E = E_{N-1} - E_N = -\varepsilon_k.
\tag{2.184}
$$

This is Koopmans' theorem. In words: to the extent of the approximation according to which the HF orbitals of an atom do not change if one of the orbitals is removed, the orbital parameter of the removed electron is equal to the (negative of the) energy necessary to remove the electron from that orbital.

This is as close as it is possible to give a physical meaning to the orbital parameter ε_k. The approximation that is necessary for Eq. (2.184) to be valid is obviously better fulfilled in a large atom with many electrons than in one with only a few electrons. There is one case in which this theorem is very meaningful. That is the case when ε_k is the orbital parameter of a valence electron. In the case of one or a few valence electrons, it is reasonable to assume that the core orbitals do not change if the valence orbital(s) is (are) removed. This case is discussed in Secs. 6.3 and 6.4.

2.7. DENSITY-MATRIX FORMULATION

In this section, we present some work of Dirac,[11] which had far-reaching effects on the development of atomic structure theory. Dirac has shown that the HF equations can be transformed into a density-matrix formalism, that is, into a formalism in which the quantity to be determined is not the set of N HF orbitals but a single-density matrix. We develop Dirac's theory here in the single-determinantal approximation.

First, let us collect the formulas that we will need. The HF equation in the single-determinantal approximation is given by Eq. (2.50):

$$H_F \varphi_k = \varepsilon_k \varphi_k \qquad (k = 1, 2, \ldots, N), \qquad (2.185)$$

where the HF Hamiltonian operator is

$$H_F = F + \sum_{j=1}^{N} U_j, \qquad (2.186)$$

with

$$U_j(1)f(1) = \int \frac{1}{r_{12}} |\varphi_j(2)|^2 \, dq_2 \, f(1) - \int \frac{1}{r_{12}} \varphi_j(1) \varphi_j^*(2) f(2) \, dq_2. \quad (2.187)$$

The operator F contains the kinetic energy and the nuclear potential. Besides these, F may contain some other external one-electron potential that does not need to be specified. The total energy of the atom is given by Eq.

(2.42), which we write down here in our new notation:

$$E_F = \sum_{i=1}^{N} \int \varphi_i^* F \varphi_i \, dq + \frac{1}{2} \sum_{i,j=1}^{N} \int \frac{1}{r_{12}} |\varphi_i(1)|^2 |\varphi_j(2)|^2 \, dq_{12}$$

$$- \frac{1}{2} \sum_{ij=1}^{N} \int \frac{1}{r_{12}} \varphi_i^*(1) \varphi_j(1) \varphi_j^*(2) \varphi_i(2) \, dq_1 \, dq_2. \qquad (2.188)$$

As the first step in the transformation of the formalism, let us introduce the HF density operator ρ, with the definition

$$\rho = \sum_{i=1}^{N} |\varphi_i\rangle\langle\varphi_i|. \qquad (2.189)$$

The representative of ρ in coordinate representation is the (first-order) HF density matrix

$$\langle 1|\rho|2\rangle \equiv \rho(1,2) = \sum_{i=1}^{N} \varphi_i(1)\varphi_i^*(2). \qquad (2.190)$$

The diagonal component of this matrix is the total HF electron density:

$$\langle 1|\rho|1\rangle \equiv \rho(1,1) \equiv \rho(1) = \sum_{i=1}^{N} |\varphi_i(1)|^2. \qquad (2.191)$$

Next, we introduce the density matrix into the HF energy expression. It is easy to see that we can write the following equations:

$$\sum_{i=1}^{N} \int \varphi_i^*(1) F_1 \varphi_i(1) \, dq_1 = \int \delta(2-1) F_2 \rho^*(1,2) \, dq_{12}; \qquad (2.192)$$

$$\frac{1}{2} \sum_{i,j=1}^{N} \int \frac{1}{r_{12}} |\varphi_i(1)|^2 |\varphi_j(2)|^2 \, dq_{12} = \frac{1}{2} \int \frac{\rho(1)\rho(2)}{r_{12}} \, dq_{12}; \qquad (2.193)$$

$$\frac{1}{2} \sum_{i,j=1}^{N} \int \frac{1}{r_{12}} \varphi_i^*(1) \varphi_j(1) \varphi_j^*(2) \varphi_i(2) \, dq_{12} = \frac{1}{2} \int \frac{1}{r_{12}} \rho(1,2)\rho^*(1,2) \, dq_{12}. \qquad (2.194)$$

In Eq. (2.192), $\delta(2-1) \equiv \delta(q_2 - q_1)$ is the Dirac delta function. Using the last three equations, we obtain for the HF energy

$$E_F = \int \delta(2 - 1) F_2 \rho^*(1, 2)\, dq_{12}$$

$$+ \frac{1}{2} \int \frac{1}{r_{12}} \rho(1) \rho(2)\, dq_{12}$$

$$- \frac{1}{2} \int \frac{1}{r_{12}} \rho(1, 2) \rho^*(1, 2)\, dq_{12}. \tag{2.195}$$

The next step is to write the HF equations in density-matrix form. In order to accomplish this, let us introduce the operators V and A. Let the representative of V in the coordinate representation be

$$\langle 1|V|2 \rangle = \delta(1 - 2) \int \frac{\langle 3|\rho|3 \rangle}{r_{13}}\, dq_3, \tag{2.196}$$

which means that if $|f\rangle$ is any ket, then the representative of $V|f\rangle$ in the coordinate representation will be

$$\langle 1|V|f \rangle = \int \langle 1|V|2 \rangle \langle 2|f \rangle\, dq_2$$

$$= \int dq_2 \left\{ \delta(1 - 2) \int \frac{1}{r_{13}} \langle 3|\rho|3 \rangle\, dq_3\, f(2) \right\}$$

$$= \int \frac{1}{r_{13}} \sum_{j=1}^{N} |\varphi_j(3)|^2\, dq_3\, f(1), \tag{2.197}$$

which is clearly the electrostatic potential in the HF equation. For the representative of A in coordinate representation we put

$$\langle 1|A|2 \rangle = \frac{\langle 1|\rho|2 \rangle}{r_{12}}, \tag{2.198}$$

which means that if $|f\rangle$ is any ket, then the representative of $A|f\rangle$ in the coordinate representation will be

$$\langle 1|A|f \rangle = \int \langle 1|A|2 \rangle \langle 2|f \rangle\, dq_2$$

$$= \int \frac{1}{r_{12}} \langle 1|\rho|2 \rangle f(2)\, dq_2$$

$$= \int \frac{1}{r_{12}} \sum_{j=1}^{N} \varphi_j(1) \varphi_j^*(2) f(2)\, dq_2, \tag{2.199}$$

which is the exchange operator in the HF equation. Using the definitions of V and A, we can write the HF equations in the form

$$H_F|\varphi_k\rangle = \varepsilon_k|\varphi_k\rangle, \tag{2.200}$$

where

$$H_F = F + V - A. \tag{2.201}$$

We can also introduce operators F, V, and A into the energy expression. For the electrostatic interaction given by Eq. (2.193), we obtain, by using the definition of the V operator,

$$\frac{1}{2}\int\frac{1}{r_{12}}\rho(1)\rho(2)\,dq_{12} = \mathrm{Tr}\left[\frac{1}{2}V\rho\right], \tag{2.202}$$

where Tr means the trace of the operator product.

The validity of this relationship can easily be proved. Using the definitions given by Eqs. (2.189) and (2.196), we obtain

$$\mathrm{Tr}\left[\frac{1}{2}V\rho\right] = \frac{1}{2}\int dq_1\,\langle 1|V\rho|1\rangle$$

$$= \frac{1}{2}\int dq_1\,dq_2\,\langle 1|V|2\rangle\langle 2|\rho|1\rangle$$

$$= \frac{1}{2}\int dq_1\,dq_2\,\delta(1-2)\int\frac{1}{r_{13}}\langle 3|\rho|3\rangle\,dq_3\,\langle 2|\rho|1\rangle$$

$$= \frac{1}{2}\int dq_1\,dq_3\,\langle 1|\rho|1\rangle\frac{1}{r_{13}}\langle 3|\rho|3\rangle$$

$$= \frac{1}{2}\sum_{i,j=1}^{N}\int\frac{1}{r_{13}}|\varphi_i(1)|^2|\varphi_j(3)|^2\,dq_1\,dq_3$$

$$= \frac{1}{2}\int\frac{1}{r_{13}}\rho(1)\rho(3)\,dq_{13}, \tag{2.203}$$

which proves the statement given in Eq. (2.202). Similarly, we obtain that $\mathrm{Tr}[A\rho/2]$ is equal to the expression given by Eq. (2.194). In order to prove this, we substitute into the trace the expressions given by Eqs. (2.198) and

(2.189) and obtain

$$
\begin{aligned}
\mathrm{Tr}\left[\frac{1}{2}A\rho\right] &= \frac{1}{2}\int dq_1 \langle 1|A\rho|1\rangle \\
&= \frac{1}{2}\int dq_1\, dq_2 \langle 1|A|2\rangle\langle 2|\rho|1\rangle \\
&= \frac{1}{2}\int dq_1\, dq_2 \frac{1}{r_{12}}\langle 1|\rho|2\rangle\langle 2|\rho|1\rangle \\
&= \frac{1}{2}\int dq_1\, dq_2 \frac{1}{r_{12}}\rho(1,2)\rho^*(1,2),
\end{aligned}
\tag{2.204}
$$

which is indeed the exchange integral given by Eq. (2.194).

The expression given by Eq. (2.192) can also be written in the form of a trace as follows:

$$
\begin{aligned}
\mathrm{Tr}\,[F\rho] &= \int dq_1 \langle 1|F\rho|1\rangle \\
&= \int dq_1\, dq_2 \langle 1|F|2\rangle\langle 2|\rho|1\rangle \\
&= \int dq_1\, dq_2 \left[\delta(2-1)F_2\rho^*(1,2)\right],
\end{aligned}
\tag{2.205}
$$

where we have used the relationship $\langle 2|\rho|1\rangle = \rho(2,1) = \rho^*(1,2)$. What we obtained is indeed equal to the expression given to Eq. (2.192). In order that Eq. (2.205) be valid, we have to define the representative of F in the coordinate representation. We put

$$
\langle 1|F|2\rangle = \delta(2-1)F_2.
\tag{2.206}
$$

This means that if $|f\rangle$ is any ket, then the representative of $F|f\rangle$ in the coordinate representation is given by the formula

$$
\begin{aligned}
\langle 1|F|f\rangle &= \int \langle 1|F|2\rangle\langle 2|f\rangle\, dq_2 \\
&= \int dq_2 \{\delta(2-1)F_2 f(2)\} = F_1 f(1),
\end{aligned}
\tag{2.207}
$$

which is the conventional expression for $F|f\rangle$. Using Eqs. (2.203), (2.204), and (2.205), we obtain for the HF energy of the atom

$$
E_F = \mathrm{Tr}\left[\left(F + \frac{1}{2}V - \frac{1}{2}A\right)\rho\right],
\tag{2.208}
$$

where the operators F, V, and A are defined by Eqs. (2.206), (2.196), and (2.198).

We transform now the HF model into a density-operator formalism. First, we recall that the HF equations in operator/ket form are given in Eqs. (2.200) and (2.201). Let us assume that the HF orbital kets can change in time because of the presence of a time-dependent potential in the operator F. Then we can replace the orbital parameters ε_k by the operator $i\hbar\,d/dt$, that is, we can write the equations in the form

$$H_F|\varphi_k\rangle = i\hbar\frac{d}{dt}|\varphi_k\rangle \qquad (k = 1, 2, \ldots, N). \tag{2.209}$$

Next, let us consider the density operator ρ as defined by Eq. (2.189):

$$\rho = \sum_{k=1}^{N} |\varphi_k\rangle\langle\varphi_k|. \tag{2.210}$$

The time derivative of this operator is

$$\dot{\rho} \equiv \frac{d}{dt}\rho = \sum_{k=1}^{N}\left(\frac{d}{dt}|\varphi_k\rangle\right)\langle\varphi_k| + \sum_{k=1}^{N}|\varphi_k\rangle\left(\langle\varphi_k|\frac{d}{dt}\right). \tag{2.211}$$

From Eq. (2.209), we have

$$\frac{d}{dt}|\varphi_k\rangle = \frac{1}{i\hbar}H_F|\varphi_k\rangle. \tag{2.212}$$

Now take the conjugate imaginary of the operator equation, Eq. (2.209). According to the rules of Dirac, we obtain

$$\langle\varphi_k|H_F = \langle\varphi_k|\frac{d}{dt}(-i\hbar), \tag{2.213}$$

and from this we have

$$\langle\varphi_k|\frac{d}{dt} = -\frac{1}{i\hbar}\langle\varphi_k|H_F. \tag{2.214}$$

Substituting Eqs. (2.212) and (2.214) into Eq. (2.211), we get

$$\dot{\rho} = \sum_{k=1}^{N}\frac{1}{i\hbar}H_F|\varphi_k\rangle\langle\varphi_k| - \sum_{k=1}^{N}\frac{1}{i\hbar}|\varphi_k\rangle\langle\varphi_k|H_F$$
$$= \frac{1}{i\hbar}(H_F\rho - \rho H_F), \tag{2.215}$$

where ρ is the HF-density operator. We note that the use of total derivatives is proper here because the eigenkets the HF equations are not functions of coordinates; only their representatives in coordinate representation are functions.

We have now obtained the equation of motion for the density operator. The equation is subjected to subsidiary conditions, which are as follows. In the HF model, the orbitals can be assumed to be orthonormal since the antisymmetric determinantal function does not change by the orthogonalization of the one-electron orbitals. Thus, we can write

$$\langle \varphi_i | \varphi_k \rangle = \delta_{ik}, \qquad (2.216)$$

and also

$$\sum_{i=1}^{N} \langle \varphi_i | \varphi_i \rangle = N. \qquad (2.217)$$

These two equations can be written in the following operator form:

$$\rho^2 = \rho, \qquad (2.218)$$

and

$$\text{Tr}\,[\rho] = N. \qquad (2.219)$$

In order to have all equations of the density-operator formalism written down in one group, we recall that

$$H_F = F + V - A, \qquad (2.220)$$

and the HF energy in operator form is given by

$$E_F = \text{Tr}\left[\left(F + \frac{1}{2}V - \frac{1}{2}A\right)\rho\right]. \qquad (2.221)$$

Our final results are given in the Eqs. (2.215), (2.218), (2.220), and (2.221). As we see from these equations, the HF model appears here as a formalism in which we have the density operator ρ and its representative in coordinate representation, the density matrix. The HF orbitals and orbital parameters disappeared from the formalism. Instead of determining the set of N HF orbitals and orbital parameters, the task here is to determine the single-density matrix and the total energy of the atom given by Eq. (2.221).

We will not further explore here these mathematically and conceptually very interesting results. We note, however, that using the density-matrix formulation, Dirac was able to derive the basic equation of the so-called Thomas–Fermi–Dirac model, that is, he was able to make the transition from a wave-function formalism to a density-functional formalism. The discussion of the density-functional formalism, along with Dirac's work related to the development of that formalism, is reserved for the second volume of this work.

In conclusion, we show on a simple example how the density-matrix equations can be used. We demonstrate that if the operator F is not time-dependent, then the HF energy is constant. Thus, let us assume that

$$\frac{dF}{dt} = 0. \tag{2.222}$$

The rate of change of the total energy is given by

$$\frac{dE_F}{dt} = \frac{d}{dt} \operatorname{Tr} \left[\left(F - \frac{1}{2}V - \frac{1}{2}A \right) \rho \right]. \tag{2.223}$$

Consider

$$\frac{d}{dt} \operatorname{Tr}[V\rho] = \frac{d}{dt} \int \langle 1|V\rho|1 \rangle \, dq_1$$

$$= \frac{d}{dt} \int \langle 1|V|2 \rangle \langle 2|\rho|1 \rangle \, dq_1 \, dq_2$$

$$= \frac{d}{dt} \int \delta(1-2) \frac{1}{r_{13}} \langle 3|\rho|3 \rangle \langle 2|\rho|1 \rangle \, dq_{123}$$

$$= \int \delta(1-2) \frac{1}{r_{13}} \langle 3|\dot{\rho}|3 \rangle \langle 2|\rho|1 \rangle \, dq_{123}$$

$$+ \int \delta(1-2) \frac{1}{r_{13}} \langle 3|\rho|3 \rangle \langle 2|\dot{\rho}|1 \rangle \, dq_{123}. \tag{2.224}$$

Integrate over q_2 in both integrals. Then we get

$$\int \frac{1}{r_{13}} \langle 3|\dot{\rho}|3 \rangle \langle 1|\rho|1 \rangle \, dq_{13}$$

$$+ \int \frac{1}{r_{13}} \langle 3|\rho|3 \rangle \langle 1|\dot{\rho}|1 \rangle \, dq_{13}$$

$$= 2 \int \frac{1}{r_{13}} \langle 3|\rho|3 \rangle \langle 1|\dot{\rho}|1 \rangle \, dq_{13}, \tag{2.225}$$

where we interchanged the dummy variables q_1 and q_3. Restoring the δ function, we get

$$\frac{d}{dt} \operatorname{Tr} \left[\frac{1}{2}V\rho \right] = \int \frac{1}{r_{13}} \delta(1-2) \langle 3|\rho|3 \rangle \langle 2|\dot{\rho}|1 \rangle \, dq_{123}$$

$$= \int \langle 1|V|2 \rangle \langle 2|\dot{\rho}|1 \rangle \, dq_1 \, dq_2$$

$$= \int \langle 1|V\dot{\rho}|1 \rangle \, dq_1 = \operatorname{Tr}[V\dot{\rho}]. \tag{2.226}$$

Likewise, we get

$$\frac{d}{dt} \, \mathrm{Tr} \left[\frac{1}{2} A\rho \right] = \mathrm{Tr} \left[A\dot{\rho} \right]. \tag{2.227}$$

Thus, we obtain for the rate of change of the HF energy

$$\frac{dE_F}{dt} = \frac{d}{dt} \, \mathrm{Tr} \left[\left(F + \frac{1}{2}V - \frac{1}{2}A \right) \rho \right]$$

$$= \mathrm{Tr} \left[\dot{F}\rho + F\dot{\rho} + V\dot{\rho} - A\dot{\rho} \right]. \tag{2.228}$$

Using the definition of H_F, we obtain

$$F\dot{\rho} + V\dot{\rho} - A\dot{\rho} = H_F\dot{\rho}, \tag{2.229}$$

and substituting from the equation of motion the expression for $\dot{\rho}$, we obtain

$$H_F\dot{\rho} = \frac{1}{i\hbar} H_F (H_F\rho - \rho H_F). \tag{2.230}$$

By putting this into the expression for the derivative of the total energy, we get

$$\frac{dE_F}{dt} = \mathrm{Tr} \left[\dot{F}\rho + \frac{1}{i\hbar} (H_F H_F \rho - H_F \rho H_F) \right]. \tag{2.231}$$

Using the relationship

$$\mathrm{Tr} \left[KLM \right] = \mathrm{Tr} \left[LMK \right],$$

we get

$$\frac{dE_F}{dt} = \mathrm{Tr} \left[\dot{F}\rho \right], \tag{2.232}$$

and if the operator F is time-independent, as we have assumed, then

$$\dot{F} = 0, \quad \frac{dE_F}{dt} = 0, \quad E_F = \mathrm{const.}, \tag{2.233}$$

which is the relationship we wanted to demonstrate.

CHAPTER 3

ATOMS WITH INCOMPLETE GROUPS

3.1. FORMULATION OF THE HARTREE – FOCK MODEL IN LS COUPLING

Dirac's general formulation for applying quantum mechanics to a system is modified for the HF model; general formulation of HF model in LS coupling; differences between Dirac's formulation and HF model formulation; LS coupling theory for a single incomplete group; application of perturbation theory; calculation of the energies by diagonal sum rule; appraisal of the significance of Slater's method.

3.2. HARTREE – FOCK ENERGIES AND EQUATIONS FOR ATOMS WITH ONE INCOMPLETE GROUP

General formula for the HF energy of an atom in LS coupling for any multiplet arising from a configuration with one incomplete group; HF equations for an orbital in the incomplete group and for an orbital in one of the complete groups; theorem about the LS independence of the potential for the orbitals in the complete groups; theorem about the orbitals always being the eigenfunctions of non-Hermitian operators.

3.3. THE THEORY OF CONFIGURATION AVERAGES

Observations of the characteristic features of the HF theory for atoms with one incomplete group; Shortley's and Slater's theory of the configuration averages; definition of the weighed mean; construction of the average energy for an atom with an arbitrary number of complete and incomplete groups; construction of a formula for the LS multiplet in terms of the average energy for atoms with one and two incomplete groups; the HF equations for configurations averages; non-Hermiticity of the HF Hamiltonian operator.

3.1. FORMULATION OF THE HF MODEL IN LS COUPLING

The single-determinantal HF theory that we presented in the preceding chapter has the attractive feature of being simple and thus being well suited for the study of the underlying physical assumptions. We have seen that the physical interpretation of the equations is quite straightforward in that formulation. Unfortunately, there are only a few atoms that can be accurately described by a single-determinantal wave function.

We proceed now to formulate the Hartree–Fock (HF) model, for an arbitrary atom, in LS coupling. We consider an atom with N electrons with the Hamiltonian

$$H = \sum_{i=1}^{N} f_i + \frac{1}{2} \sum_{i,j=1}^{N} \frac{1}{r_{ij}}, \tag{3.1}$$

where f_i is the kinetic-energy operator plus the nuclear potential, and r_{ij} is the distance electrons i and j. Let $\mathbf{S}_k$ be the operator of the spin of the kth electron. We define the total spin operator as

$$\mathbf{S} = \sum_{k=1}^{N} \mathbf{S}_k. \tag{3.2}$$

The z-component of this operator is

$$S_z = \sum_{k=1}^{N} S_{kz}. \tag{3.3}$$

Likewise, let the angular momentum operator of the kth electron be L_k; then the total angular momentum is

$$\mathbf{L} = \sum_{k=1}^{N} \mathbf{L}_k, \tag{3.4}$$

with the z-component given by

$$L_z = \sum_{k=1}^{N} L_{kz}. \tag{3.5}$$

According to the general formulation of Dirac, the quantum mechanics description of a system consists of selecting a complete set of commuting observables (Hermitian operators that possess complete sets of eigenfunctions) and determining the joint eigenfunction of these observables. The measurable quantities associated with the system will be the eigenvalues of the observables belonging to the joint eigenfunction. The set of observables will be complete if there is only one joint eigenfunction belonging to each set of eigenvalues.

Since the HF approximation is a *model*, we have to modify Dirac's formulation. No matter how we choose the set of commuting observables, the Hamiltonian must be one of them. In the HF approximation, however, we cannot seek eigenfunctions that are exact eigenfunctions of the Hamiltonian because, in the HF approximation, we work with one-electron functions and the exact eigenfunctions of the Hamiltonian will have electron-correlation built into them and thus cannot be represented by a single-determinantal wave function or by a finite set of determinantal wave functions.

We formulate the HF model in LS coupling as follows. For each multiplet, that is, for each energy level of the atom with definite L and S, we construct wave functions that are joint eigenfunctions of L^2, S^2, L_z, and S_z (the choice of these operators is dictated by the fact that they commute with the Hamiltonian). The joint eigenfunctions will be built from determinantal wave functions, which, in turn, will be constructed from central-field type one-electron orbitals. Then, the expectation value of the Hamiltonian, Eq. (3.1), will be formed using the joint eigenfunctions. The expectation value, that is, the total energy of the atom, will be varied with respect to the radial parts of the one-electron orbitals. The HF equations for the radial parts of the orbitals will be obtained by this variation. The solutions of the HF equations will define the best one-electron orbitals that can be obtained with the joint eigenfunction. The HF energy of the atom will be identified with the minimum of the total energy obtained by the variation procedure.

As we see, this formulation, which is the conventional formulation of the HF model in LS coupling, is different from Dirac's general formulation. We will construct wave functions that are joint eigenfunctions of operators, but the requirement that a joint eigenfunction be an eigenfunction of the Hamiltonian must be replaced by the requirement that it should minimize the expectation value of the Hamiltonian. Due to the restrictions placed on the form of the joint eigenfunction, the minimum of the expectation value will not be equal to an exact eigenvalue of the Hamiltonian and the joint eigenfunction will not be an exact eigenfunction of the Hamiltonian.

Before building up the HF model according to the previous formulation, we remind the reader of the strategy outlined in the first chapter of the book. As we have outlined it, the goal of this book is to present a comprehensive theory of the Hamiltonian operator, or, more accurately, to present a comprehensive survey of that part of atomic structure theory that rests on the Hamiltonian operator. In other words, we will concentrate on the question of how approximate eigenfunctions and eigenvalues of the Hamiltonian operator can be obtained and we will not go into the details of how the approximate eigenfunctions of the Hamiltonian can be made simultaneous eigenfunctions of angular momentum and spin operators. That part of the atomic structure theory that deals with the construction of angular momentum and spin eigenfunctions will only be *summarized* in this book.

TABLE 3.1

Atom	Z	Ground state	1s	2s	2p	3s	3p	3d	4s	4p	4d	4f	5s	5p	5d	5f	6s	6p	6d	7s
H	1	$^2S_{1/2}$	1																	
He	2	$^1S_0$	2																	
Li	3	$^2S_{1/2}$	2	1																
Be	4	$^1S_0$	2	2																
B	5	$^2P_{1/2}$	2	2	1															
C	6	$^3P_0$	2	2	2															
N	7	$^4S_{3/2}$	2	2	3															
O	8	$^3P_2$	2	2	4															
F	9	$^2P_{3/2}$	2	2	5															
Ne	10	$^1S_0$	2	2	6															
Na	11	$^2S_{1/2}$	2	2	6	1														
Mg	12	$^1S_0$	2	2	6	2														
Al	13	$^2P_{1/2}$	2	2	6	2	1													
Si	14	$^3P_0$	2	2	6	2	2													
P	15	$^4S_{3/2}$	2	2	6	2	3													
S	16	$^3P_2$	2	2	6	2	4													
Cl	17	$^2P_{3/2}$	2	2	6	2	5													
Ar	18	$^1S_0$	2	2	6	2	6													
K	19	$^2S_{1/2}$	2	2	6	2	6		1											
Ca	20	$^1S_0$	2	2	6	2	6		2											
Sc	21	$^2D_{3/2}$	2	2	6	2	6	1	2											
Ti	22	$^3F_2$	2	2	6	2	6	2	2											
V	23	$^4F_{3/2}$	2	2	6	2	6	3	2											
Cr	24	$^7S_3$	2	2	6	2	6	5	1											
Mn	25	$^6S_{5/2}$	2	2	6	2	6	5	2											
Fe	26	$^5D_4$	2	2	6	2	6	6	2											
Co	27	$^4F_{9/2}$	2	2	6	2	6	7	2											
Ni	28	$^3F_4$	2	2	6	2	6	8	2											
Cu	29	$^2S_{1/2}$	2	2	6	2	6	10	1											
Zn	30	$^1S_0$	2	2	6	2	6	10	2											
Ga	31	$^2P_{1/2}$	2	2	6	2	6	10	2	1										
Ge	32	$^3P_0$	2	2	6	2	6	10	2	2										
As	33	$^4S_{3/2}$	2	2	6	2	6	10	2	3										
Se	34	$^3P_2$	2	2	6	2	6	10	2	4										
Br	35	$^2P_{3/2}$	2	2	6	2	6	10	2	5										
Kr	36	$^1S_0$	2	2	6	2	6	10	2	6										
Rb	37	$^2S_{1/2}$	2	2	6	2	6	10	2	6			1							
Sr	38	$^1S_0$	2	2	6	2	6	10	2	6			2							
Y	39	$^2D_{3/2}$	2	2	6	2	6	10	2	6	1		2							
Zr	40	$^3F_2$	2	2	6	2	6	10	2	6	2		2							
Nb	41	$^6D_{1/2}$	2	2	6	2	6	10	2	6	4		1							
Mo	42	$^7S_3$	2	2	6	2	6	10	2	6	5		1							
Tc	43	$^6S_{5/2}$	2	2	6	2	6	10	2	6	5		2							
Ru	44	$^5F_5$	2	2	6	2	6	10	2	6	7		1							
Rh	45	$^4F_{9/2}$	2	2	6	2	6	10	2	6	8		1							
Pd	46	$^1S_0$	2	2	6	2	6	10	2	6	10									

TABLE 3.1 *(Continued)*

Atom	Z	Ground state	1s	2s	2p	3s	3p	3d	4s	4p	4d	4f	5s	5p	5d	5f	6s	6p	6d	7s
Ag	47	$^2S_{1/2}$	2	2	6	2	6	10	2	6	10		1							
Cd	48	$^1S_0$	2	2	6	2	6	10	2	6	10		2							
In	49	$^2P_{1/2}$	2	2	6	2	6	10	2	6	10		2	1						
Sn	50	$^3P_0$	2	2	6	2	6	10	2	6	10		2	2						
Sb	51	$^4S_{3/2}$	2	2	6	2	6	10	2	6	10		2	3						
Te	52	$^3P_2$	2	2	6	2	6	10	2	6	10		2	4						
I	53	$^2P_{3/2}$	2	2	6	2	6	10	2	6	10		2	5						
Xe	54	$^1S_0$	2	2	6	2	6	10	2	6	10		2	6						
Cs	55	$^2S_{1/2}$	2	2	6	2	6	10	2	6	10		2	6			1			
Ba	56	$^1S_0$	2	2	6	2	6	10	2	6	10		2	6			2			
La	57	$^2D_{3/2}$	2	2	6	2	6	10	2	6	10		2	6	1		2			
Ce	58	$^1G_4$	2	2	6	2	6	10	2	6	10	1	2	6	1		2			
Pr	59	$^4I_{9/2}$	2	2	6	2	6	10	2	6	10	3	2	6			2			
Nd	60	$^5I_4$	2	2	6	2	6	10	2	6	10	4	2	6			2			
Pm	61	$^6H_{5/2}$	2	2	6	2	6	10	2	6	10	5	2	6			2			
Sm	62	$^7F_0$	2	2	6	2	6	10	2	6	10	6	2	6			2			
Eu	63	$^8S_{7/2}$	2	2	6	2	6	10	2	6	10	7	2	6			2			
Gd	64	$^9D_2$	2	2	6	2	6	10	2	6	10	7	2	6	1		2			
Tb	65	$^6H_{15/2}$	2	2	6	2	6	10	2	6	10	9	2	6			2			
Dy	66	$^5I_8$	2	2	6	2	6	10	2	6	10	10	2	6			2			
Ho	67	$^4I_{15/2}$	2	2	6	2	6	10	2	6	10	11	2	6			2			
Er	68	$^3H_6$	2	2	6	2	6	10	2	6	10	12	2	6			2			
Tm	69	$^2F_{7/2}$	2	2	6	2	6	10	2	6	10	13	2	6			2			
Yb	70	$^1S_0$	2	2	6	2	6	10	2	6	10	14	2	6			2			
Lu	71	$^2D_{3/2}$	2	2	6	2	6	10	2	6	10	14	2	6	1		2			
Hf	72	$^3F_2$	2	2	6	2	6	10	2	6	10	14	2	6	2		2			
Ta	73	$^4F_{3/2}$	2	2	6	2	6	10	2	6	10	14	2	6	3		2			
W	74	$^5D_0$	2	2	6	2	6	10	2	6	10	14	2	6	4		2			
Re	75	$^6S_{5/2}$	2	2	6	2	6	10	2	6	10	14	2	6	5		2			
Os	76	$^5D_4$	2	2	6	2	6	10	2	6	10	14	2	6	6		2			
Ir	77	$^4F_{9/2}$	2	2	6	2	6	10	2	6	10	14	2	6	7		2			
Pt	78	$^3D_3$	2	2	6	2	6	10	2	6	10	14	2	6	9		1			
Au	79	$^2S_{1/2}$	2	2	6	2	6	10	2	6	10	14	2	6	10		1			
Hg	80	$^1S_0$	2	2	6	2	6	10	2	6	10	14	2	6	10		2			
Tl	81	$^1P_{1/2}$	2	2	6	2	6	10	2	6	10	14	2	6	10		2	1		
Pb	82	$^3P_0$	2	2	6	2	6	10	2	6	10	14	2	6	10		2	2		
Bi	83	$^3S_{3/2}$	2	2	6	2	6	10	2	6	10	14	2	6	10		2	3		
Po	84	$^3P_2$	2	2	6	2	6	10	2	6	10	14	2	6	10		2	4		
At	85	$^2P_{3/2}$	2	2	6	2	6	10	2	6	10	14	2	6	10		2	5		
Rn	86	$^1S_0$	2	2	6	2	6	10	2	6	10	14	2	6	10		2	6		
Fr	87	$^2S_{1/2}$	2	2	6	2	6	10	2	6	10	14	2	6	10		2	6		1
Ra	88	$^1S_0$	2	2	6	2	6	10	2	6	10	14	2	6	10		2	6		2
Ac	89	$^2D_{3/2}$	2	2	6	2	6	10	2	6	10	14	2	6	10		2	6	1	2
Th	90	$^3F_2$	2	2	6	2	6	10	2	6	10	14	2	6	10		2	6	2	2
Pa	91	$^4I_{9/2}$	2	2	6	2	6	10	2	6	10	14	2	6	10	2	2	6	1	2
U	92	$^5L_6$	2	2	6	2	6	10	2	6	10	14	2	6	10	3	2	6	1	2

In this section, we summarize how determinantal wave functions can be constructed that are eigenfunctions of certain sets of angular momentum and spin operators and how the energy expressions associated with these eigenfunctions can be obtained. In the remaining parts of the chapter, we show how the approximate eigenvalues of the Hamiltonian can be obtained using these wave functions and energy expressions.

We characterize an atom in its ground state or in any excited state by its electron configuration. The electron configuration is defined by giving the occupation numbers of one-electron states with the principal quantum number n and the azimuthal quantum number l. Thus, a configuration has this form:

$$(n_1 l_1)^{q(n_1 l_1)} (n_2 l_2)^{q(n_2 l_2)} \cdots (n_N l_N)^{q(n_N l_N)},$$

where $q(nl)$ is the occupation number of the state (nl). In Table 3.1, we list the electron configurations of neutral atoms in the ground state. The electrons of an atom can be divided into core electrons and valence electrons. The former always occupy complete groups and the latter may form complete or incomplete groups. The joint eigenfunction of complete groups is always a single-determinantal wave function and the energy expression for that case was given in Sec. 2.3.

In this section, we present LS coupling for a single incomplete group. Generally, we think of this group as being a group of valence electrons, but this assumption is not necessary for the validity of the presentation.

Let us consider the single incomplete group characterized by the quantum numbers (nl). In the Hartree model, Sec. 2.1, we have treated the problem in such a way that we have obtained one orbital P_{nl} and one orbital energy ε_{nl} for this group. But it is known from quantum mechanics that an atom with such a configuration can have several multiplets, that is, several different energy levels depending on the sets of L and S values that are possible for the configuration. The multiplet structure does not appear in the Hartree model; in that model, the energy of the atom is independent from L and S.

In order to introduce the multiplet structure into the HF theory, we proceed according to the method of Slater.[12] This method corresponds in almost all details to the general formulation given before except that in the discussion presented here, the step of constructing eigenfunctions of L^2 and S^2 will be bypassed and the energy expression will be obtained without deriving the explicit form of these eigenfunctions.

For the incomplete group characterized by (nl), we have a total of $2(2l + 1)$ spin orbitals of the form

$$\varphi_{nlm_l m_s} = \frac{P_{nl}(r)}{r} Y_{lm_l}(\vartheta, \varphi) \eta_{m_s}(\sigma). \tag{3.6}$$

Because we have less than $2(2l + 1)$ electrons in the incomplete group, we can select from these orbitals $q(nl)$ functions that differ from each other in the m_l and m_s quantum numbers. From these orbitals, we can construct determinantal wave functions, $\phi_1, \phi_2, \ldots, \phi_\alpha$, each of which contains one of the selected sets of one-electron orbitals. Each of these determinants will be an eigenfunction of the z-component operators, Eqs. (3.3) and (3.5). Let us denote the quantum numbers associated with L_z and S_z by M_L and M_S. Clearly, for each determinantal wave function, we have

$$M_L = \sum_k m_{lk},$$ (3.7)

and

$$M_S = \sum_k m_{sk}$$ (3.8)

We note that the index α in ϕ_α is not identical with (M_L, M_S) because among the determinantal wave functions, there might be some that are different but have the same (M_L, M_S). We also note that, since each determinant differs from all the others in at least one of the one-electron orbitals, the determinants will be orthogonal.

Next, we establish the connection between the possible multiplets and the determinantal wave functions. The possible L and S values are determined by the quantum mechanics rules. For two electrons with quantum numbers (l_1, s_1) and (l_2, s_2), we have the following range for L and S:

$$L = |l_1 - l_2|, |l_1 - l_2| + 1, \ldots, (l_1 + l_2),$$ (3.9a)

and

$$S = |s_1 - s_2|, |s_1 - s_2| + 1, \ldots, (s_1 + s_2).$$ (3.9b)

The equations are similar for more than two electrons. This calculation does not take into account the Pauli exclusion principle, which may eliminate some of the multiplets.

In order to take the Pauli exclusion principle into account, we draw up a list of the (M_L, M_S) values of all the determinantal wave functions that we have constructed. Let $n(M_L, M_S)$ be the number of determinants with (M_L, M_S). For each multiplet with the quantum numbers (L_k, S_k), we must have determinantal wave functions for each combination of (M_L, M_s) that fit into the range

$$-L_k \leq M_L \leq L_k,$$ (3.10a)

$$-S_k \leq M_S \leq S_k.$$ (3.10b)

Thus, for each (L_k, S_k), we must have $(2L_k + 1)(2S_k + 1)$ determinantal wave functions.

The determinantal wave function with (M_L, M_S) may be a member of the necessary set for several different multiplets. The general rule is this: The Pauli exclusion principle permits those multiplets for which complete sets of determinantal wave functions can be constructed. Let the number of multiplets that require a determinantal wave function with (M_L, M_S) be $n'(M_L, M_S)$. Then the permitted multiplets are identified by the equations[†]

$$n(M_L, M_S) = n'(M_L, M_S), \tag{3.11}$$

which must be satisfied for all (M_L, M_S) combinations that occur in our list mentioned before. We can state now that our set of determinantal wave functions, $\phi_1, \ldots, \phi_\alpha$, will contain the necessary $(2L_k + 1)(2S_k + 1)$ functions for each permitted multiplet with (L_k, S_k).

Next we construct a fictitious perturbation problem as follows. We assume that our one-electron orbitals are solutions of the Schroedinger equation:

$$H\varphi_k = \varepsilon_k \varphi_k, \tag{3.12}$$

where the Hamiltonian H, which is the same for all orbitals in the incomplete group, is given as

$$H = t + g + V, \tag{3.13}$$

where t is the kinetic-energy operator, g is the nuclear potential plus the fixed potential of the core electrons. This potential is included in this term because here we are talking about a single incomplete group outside of the core, which contains complete groups. V is a central potential representing the potential of the (nl) electrons in the incomplete group. Physically, it is plausible to say that V is the Hartree potential (Sec. 2.1), and, in fact, this choice is made in many textbooks in which it is remarked that the procedure rests on the Hartree method. However, as we shall see presently, the procedure is independent of the choice of V and, in fact, we can put $V = 0$. Thus, the assumption that the method rests on the Hartree model is not necessary.

Next, let us establish the notation for the perturbation theory. Let

$$H_0 = F + V_0, \tag{3.14a}$$

where

$$F = \sum_k (t_k + g_k), \tag{3.14b}$$

[†]Slater constructed a marvelous graphical method for the speedy handling of these equations [Slater I, 296].

and

$$V_0 = \sum_k V(k).$$ (3.14c)

Here t and g are the same as in Eq. (3.13).

The operators H_0, F, and V_0 are defined for the incomplete group, that is, the summation over k is over the electrons in the group. From the definitions, we see that

$$H_0 = \sum_k H(k),$$ (13.4d)

where H is given by Eq. (3.13). Also, let us define the Coulomb interaction energy for the incomplete group:

$$Q = \frac{1}{2} \sum_{i,j} \frac{1}{r_{ij}},$$ (3.14e)

where the summation is again over the electrons in the group. Finally, let

$$W = Q - V_0.$$ (3.14f)

With these notations, we get for the exact Hamiltonian of the electrons in the incomplete group:

$$H = \sum_k (t_k + g_k) + \frac{1}{2} \sum_{i,j} \frac{1}{r_{ij}} = F + Q.$$ (3.15)

Since the one-electron orbitals are the solutions of Eq. (3.12), the determinantal wave functions are solutions of the equation

$$H_0 \phi_\lambda = E_0 \phi_\lambda \quad (\lambda = 1, 2, \ldots, \alpha),$$ (3.16)

where

$$E_0 = \sum_k \varepsilon_k.$$ (3.17)

To each eigenvalue E_0, we have α determinantal wave functions. Let us now apply the degenerate perturbation theory and let us determine the energy up to the first order. Our "unperturbed" energy is E_0 and let us denote the first-order correction by E_1. We seek the perturbed wave function in the form

$$\Psi = \sum_{\lambda=1}^{\alpha} C_\lambda \phi_\lambda,$$ (3.18)

that is, we form a linear combination of all our determinantal wave functions. We define as the perturbation the operator W, that is, we put the exact Hamiltonian in the form

$$H = F + Q = F + V_0 + Q - V_0$$
$$= H_0 + W. \tag{3.19}$$

According to perturbation theory, we get the first-order correction to the energy from the scalar equation,

$$\det[W_{\kappa\lambda} - E_1\delta_{\kappa\lambda}] = 0 \qquad (\kappa, \lambda = 1, \ldots, \alpha), \tag{3.20}$$

where $W_{\kappa\lambda}$ is the matrix component of the perturbation operator.

Looking at the secular equation, we realize at once that it can be simplified considerably. From Eq. (3.14f), we see that the operator W will commute with the L_z and S_z operators. This being the case, we get

$$W_{\kappa\lambda} = \langle \phi_\kappa | W | \phi_\lambda \rangle = 0, \tag{3.21}$$

unless ϕ_λ and ϕ_κ belong to the same (M_L, M_S). Thus, by grouping together the determinantal functions with the same (M_L, M_S), we can put the secular determinant in the so-called block-diagonal form:

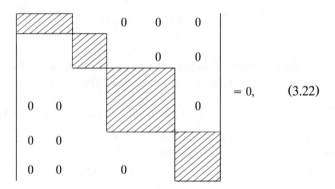

$$= 0, \tag{3.22}$$

where only the subdeterminants, indicated by the shaded areas, will be different from zero. We can make the secular determinant equal to zero by putting each of the subdeterminants in the shaded areas equal to zero.

The last step makes the solution of the secular equation very much easier, but Slater has shown that in many cases, the problem becomes even simpler by the application of the diagonal sum rule. According to this rule, we have, for the secular equation, Eq. (3.20), the following relationship:

$$\sum_{\lambda=1}^{\alpha} E_1^{(\lambda)} = \sum_{\lambda=1}^{\alpha} W_{\lambda\lambda}, \tag{3.23}$$

where $E_1^{(1)}, E_1^{(2)}, \ldots, E_1^{(\alpha)}$ are the roots of the secular equation. The rule says that the sum of the roots is equal to the sum of the diagonal terms of the $W_{\kappa\lambda}$ matrix. Thus, in many cases, we do not need to calculate even the subdeterminants of the secular problem, but, instead, we can obtain the roots by this rule.

We now show that the secular determinant can be brought to mathematically more convenient forms. First, we show that the choice of V in Eq. (3.13) is irrelevant. Let

$$E^{(\lambda)} = E_0 + E_1^{(\lambda)}. \tag{3.24}$$

Then we obtain,

$$\begin{aligned}
\sum_{\lambda=1}^{\alpha} E^{(\lambda)} &= \sum_{\lambda=1}^{\alpha} E_1^{(\lambda)} + \sum_{\lambda=1}^{\alpha} \langle \lambda | E_0 | \lambda \rangle \\
&= \sum_{\lambda=1}^{\alpha} \left\{ \langle \lambda | W | \lambda \rangle + \langle \lambda | E_0 | \lambda \rangle \right\} \\
&= \sum_{\lambda=1}^{\alpha} \left\{ \langle \lambda | Q - V_0 + F + V_0 | \lambda \rangle \right\} = \sum_{\lambda=1}^{\alpha} \langle \lambda | H | \lambda \rangle, \quad (3.25)
\end{aligned}$$

where we have used, besides the diagonal sum rule, Eqs. (3.14a), (3.14f), and (3.16). Thus, we can write the diagonal sum rule in the form

$$\sum_{\lambda=1}^{\alpha} E^{(\lambda)} = \sum_{\lambda=1}^{\alpha} H_{\lambda\lambda}. \tag{3.26}$$

The sum of the diagonal matrix components of the complete Hamiltonian, given by Eq. (3.15), is equal to the sum of the total energies of the system up to the first-order correction. Since the perturbation operator V_0 dropped out of this equation, it is proved that the choice of the potential in Eq. (3.13) is irrelevant.

We can use this theorem to simplify the secular equation. If we are confronted with a problem in which the diagonal sum rule cannot be used, we have to calculate the off-diagonal components. In order to eliminate the fictitious potential from the calculations, we simply put $V_0 = 0$ and get from the off-diagonal matrix components in the secular equation

$$W_{\kappa\lambda} = Q_{\kappa\lambda}, \tag{3.27}$$

where Q is the Coulomb interaction of the electrons in the incomplete group.

For the diagonal components, we get

$$W_{\lambda\lambda} - E_1 = W_{\lambda\lambda} + E_0 - E$$
$$= H_{\lambda\lambda} - E, \qquad (3.28)$$

where $E = E_0 + E_1$, and we have again used Eqs. (3.14a), (3.14f), and (3.16). Thus, the diagonal components will be the matrix components of the total Hamiltonian and the off-diagonal components will contain only the interaction operator Q.

According to Slater, the diagonal sum rule cannot be used for the calculation of energies if the analysis shows that for the given configuration, there is more than one multiplet with the same (LS). In that case, a secular equation must be solved.

We now summarize Slater's method for the formulation of the HF model in LS coupling. We constructed determinantal wave functions built from one-electron spin-orbitals. The determinantal wave functions possess the property that linear combinations of them are eigenfunctions of the angular momentum and spin operators L^2, S^2, L_z, and S_z. The radial parts of the spin-orbitals are, as yet, unspecified, that is, the wave functions constructed satisfy the angular-momentum conditions regardless of the special form of the radial orbitals. The angular-momentum conditions are satisfied for a set of simple analytic radial functions as well as for the radial functions obtained from the HF equations.

As we have mentioned before, Slater's method does not correspond exactly to the general formulation of the HF model in LS coupling. In the general formulation, we specified the construction of wave functions that satisfy the angular-momentum conditions as the first step. In Slater's method, the determinantal wave functions, $\phi_1, \ldots, \phi_\alpha$ are eigenfunctions of L_z and S_z but not of L^2 and S^2. Using perturbation theory, Slater bypassed the step in which the joint eigenfunctions are constructed and obtained the correct energy expressions directly. The method is especially straightforward if the diagonal-sum rule can be used. We note that the joint eigenfunctions can be constructed subsequently by forming the appropriate linear combinations of the determinantal functions $\phi_1, \ldots, \phi_\alpha$.

It should be noted that this method is one of the most interesting applications of the perturbation theory. Contrary to conventional usage, the formalism is not used here for the improvement of the energy in successive approximations. Rather, it is used for the construction of energy expressions that rest on the correct angular-momentum properties. In other words, it is used to resolve the degeneracy of the Hartree formulation in which the energy is (LS)-independent. For the case where the diagonal-sum rule is applicable, our results are embodied in Eq. (3.26). In that equation, the index λ is the serial number of the roots of the secular equation. The roots will be the energies of the various LS multiplets. (There might be, of course, multiple roots.) Thus we can write that Slater's procedure results in obtaining the energy expression for the various multiplets, that is, in obtaining for a given configuration the energy expressions

$$E_i = E(L_i, S_i) \qquad (i = 1, \ldots, k), \tag{3.29}$$

where k is the number of multiplets for the given configuration.

As we have mentioned before, in the procedure presented here, the radial orbitals remained unspecified. In order to obtain the HF model, we have yet to derive the equations for the radial orbitals. This will be done in subsequent sections. We see clearly here the division between the "angular-momentum" part of the theory, which deals with the construction of wave functions and energy expressions that satisfy the angular-momentum conditions, and the "Hamiltonian" part, which deals with the calculation of approximate eigenvalues of the Hamiltonian operator. Slater's method comprised the "angular-momentum" part of the theory. The derivation of the HF equations, from the minimization of the expectation values of the Hamiltonian, will comprise the "Hamiltonian" part.

In this section, the LS coupling was formulated only for atoms with one incomplete group. How far can we go with this formulation? In order to answer this question, let us take a look at Table 3.1 in which we have the electron configurations of neutral atoms in the ground state. We see the following:

1. There is a small number of atoms with complete groups, for example, He, Ne, A, Kr, Xe, and Rn and also Mg, Ca, Zn, Sr, Cd, Ba, etc. These atoms and those with a single electron outside complete groups can be treated with the single-determinantal theory presented in Sec. 2.3.

2. There is a large number of atoms with a single incomplete group. These are first the atoms with incomplete "p" groups like the series beginning with B ($Z = 5$), Al ($Z = 13$), Ga ($Z = 31$), In ($Z = 49$), and Tl ($Z = 81$). Then we have atoms with incomplete "d" groups like the atoms between Sc ($Z = 21$) and Ni ($Z = 28$) (with the exception of Cr ($Z = 24$), which has, in addition to the incomplete d group, an incomplete s group), and the atoms between Lu ($Z = 71$) and Pt ($Z = 78$). Finally, we have some atoms with an incomplete "f" group like the atoms between Ce ($Z = 58$) and Tu ($Z = 69$).[†] These are all atoms with a single incomplete group. Actually, the alkali atoms also belong to this type, but they can be represented with a single-determinantal wave function.

3. Finally, there is a relatively small number of atoms that have more than one incomplete group in the ground state. Examples are the atoms between Nb ($Z = 41$) and Rh ($Z = 45$) and the individual atom Cr, mentioned before, and Ce and Gd. As we see, the large majority of neutral atoms in the ground state possess either configurations with

---

[†]The atoms Ce ($Z = 58$) and Gd ($Z = 64$) also have a 5d electron, so that these atoms have two incomplete groups. It should also be noted that for many rare earth atoms in this series, the LS coupling is not suitable. See Sec. 6.1.

complete groups or configurations with one incomplete group. Thus, the LS coupling theory developed up to now will cover almost all neutral atoms in the ground state.

We should note, however, that the vast body of atomic spectra involves in most cases atoms in their excited states, which are results of configurations with numerous incomplete groups. For these cases, the LS coupling must be formulated for more than one incomplete group. We will present a summary of that formulation in Section 6.2, where we consider the application of the HF model to excited states.

Finally, we note that a list of multiplets that can arise from a single incomplete group of electrons is given by Slater [Slater I, 304]. How the energy expressions and HF equations belonging to these multiplets can be constructed is discussed in the following two sections.

3.2. HF ENERGIES AND EQUATIONS FOR ATOMS WITH ONE INCOMPLETE GROUP

We have seen in Sec. 2.3 that the Hartree–Fock energy of an atom with complete groups consists of terms that can be characterized as either intra-group energies or intergroup interactions. This property is the result of the HF total wave function being built from products of one-electron orbitals. We can assume that the HF energy expression for an atom with one incomplete group will have a similar structure. Thus, we divide the electrons of the atom into those that belong to complete groups and those that belong to the single incomplete group and we assume that the HF energy is of the form:

$$E = E_C + E_{CI} + E_I, \qquad (3.30)$$

where the first right-hand term is the energy of the electrons in the complete groups, the last term is the energy of the incomplete group, and the second term is the complete/incomplete group interaction. We now derive a general formula for E.

We can take the energy of the complete groups, E_C, from Sec. 2.3. Thus, for E_C, we already have the general formula; it is given by Eq. (2.73). The energy of the single incomplete group, E_I, must be constructed by the rules described in the preceding section. Thus, what we have to undertake here is the construction of the complete/incomplete interaction, E_{CI}.

In order to determine the form of E_{CI}, let us consider the interaction of two electrons, i and j, with i belonging to the incomplete group and j to a complete group. In the determinantal approximation, this interaction has the form

$$E_{CI} \approx \left\langle ij \left| \frac{1}{r_{ij}} \right| ij \right\rangle - \left\langle ij \left| \frac{1}{r_{ij}} \right| ji \right\rangle, \qquad (3.31)$$

and if we sum over the electrons in the complete groups, we get

$$E_{CI} \approx \sum_j \left[\left\langle ij \left| \frac{1}{r_{12}} \right| ij \right\rangle - \left\langle ij \left| \frac{1}{r_{12}} \right| ji \right\rangle \right]. \tag{3.32}$$

In order to see the nature of this expression, we write it in the form

$$E_{CI} \approx \langle i|V|i \rangle - \langle i|A|i \rangle, \tag{3.33}$$

where

$$V(1) = \sum_j \int \frac{|\varphi_j(2)|^2}{r_{12}} \, dq_2, \tag{3.34a}$$

and

$$A(1)\varphi_i(1) = \sum_j \int \frac{1}{r_{12}} \varphi_j(1)\varphi_j^*(2)\varphi_i(2) \, dq_2. \tag{3.34b}$$

In Sec. 2.3 and in App. D, we show that if the summation in Eqs. (3.34a) and (3.34b) is over complete groups, then the quantity

$$(V - A)\varphi_i \tag{3.35}$$

will have the same angular and spin part as φ_i. (Actually, the proof we present refers only to the angular part, but the statement about the spin part is trivial for complete groups.) Now in order to construct E_{CI}, we must form the matrix components of $(V - A)$ with respect to linear combinations of the determinantal functions $\phi_1, \ldots, \phi_\alpha$. These are the determinants that we constructed to obtain (LS) coupling. Because these determinants are built from one-electron orbitals, the matrix components break down into terms like

$$\langle \varphi_\alpha|V - A|\varphi_\beta \rangle, \tag{3.36}$$

where φ_α and φ_β are two of the one-electron orbitals used to construct the determinantal functions. The one-electron orbitals differ in their (m_l, m_s) numbers.

As we have seen, $(V - A)\varphi_\beta$ has the same angular and spin parts as φ_β; therefore, because the one-electron orbitals are orthogonal with respect to m_l and m_s, we find that E_{CI} is of the form

$$E_{CI} \approx \sum_\beta \langle \varphi_\beta|V - A|\varphi_\beta \rangle, \tag{3.37}$$

where $\langle \varphi_\beta|V - A|\varphi_\beta \rangle$ is the interaction energy of the orbital φ_β with the orbitals in the complete groups. Because $(V - A)$ does not affect the angular and spin parts of the orbital on which it operates, we can state that

The interaction energy of an orbital in the incomplete group with the orbitals of
the complete groups is the same for all orbitals in the incomplete group .

From this theorem, it is easy to construct E_{CI}. We know from Eq. (2.73)
that the interaction between two complete groups, with quantum numbers
(nl) and $(n'l')$, is given by

$$2(2l + 1)2(2l' + 1)F_0(nl, n'l')$$
$$- \sum_k 2\sqrt{(2l + 1)(2l' + 1)}\, c^k(l0, l'0)G_k(nl, n'l'), \qquad (3.38)$$

where we have substituted for Hartree's notation the notation of Slater,
according to Eq. (2.101b).

Let (nl) denote the incomplete group with the occupation number $q(nl)$.
We obtain the interaction energy between *one* electron in this group and the
whole complete $(n'l')$ group by dividing Eq. (3.38) by the number of the
electrons in the complete (nl) group, $2(2l + 1)$. Thus, the total interaction
energy between the incomplete group and all complete groups $(n'l')$ is

$$E_{CI} = \sum_{(n'l')} \frac{q(nl)}{2(2l + 1)}\left[2(2l + 1)2(2l' + 1)F_0(nl, n'l') \right.$$

$$\left. - \sum_k 2\sqrt{(2l + 1)(2l' + 1)}\, c^k(l0, l'0)G_k(nl, n'l') \right]$$

$$= \sum_{(n'l')} \left[q(nl)2(2l' + 1)F_0(nl, n'l') \right.$$

$$\left. -q(nl) \sum_k \sqrt{\frac{(2l' + 1)}{(2l + 1)}}\, c^k(l0, l'0)G_k(nl, n'l') \right], \qquad (3.39)$$

where, of course, the first term is the electrostatic interaction, and the second
is the exchange interaction.

As we have mentioned, the expression for E_I, the energy of the incomplete group, must be constructed on the basis of the preceding section.
Nevertheless, even without actually constructing E_I, we are able to write
down its general form by analyzing Eq. (2.73). We see from that equation
that the intragroup energies are given by the first, second, and fourth terms.
The first term arises from the one-electron operators $(t + g)$. Obviously, the
form of this term is the same for an incomplete group as for a complete. The
second term is the electrostatic interaction within the group and the fourth is
the exchange interaction. Thus, the total intragroup interaction is given by
$F_0(nl, nl)$ integrals and by $F_k(nl, nl)$ integrals with $k > 0$. This must be also
true for an incomplete group because any G_k integral that represents

intragroup interaction can be transformed into an F_k integral using the relationship

$$G_k(nl, nl) = F_k(nl, nl). \qquad (3.40)$$

Thus, the expression for E_I must have the form

$$E_I = q(nl)I(nl) + \frac{q(nl)}{2}[q(nl) - 1]F_0(nl, nl)$$
$$- \sum_{k>0} \tilde{A}_{lk} F_k(nl, nl). \qquad (3.41)$$

We are now in the position to construct the general formula for the energy of an atom with one incomplete group. The incomplete group is denoted by (nl) and its occupation number by $q(n, l)$. We use Eq. (2.73) for the energy of the complete groups, with the notation of Slater given by Eqs. (2.100b) and (2.101b). For E_{CI}, we use Eq. (3.39), and for the energy of the incomplete group, Eq. (3.41).

In this way, we obtain

$$E_{HF}(LS)$$
$$= E_C + E_{CI} + E_I(LS)$$
$$= \sum_{(n'l')} 2(2l' + 1)I(n'l')$$
$$+ \sum_{(n'l')} \frac{2(2l' + 1)}{2}[2(2l' + 1) - 1]F_0(n'l', n'l')$$
$$+ \sum_{(n'l')} \sum_{\substack{(n''l'') \\ (n''l'') \neq (n'l')}} 2(2l' + 1)2(2l'' + 1)F_0(n'l', n''l'')$$
$$- \sum_{(n'l')} \sum_{k>0} (2l' + 1)c^k(l'0, l'0)F_k(n'l', n'l')$$
$$- \sum_{(n'l')} \sum_{\substack{(n''l'') \\ (n''l'') \neq (n'l')}} \sum_k 2\sqrt{(2l' + 1)(2l'' + 1)}\, c^k(l'0, l''0)G_k(n'l', n''l'')$$
$$+ \sum_{(n'l')} q(nl)2(2l' + 1)F_0(nl, n'l')$$
$$- \sum_{(n'l')} \sum_k q(nl)\sqrt{\frac{(2l' + 1)}{(2l + 1)}}\, c^k(l0, l'0)G_k(nl, n'l')$$
$$+ q(nl)I(nl) + \frac{q(nl)}{2}[q(nl) - 1]F_0(nl, nl)$$
$$- \sum_{k>0} \tilde{A}_{lk}(LS)F_k(nl, nl). \qquad (3.42)$$

In this expression, the complete groups are indicated by $(n'l')$ and $(n''l'')$ and the double summations are for all pairs of groups. We note that this is a general formula in the sense that all terms can be written down immediately, for any configuration, except the last term, which must be constructed separately for each multiplet with the methods described in the preceding section. Apart from the constants $\tilde{A}_{lk}$, which are different for each (LS), the only constants that are needed are the c^k's of Slater. The $\tilde{A}_{lk}$ constants were given by Hartree [Hartree, 111] for the ground states of atoms with p^q and d^q incomplete groups. Slater has given these constants for many more multiplets but since Slater's tables involve the average of configurations, his constants are discussed in the next section.

The formula given by Eq. (3.42) possesses a plausible physical interpretation. The first five terms are representing the energy of the complete groups; the physical meaning of these terms was analyzed following Eq. (2.73). The remaining terms are the results of the addition of an incomplete group to the complete groups. The sixth and seventh terms are the electrostatic and exchange interactions between the incomplete and complete groups. The eighth term is the kinetic plus nuclear potential energy of the incomplete group. The ninth and tenth terms are representing the electrostatic and exchange interaction energies within the incomplete group.

Our next task is to derive the HF equations by applying the variation principle. First, we derive the equation for an orbital P_{nl} in the incomplete group. Let δ be the variation with respect to P_{nl}. The subsidiary conditions are the normalization and orthogonality to all orbitals in the complete groups. We denote the Lagrangian multipliers by $\bar{\varepsilon}_{nl}$ and $\bar{\lambda}_{nl, n'l'}$. The subsidiary conditions are

$$E_0 = -\bar{\varepsilon}_{nl}\langle P_{nl}|P_{nl}\rangle - \sum_{n'}\bar{\lambda}_{nl, n'l}\langle P_{nl}|P_{n'l}\rangle, \qquad (3.43)$$

and our variational principle becomes

$$\delta(E_{HF}(LS) + E_0) = 0. \qquad (3.44)$$

We carry out the variation taking into account Eqs. (2.78a) to (2.78d). After the variation, we divide the resulting equation by $2q(nl)$, in order to obtain it in the conventional form, that is, in the form in which the second derivative is multiplied by $-\frac{1}{2}$. Before carrying out these steps, let us take a look at Eq. (3.42). The first five terms do not contain P_{nl}. This function occurs only in the last five terms, so it is only these that we have to consider.

In this way, we obtain

$$
\left\{ -\frac{1}{2}\frac{d^2}{dr^2} + \frac{l(l+1)}{2r^2} - \frac{Z}{r} + [q(nl) - 1]\frac{1}{r}Y_0(nl, nl|r) \right.
$$

$$
+ \sum_{\substack{(n'l') \\ (n'l') \neq (nl)}} 2(2l' + 1)\frac{1}{r}Y_0(n'l', n'l'|r) - \sum_k \tilde{\alpha}_{lk}(LS)\frac{1}{r}Y_k(nl, nl|r) \Bigg\} P_{nl}(r)
$$

$$
- \sum_{\substack{(n'l') \\ (n'l') \neq (nl)}} \sum_k \sqrt{\frac{2l'+1}{2l+1}}\, c^k(l0, l'0)\frac{1}{r}Y_k(nl, n'l'|r)P_{n'l'}(r)
$$

$$
= \varepsilon_{nl}P_{nl} + \sum_{\substack{n' \\ (n' \neq n)}} \lambda_{nl, n'l}P_{n'l}(r). \tag{3.45}
$$

In this equation, we have

$$
\tilde{\alpha}_{lk}(LS) = \frac{4\tilde{A}_{lk}(LS)}{2q(nl)}. \tag{3.46}
$$

Apart from the off-diagonal Lagrangian multipliers, this equation is formally the same as Eq. (2.83), which gives the HF equation for an orbital in complete groups. In fact, the last term on the left side of Eq. (3.45) is not only formally but exactly identical with the corresponding term in Eq. (2.83); we recognize this if we look at Eq. (2.101a). The reason for this is simple: that term represents the exchange interaction of the orbital P_{nl} with all the orbitals in the complete groups and we have seen that this quantity does not depend on whether the orbital P_{nl} is in a complete or in an incomplete group.

Let us write Eq. (3.45) in the same simple form as Eq. (2.88). We put

$$
\left\{ -\frac{1}{2}\frac{d^2}{dr^2} + \frac{l(l+1)}{2r^2} - \frac{Z}{r} + U(nl, LS|r) \right\} P_{nl}
$$

$$
= \varepsilon_{nl}P_{nl} + \sum_{n'} \lambda_{nl, n'l}P_{n'l}, \tag{3.47}
$$

where the HF potential is written in localized form:

$$U(nl, LS|r) = \sum_{(n'l')} q(n'l') \frac{1}{r} Y_0(n'l', n'l'|r)$$

$$- \frac{1}{r} Y_0(nl, nl) - \sum_{\substack{(n'l') \\ (n'l') \neq (nl)}} \sum_k \sqrt{\frac{2l' + 1}{2l + 1}}$$

$$\times c^k(l0, l'0) \frac{1}{r} Y_k(nl, n'l'|r) \frac{P_{n'l'}(r)}{P_{nl}(r)}$$

$$- \sum_{k>0} \tilde{\alpha}_{lk}(LS) \frac{1}{r} Y_k(nl, nl|r). \tag{3.48}$$

The potential is, of course, LS-dependent through the last term. The summation in the first term is for *all* groups.

Next, we derive the HF equation for an orbital in one of the complete groups with one incomplete group present. Our energy expression is Eq. (3.42). Now we can disregard the last three terms because those contain only the orbital of the incomplete group. Also, we must remember that the first five terms of Eq. (3.42) are identical with the expression given in Eq. (2.73) because these five terms are giving the energy of the complete groups. Thus, the variation of these five terms gives the potential shown by Eq. (2.87). The only difference between the potential obtained in the presence of one incomplete group and the potential obtained for complete groups is the terms arising from the complete/incomplete interaction energies in Eq. (3.42). These are given by the sixth and seventh terms.

Let δ denote variation with respect to $P_{n'l'}$. Then from the variation of the sixth and seventh terms, we get

$$q(nl) \frac{1}{r} Y_0(nl, nl|r) P_{n'l'} - \frac{q(nl)}{2(2l' + 1)}$$

$$\times \sqrt{\frac{2l' + 1}{2l + 1}} \sum_k c^k(l'0, l0) \frac{1}{r} Y_k(n'l', nl|r) P_{nl}(r). \tag{3.49}$$

These expressions are obtained by dividing the result of the variation by $2[2(2l' + 1)]$, so that we get the equations in standard form.

Now we can write down the HF equation for the orbital $P_{n'l'}$ that is in one of the complete groups. We get

$$\left\{ -\frac{1}{2} \frac{d^2}{dr^2} + \frac{l'(l' + 1)}{2r^2} - \frac{Z}{r} + U(n'l', LS|r) \right\} P_{n'l'}$$

$$= \varepsilon_{n'l'} P_{n'l'} + \sum_{\substack{n'' \\ (n'' \neq n')}} \lambda_{n'l', n''l'} P_{n''l'}, \tag{3.50}$$

where the (localized) HF potential is given by

$$U(n'l', LS|r) = \sum_{(n''l'')} q(n''l'') \frac{1}{r} Y_0(n''l'', n''l''|r)$$

$$- \frac{1}{r} Y_0(n'l', n'l'|r) - \sum_{\substack{(n''l'') \\ (n''l'') \neq (n'l')}} \sum_k \sqrt{\frac{2l''+1}{2l'+1}} c^k(l'0, l''0)$$

$$\times \frac{1}{r} Y_k(n'l', n''l''|r) \frac{P_{n''l''}}{P_{n'l'}} - \sum_{k>0} c^k(l'0, l'0)$$

$$\times \frac{1}{r} Y_k(n'l', n'l'|r) + q(nl) \frac{1}{r} Y_0(nl, nl|r)$$

$$- q(nl) \sum_k \frac{c^k(l'0, l0)}{2\sqrt{(2l'+1)(2l+1)}}$$

$$\times \frac{1}{r} Y_k(n'l', nl|r) \frac{P_{nl}}{P_{n'l'}}. \tag{3.51}$$

We note again that (nl) indicates the incomplete group and $(n''l'')$ means a complete group. The summations in the first and third terms are over the complete groups.

We notice something very interesting in Eq. (3.51). The potential does not explicitly depend on the constants $\tilde{A}_{lk}(LS)$, which made the energy, given by Eq. (3.42), LS-dependent. Of course, the potential depends on the orbital of the incomplete group, P_{nl}, and this orbital is different for different multiplets. Thus, we can formulate the following theorem:

In an atom with one incomplete group, the HF potential for an orbital in a complete group does not explicitly depend on the multiplet structure, that is, the potential is formally the same for all multiplets that can be generated by the configuration. The potential depends implicitly on the multiplet structure through the orbital P_{nl} of the incomplete group, which is different for different multiplets.

Finally, we want to discuss the role of the Lagrangian multipliers. We have seen that for atoms with complete groups, where the total wave function is a single determinant, the Lagrangians can be eliminated by a unitary transformation. Here the wave function is not a single determinant, so we cannot *a priori* eliminate all the multipliers. We now show, however, that the multipliers *can* be eliminated from the complete-group equations.

At this point, we recall the discussion of the behavior of the HF equations under a unitary transformation, which is presented in App. B. In analogy to Eq. (B.12), we subject the solutions of Eq. (3.50) to a unitary transformation that, in the case of radial wave functions, has the form

$$\hat{P}_{n'l'} = \sum_{n''} C_{n'l', n''l'} P_{n''l'}. \tag{3.52}$$

(We note that the l' index is fixed in all quantities.) In the case of complete groups, we show in App. B that the HF potential is invariant under the transformation. We now show that the HF Hamiltonian operator of Eq. (3.50) is also invariant under a unitary transformation like Eq. (3.52).

We see from the formulas that the HF Hamiltonian depends on l' through the $l'(l' + 1)/2r^2$ term. The first thing about Eq. (3.52) is that the transformation does not affect l'. Next, we observe that the first four terms in the potential given by Eq. (3.51) constitute an operator that is the same as the operator in Eq. (2.98); these terms represent the potential operator generated by the complete groups. As we see from Eq. (2.98), this operator depends on l_i, which means the dependence on l' for Eq. (3.51). That means that the first four terms of Eq. (3.51) are invariant under the transformation just as the HF potential for complete groups is invariant under the transformation given by Eq. (B.12). Our argument hinges on the question whether the last two terms of Eq. (3.51) are also invariant under our unitary transformation, Eq. (3.52).

The term before the last does not contain $P_{n'l'}$; therefore, it is not affected. We write the last term in operator form:

$$\left[-q(nl) \sum_k \frac{c^k(l'0, l0)}{2\sqrt{(2l' + 1)(2l + 1)}} \frac{1}{r} Y_k(n'l', nl|r) \frac{P_{nl}}{P_{n'l'}} \right] P_{n'l'}$$

$$= -q(nl) \sum_k \frac{c^k(l'0, l0)}{2\sqrt{(2l' + 1)(2l + 1)}} \frac{1}{r} P_{nl}(r)$$

$$\times \int P_{nl}(r') L_k(r, r') P_{n'l'}(r') \, dr'$$

$$= U_{op}^{l'}(r) P_{n'l'}(r), \tag{3.53}$$

where $U_{op}^{l'}(r)$ is a linear operator with the kernel

$$K^{l'}(r, r') = -q(nl) \sum_k \frac{c^k(l'0, l0|r)}{2\sqrt{(2l' + 1)(2l + 1)}}$$

$$\times \frac{1}{r} P_{nl}(r) P_{nl}(r') L_k(r, r'). \tag{3.53a}$$

Thus, we see that $U_{op}^{l'}$ is the exchange operator representing the complete/incomplete exchange interaction. This operator depends on l' and is not affected by the transformation, Eq. (3.52). Therefore,

(a) the l'-dependent term, $l'(l' + 1)/2r^2$, is not affected by the transformation;
(b) the first four terms of Eq. (3.51) are invariant under the transformation because they represent complete groups;
(c) the last two terms, if written into operator form, do not depend on $P_{n'l'}$; therefore, they are also invariant under the transformation.

Thus, we conclude that the HF Hamiltonian operator of the Eq. (3.50) is invariant under the unitary transformation, Eq. (3.52). From this, it follows that the transformation can be used to diagonalize the Lagrangian matrix and Eq. (3.50) can be transformed into the form

$$\left\{ -\frac{1}{2}\frac{d^2}{dr^2} + \frac{l'(l' + 1)}{2r^2} - \frac{Z}{r} + U(n'l', LS|r) \right\} P_{n'l'} = \varepsilon_{n'l'} P_{n'l'}, \quad (3.54)$$

where U is given by Eq. (3.51).

Next, we turn to Eq. (3.47), which is the equation for the orbital in the incomplete group. Unfortunately, the Lagrangians cannot be eliminated from this equation. Here we are confronted with the fact that the HF orbital, P_{nl}, is the solution of an operator that is not Hermitian. We can see this by multiplying Eq. (3.47) from the left by $P_{n'l}$ ($n' \neq n$) and integrating. We obtain

$$\lambda_{nl, n'l} = \langle P_{n'l} | H_F | P_{nl} \rangle, \quad (3.55)$$

where H_F is the operator in the curly brackets of Eq. (3.47). Using this relationship, we can write Eq. (3.47) in the form

$$H_F P_{nl} = \varepsilon_{nl} P_{nl} + \sum_{n'} |P_{n'l}\rangle\langle P_{n'l}| H_F |P_{nl}\rangle. \quad (3.56)$$

Let us introduce the projection operators Ω and Π with the definitions

$$\Omega \equiv \sum_{n'} |P_{n'l}\rangle\langle P_{n'l}|, \quad (3.57a)$$

$$\Pi \equiv 1 - \Omega. \quad (3.57b)$$

The summation is over all complete groups. Using these operators, we can write the HF equation for the orbital in the incomplete group in the form

$$H'_F P_{nl} = \varepsilon_{nl} P_{nl}, \quad (3.58)$$

where

$$H'_F = \Pi H_F. \tag{3.59}$$

From Eq. (3.58), it is clear that the orbital P_{nl}, the wave function in the incomplete group, is an eigenfunction of a non-Hermitian operator; the operators H_F and Π are Hermitian, but they do not commute; therefore, H'_F is not Hermitian.

Finally, we note that, strictly speaking, the summation over the Lagrangians in Eq. (3.50) should include the orbital in the incomplete group. We have omitted this term because the orbital is orthogonal to the complete-group orbitals anyhow by virtue of the presence of the Lagrangians in Eq. (3.47). Therefore, it is not necessary to include this term in Eq. (3.50).

3.3. THE THEORY OF CONFIGURATION AVERAGES

In the preceding section, we have described the HF theory of atoms, with one incomplete group in LS coupling. A closer examination of this theory reveals that there are certain features in the formalism that permit considerable conceptual and mathematical simplifications.

First, let us take a look at the energy expression given by Eq. (3.42). An interesting feature of that expression is that only the last term is LS-dependent. As we have seen, the whole expression can be divided into three well-defined parts; the last three terms represent the intragroup interaction of the incomplete group. Now we recognize that the explicit LS-dependence is restricted to the third term of the intragroup interaction. Thus, it is evident that the energy expression for the different multiplets will differ in only a very few terms.

Next let us take a look at the HF potential, for the orbital P_{nl} in the incomplete group, given by Eq. (3.48). Here we see that the explicit LS-dependence is again restricted to one term. As we see, the electrostatic potential of all electrons, including those in the incomplete group, and the exchange potential representing the interaction of P_{nl} with the orbitals in the complete groups are LS-independent. The explicit LS-dependence is exhibited only by the term representing the exchange interaction between the orbital P_{nl} and the other orbitals in the incomplete group.

Finally, we look at Eq. (3.51), which is the potential of the orbitals in the complete groups. As we have emphasized in a theorem, this potential does not have any explicit LS-dependence.

In a consistent application of the theory, we would have to calculate all orbitals for each of the multiplets arising from a configuration. It is clear, however, that most of the orbitals would not change much as we move from one multiplet to another. The orbital in the incomplete group would change, of course, but the change would not be excessive because the potential does not change much. The orbitals in the complete groups would change even less in view of the lack of explicit LS-dependence in their potential.

These circumstances induced Slater to introduce the concept of the average energy of a configuration and to show that much of the atomic structure theory can be simplified by the introduction of this concept and by the development of its theory. Slater's theory, which incorporated some earlier work of Shortley,[13] is presented here [Slater I, 322].

We consider, first, one incomplete group. The average energy of a configuration, or weighted mean energy, is defined as follows:

$$E_{av} = \frac{\sum\limits_{k} g_k E(L_k S_k)}{\sum\limits_{k} g_k}. \tag{3.60}$$

Here $E(L_k S_k)$ is the energy of the multiplet with (L_k, S_k), g_k is a weight factor, and the summations are over all multiplets of the configuration. For the weight factor, we put

$$g_k = (2L_k + 1)(2S_k + 1), \tag{3.61}$$

because, as we have seen in Sec. 3.1, each multiplet has this many determinantal wave functions associated with it. With this choice, we get

$$\sum_k g_k = \sum_k (2L_k + 1)(2S_k + 1) = K, \tag{3.62}$$

where K is the total number of determinantal wave functions that can be constructed for this configuration.

Using the diagonal-sum rule, Eq. (3.26), we obtain

$$E_{av} = \frac{\sum\limits_{i=1}^{K} \langle \phi_i | H | \phi_i \rangle}{K}, \tag{3.63}$$

where the set of determinantal wave functions, $\phi, \ldots, \phi_K$, is what we have constructed in Sec. 3.1 to obtain the multiplet energies.

Let us consider that part of the diagonal matrix components that contain the interaction operator Q given by Eq. (3.14e). For a particular diagonal matrix component, we obtain, using Eq. (A.34) from App. A,

$$\langle \phi_k | Q | \phi_k \rangle = \frac{1}{2} \sum_{i,j} \left\{ \left\langle ij \left| \frac{1}{r_{12}} \right| ij \right\rangle - \left\langle ij \left| \frac{1}{r_{12}} \right| ji \right\rangle \right\}, \qquad (3.64)$$

where the double summation is over all orbitals that occur in the determinant ϕ_k. If we now write down such an equation for each determinant, then we get, for the interaction part of the diagonal matrix component, the expression

$$\sum_k \langle \phi_k | Q | \phi_k \rangle = \sum_k \frac{1}{2} \sum_{ij}^{(k)} \left\{ \left\langle \left| \frac{1}{r_{12}} \right| ij \right\rangle - \left\langle ij \left| \frac{1}{r_{12}} \right| ji \right\rangle \right\}. \qquad (3.65)$$

The superscript (k) means that in each summation over (ij), we have to consider those orbitals that occur in ϕ_k.

The summations in this equation can be carried out differently. Originally, we have a total of $N_0 = 2(2l + 1)$ spin orbitals for this group. From these orbitals, we selected a set of $N_1 = q(nl)$ orbitals and formed determinantal wave functions from all possible sets that can be selected. The summations in Eq. (3.65) can also be carried out in such a way that, first, we select a pair, (ij), from all the available N_0 orbitals. Then we multiply this by the number of determinants in which (ij) occurs and sum up over all (ij) pairs. Thus,

$$\begin{aligned}
\sum_k \langle \phi_k | Q | \phi_k \rangle &= \sum_k \frac{1}{2} \sum_{ij}^{(k)} \left\{ \left\langle ij \left| \frac{1}{r_{12}} \right| ij \right\rangle - \left\langle ij \left| \frac{1}{r_{12}} \right| ji \right\rangle \right\} \\
&= \frac{1}{2} \sum_{ij}^{(N_0)} \left[\left\{ \left\langle ij \left| \frac{1}{r_{12}} \right| ij \right\rangle - \left\langle ij \left| \frac{1}{r_{12}} \right| ji \right\rangle \right\} \right. \\
&\qquad\qquad \left. \times \left(\begin{array}{c} \text{Number of determinants} \\ \text{with } (ij) \end{array} \right) \right], \qquad (3.66)
\end{aligned}$$

where now the summation is over all pairs in N_0. On putting this expression into Eq. (3.63), we obtain

$$E_{av} = \langle \phi_\alpha | F | \phi_\alpha \rangle$$

$$+ \frac{\dfrac{1}{2} \sum_{ij}^{(N_0)} \left[\left\{ \left\langle ij \left| \dfrac{1}{r_{12}} \right| ij \right\rangle \cdots \right\} \left(\begin{array}{c} \text{number of} \\ \text{determinants with} \\ (ij) \end{array} \right) \right]}{K} \qquad (3.67)$$

In this equation, ϕ_α is any one of the determinantal functions because the diagonal matrix components of the operator F, that is given by Eq. (3.14b), are the same with respect to any of the determinantal functions, $\phi_1, \phi_2, \ldots, \phi_K$.

The constants in the previous equation are easily obtained. The determinants are $(N_1 \times N_1)$. The number of determinants that will contain (ij) will be equal to the number of ways in which we can select $N_1 - 2$ orbitals out of the $N_0 - 2$ that remained from the N_0 after we selected the two, i and j. Thus, this number is

$$\begin{pmatrix} N_0 - 2 \\ N_1 - 2 \end{pmatrix} = \frac{(N_0 - 2)!}{(N_1 - 2)!(N_0 - 2 - N_1 + 2)!}$$

$$= \frac{(N_0 - 2)!}{(N_1 - 2)!(N_0 - N_1)!}. \tag{3.68}$$

Because this is the same for all (ij), we can take the factor out of the summation of Eq. (3.67) and combine it with K. This quantity is the total number of determinants, that is,

$$K = \begin{pmatrix} N_0 \\ N_1 \end{pmatrix} = \frac{N_0!}{N_1!(N_0 - N_1)!}, \tag{3.69}$$

and, thus

$$\begin{pmatrix} \text{number of determinants} \\ \text{with } (ij) \\ \hline K \end{pmatrix} = \frac{(N_0 - 2)!}{(N_1 - 2)!(N_0 - N_1)!} \bigg/ \frac{(N_0)!}{(N_1)!(N_0 - N_1)!}$$

$$= \frac{(N_0 - 2)!(N_1)!(N_0 - N_1)!}{(N_1 - 2)!(N_0 - N_1)!(N_0)!}$$

$$= \frac{(N_1)!}{(N_1 - 2)!} \bigg/ \frac{(N_0)!}{(N_0 - 2)!} = \frac{\begin{pmatrix} N_1 \\ 2 \end{pmatrix}}{\begin{pmatrix} N_0 \\ 2 \end{pmatrix}}. \tag{3.70}$$

Substituting this result into Eq. (3.67), we obtain

$$E_{av} = q(nl)I(nl) + \frac{\binom{N_1}{2}\frac{1}{2}\sum_{ij}^{(N_0)}\left\{\left\langle ij\left|\frac{1}{r_{12}}\right|ij\right\rangle - \left\langle ij\left|\frac{1}{r_{12}}\right|ji\right\rangle\right\}}{\binom{N_0}{2}}$$

$$= q(nl)I(nl) + \frac{q(nl)}{2}[q(nl) - 1]$$

$$\times \left[\left\langle ij\left|\frac{1}{r_{12}}\right|ij\right\rangle - \left\langle ij\left|\frac{1}{r_{12}}\right|ji\right\rangle\right]_{av}, \tag{3.71}$$

where the average of the two-electron interaction is defined as

$$\left[\left\langle ij\left|\frac{1}{r_{12}}\right|ij\right\rangle - \left\langle ij\left|\frac{1}{r_{12}}\right|ji\right\rangle\right]_{av}$$

$$= \frac{\frac{1}{2}\sum_{ij}^{(N_0)}\left\{\left\langle ij\left|\frac{1}{r_{12}}\right|ij\right\rangle - \left\langle ij\left|\frac{1}{r_{12}}\right|ji\right\rangle\right\}}{\binom{N_0}{2}}. \tag{3.72}$$

In writing down Eq. (3.71), we transformed the diagonal matrix component of F into the conventional expression and we also wrote

$$\binom{N_1}{2} = \frac{q(nl)}{2}[q(nl) - 1], \tag{3.73}$$

where $q(nl)$ is the number of electrons in the incomplete group.

The calculation of the average of the two-electron interaction is easy because, in Eq. (2.116), we have the total intragroup interaction, electrostatic plus exchange, for the complete group with (nl):

$$\frac{1}{2}\sum_{ij}^{(N_0)}\left[\left\langle ij\left|\frac{1}{r_{12}}\right|ij\right\rangle - \left\langle ij\left|\frac{1}{r_{12}}\right|ji\right\rangle\right]$$

$$= \frac{1}{2}[4l + 2]^2 F_0(nl, nl) - \frac{4l + 2}{2}\sum_{k=0}^{2l} c^k(l0, l0) F_k(nl, nl). \tag{3.74}$$

For a complete group, we get

$$
\binom{N_0}{2} = \binom{2(2l+1)}{2} = \binom{4l+2}{2}
$$

$$
= \frac{(4l+2)!}{2!(4l+2-2)!} = \frac{(4l+1)(4l+2)}{2}, \tag{3.75}
$$

and combining the last two equations, we obtain

$$
\left[\left(\left\langle ij \left| \frac{1}{r_{12}} \right| ij \right\rangle - \left\langle ij \left| \frac{1}{r_{12}} \right| ji \right\rangle \right) \right]_{av}
$$

$$
= \left(\frac{1}{2}[4l+2]^2 F_0(nl, nl) - \frac{4l+2}{2} F_0(nl, nl) \right.
$$

$$
\left. - \frac{4l+2}{2} \sum_{k>0}^{2l} c^k(l0, l0) F_k(nl, nl) \right) \Big/ \frac{(4l+1)(4l+2)}{2}
$$

$$
= F_0(nl, nl) - \frac{1}{4l+1} \sum_{k>0}^{2l} c^k(l0, l0) F_k(nl, nl). \tag{3.76}
$$

By putting this into Eq. (3.71), we obtain the average energy of the configuration of an incomplete group with the occupation number $q(nl)$:

$$
E_{av} = q(nl)I(nl) + \frac{q(nl)}{2}[q(nl) - 1]
$$

$$
\times \left\{ F_0(nl, nl) - \frac{1}{4l+1} \sum_{k>0}^{2l} c^k(l0, l0) F_k(nl, nl) \right\}. \tag{3.77}
$$

Next, let us consider two partly filled groups with the occupation numbers $N_1 = q(nl)$ and $N_1' = q(n'l')$. The complete groups would hold N_0 and N_0', respectively. In a straightforward fashion, we obtain the formula, analogous to the second term of Eq. (3.71),

$$
E_{av} = q(nl)q(n'l') \left[\left(\left\langle ij \left| \frac{1}{r_{12}} \right| ij \right\rangle - \left\langle ij \left| \frac{1}{r_{12}} \right| ji \right\rangle \right) \right]_{av}, \tag{3.78}
$$

where in the average of the two-electron interaction, i is in one of groups and

j in the other. We have

$$\left[\left\langle ij \left| \frac{1}{r_{12}} \right| ij \right\rangle - \left\langle ij \left| \frac{1}{r_{12}} \right| ji \right\rangle \right]$$

$$= \frac{\sum_i^{(N_0)} \sum_j^{(N_0')} \left\{ \left\langle ij \left| \frac{1}{r_{12}} \right| ij \right\rangle - \left\langle ij \left| \frac{1}{r_{12}} \right| ji \right\rangle \right\}}{N_0 N_0'}. \qquad (3.79)$$

From Eq. (2.116), we obtain

$$\sum_i^{(N_0)} \sum_j^{(N_0')} \left\{ \left\langle ij \left| \frac{1}{r_{12}} \right| ij \right\rangle - \left\langle ij \left| \frac{1}{r_{12}} \right| ji \right\rangle \right\}$$

$$= \left\{ [4l + 2][4l' + 2] F_0(nl, n'l') \right.$$

$$\left. - 2\sqrt{(2l + 1)(2l' + 1)} \sum_{k=0}^{l+l'} c^k(l0, l'0) G_k(nl, n'l') \right\},$$

$$(n'l') \neq (nl). \qquad (3.80)$$

Also we have

$$N_0 N_0' = 2(2l + 1)2(2l' + 1), \qquad (3.81)$$

and by substituting the two last expressions into Eq. (3.79) and the resulting quantity into Eq. (3.78), we obtain

$$E_{av} = q(nl)q(n'l') \left\{ F_0(nl, n'l') - \frac{1}{2\sqrt{(2l + 1)(2l' + 1)}} \right.$$

$$\left. \times \sum_{k=0}^{l+l'} c^k(l0, l'0) G_k(nl, n'l') \right\}. \qquad (3.82)$$

Finally, let us consider an atom with an arbitrary number of complete and incomplete groups. The average energy of the configuration is the sum of expressions like in Eq. (3.77) for individual groups and expressions like in Eq.

(3.82) for pairs of groups. Thus we obtain Slater's general formula,

$$
\begin{aligned}
E_{av} = \sum_{(nl)} \Bigg\{ & q(nl) I(nl) + \frac{q(nl)}{2} [q(nl) - 1] \\
& \times \Bigg[F_0(nl, nl) - \frac{1}{4l + 1} \sum_{k>0}^{2l} c^k(l0, l0) F_k(nl, nl) \Bigg] \Bigg\} \\
+ \frac{1}{2} \sum_{\substack{(nl) \ (n'l') \\ (n'l') \neq (nl)}} & q(nl) q(n'l') \Bigg\{ F_0(nl, n'l') - \frac{1}{2\sqrt{(2l + 1)(2l' + 1)}} \\
& \times \sum_{k=0}^{l+l'} c^k(l0, l'0) G_k(nl, n'l') \Bigg\}, \quad (3.83)
\end{aligned}
$$

where in the double summation we have to sum over all groups independently.

We now begin the analysis of the average-energy expression. This is a general formula for an arbitrary atom in an arbitrary configuration with complete and incomplete groups. It is a general formula because it can be written down at once for any configuration, that is, it does not require the construction of LS-dependent terms. As the first fact, we note that E_{av} becomes identical with the HF energy for complete groups. Such groups generate only one multiplet; therefore, the average energy is equal to the exact. It is clear also that E_{av} becomes identical with the exact expression for atoms with one electron outside complete groups.

Next, we show that the exact HF energy for a multiplet deviates from E_{av} only in an expression that contains very few terms and that we can easily derive a formula for this expression. We consider an atom with one incomplete group. For that case, we have the exact HF energy in Eq. (3.42). We show that the connection between the exact and average energies can be put into the form

$$ E_{HF}(LS) = E_{av} + X(LS), \qquad (3.84) $$

and we derive a formula for $X(LS)$. We have, first,

$$ X(LS) = E_{HF}(LS) - E_{av}. \qquad (3.85) $$

For $E_{HF}(LS)$, we use Eq. (3.42) and E_{av} is given by Eq. (3.83). In order to compare the two formulas, we change the double summations in E_{av} in such a way that they are summations for all pairs of groups, that is, we omit the $\frac{1}{2}$ from these terms. Next, we recall that the average energy gives the exact result for complete groups. This means that the expression given by E_{av} for complete groups cancels the first five terms in Eq. (3.42). What remains from both expressions is the interaction between the incomplete group and the complete groups plus the intragroup interactions of the incomplete group. From the latter, the terms with $I(nl)$ and $F_0(nl, nl)$ are identical in the two formulas; thus, they drop out. We obtain

$$
\begin{aligned}
X(LS) = &\sum_{(n'l')} q(nl) 2(2l' + 1) F_0(nl, n'l') \\
&- \sum_{(n'l')} q(nl) \sqrt{\frac{2l' + 1}{2l + 1}} \sum_{k=0}^{l+l'} c^k(l0, l'0) G_k(nl, n'l') \\
&- \sum_{k>0} \tilde{A}_{lk}(LS) F_k(nl, nl) \\
&- \sum_{(n'l')} q(nl) q(n'l') \left\{ F_0(nl, n'l') - \frac{1}{2\sqrt{(2l + 1)(2l' + 1)}} \right. \\
&\hspace{3cm} \left. \times \sum_{k=0}^{l+l'} c^k(l0, l'0) G_k(nl, n'l') \right\} \\
&+ \frac{q(nl)}{2} [q(nl) - 1] \frac{1}{4l + 1} \\
&\times \sum_{k>0}^{2l} c^k(l0, l0) F_k(nl, nl).
\end{aligned}
\tag{3.86}
$$

We see immediately that the first and fourth terms cancel, because in the fourth term, the summation over $(n'l')$ is for complete groups and thus $q(n'l') = 2(2l' + 1)$. Using this identity in the fifth term, we get the factor of that term in the form

$$
q(n'l') \frac{1}{2\sqrt{(2l + 1)(2l' + 1)}} = \frac{2(2l' + 1)}{2\sqrt{(2l + 1)(2l' + 1)}}
$$
$$
= \sqrt{\frac{2l' + 1}{2l + 1}},
\tag{3.87}
$$

and now it is evident that the second and fifth terms cancel. Thus, we obtain

$$
\begin{aligned}
X(LS) = &- \sum_{k>0} \tilde{A}_{lk}(LS) F_k(nl, nl) + \frac{q(nl)}{2} [q(nl) - 1] \\
&\times \frac{1}{4l + 1} \sum_{k>0}^{2l} c^k(l0, l0) F_k(nl, nl).
\end{aligned}
\tag{3.88}
$$

The upper limit of the first summation is the same as in the second. Thus, we can write

$$X(LS) = \sum_{k>0}^{2l} f_k(LS)F_k(nl, nl),\qquad(3.89)$$

where

$$f_k(LS) = \frac{q(nl)}{2}[q(nl) - 1]\frac{1}{4l + 1}c^k(l0, l0)$$
$$- \tilde{A}_{lk}(LS) \qquad (0 \le k \le 2l).\qquad(3.90)$$

These are the coefficients tabulated by Slater [Slater II, 287, 294] for p^q and d^q configurations.[†] As we see from those tables, the sum over k contains one term for p^q configurations and two for d^q configurations.

A by-product of this derivation is that the average energy gives the exact result not only for complete groups, but also for the complete/incomplete interaction plus the intragroup electrostatic energy of incomplete groups. Thus, what is being averaged is really only the intragroup exchange energy represented by the $F_k(nl, nl)$, $k > 0$, integrals.

Next, we derive the exact formula for two incomplete groups, one of which consists of one electron. We again have

$$X = E_{HF} - E_{av}.\qquad(3.91)$$

Obviously, we do not have to bother about the complete groups, about the interaction energy between any of the incomplete groups and complete groups, and about the intragroup electrostatic interaction of the incomplete groups. (This last is zero for a group with one electron.) Both in E_{HF} and in E_{av}, we will have the terms representing the intragroup exchange energy for the group with more than one electron. These terms lead to the expression given by Eqs. (3.89) and (3.90). The presence of another incomplete group with one electron will not influence this term. Besides this term, we will have in X the electrostatic and exchange interactions between the two incomplete groups. The general form of that can easily be determined. Let the spin orbital φ_i be associated with the single electron in the incomplete group and let φ_j be one of the spin orbitals of the other incomplete group. Using Eqs. (2.70b) and (2.70c), for the interaction energy of the two spin orbitals, we

[†]More complete tables including the f^q configurations were published by Nielson and Koster, Ref. 14.

obtain

$$\left\langle \varphi_i \varphi_j \left| \frac{1}{r_{12}} \right| \varphi_i \varphi_j \right\rangle - \left\langle \varphi_i \varphi_j \left| \frac{1}{r_{12}} \right| \varphi_j \varphi_i \right\rangle$$

$$= \sum_{k=0}^{l_m} a_k(l_i m_{li}, l_j m_{lj}) F_k(n_i l_i, n_j l_j)$$

$$- \sum_{k=0}^{l_i+l_j} \delta(m_{si} m_{sj}) b_k(l_i m_{li}, l_j m_{lj}) G_k(n_i l_i, n_j l_j). \quad (3.92)$$

On the basis of this formula, it is evident that X will contain the following type of expression:

$$\sum_{k=0}^{l_m} g_k F_k(nl, n''l'') + \sum_{k=0}^{l+l''} h_k G_k(nl, n''l''), \quad (3.93)$$

where (nl) characterizes the incomplete group with one electron and $(n''l'')$ the other incomplete group, $(l_m$ is the lesser of $(2l)$ and $(2l''))$. The g_k and h_k constants will depend on the multiplet and on the configuration.

We can summarize the argument by writing down the explicit formulas. For the *HF* energy, we get

$$E_{HF} = E_{av} + X, \quad (3.94)$$

where

$$X = \sum_{k>0}^{2l''} f_k F_k(n''l'', n''l'')$$

$$+ \sum_{k=0}^{l_m} g_k F_k(nl, n''l'') + \sum_{k=0}^{l+l''} h_k G_k(nl, n''l''). \quad (3.95)$$

The constants occurring in this formula are tabulated by Slater for numerous multiplets [Slater II, Appendix 21, 286].[†] The limits of the summations show that in most cases, X contains only few terms. In fact, as we can see from Slater's tables, in most cases, the formulas for X are even shorter because some of the k values do not occur. As we have seen, for a d^q configuration, there are only two terms in the summation. For a $d^h p$ configuration, the g_k summation contains only one term and the h_k summation only two terms. Thus, for a $d^h p$ configuration, there is a total of five terms in X.

[†]See also the explicit formulas, derived using Racah's method, for these constants in Sec. 6.2.

We now summarize the properties of E_{av} that we have discussed. We have established the following:

1. E_{av} gives the exact expression for complete groups and for atoms with one electron outside complete groups.
2. E_{av} differs from the exact *HF* energy in only a few terms for configurations that contain one incomplete group. In fact, for p^q configurations, the difference is one term, and d^q configurations, the difference is two. Configurations with complete groups and with one incomplete group are describing the vast majority of neutral atoms in the ground state.
3. For configurations with two incomplete groups, one of which consists of one electron, we obtained a general formula for the *HF* energy and we have shown that E_{HF} differs from the average energy in only a few terms.

Based on the preceding conclusions, we now reformulate the *HF* model in terms of the average of configurations. Instead of minimizing the energy for each multiplet, that is, calculating different orbitals for each multiplet, we calculate only one set of orbitals for each configuration. The wave functions, total energies, and expectation values of operators for each multiplet are computed by using the single set of orbitals that is derived for the average of the configuration. As we will see presently, the structure of the equations will show that the single set of orbitals provides a very good approximation for all multiplets that can be generated by the configuration.

Following these guidelines, we now derive the *HF* equations for the average of the configuration. For the reader's convenience, we again write down the formula for the total energy:

$$
E_{av} = \sum_{(nl)} q(nl) I(nl) + \sum_{(nl)} \frac{q(nl)}{2} [q(nl) - 1]
$$

$$
\times \left\{ F_0(nl, nl) - \frac{1}{4l + 1} \sum_{k>0} c^k(l0, l0) F_k(nl, nl) \right\}
$$

$$
+ \sum_{\substack{(nl)\ (n'l') \\ (n'l') \neq (nl)}} q(nl) q(n'l')
$$

$$
\times \left\{ F_0(nl, n'l') - \frac{1}{2\sqrt{(2l + 1)(2l' + 1)}} \right.
$$

$$
\left. \times \sum_{k=0}^{l+l'} c^k(l0, l'0) G_k(nl, n'l') \right\}.
$$

(3.96)

We quote the equations describing the variation of the F_k and G_k integrals (Eqs. (2.78a) to (2.78d)):

$$\delta F_k(nl, n'l') = 2\left\langle \delta P_{nl} \left| \frac{1}{r} Y_k(n'l', n'l'|r) \right| P_{nl} \right\rangle; \qquad (3.97a)$$

$$\delta F_k(nl, nl) = 4\left\langle \delta P_{nl} \left| \frac{1}{r} Y_k(nl, nl|r) \right| P_{nl} \right\rangle; \qquad (3.97b)$$

$$\delta G_k(nl, n'l') = 2\left\langle \delta P_{nl} \left| \frac{1}{r} Y_k(nl, n'l'|r) \right| P_{n'l'} \right\rangle. \qquad (3.97c)$$

The subsidiary conditions in the variations are the normalization and orthogonality of the P_{nl}'s. We put

$$E_0 = - \sum_{(nl)} \bar{\varepsilon}_{nl} \langle P_{nl}|P_{nl} \rangle - \sum_{(nl)} \sum_{\substack{n' \\ (n' \neq n)}} \bar{\lambda}_{nl, n'l} \langle P_{nl}|P_{n'l} \rangle, \qquad (3.98)$$

where $\bar{\varepsilon}_{nl}$ and $\bar{\lambda}_{nl, n'l}$ are Lagrangian multipliers. Our variation principle is then

$$\delta(E_{av} + E_0) = 0. \qquad (3.99)$$

We carry out the variation and we divide the resultant equation by $2q(nl)$ in order to bring it to the standard form. In this way, we obtain

$$\begin{aligned}
&\left\{ -\frac{1}{2}\frac{d^2}{dr^2} + \frac{l(l+1)}{2r^2} - \frac{Z}{r} + \sum_{\substack{(n'l') \\ \text{(all groups)}}} q(n'l')\frac{1}{r}Y_0(n'l', n'l'|r) \right. \\[2mm]
&\quad - \frac{1}{r}Y_0(nl, nl|r) - \frac{q(nl)-1}{4l+1} \sum_{k>0} c^k(l0, l0)\frac{1}{r}Y_k(nl, nl|r) \\[2mm]
&\quad - \sum_{\substack{(n'l') \\ (n'l') \neq (nl)}} \frac{q(n'l')}{2(2l'+1)}\sqrt{\frac{2l'+1}{2l+1}} \\[2mm]
&\quad \left. \times \sum_k c^k(l0, l'0)\frac{1}{r}Y_k(nl, n'l'|r)\frac{P_{n'l'}(r)}{P_{nl}(r)} \right\} P_{nl} \\[2mm]
&= \varepsilon_{nl}P_{nl} + \sum_{\substack{n' \\ (n' \neq n)}} \lambda_{nl, n'l}P_{n'l}. \qquad (3.100)
\end{aligned}$$

In this equation, we have formally localized the exchange operator, given by the last term on the left side, which represents the exchange interaction between P_{nl} and the orbitals in the other groups.

We now proceed to the analysis of this equation. First, we must verify that it will become identical with the exact equation for complete groups. The *HF* potential for complete groups is given by Eq. (2.98), which we write down here:

$$U(r) = \sum_{\substack{(n'l') \\ \text{(all groups)}}} 2(2l' + 1)\frac{1}{r}Y_0(n'l', n'l'|r)$$

$$- \sum_{(n'l')} \sum_k \sqrt{\frac{2l' + 1}{2l + 1}}\, c^k(l0, l'0)\frac{1}{r}Y_k(nl, n'l'|r)\frac{P_{n'l'}}{P_{nl}}, \quad (3.101)$$

where we have written the exchange interaction in local form. Consider now the fifth, sixth, and seventh terms in Eq. (3.100). For complete groups, we put

$$q(nl) = 2(2l + 1),$$

$$q(n'l') = 2(2l' + 1), \quad (3.102)$$

and then the fifth term can be incorporated into the sixth with $k = 0$. On the other hand, the expression in the seventh term gives, for $(n'l') = (nl)$,

$$-\frac{2(2l + 1)}{2(2l + 1)}\sum_{k=0}^{2l} c^k(l0, l0)\frac{1}{r}Y_k(nl, nl|r), \quad (3.103)$$

which is identical with the sixth term. Thus, for complete groups, the fifth, sixth, and seventh terms become identical with the exchange potential in Eq. (3.101).

Next, let us consider an atom with one incomplete group. First, we want the *HF* equation for an orbital in the incomplete group. In order to get that, we put $q(n'l') = 2(2l' + 1)$ in the last term on the left side of Eq. (3.100). Then the equation becomes

$$\left\{-\frac{1}{2}\frac{d^2}{dr^2} + \frac{l(l + 1)}{2r^2} - \frac{Z}{r} + U_{av}(nl|r)\right\}P_{nl}$$

$$= \varepsilon_{nl}P_{nl} + \sum_{n'}\lambda_{nl, n'l}P_{n'l}, \quad (3.104)$$

where the *HF* potential is given by

$$U_{av}(nl|r)$$

$$= \sum_{(n'l')} q(n'l') \frac{1}{r} Y_0(n'l', n'l'|r)$$

$$- \frac{1}{r} Y_0(nl, nl|r) - \sum_{\substack{(n'l') \\ (n'l') \neq (nl)}} \sqrt{\frac{2l'+1}{2l+1}} \sum_k c^k(l0, l'0)$$

$$\times \frac{1}{r} Y_k(nl, n'l'|r) \frac{P_{n'l'}}{P_{nl}}$$

$$- \frac{q(nl)-1}{4l+1} \sum_{k>0} c^k(l0, l0) \frac{1}{r} Y_k(nl, nl|r). \qquad (3.105)$$

The *HF* potential for the *LS* multiplet was given by Eq. (3.48). Using that equation, for the difference of the two potentials, we obtain

$$U_d = U(nl, LS|r) - U_{av}(nl|r)$$

$$= - \sum_{k>0} \tilde{a}_{lk}(LS) \frac{1}{r} Y_k(nl, nl|r)$$

$$+ \sum_{k>0} \frac{q(nl)-1}{4l+1} c^k(l0, l0) \frac{1}{r} Y_k(nl, nl|r)$$

$$+ \sum_{k>0} \left[\frac{q(nl)-1}{4l+1} c^k(l0, l0) - \frac{4\tilde{A}_{lk}(LS)}{2q(nl)} \right]$$

$$\times \frac{1}{r} Y_k(nl, nl|r), \qquad (3.106)$$

where we have used Eq. (3.46). Comparing this with Eq. (3.90), we see that the difference between the exact *HF* potential and the average potential is given by

$$U_d = \sum_{k>0}^{2l} f'_k(LS) \frac{1}{r} Y_k(nl, nl|r), \qquad (3.107)$$

where the constants are related to Slater's constants by the relationship

$$f'_k(LS) = \frac{2}{q(nl)} f_k(L, S), \qquad (3.108)$$

and $f_k(LS)$ is the constant given by Eq. (3.90), that is,

$$f_k(LS) = \frac{q(nl)}{2}[q(nl) - 1]\frac{1}{4l+1}c^k(l0,l0)$$

$$- \tilde{A}_{lk}(LS) \qquad (0 \leq k \leq 2l). \tag{3.109}$$

Now we can write

$$U(nl, LS|r) = U_{av}(nl) + \sum_{k>0}^{2l} f'_k(LS)\frac{1}{r}Y_k(nl, nl|r). \tag{3.110}$$

The physical meaning of this equation is this: In order to get the potential for the LS multiplet, we must take the potential for the average of the configuration and add to it the second term, which is nothing else than what we get from X, Eq. (3.84), by varying it with respect to P_{nl} and dividing it by $2q(nl)$. Thus, in order to get the HF potential for any multiplet, we must add to the average potential an expression that is easily obtainable from the expressions tabulated by Slater.

This is hardly surprising because we emphasized that the average energy differs from the exact energy by those terms tabulated by Slater; we also pointed out that X consists only of a few terms in most cases. Nevertheless, the relationship given by Eq. (3.110) represents a significant conceptual simplification of multiplet theory.

We are still discussing an atom with one incomplete group. Having obtained the equation for the orbital in the incomplete group, we now derive the equation for one of the orbitals in the complete groups. Let the quantum numbers of the selected orbital be $(n'l')$, the quantum numbers of the incomplete group (nl), and the summations for complete groups indicated by $(n''l'')$. Then, in Eq. (3.100), the seventh term, which represents the exchange interaction between $(n'l')$ and all the other complete groups and between $(n'l')$ and (nl), becomes

$$-\sum_{\substack{(n''l'') \\ (n''l'' \neq n'l')}} \frac{2(2l''+1)}{2(2l'+1)}\sqrt{\frac{2l''+1}{2l'+1}}\sum_k c^k(l'0,l''0)$$

$$\times \frac{1}{r}Y_k(n'l', n''l''|r)\frac{P_{n''l''}}{P_{n'l'}}$$

$$-\frac{q(nl)}{2(2l+1)}\sqrt{\frac{2l+1}{2l'+1}}\sum_k c^k(l'0,l0)$$

$$\times \frac{1}{r}Y_k(n'l', nl|r)\frac{P_{nl}}{P_{n'l'}}, \tag{3.111}$$

where the first term is the exchange interaction with the complete groups and the second is the exchange interaction with the incomplete group. In the sixth term of Eq. (3.100), we put $q(nl) \to q(n'l') = 2(2l' + 1)$ since $(n'l')$ is a complete group. In this way we obtain the *HF* potential for the orbital $P_{n'l'}$, which is in a complete group:

$$
\begin{aligned}
U_{av} = &\sum_{(n''l'')} q(n''l'') \frac{1}{r} Y_0(n''l'', n''l''|r) \\
&- \frac{1}{r} Y_0(n'l', n'l') - \sum_{k>0} c^k(l'0, l'0) \frac{1}{r} Y_k(n'l', n'l'|r) \\
&- \sum_{(n''l'')} \sum_k \sqrt{\frac{2l''+1}{2l'+1}} c^k(l'0, l''0) \frac{1}{r} Y_k(n'l', n''l''|r) \\
&\quad (n''l'') \neq (n'l') \\
&\times \frac{P_{n''l''}}{P_{n'l'}} - \frac{q(nl)}{2(2l+1)} \sqrt{\frac{2l+1}{2l'+1}} \sum_k c^k(l'0, l0) \\
&\times \frac{1}{r} Y_k(n'l', nl|r) \frac{P_{nl}}{P_{n'l'}}.
\end{aligned}
\tag{3.112}
$$

A look at Eq. (3.51) reveals that this expression is identical with the expression that is the exact *HF* potential for the *LS* multiplet. (The fifth term of Eq. (3.51) is incorporated into the first term of Eq. (3.112).) Thus, for an orbital in a complete group, the *HF* equation for the average configuration gives the (formally) exact result. This is, of course, just another way of saying that the average energy of a configuration gives the exact result for the interactions between complete and incomplete groups.

The interpretations of Eq. (3.110) and of Eq. (3.112) form the theoretical justification for the introduction of the average energy of configurations. Let us consider first Eq. (3.110). We have seen that the potential for the orbital in the incomplete group differs only in a few terms from the average potential. Thus, it is reasonable to assume that an orbital computed from the average-configuration *HF* equation will be a very good approximation to the orbitals that we would obtain by using the correct, exact potential for each multiplet. This assumption is reasonable not only because the two potentials differ in only a few terms, but also because of the nature of those terms that make up the difference. As we see from the derivation, the difference consists of potential terms representing the intragroup exchange interaction. The effect of these terms is certainly smaller than the effect of the intragroup electrostatic interaction, not to mention the effect of the total potential of the complete groups, electrostatic plus exchange, which is the same in the two potentials. Another argument in favor of the average-of-configuration method is provided by Eq. (3.112). This equation shows that the average potential for the orbitals in the complete groups is formally identical with the exact potential for any of the multiplets.

"Formally identical" means that the functional dependence of the average potential on the orbital in the incomplete group is the same as in the exact potential. Thus, by replacing the exact potential by the potential of the average configuration, we do not replace one potential function by another completely different potential function. The change means only that in the two potentials, *one* of the orbitals will be different slightly. It is physically plausible that the effect of this difference is not very great. This follows because the orbital of the incomplete group is contained in those terms of the potential that represent the interaction between the incomplete group and the selected complete group. The incomplete group is usually the outermost group of the valence electrons and the selected complete group is one of the inner groups. It is well known that the effect of the outermost group on the electron density of inner groups is small, except, perhaps, for the group immediately below the valence electrons. Thus, a slight change in the orbital P_{nl} has almost certainly only a small effect on the potential and electron density of the inner groups.

Thus, the analysis of both equations, Eq. (3.110) and Eq. (3.112), provides arguments in favor of the average-of-configuration method. We adopt these arguments and we will treat this approximation, in the rest of the book, as a useful extension of the *HF* model.

In order to complete the discussion, we note that an equation analogous to Eq. (3.110) can be constructed for any number of incomplete groups. For example, on the basis of Eq. (3.95), we can construct equations for two incomplete groups, one of which consists of one electron. We obtain for the orbital of the one-electron incomplete group, P_{nl},

$$U(nl, LS|r) = U_{av} + \sum_{k=0}^{l_m} g_k \frac{1}{r} Y_k(n''l'', n''l''|r)$$

$$+ \sum_{k=0}^{l+l''} h_k \frac{1}{r} Y_k(nl, n''l''|r) \frac{P_{n''l''}}{P_{nl}}, \qquad (3.113)$$

where g_k and h_k are the same as in Eq. (3.95). For the orbital $P_{n''l''}$ of the incomplete group with more than one electron, we obtain

$$U(n''l'', LS|r) = U_{av} + \sum_{k>0}^{2l''} f'_k \frac{1}{r} Y_k(n''l'', n''l'|r) + \sum_{k=0}^{l_m} g'_k \frac{1}{r} Y_k(nl, nl|r)$$

$$+ \sum_{k=0}^{l+l''} h'_k \frac{1}{r} Y_k(nl, n''l''|r) \frac{P_{nl}}{P_{n''l''}}, \qquad (3.114)$$

where

$$f'_k = \frac{2f_k}{q(n''l'')}, \tag{3.115a}$$

$$g'_k = \frac{g_k}{q(n''l'')}, \tag{3.115b}$$

and

$$h'_k = \frac{h_k}{q(n''l'')}. \tag{3.115c}$$

We close the section by proving that in the average-of-configuration model, the *HF* orbitals are again eigenfunctions of non-Hermitian operators. The *HF* equation is Eq. (3.100). Let us denote the potential in that equation by U_{av}. We obtain

$$
U_{av} = \sum_{(n'l')} q(n'l') \frac{1}{r} Y_0(n'l', n'l'|r)
$$

$$
- \frac{1}{r} Y_0(nl, nl|r)
$$

$$
- \frac{q(nl) - 1}{4l + 1} \sum_{k>0} c^k(l0, l0) \frac{1}{r} Y_k(nl, nl|r)
$$

$$
- \sum_{\substack{(n'l') \\ (n'l') \neq (nl)}} \frac{q(n'l')}{2(2l' + 1)} \sqrt{\frac{2l' + 1}{2l + 1}} \sum_k c^k(l0, l'0)
$$

$$
\times \frac{1}{r} Y_k(nl, n'l'|r) \frac{P_{n'l'}}{P_{nl}}. \tag{3.116}
$$

This is the potential for the orbital P_{nl} in the group (nl) with the other groups indicated by $(n'l')$. We note that the potentials in Eqs. (3.105) and (3.112), also denoted by U_{av} are special cases of this potential for an orbital in an incomplete group and for an orbital in one of the complete groups. Using the notation of Eq. (3.116), we can write the *HF* equation in the form

$$
\left\{ -\frac{1}{2} \frac{d^2}{dr^2} + \frac{l(l + 1)}{2r^2} - \frac{Z}{r} + U_{av} \right\} P_{nl}
$$

$$
= \varepsilon_{nl} P_{nl} + \sum_{\substack{n' \\ (n' \neq n)}} \lambda_{nl, n'l} P_{n'l}. \tag{3.117}
$$

It is easy to see, using the same method that we have used before, that this equation can be written in the form

$$H'_{av} P_{nl} = \varepsilon_{nl} P_{nl},$$ (3.118)

where

$$H'_{av} = \Pi H_{av},$$ (3.119a)

$$H_{av} = -\frac{1}{2}\frac{d^2}{dr^2} + \frac{l(l+1)}{2r^2} - \frac{Z}{r} + U_{av},$$ (3.119b)

$$\Pi = 1 - \Omega,$$ (3.119c)

$$\Omega = \sum_{\substack{n' \\ (n' \neq n)}} |P_{n'l}\rangle\langle P_{n'l}|.$$ (3.119d)

Thus, we see from Eq. (3.118) that the orbitals are the eigenfunctions of non-Hermitian operators. The non-Hermiticity follows because H_{av} and Π do not commute. Nevertheless, it is easy to show that the eigenvalues of Eq. (3.118) are always real; therefore, the basic principles of the quantum mechanics are not violated.

CHAPTER 4

RELATIVISTIC HARTREE – FOCK
THEORY

4.1. INTRODUCTION: IS RELATIVITY THEORY IMPORTANT FOR ATOMIC STRUCTURE?

Discussion of the conceptual and practical reasons for a relativistic formulation of atomic structure theory.

4.2. RELATIVISTIC MANY-ELECTRON HAMILTONIANS

A. The nonrelativistic Hamiltonian in the presence of an external electromagnetic field. B. Darwin's relativistic but nonquantum-mechanical Hamiltonian and its comparison with the nonrelativistic expression. C. Summary of Dirac's exact, relativistic, one-electron theory; a discussion of the approximate, two-component equation and its comparison with the nonrelativistic theory. D. Breit's method for constructing the relativistic many-electron Hamiltonian by the correspondence principle; reduction of the exact equation to a two-component equation; comparison of this equation with Darwin's and Dirac's expressions. E. The physical meaning of Breit's operator and comparison with quantum electrodynamics.

4.3. THE HARTREE – FOCK – DIRAC (HFD) MODEL

A. The basic formulas of the HFD model; formulas for the Hamiltonian, for the determinantal wave function of the atom, and for the one-electron wave functions of the electrons; the matrix components of the Hamiltonian with respect to the one-electron functions; the energy expression for an atom with complete groups; comparison with the nonrelativistic expression. B. Derivation of the HFD equations for atoms with complete groups. C. The relativistic energy expression for the average of a configuration with incomplete groups; the HFD equations for the configuration average.

4.1. INTRODUCTION: IS RELATIVITY THEORY IMPORTANT FOR ATOMIC STRUCTURE?

The Hartree–Fock (HF) model that we have presented in the preceding two chapters is very elegant mathematically and provides a great deal of accurate information about atomic structure. Nevertheless, the model is not quite satisfactory for conceptual as well as for practical reasons. First, it is the principle of the special theory of relativity that all physical theories (and models) must be formulated in such a way as to be *covariant*, that is, invariant to a Lorentz transformation between two uniformly moving coordinate systems. The HF model, as we have developed it in the preceding chapters, does not satisfy this requirement. Second, the model does not explain the very important spin effects: these are the fine structure of the optical spectra and the doublet structure of the x-ray spectra.

Thus, we see that for conceptual as well as for practical reasons, it is necessary that atomic structure theory should be formulated relativistically. Here we want to emphasize especially the conceptual reasons. According to Dirac, a quantum mechanical problem should be formulated in terms of a complete set of commuting observables, that is, a complete set of commuting Hermitian operators whose eigenvalues would be equal to the results of the corresponding measurements. The HF model developed before rested on the Hamiltonian operator written down in Eq. (2.34). It is obvious that this operator does not satisfy the requirement of Lorentz invariance. Therefore, the exact eigenvalues of this operator (even if we would be able to determine them!) could not be expected to agree with experimental data.

Regardless of how much useful information can be gained from a model that rests on this Hamiltonian, there is something profoundly unsatisfactory about a theory that could not yield, even "in principle," numbers that would be equal to empirical data.

The situation is unsatisfactory for practical reasons as well. Included in the most important data about atoms are the x-ray line spectra. The energy spectrum of the core electrons, as described by the nonrelativistic HF model, does not resemble the empirical spectrum because the model yields only one energy level for each (nl) group. Similarly, the fine-structure splitting is absent from the optical spectrum provided by the HF model.

Thus, the answer to the question posed in the section title, "is relativity theory important for atomic structure?," is strongly in the affirmative. Indeed, relativity theory is not only important for atomic structure, but it is absolutely essential that a comprehensive atomic theory should be formulated relativistically.

In this chapter, our main goal is to present the relativistic formulation of the HF model. This will be done in two steps. First, in the next section, we study the methods by which the relativistic Hamiltonian for an N-electron system can be set up. Having obtained the Hamiltonian, we formulate the relativistic HF model in Section 4.3. The evaluation of results obtained with this model (as well as with the nonrelativistic model) follows in Chapter 5.

4.2. RELATIVISTIC MANY-ELECTRON HAMILTONIANS

In this section, we present the pyramidlike step-by-step process by which a relativistic Hamiltonian for an interacting system of N electrons can be set up. As a first step in this process, we review the nonrelativistic formulas. Everywhere in the section, we write the Hamiltonians in such a way as to include an arbitrary external electromagnetic field. In order to keep the formulas relatively simple, we write down the many-electron Hamiltonians only for two particles. Generalizations for N electrons can be written down immediately.

A. Nonrelativistic Hamiltonians

Let us consider first a single electron moving in an electromagnetic field that is described by the scalar potential $V = V(\mathbf{r}, t)$ and the vector potential $\mathbf{A} = \mathbf{A}(\mathbf{r}, t)$. The vectors of the electric field strength, $\mathscr{E}$, and the magnetic field strength $\mathscr{H}$, are related to the potential as follows:

$$\mathscr{E} = -\nabla V - \frac{1}{c} \frac{\partial \mathbf{A}}{\partial t}, \tag{4.1}$$

and

$$\mathscr{H} = \operatorname{curl} \mathbf{A} = \nabla \times \mathbf{A}. \tag{4.2}$$

The potentials are also subjected to the Lorentz condition

$$\operatorname{div} \mathbf{A} + \frac{1}{c} \frac{\partial V}{\partial t} = 0. \tag{4.3}$$

It can be shown that if the current densities are zero, $\mathbf{j} = \rho = 0$, then the potential can be subjected to a gauge transformation after which

$$\operatorname{div} \mathbf{A} = 0, \tag{4.4}$$

for all $\mathbf{r}$ and t. Naturally, this equation is valid anyhow if the scalar potential is time-independent; we assume that its validity can be taken for granted in all cases that will be considered in this work.

   If Eq. (4.4) is valid, the operator of the momentum commutes with the vector potential. Because this fact simplifies the formulas, we prove it here. Let p_k be one of the cartesian components of the momentum and A_k the component of the vector potential. Then

$$p_k A_k - A_k p_k = -i\hbar \frac{\partial}{\partial x_k} A_k(x_1 x_2 x_3) + i\hbar A_k(x_1 x_2 x_3) \frac{\partial}{\partial x_k}$$

$$= i\hbar A_k(x_1 x_2 x_3) \frac{\partial}{\partial x_k} - i\hbar A_k(x_1 x_2 x_3) \frac{\partial}{\partial x_k} - i\hbar \frac{\partial A_k}{\partial x_k}$$

$$= -i\hbar \frac{\partial A_k}{\partial x_k}. \tag{4.5}$$

From this, we obtain

$$\mathbf{p} \cdot \mathbf{A} - \mathbf{A} \cdot \mathbf{p} = -i\hbar \, \mathrm{div} \, \mathbf{A}, \tag{4.6}$$

and, in view of Eq. (4.4),

$$\mathbf{p} \cdot \mathbf{A} = \mathbf{A} \cdot \mathbf{p}. \tag{4.7}$$

The classical Hamiltonian for an electron that is moving in an electromagnetic field is

$$H = \frac{1}{2m}\left(\mathbf{p} + \frac{e}{c}\mathbf{A}\right)^2 - eV, \tag{4.8}$$

and we obtain the Hamilton operator from this by replacing $\mathbf{p}$ with the corresponding operator, that is, by putting

$$\mathbf{p} = -i\hbar \nabla. \tag{4.9}$$

Let us assume that the electron is moving in an atom. Then we put

$$V = \frac{Ze}{r} + \varphi(\mathbf{r}, t), \tag{4.10}$$

where the first term is the potential of the nucleus, and the second term is an arbitrary, external potential that may be time-dependent. Our Hamiltonian becomes

$$H = \frac{P^2}{2m} - \frac{Ze^2}{r} + \frac{e}{mc}\mathbf{A} \cdot \mathbf{p} + \frac{e^2}{2mc^2}A^2 - e\varphi, \tag{4.11}$$

which is the general nonrelativistic Hamiltonian for an electron in an electromagnetic field. We note that in the formulation, we used Eqs. (4.4) and (4.7).

We generalize the results to two electrons. The classical Hamiltonian is

$$H = \frac{1}{2m}\left(\mathbf{p}_1 + \frac{e}{c}\mathbf{A}_1\right)^2 - eV(1)$$

$$+ \frac{1}{2m}\left(\mathbf{p}_2 + \frac{e}{c}\mathbf{A}_2\right)^2 - eV(2) + \frac{e^2}{r_{12}}. \tag{4.12}$$

Clearly, this function is simply the sum of two one-electron Hamiltonians like in Eq. (4.8), plus the ordinary electrostatic interaction energy e^2/r_{12}, with r_{12} being the distance between the electrons. Making the operator substitution given by Eq. (4.9), we obtain the Hamiltonian

$$H = H_0(1) + H_0(2) + H_1'(1) + H_1'(2) + \frac{e^2}{r_{12}}. \tag{4.13}$$

Here we have introduced the notations

$$H_0(i) = \frac{p_i^2}{2m} - eV(i), \tag{4.14}$$

and

$$H_1'(i) = \frac{e}{mc}\mathbf{A}_i \cdot \mathbf{p}_i + \frac{e^2}{2mc^2}A_i^2. \tag{4.15}$$

As we see, H_0 contains the kinetic energy plus the scalar potentials and H_1' contains the magnetic interaction with the external field.

If there is no external electromagnetic field, then $\mathbf{A} = \varphi = 0$ and we get

$$H_0(i) = \frac{p_i^2}{2m} - \frac{Ze^2}{r_i}, \tag{4.16}$$

and our Hamiltonian becomes

$$H = \frac{p_i^2}{2m} - \frac{Ze^2}{r_1} + \frac{p_2^2}{2m} - \frac{Ze^2}{r_2} + \frac{e^2}{r_{12}}. \tag{4.17}$$

This is the Hamiltonian that we wrote down in Eqs. (2.34) and (3.1), that is, this is the Hamiltonian that serves as the basis for the nonrelativistic HF model.

B. Darwin's Classical Relativistic Hamiltonian

Having reviewed the nonrelativistic formulas, we now proceed to the discussion of Darwin's work. Darwin's goal was to set up a Hamiltonian for an interacting system of N electrons. Darwin formulated the Hamiltonian relativistically, but his work is not yet a quantum mechanical treatment. As we will see in the forthcoming discussion, Darwin's work is of pivotal significance for relativistic many-electron theory. For this reason, we have reconstructed the somewhat cryptic original derivation and present it, in its entirety, in App. F. Here we summarize the results.

Darwin's main point was that in determining the electromagnetic interaction between two electrons, there are effects that must be considered besides the electrostatic energy e^2/r_{12}. The first is the magnetic interaction. An electron that is moving constitutes a current; therefore, it creates a magnetic field. Thus, the electron will be surrounded, besides its electrostatic Coulomb field, by a magnetic field. Therefore, the interaction of two electrons is not only electrostatic, but also a magnetic interaction.

The second effect that must be taken into account is the retardation of potential. According to the theory of relativity, the electrostatic potentials created by an electron are not felt instantaneously by another electron. The electromagnetic potentials propagate with the finite speed of light and the time needed for them to proceed from the location of one electron to the location of another must be taken into account in the calculation of the interaction energy. This phenomenon is called the retardation effect.

Darwin derived a Hamiltonian in which both of these effects were taken into account. In addition, his Hamiltonian also contains a term that takes into account the energy resulting from the relativistic increase of the electron mass at high speeds. The whole derivation, which the reader finds in App. F, is done in an approximation that includes terms up to $1/c^2$, where c is the speed of light.

For two particles, Darwin's expression reads as follows (Eq. (F.48) from App. F):

$$
\begin{aligned}
H = {} & \frac{p_1^2}{2m} - eV(1) + \frac{p_2^2}{2m} - eV(2) \\
& + \frac{e}{mc}\mathbf{A}_1 \cdot \mathbf{p}_1 + \frac{e}{mc}\mathbf{A}_2 \cdot \mathbf{p}_2 \\
& - \frac{p_1^4}{8c^2m^3} - \frac{p_2^4}{8c^2m^3} + \frac{e^2}{r_{12}} \\
& - \frac{e^2}{2m^2c^2}\left\{ \frac{\mathbf{p}_1 \cdot \mathbf{p}_2}{r_{12}} + \frac{(\mathbf{p}_1 \cdot \mathbf{r}_{12})(\mathbf{p}_2 \cdot \mathbf{r}_{21})}{r_{12}^3} \right\}.
\end{aligned}
\tag{4.18}
$$

We have used all the symbols before except vector $\mathbf{r}_{21}$ which is defined as

$$\mathbf{r}_{21} = \mathbf{r}_2 - \mathbf{r}_1, \tag{4.19}$$

and

$$r_{21} = r_{12} = |\mathbf{r}_2 - \mathbf{r}_1|. \tag{4.20}$$

The previous Hamiltonian describes the relativistic interaction of two electrons that are moving in the external electromagnetic field characterized by the scalar potential V and the vector potential $\mathbf{A}$. The physical interpretation of this expression is straightforward. The first two terms are the same as in Eq. (4.14). These represent the kinetic energy of "electron one" and the potential energy in the external field. The next two terms are the same for "electron two." The fifth term is the interaction of "electron one" with the external magnetic field and it is the same as the first term of Eq. (4.15). In Darwin's expression, we do not have the A^2 term as in Eq. (4.15). The sixth term is the magnetic interaction of "electron two" and the seventh and eighth terms are the corrections of the kinetic energies due to the relativistic increase in the electron mass at higher velocities. The next term is the electrostatic interaction energy of the two electrons and the last, tenth term, represents the magnetic and retardation effects. As we show in App. F, the electrostatic and magnetic interaction energies are given by the expression

$$g_0(1,2) = \frac{e^2}{r_{12}}\left[1 - \frac{1}{m^2 c^2}\mathbf{p}_1 \cdot \mathbf{p}_2\right], \tag{4.21}$$

and the retardation effects are described by the formula

$$g_1(1,2) = \frac{e^2}{2m^2 c^2 r_{12}}\left[\mathbf{p}_1 \cdot \mathbf{p}_2 - \frac{(\mathbf{p}_1 \cdot \mathbf{r}_{21})(\mathbf{p}_2 \cdot \mathbf{r}_{21})}{r_{12}^2}\right]. \tag{4.22}$$

The last two terms of the Hamiltonian in Eq. (4.18) are the sum of these two expressions.

Thus, we see that the Hamiltonian constructed by Darwin contains the two effects that are required by relativity theory and are missing from the nonrelativistic Hamiltonian: Darwin's expression contains the magnetic interaction between two moving charges and it contains the retardation of the potentials. In addition, there is a relativistic kinetic-energy correction for high velocities. We can turn Darwin's expression into an operator immediately by making the substitution $\mathbf{p} = -i\hbar\nabla$. As we will see presently, this is not the only way to turn Darwin's Hamiltonian into an operator. At this point, however, let us transform the function into an operator, that is, let us assume that the momentum vector is everywhere replaced by the momentum operator. Then, let us introduce the notations

$$H_2(i) = -\frac{p_i^4}{8c^2m^3},$$ (4.23)

$$H_3(1,2) = -\frac{e^2}{m^2c^2r_{12}}\mathbf{p}_1 \cdot \mathbf{p}_2,$$ (4.24)

and

$$H_4(1,2) = \frac{e^2}{2m^2c^2r_{12}}\left[\mathbf{p}_1 \cdot \mathbf{p}_2 - \frac{(\mathbf{p}_1 \cdot \mathbf{r}_{21})(\mathbf{p}_2 \cdot \mathbf{r}_{21})}{r_{12}^2}\right].$$ (4.25)

Using these notations and the symbols defined in Eqs. (4.14) and (4.15), we can write Darwin's Hamiltonian in the form

$$H = H_0(1) + H_0(2) + H_1'(1) + H_2'(2) + \frac{e^2}{r_{12}}$$

$$+ H_2(1) + H_2(2) + H_3(1,2) + H_4(1,2).$$ (4.26)

It is understood that in this expression, H_1' does not contain the A^2 term as in Eq. (4.15). Using this formula, we can write

$$H(\text{Darwin}) = H(\text{nonrel.}) + H_2(1) + H_2(2) + H_3(1,2) + H_4(1,2),$$ (4.27)

where $H(\text{nonrel.})$ is the operator given by Eq. (4.13). Thus, it is clear that Darwin's operator differs from the nonrelativistic Hamiltonian by the addition of the operators H_2, H_3, and H_4.

C. Dirac's Relativistic One-Electron Theory

Although Darwin's Hamiltonian does contain relativistic effects, it is not satisfactory because it is not Lorentz-invariant. Omitting the two-electron terms from Eq. (4.26), we obtain the sum of 2 one-electron operators. These are based on the classical nonrelativistic Hamiltonian given by Eq. (4.8); in addition, they contain the relativistic mass corrections. Clearly, a wave equation based on these operators would not be Lorentz-invariant.

The exact, that is, fully Lorentz-invariant theory of an electron was formulated by Dirac.[16] The relevant equations of Dirac's theory are summarized in App. G. Dirac's Hamiltonian for a single electron that is moving in an external electromagnetic field is given by Eq. (G.4):

$$H_D = -c\boldsymbol{\alpha} \cdot \mathbf{P} - \beta mc^2 - eV.$$ (4.28)

In this operator, we have the Dirac matrix vectors $\boldsymbol{\alpha}$ and $\boldsymbol{\beta}$, which are given in Eqs. (G.12). The operator $\mathbf{P}$ is given by Eq. (G.3); it is also in the first term of our expression in Eq. (4.8). The external potential V is given by Eq. (4.10).

Because the quantities $\boldsymbol{\alpha}$ and $\boldsymbol{\beta}$ must satisfy the commutation relations given by Eqs. (G.8) and (G.9), the wave equation constructed with H_D is a four-component equation, that is, the eigenfunctions of H_D are four-component functions like in Eq. (G.13), and the Dirac equation itself is a set of four equations that are in Eq. (G.20). Because the eigenfunctions of H_D play an important role in the relativistic HF theory, we have summarized the relevant properties of these functions in App. H.

We are assuming that the reader is familiar with the main features of Dirac's theory. Here we recall that this theory is fully Lorenz-invariant and describes accurately the fine structure of the H-like atoms. This last property of the theory follows from the ability of the Dirac equation to account for the existence of the electron spin. When we introduced the electron spin into the HF model by adopting Slater's idea of the spin-orbitals, this was done in a purely ad hoc way. We have introduced spin functions by constructing spin-orbitals and constructed Slater determinants from the spin-orbitals. The scope of the Hartree model was significantly, indeed decisively, enlarged by the introduction of Slater determinants. On the other hand, the introduction of spin-orbitals did not rest on any property of the nonrelativistic Hamiltonian operator, that is, it did not follow from the theory in a natural way.

In contrast, the strength of the Dirac theory rests on the fact that the Dirac Hamiltonian operator given by Eq. (4.28) commutes with the square of the total angular-momentum operator, that is, with the operator that is the sum of the orbital and spin angular momenta. From this, it follows that the electron possesses an intrinsic angular momentum called spin. This is one of the *results* of the Dirac theory; the spin had to be introduced artificially into the nonrelativistic theory.

The Dirac Hamiltonian in its exact form does not directly show the presence of the spin effects. These important effects become visible if we reduce the Dirac equation to its approximate two-component form. It was Darwin who demonstrated that the exact four-component Dirac equation can be reduced, in a good approximation, to a two-component form.[17, 18] This reduction is presented in detail in App. G. Here we review the main steps of the reduction and analyze the results.

The first step in the reduction process is to show that from the four components of the Dirac eigenfunction, two can be classified as "large" and two as 'small" components. Symbolically, we put

$$\begin{pmatrix} u_1 \\ u_2 \end{pmatrix} \ll \begin{pmatrix} u_3 \\ u_4 \end{pmatrix}, \tag{4.29}$$

where u_1, u_2, u_3, and u_4 are the four components of the exact equation. This relationship can be proved by expressing u_1 and u_2 approximately in terms of

u_3 and u_4. These approximate relationships are given in Eqs. (G.44) and (G.46) and show that u_1 and u_2 are smaller than u_3 and u_4 by a factor of v_k/c, where v_k is any component of the electron velocity. Except for extreme relativistic velocities, this factor is small. The next step is to eliminate u_1 and u_2 from the Dirac equations by means of the approximate relationships, which give them in terms of u_3 and u_4. By this process, a system of four equations is constructed of which the third and fourth equations, the equations for the large components u_3 and u_4, do not contain the small components any more. By simply omitting the equations for the small components, we obtain a set of two equations, which means that the original four-component system is reduced to a two-component system.

The result of this reduction process is the two-component equation given by Eq. (G.56). We write the equation here in such a way as to use the notations that we introduced for the nonrelativistic equations. Let the operators H_0, H_1', and H_2 be defined by Eqs. (4.14), (4.15), and (4.23). In addition, let

$$H_1'' = \frac{e\hbar}{2mc}(\boldsymbol{\sigma} \cdot \mathscr{H}), \qquad (4.30)$$

$$H_5 = -\frac{ie\hbar}{4m^2c^2}(\mathscr{E} \cdot \mathbf{p}), \qquad (4.31)$$

and

$$H_6 = -\frac{e\hbar}{2m^2c^2}[\mathscr{E} \cdot (\boldsymbol{\sigma} \times \mathbf{p})]. \qquad (4.32)$$

The 2-component Dirac Hamiltonian contained in Eq. (G.56) then reads

$$H_D = H_0 + H_1' + H_1'' + H_2 + H_6. \qquad (4.33)$$

Each term can be interpreted in a clear-cut fashion and can also be related to the nonrelativistic equations. Let us consider the operators term by term.

H_0 is the nonrelativistic part of the Hamiltonian, that is,

$$H_0 = \frac{p^2}{2m} - eV, \qquad (4.34)$$

where $\mathbf{p}$ is the momentum operator, and V is given by Eq. (4.10). Operators H_1' and H_1'' are the interactions of the electron with the external magnetic field. We have

$$H_1' + H_1'' = \frac{e}{mc}\mathbf{A} \cdot \mathbf{p} + \frac{e^2}{2mc^2}A^2 + \frac{e\hbar}{2mc}(\boldsymbol{\sigma} \cdot \mathscr{H}). \qquad (4.35)$$

The first two terms arise because the electron, as it is moving in a magnetic field, constitutes a current that experiences a force acting on it because of the presence of the magnetic field. The last term is the energy of the electron's spin magnetic moment in the magnetic field. Indeed, we recognize that

$$\mu = \frac{e\hbar}{2mc}, \tag{4.36}$$

is the Bohr magneton and that the intrinsic magnetic moment of the electron is given by

$$\boldsymbol{\mu}_s = -\mu\boldsymbol{\sigma}, \tag{4.37}$$

and so the last term of Eq. (4.35) can be written as

$$H_1'' = -(\boldsymbol{\mu}_s \cdot \mathcal{H}), \tag{4.38}$$

which is the energy of the magnetic moment $\boldsymbol{\mu}_s$ in the magnetic field. Thus, we can introduce the symbol

$$H_1 = H_1' + H_1'', \tag{4.39}$$

and it is clear that H_1 represents the full interaction of the electron with the external magnetic field.

Let us return now to Eq. (4.33). The next term, H_2, is the correction for high velocities:

$$H_2 = -\frac{p^4}{8m^3c^2}. \tag{4.40}$$

Finally, the last two terms, H_5 and H_6, are the interactions of the electron with the external electric field. First, we have

$$H_5 = -\frac{ie\hbar}{4m^2c^2}(\mathcal{E} \cdot \mathbf{p}). \tag{4.41}$$

In App. G, we called this the Darwin term and it is a typical product of Dirac's theory. It is arising from the motion of the electron in the electric field. The interaction of a static electron with the electric field is given by $-eV$, which is contained in the operator H_0. The energy represented by Eq. (4.41) is zero if the electron is not moving, that is, if $\mathbf{p} = 0$.

The last term, H_6, is the most important result of the Dirac theory. This term represents the spin-orbit interaction. As we show in App. G, this term can be brought to the form

$$H_6 = -\frac{e\hbar}{4m^2c^2}[\mathscr{E} \cdot (\boldsymbol{\sigma} \times \mathbf{p})]$$

$$= -\frac{e\hbar}{4m^2c^2}\frac{1}{r}\frac{dV}{dr}\boldsymbol{\sigma} \cdot (\mathbf{r} \times \mathbf{p})$$

$$= -\frac{e}{2m^2c^2}\frac{1}{r}\frac{dV}{dr}(\mathbf{s} \cdot \mathbf{L}). \qquad (4.42)$$

Equation (4.42) is valid if the external vector potential is zero, that is, if

$$\mathscr{E} = -\nabla V, \qquad (4.43)$$

and it is also assumed that V is a central field and we have

$$\mathscr{E} = -\nabla V = -\frac{dV}{dr} \cdot \frac{\mathbf{r}}{r}. \qquad (4.44)$$

In Eq. (4.42), the spin and angular-momentum vectors are defined as

$$\mathbf{s} = \frac{\hbar}{2}\boldsymbol{\sigma}, \qquad (4.45)$$

and

$$\mathbf{L} = \mathbf{r} \times \mathbf{p}. \qquad (4.46)$$

H_6 is the same as Thomas's semiclassical formula for the spin-orbit energy except that we have, in our formulas, the factor $-e$, which is there because in out formula V is the potential; if V would be the potential energy, that is, $-eV$, then our formula would be identical with Thomas's.

In order to clarify the meaning of H_6 more clearly, we want to make the terminology more exact. In preceding Eqs. (4.41) and (4.42), we stated that H_5 and H_6 represent the interactions of the electron with the external electric field. As we stated in Eq. (4.10), the definition of the potential is

$$V = \frac{Ze}{r} + \varphi, \qquad (4.47)$$

where the first term is the potential of the nucleus, and the second is an arbitrary additional potential. The connection between $\mathscr{E}$ and V is, in general,

$$\mathscr{E} = -\nabla V - \frac{1}{c}\frac{\partial \mathbf{A}}{\partial t} = -\nabla\left(\frac{Ze}{r} + \varphi\right) - \frac{1}{c}\frac{\partial \mathbf{A}}{\partial t}. \qquad (4.48)$$

As we have stated, the derivation of Eq. (4.42) for H_6 is valid if $\mathbf{A} = 0$ and the potential V is central.

Now the use of the word "external" is misleading for the following reason. From the point of view of the Dirac formalism, the whole V potential is external, and if that definition is maintained, then it is correct to say that H_5 and H_6 represent the electron's interaction with the external electric field. However, physically it is more meaningful to call only the φ-potential "external"; the potential of the nucleus is part of the atom.[†]

The difference between the nuclear potential and φ becomes evident if we put the "really external" potential equal to zero, that is, if we consider an atom in the absence of an external electromagnetic field. Then we get

$$V = \frac{Ze}{r}, \tag{4.49}$$

and

$$\frac{dV}{dr} = -\frac{Ze}{r^2}. \tag{4.50}$$

H_6 then becomes

$$H_6 = \frac{1}{2m^2c^2}\frac{Ze^2}{r^3}(\mathbf{s}\cdot\mathbf{L}). \tag{4.51}$$

The point of this argument is that H_6 does not vanish if we put the external potential φ equal to zero. In fact, H_6 represents the interaction of the intrinsic magnetic moment of the electron with the magnetic field created by the electron itself. The magnetic field is generated by the electron because the moving electron constitutes a current. The nuclear potential enter H_6 because the motion of the electron, that is, the current, depends on the nuclear potential. It is clear that this interaction will be present even if the "external" potential φ is zero.

We are now able to compare the Dirac approximate Hamiltonian, Eq. (4.33), with the nonrelativistic Hamiltonian, Eq. (4.11). By using the operators H_0 and H_1', the nonrelativistic Hamiltonian can be written in the form

$$H(\text{nonrel.}) = H_0 + H_1', \tag{4.52}$$

and so we obtain from Eq. (4.33)

$$H_D = H(\text{nonrel.}) + H_1'' + H_2 + H_5 + H_6. \tag{4.53}$$

[†]The argument presented here is for a one-electron atom. For an atom with many electrons, φ will consist of the external potential plus the potential for the other electrons. Thus, in that case, putting the external potential equal to zero will not mean $\varphi = 0$. See Section 9.4.

In this equation, H_2 is the high-velocity correction, Eq. (4.23). Operator H_5 given by Eq. (4.31) is the interaction of the moving electron with the electric field. The remaining terms, H_1'' and H_6, are the spin effects. The first of these, H_1'', is the interaction of the intrinsic magnetic moment with the external magnetic field; the second, H_6, is the interaction of the intrinsic magnetic moment with the magnetic field created by the electron itself. The fact that H_1'' and H_6 represent the spin effects can be made even more visible by introducing into the formulas the intrinsic magnetic moment vector, which we have defined by Eqs. (4.36) and (4.37). By introducing these symbols, H_1'' takes the form given by Eq. (4.38). For H_6, we obtain, by taking into account Eqs. (G.59), (4.36), and (4.37),

$$
\begin{aligned}
H_6 &= \frac{e\hbar}{4m^2c^2} \left[\boldsymbol{\sigma} \cdot (\mathscr{E} \times \mathbf{p}) \right] \\
&= \frac{\mu}{2mc} \left[\boldsymbol{\sigma} \cdot (\mathscr{E} \times \mathbf{p}) \right] \\
&= -\frac{1}{2mc} \left[\boldsymbol{\mu}_s \cdot (\mathscr{E} \times \mathbf{p}) \right].
\end{aligned}
\tag{4.54}
$$

We have now summarized the reduction of Dirac's equation to a two-component form. The resulting formula, given by Eq. (4.53), differs from the nonrelativistic Hamiltonian by the presence of four terms, two of which are the spin effects. Thus, it is clearly demonstrated that in the Dirac theory, the spin appears as the result of the theory being fully Lorentz-invariant. The spin is an integral part of the theory and it does not need to be introduced artificially by adding terms to the Hamiltonian.

D. Breit's Relativistic Many-Electron Hamiltonian

We return now to our central problem, which is to construct a relativistic many-electron Hamiltonian. One would think that Darwin's formula, Eq. (4.18), would be a satisfactory solution. The classical formula of Darwin can be turned easily into an operator by replacing everywhere the momentum vector $\mathbf{p}$ by the corresponding quantum mechanical operator. The result is Eq. (4.27) and it is evident immediately that this operator is not a satisfactory solution to our problem. We can see that by comparing Eq. (4.27) with Dirac's operator, Eq. (4.33).

The main difference between the two operators is that the all-important spin effects are missing from the one-electron part of Darwin's expression. Therefore, the replacement of the momentum vector by the corresponding quantum mechanical operator does not turn Darwin's classical Hamiltonian function into an acceptable relativistic Hamiltonian.

A more satisfactory theory was formulated by Breit.[19] In constructing the many-electron Hamiltonian function Breit was guided by the correspondence principle. The correspondence principle of quantum mechanics states that in the classical limit, the Hamiltonian operator should become identical with the corresponding classical Hamiltonian. For the classical (relativistic, but non-quantum-mechanical) Hamiltonian we accept Darwin's expression, Eq. (4.18), as the correct formula. Thus, in the classical limit, the Hamiltonian to be constructed must reduce to this expression. It is clear, however, that it is not enough simply to replace the momentum in Darwin's expression by the corresponding operators. This procedure leads to Eq. (4.27), which is not satisfactory.

Breit turned to the matrix formulation of quantum mechanics and investigated the equation of motion of the electron in the Dirac theory. Let H_D be the exact Dirac Hamiltonian for an electron moving in an electromagnetic field, that is, let

$$H_D = -c\boldsymbol{\alpha} \cdot \mathbf{P} - \beta mc^2 - eV. \tag{4.55}$$

Let the electron's position be $\mathbf{r}$ and its momentum $\mathbf{p}$. The operator $\mathbf{P}$ is defined as

$$\mathbf{P} = \mathbf{p} + \frac{e}{c}\mathbf{A}. \tag{4.56}$$

V and $\mathbf{A}$ are the scalar and vector potentials, respectively, and $\boldsymbol{\alpha}$ and β are the Dirac matrix vectors. Using H_D, we obtain, for the velocity in the Heisenberg picture,

$$\begin{aligned}
\mathbf{v} = \frac{d\mathbf{r}}{dt} &= \frac{\partial \mathbf{r}}{\partial t} + \frac{1}{i\hbar}[\mathbf{r}H_D - H_D\mathbf{r}] \\
&= -c\boldsymbol{\alpha},
\end{aligned} \tag{4.57}$$

and for the equation of motion, we get

$$\frac{d\mathbf{P}}{dt} = -e\mathscr{E} + e[\boldsymbol{\alpha} \times \mathscr{H}]. \tag{4.58}$$

where $\mathscr{E}$ and $\mathscr{H}$ are the electric and magnetic fields, respectively. Breit's derivations are presented in detail in App. I, where the result for $d\mathbf{r}/dt$ is given by Eq. (I.37) and the formula for $d\mathbf{P}/dt$ is Eq. (I.49).

Now in classical mechanics, we get, using the Hamiltonian formalism,

$$\frac{d\mathbf{P}}{dt} = -e\mathscr{E} - \frac{e}{c}[\mathbf{v} \times \mathscr{H}], \tag{4.59}$$

where the right side is the well-known formula for the Lorentz force. The reader finds the derivation of this relationship in App. I, where this formula is Eq. (I.51).

The comparison of the classical formula, Eq. (4.59), with the quantum mechanical formulas, Eqs. (4.57) and (4.58), is very instructive. From Eq. (4.57), we obtain that the correspondence principle for the Dirac theory yields the relationship

$$\mathbf{v} = -c\boldsymbol{\alpha}. \tag{4.60}$$

Comparing Eq. (4.58) with Eq. (4.59), we see that exact correspondence between them can be established if we adopt the identity given by Eq. (4.60). The conclusion is that in the Dirac theory, the quantum mechanical equation of motion reduces to the classical equation of motion if the substitution

$$\boldsymbol{\alpha} = -\frac{1}{c}\mathbf{v}, \tag{4.61}$$

is carried out.

From this relationship, Breit concluded that the classical Hamiltonian of Darwin can be turned into the correct Hamiltonian operator by making the substitution given by Eq. (4.60). More accurately, Breit divided the argument into two parts. The many-electron Hamiltonian must have a one-electron part and a two-electron interaction term. The correct operator for the one-electron part is Dirac's operator, H_D. For the two-electron part, Breit used the two-electron parts of Darwin's expression, which are given by the Eqs. (F.52) and (F.53). Making the substitution of Eq. (4.61) in the interaction operators, Breit obtained the expression for the N-electron system:

$$H = \sum_{k=1}^{N} H_D(k) + \sum_{i<k} \left[g_0(i,k) + g_1(i,k) \right], \tag{4.62}$$

where H_D is the exact Dirac one-electron Hamiltonian and the interaction terms are

$$g_0(1,2) = \frac{e^2}{r_{12}} \left[1 - \boldsymbol{\alpha}(1) \cdot \boldsymbol{\alpha}(2) \right], \tag{4.63}$$

and

$$g_1(1,2) = \frac{e^2}{2r_{12}} \left[\boldsymbol{\alpha}(1) \cdot \boldsymbol{\alpha}(2) - \frac{(\boldsymbol{\alpha}(1) \cdot \mathbf{r}_{21})(\boldsymbol{\alpha}(2) \cdot \mathbf{r}_{21})}{r_{12}^2} \right]. \tag{4.64}$$

Breit was able to show that the Hamiltonian given by Eq. (4.62) reduces, in the classical non-quantum-mechanical limit, to the Hamiltonian function of Darwin, that is, to the function given in Eq. (4.18). The argument leading to this conclusion is summarized at the end of App. I. Thus, the operator given by Eq. (4.62) can be considered to be a meaningful expression for the relativistic Hamiltonian of an N-electron system.

Some mathematical properties of this operator will be considered in Section 4.2/E. Here we proceed to the physical interpretation. To see the physical meaning of this operator is even more difficult than in the case of the exact Dirac Hamiltonian, which we analyzed in Sec. 4.2/C. We have seen that in order to understand the physical meaning of the Dirac Hamiltonian, we had to reduce it to an approximate 2-component form. A similar approximation is needed here. In fact, Breit has shown that the original theory can be simplified similarly as in the case of Dirac's formalism.

Let us consider Breit's Hamiltonian for two electrons. The Schroedinger equation then becomes

$$\{H_D(1) + H_D(2) + g_0(1,2) + g_1(1,2)\}\psi = E\psi. \qquad (4.65)$$

The solutions of this equation will be 16-component wave functions of the form

$$\psi = \begin{vmatrix} \psi_{11}(1,2) & \psi_{12}(1,2) & \psi_{13}(1,2) & \psi_{14}(1,2) \\ \psi_{21}(1,2) & \psi_{22}(1,2) & \psi_{23}(1,2) & \psi_{24}(1,2) \\ \psi_{31}(1,2) & \psi_{32}(1,2) & \psi_{33}(1,2) & \psi_{34}(1,2) \\ \psi_{41}(1,2) & \psi_{42}(1,2) & \psi_{43}(1,2) & \psi_{44}(1,2) \end{vmatrix} \qquad (4.66)$$

All wave functions in this matrix are two-electron functions. From the subscripts, the first refers to "particle one" and the second to "particle two". We can understand the structure of ψ if we consider one of the typical terms in g_0. Let $\alpha_k(1)$ be one of the Dirac matrix-vector components that operates on "electron one" and $\alpha_j(2)$ one that operates on "electron two." Then, a typical product like $\alpha_k(1)\alpha_j(2)$ operates on ψ as follows:

$$\left(\alpha_k(1)\alpha_j(2)\psi\right)_{mn} = \sum_{s,t}\left(\alpha_k(1)\right)_{ms}\left(\alpha_j(2)\right)_{nt}\psi_{st}(1,2). \qquad (4.67)$$

The results of this is another 16-component wave functions composed from the ψ_{st}'s.

Breit was able to show that from the 16 components, four can be classified as "large" and the rest as "small" components. The large components are in the lower right corner, indicated by a box in Eq. (4.66). The argument used to show this was the same argument that was used by Darwin in reducing the Dirac equation. Because we have presented that argument in detail in App. G, there is no need to repeat it here. We proceed now to write down the reduced Hamiltonian that operates only on the large components. We note that the relationship of this operator to the exact Hamiltonian, which is given by Eq. (4.65), is the same as the relationship of the reduced Dirac operator, Eq. (4.53), to the exact Dirac Hamiltonian, Eq. (4.28).

In order to write down Breit's reduced operator, we use the notations of Eqs. (4.14), (4.15), (4.23), (4.24), (4.25), (4.30), (4.31), and (4.32). In addition, let

$$H_7(1,2) = \frac{e\hbar}{4m^2c^2}\frac{2e}{r_{12}^3}\{(\mathbf{r}_{12} \times \mathbf{p}_2) \cdot \boldsymbol{\sigma}_1 + (\mathbf{r}_{21} \times \mathbf{p}_1) \cdot \boldsymbol{\sigma}_2\}, \quad (4.68)$$

and

$$H_8(1,2) = \left(\frac{e\hbar}{2mc}\right)^2 \left\{ -\frac{8\pi}{3}(\boldsymbol{\sigma}_1 \cdot \boldsymbol{\sigma}_2)\delta(\mathbf{r}_{12}) \right.$$
$$\left. + \frac{1}{r_{12}^3}\left[\boldsymbol{\sigma}_1 \cdot \boldsymbol{\sigma}_2 - \frac{3(\boldsymbol{\sigma}_1 \cdot \mathbf{r}_{12})(\boldsymbol{\sigma}_2 \cdot \mathbf{r}_{12})}{r_{12}^2}\right]\right\}. \quad (4.69)$$

In these formulas, $\mathbf{r}_{12} = \mathbf{r}_1 - \mathbf{r}_2$, the $\boldsymbol{\sigma}$'s are the Pauli pin operators operating on the coordinates of electrons "one" and "two," respectively, and δ is the Dirac delta function:

$$\delta(\mathbf{r}_{12}) = \delta(x_1 - x_2)\delta(y_1 - y_2)\delta(z_1 - z_2). \quad (4.70)$$

The physical meaning of H_7 and H_8 is as follows. H_7 is the spin–other–orbit interaction, that is, the interaction energy of the intrinsic magnetic moment of electron "one" with the magnetic field created by the motion of electron "two"—and vice versa. The form of the operator reveals this clearly, especially if we again introduce the intrinsic magnetic-moment vector from Eq. (4.37). Using that formula, we obtain

$$H_7 = -\frac{1}{2mc}\frac{2e}{r_{12}^3}\{(\mathbf{r}_{12} \times \mathbf{p}_2) \cdot \boldsymbol{\mu}_s(1) + (\mathbf{r}_{21} \times \mathbf{p}_1) \cdot \boldsymbol{\mu}_s(2)\}, \quad (4.71)$$

where we have used Eq. (4.36).

Operator H_8 is the spin–spin interaction, that is, the interaction energy of the two intrinsic magnetic moments. We can write

$$H_8(1,2) = -\frac{8\pi}{3}[\boldsymbol{\mu}_s(1) \cdot \boldsymbol{\mu}_s(2)]\delta(\mathbf{r}_{12})$$
$$+ \frac{1}{r_{12}^3}\left[\boldsymbol{\mu}_s(1) \cdot \boldsymbol{\mu}_s(2) - \frac{3(\boldsymbol{\mu}_s(1) \cdot \mathbf{r}_{12})(\boldsymbol{\mu}_s(2) \cdot \mathbf{r}_{12})}{r_{12}^2}\right]. \quad (4.72)$$

We write down now Breit's reduced operator for two electrons:

$$H(1,2) = H_0(1) + H_1(1) + H_2(1) + H_5(1) + H_6(1)$$
$$+ H_0(2) + H_1(2) + H_2(2) + H_5(2) + H_6(2) + \frac{e^2}{r_{12}}$$
$$+ H_3(1,2) + H_4(1,2) + H_7(1,2) + H_8(1,2). \qquad (4.73)$$

The structure of this formula is very instructive. The one-electron part is the same as the reduced Dirac operator, Eq. (4.53). Indeed, if we write down Eq. (4.53) and take into account Eq. (4.39), we get

$$H_D = H_0 + H_1 + H_2 + H_5 + H_6, \qquad (4.74)$$

which is identical with the one-electron part of Eq. (4.73). The two-electron part consists of two sections. The terms e^2/r_{12}, H_3, and H_4 are the same operators that we get if we replace the momentum vector in Darwin's operator by the quantum-mechanical operator $-i\hbar\nabla$. We see this from Eq. (4.26), where the two-electron part is $e^2/r_{12} + H_3 + H_4$. The terms H_7 and H_8 in Eq. (4.73) are the spin terms, which do not occur in Darwin's expression.

Even more interesting is to compare Eq. (4.73) directly with Eq. (4.26). We can write Breit's operator, Eq. (4.73), in the form

$$H_B(1,2) = H_D(1,2) + H_1''(1) + H_1''(2)$$
$$+ H_5(1) + H_5(2) + H_6(1) + H_6(2)$$
$$+ H_7(1,2) + H_8(1,2). \qquad (4.75)$$

In this formula, $H_D(1,2)$ is Darwin's operator given by Eq. (4.26), that is, $H_D(1,2)$ is the operator that we get by replacing the momentum vector in Darwin's Hamiltonian by $-i\hbar\nabla$. Thus, Eq. (4.75) shows the terms that we *do not* get by the replacement of **p** by $-i\hbar\nabla$. Looking at the terms that must be added to Darwin's operator to get Breit's Hamiltonian, we see that H_5, Eq. (4.31), is the interaction of the moving electron with the electric field. This term is typical of the Dirac theory and does not occur in Darwin's operator. The rest of the terms, namely, H_1'', H_6, H_7, and H_8 are all *spin effects*, a fact that we see immediately by looking at the equations defining these operators, that is, at Eqs. (4.30), (4.32), (4.68), and (4.69). Therefore, we see clearly that Breit's reduced operator differs from Darwin's expression primarily by the presence of the important interaction terms generated by the electron spin.

Let us summarize now how Breit's operator, Eq. (4.62), is constructed. Darwin's classical Hamiltonian is accepted as the accurate classical (relativistic but non-quantum-mechanical) expression. The correct Hamiltonian is generated by adding to the Dirac one-electron operators the electron–electron interaction terms of Darwin's expression. In these terms, we do not just replace **p** by $-i\hbar\nabla$. Rather, we rely on the correspondence principle and use the relationship given by Eq. (4.61). The result is an electron–electron operator that contains, besides the magnetic and retardation effects of Darwin's expression, the important spin interaction terms. Thus, Breit's operator is constructed in such a way that the one-electron part is identical with Dirac's exact one-electron operator and the two-electron part reduces to Darwin's expression plus the interaction terms generated by the electron spin.

E. A Discussion of Some Properties of Breit's Theory

The derivation of the Breit equation, as we have outlined it in the preceding section, suffers from several shortcomings. First, because the equation is based on Darwin's expression, it is an approximation in which terms up to $1/c^2$ have been included (see App. F). Second, the derivation by the correspondence principle is elegant but does not necessarily give a unique quantum-mechanical formulation.[20] In addition to these points, we should ask the question: What is the meaning of the exact solution of the Schroedinger equation, Eq. (4.65)? We must ask this question because our motivation in introducing relativity into atomic theory was to find operators whose eigenvalues would agree with empirical data. We can compare the eigenvalues of operators with empirical data only if exact solutions of the Schroedinger equation in fact exist and are meaningful.

We follow here the discussion presented by Bethe and Salpeter,[21] which was developed for two-electron systems; the generalization of the basic conclusions to N-electron systems is plausible. In order to establish the validity of Eq. (4.65), it is necessary to calculate the electron–electron interaction by quantum electrodynamics. In quantum electrodynamics, the electron–electron interaction is described as a process in which a virtual photon is emitted by one of the electrons and absorbed by the other. The energy change, ΔE, due to this emission–absorption process, is evaluated in second-order perturbation theory. The unperturbed equation is

$$H_0\psi = \left(H_D(1) + H_D(2) + \frac{e^2}{r_{12}}\right)\psi = E\psi, \qquad (4.76)$$

which is obtained from Eq. (4.65) by omitting the magnetic and retardation terms, that is, by representing the electron–electron interaction by the instantaneous term e^2/r_{12}. It is assumed that the exact solutions of the pseudorelativistic equation, Eq. (4.76), exist and they serve as the unperturbed solutions. This equation has the merit that it is fully Lorentz-invariant if the interaction term is omitted; also, it reduces to the ordinary two-electron Schroedinger equation in the nonrelativistic limit.

The energy change, ΔE, is then calculated with this method. Now, let us write Eq. (4.65) in the form

$$(H_0 + H')\psi = E\psi, \tag{4.77}$$

where H_0 is that Hamiltonian given by Eq. (4.76), and H' is the Breit operator:

$$H' = -\frac{e^2}{2r_{12}} \left[\boldsymbol{\alpha}(1) \cdot \boldsymbol{\alpha}(2) + \frac{(\boldsymbol{\alpha}(1) \cdot \mathbf{r}_{12})(\boldsymbol{\alpha}(2) \cdot \mathbf{r}_{12})}{r_{12}^2} \right]. \tag{4.78}$$

As we see from Eqs. (4.63) and (4.64), H' is not equal to $g_0 + g_1$; the letter contains e^2/r_{12}, which, in Eq. (4.77), is included in H_0.

Let us assume again that the eigenfunctions and eigenvalues of H_0 exist, that is, we have the solutions of Eq. (4.76). Considering now H_0 to be the unperturbed part of Eq. (4.77), we can calculate the solutions of that equation by perturbation theory, viewing H' as the perturbation. If we do this in first order, then it turns out that the first-order energy gives approximately the correct answer, that is, we have the approximate relationship

$$\Delta E \approx \int \psi^* H' \psi \, dv, \tag{4.79}$$

where ψ is the solution of Eq. (4.76), and ΔE is the energy shift caused by the virtual emission–absorption process.

Thus, it can be said that *if the first-order perturbation theory is used*, then H', the Breit operator, describes the relativistic electron–electron interaction in good approximation. The trouble is that if we attempt to improve this calculation by calculating the second-order energy, by using H' as the perturbation, the results will be of the wrong order of magnitude and too large. Mathematically, the reason for this is that Eq. (4.76) possesses, just like the Dirac one-electron equation, solutions with negative total energy. These negative energies will make the second-order energy correction too large.

The conclusion of this argument is that we are not able to define the exact solution of the Breit equation by approaching it with perturbation theory. According to Bethe and Salpeter, "The Breit equation, Eq. (4.65), gives the leading term for the relativistic corrections to the interaction between the two electrons, if the Breit operator, H', is treated by first-order perturbation theory."

This statement does not mean that the higher-order corrections cannot be calculated. They can be calculated by considering the higher-order corrections to the quantum electrodynamical description of the interaction process. However, if these higher-order corrections were considered, the description of the relativistic interaction would have to be made by an operator other than Breit's expression.

The argument can be summarized by paraphrasing Bethe and Salpeter in such a way that is closer to the terminology of atomic structure theory. We put this into the form of a theorem:

> The Breit interaction operator is a *model* of the actual relativistic interaction. In calculations, either one has to calculate the expectation value of this operator or one has to use first-order perturbation theory.

The question now arises about the meaning of the reduced Breit operator given by Eq. (4.73). Clearly, that is only an approximation to the exact operator, given by Eq. (4.65), but it has the advantage that this operator does not have eigenvalues with negative total energies. Therefore, the second-order perturbation energy need not yield unreasonable results in this case. Thus, it is not unreasonable to talk about the exact solutions of the Schroedinger equation in which the Hamiltonian is given by Eq. (4.73). However, we should remember at all times that the *exact* solutions of such a Schroedinger equation would contain only an *approximation* to the relativistic electron–electron interaction operator. Thus, the exact eigenvalues of the Hamiltonian given by Eq. (4.73) will still be only approximations to the corresponding empirical data.

Finally, we provide the reader who is interested in the quantum electrodynamical background of many electron theory with some more recent references that can be used as starting points for further study. The review article by Sucher[22] discusses the foundations of relativistic many-electron theory. The same topic is discussed, from a nontraditional point of view, by Detrich and Roothaan.[23] The connection between quantum electrodynamics and some aspects of atomic structure theory are discussed in the recent conference report of Sapirstein.[24]

4.3. THE HARTREE – FOCK – DIRAC (HFD) MODEL

A. The HFD Energy of an Atom with Complete Groups

A relativistic Hartree–Fock theory was first formulated by Swirles.[25] Although Swirles' work contained almost all basic ideas of the theory, it did not receive much attention, probably because of the complexity of the formalism as well as because of the lack of proper computing facilities at the time of its publication. New impetus was given to relativistic atomic theory some years after the publication of Swirles' work by the studies of Grant.[26] His work is essentially a clarified and extended version of Swirles' formulation. Using Racah's theory,[27] Grant succeeded to simplify Swirles' formalism considerably and put it into a clearly computable form. Grant also improved Swirles' work by including the magnetic effects in the formalism; Swirles considered only the instantaneous electrostatic interaction in the electron–electron interaction part of the many-electron Hamiltonian. The presentation in this section follows closely the work of Grant.

In order to show that the relativistic Hartree–Fock model, which we will call the Hartree–Fock–Dirac (HFD) theory, is closely analogous to the nonrelativistic formalism, we summarize here the basic equations of the latter and then proceed to the formulation of the HFD model. At first, we consider atoms with complete groups. For these atoms, their HF model is a single-determinantal theory, which we presented in Sec. 2.2.

In building up the HF model, we started with the Hamiltonian given by Eq. (2.34):

$$H = -\sum_{i=1}^{N} \frac{1}{2}\Delta_i - \sum_{i=1}^{N} \frac{Z}{r_i} + \frac{1}{2}\sum_{\substack{i,j=1 \\ (i \neq j)}}^{N} \frac{1}{r_{ij}}. \tag{4.80}$$

This operator is in atomic units. The normalized determinantal wave function for an N-electron atom is

$$\Psi_T = (N!)^{-1/2} \det[\varphi_1(1)\varphi_2(2) \cdots \varphi_N(N)], \tag{4.81}$$

where the spin-orbitals are defined as

$$\varphi_k(i) = \varphi_k(q_i) = \psi_k(\mathbf{r}_i)\eta_k(\sigma_i). \tag{4.82}$$

ψ_k and η_k are the spatial and spin functions, respectively, and $\mathbf{r}_i$ and σ_i are the spatial and spin coordinates, respectively. The one-electron spin-orbitals are orthonormal, Eq. (2.41):

$$\langle \varphi_i | \varphi_j \rangle = \delta_{ij} \qquad (i, j = 1, \ldots, N). \tag{4.83}$$

The total energy of the atom is the expectation value of the Hamiltonian operator, Eq. (4.80), with respect to the determinantal wave function, Eq. (4.81). That is, according to Eq. (2.42), we get

$$E_F = \langle \Psi_T | H | \Psi_T \rangle. \tag{4.84}$$

The HF equations are derived by varying E_F with respect to the one-electron spin-orbitals under the subsidiary conditions given by Eqs. (4.83).

In formulating the HFD model, we start with the Breit Hamiltonian operator. We have written down that operator in Eq. (4.65). Generalizing it for N electrons, we obtain

$$H = \sum_{k=1}^{N} H_D(k) + \sum_{k,j=1}^{N} \{g_0(k,j) + g_1(k,j)\}. \tag{4.85}$$

In this formula, H_D is the Dirac one-electron Hamiltonian:

$$H_D(k) = ic\boldsymbol{\alpha}(k) \cdot \nabla_k - \beta(k)c^2 - V(k), \tag{4.86}$$

which we obtain from Eq. (4.55) by writing it down in the absence of an external electromagnetic field and putting it into atomic units. The last step means to write $\hbar = m = e = 1$. The numerical value of c in atomic units is

$$c = 137.037. \tag{4.87}$$

In H_D, the potential is given by $V = Z/r$.

The interaction terms of the Hamiltonian were defined by Eqs. (4.63) and (4.64). According to them,

$$g_0(1,2) = \frac{1}{r_{12}}[1 - \boldsymbol{\alpha}(1) \cdot \boldsymbol{\alpha}(2)], \tag{4.88a}$$

and

$$g_1(1,2) = \frac{1}{2r_{12}}\left[\boldsymbol{\alpha}(1) \cdot \boldsymbol{\alpha}(2) - \frac{(\boldsymbol{\alpha}(1) \cdot \mathbf{r}_{12})(\boldsymbol{\alpha}(2) \cdot \mathbf{r}_{12})}{r_{12}^2}\right]. \tag{4.88b}$$

These are written now in atomic units and we recall that g_0 represents the electrostatic and magnetic interactions and g_1 the retardation effects.

The complete Breit Hamiltonian is given by the formula

$$H = \sum_{k=1}^{N}\left[ic\boldsymbol{\alpha}(k) \cdot \nabla_k - \beta(k)c^2 - \frac{Z}{r_k}\right] + \sum_{k,j=1}^{N}\frac{1}{r_{kj}}[1 - \boldsymbol{\alpha}(k) \cdot \boldsymbol{\alpha}(j)]$$

$$+ \sum_{k,j=1}^{N}\frac{1}{2r_{kj}}\left[\boldsymbol{\alpha}(k) \cdot \boldsymbol{\alpha}(j) - \frac{(\boldsymbol{\alpha}(k) \cdot \mathbf{r}_{jk})(\boldsymbol{\alpha}(j) \cdot \mathbf{r}_{jk})}{r_{kj}^2}\right]. \tag{4.89}$$

The total wave function of the atom is built up from one-electron functions. We denote these by $\varphi_A, \varphi_B, \dots$, and so on. Here the capital letters $A, B, \dots$ mean a set of quantum numbers that characterize the one-electron wave functions similarly as the letter k characterized the nonrelativistic spin-orbital φ_k in Eq. (4.82). The total wave function of the atom is the determinant

$$\Psi_T = (N!)^{-1/2}\begin{vmatrix} \varphi_1(1) & \varphi_1(2) & \cdots & \varphi_1(N) \\ \varphi_2(1) & \varphi_2(2) & \cdots & \varphi_2(N) \\ \vdots & \vdots & & \vdots \\ \varphi_N(1) & \varphi_N(2) & \cdots & \varphi_N(N) \end{vmatrix}, \tag{4.90}$$

where $\varphi_1, \varphi_2, \ldots$ are the one-electron functions, and the normalization constant is the same as in the HF theory. The one-electron functions are orthonormal:

$$\langle \varphi_A | \varphi_B \rangle = \delta_{AB}. \tag{4.91}$$

Next, we clarify the meaning of the index A in the one-electron function φ_A. In the nonrelativistic case, that is, in the case of the wave function φ_k given by Eq. (4.82), the index k meant the following set of quantum numbers:

$$k = (n_k, l_k, m_{lk}, m_{sk}). \tag{4.92a}$$

In the case of the relativistic one-electron functions, several different equivalent sets can be used. In the relativistic one-electron theory, m_l and m_s are not good quantum numbers. Instead of the previous set, we can use

$$A = (n_A, l_A, j_A, m_{jA}). \tag{4.92b}$$

Here j and m_j are the quantum numbers associated with the total angular momentum, Eq. (H.2a), and its z-component, Eq. (H.2b). According to the quantum theory of angular momentum, each j value can be generated by two different azimuthal quantum numbers l and $\bar{l}$ according to the formulas

$$j = l + \tfrac{1}{2} \quad \text{or} \quad j = \bar{l} - \tfrac{1}{2}. \tag{4.93}$$

Let us introduce parameter a with the definition

$$
\begin{aligned}
a &= +1, \quad \text{if} \quad j = l + \tfrac{1}{2}; \\
a &= -1, \quad \text{if} \quad j = \bar{l} - \tfrac{1}{2}.
\end{aligned}
\tag{4.94}
$$

Thus, a is equal to $+1$ for those cases in which j is generated from the azimuthal quantum number l with the formula $j = l + \tfrac{1}{2}$, and it is equal to -1 for those cases in which $j = \bar{l} - \tfrac{1}{2}$. Therefore, instead of the set given by Eq. (4.92b), we can use

$$A = (n_A, j_A, m_{j_A}, a_A), \tag{4.92c}$$

where the value of a_A determines the value of l_A for a given j_A.

Next, let us introduce κ by the definition

$$
\begin{aligned}
\kappa &= -\left(j + \tfrac{1}{2}\right) = -(l + 1), \quad \text{for} \quad a = +1, \\
\kappa &= j + \tfrac{1}{2} = \bar{l}, \quad \text{for} \quad a = -1.
\end{aligned}
\tag{4.95}
$$

From these definitions, we see that κ can take on the values $\kappa = 1, 2, 3 \ldots$ because of the $\kappa = \bar{l}$ relationship. The value $\kappa = 0$ is not possible because the $\kappa = \bar{l}$ relationship holds only for $a = -1$, that is, for $j = \bar{l} - \frac{1}{2}$ and that is not permitted for $\bar{l} = 0$. We see also that κ can take on the values of $\kappa = -1, -2, \ldots$ because of $\kappa = -(l + 1)$. Further, we obtain the relationship

$$\kappa = -\left(j + \tfrac{1}{2}\right)a, \tag{4.96}$$

because for

$$a = 1, \qquad \kappa = -\left(j + \tfrac{1}{2}\right);$$
$$a = -1, \qquad \kappa = j + \tfrac{1}{2}. \tag{4.97}$$

There are the same formulas as in Eq. (4.95); thus, Eq. (4.96) is equivalent to the equations in Eq. (4.95). Finally, from Eq. (4.97), we get

$$|\kappa| = j + \tfrac{1}{2}. \tag{4.98}$$

This equation shows that $|\kappa|$ determines j because

$$j = |\kappa| - \tfrac{1}{2}. \tag{4.98a}$$

The sign of κ determines whether j is generated by the formula $j = l + \frac{1}{2}$ or by $j = \bar{l} - \frac{1}{2}$. Indeed, by looking at Eq. (4.95), we see that κ will be positive for $j = \bar{l} - \frac{1}{2}$, that is, for $a = -1$; κ will be negative for $j = l + \frac{1}{2}$, that is, for $a = +1$. Therefore, we can summarize the properties of κ by the following statements:

1. $|\kappa|$ determines the value of j.
2. The sign of κ determines whether j is associated with l or with $\bar{l}$.

From this discussion, it is clear that κ is equivalent to the *two* quantum numbers (jl). Thus, instead of the set given by Eq. (4.92b), we can use the set

$$A = \left(n_A, \kappa_A, m_{j_A}\right), \tag{4.92d}$$

which is equivalent to four quantum numbers.

Finally, the index may be defined in such a way that instead of using j, a, or κ, we simply use l and $\bar{l}$. Thus, we can put

$$A = \left(n_A, l_A, m_{j_A}\right), \tag{4.92e}$$

which means $j_A = l_A + \frac{1}{2}$, or we can write

$$A = \left(n_A, \bar{l}_A, m_{j_A} \right), \tag{4.92f}$$

which means $j_A = \bar{l}_A - \frac{1}{2}$.

The relationship between the principal quantum number n and the value of j or $|\kappa|$ is given by the formula

$$n = n_r + |\kappa|, \tag{4.99}$$

where n_r is the number of nodes in the radial part of the one-electron wave function.

In Table 4.1, we demonstrate the connection between the various quantum numbers. The meaning of various combinations can be seen immediately by looking at any of the horizontal lines that contain the various possible combinations for the choice of the quantum number index A. As the preceding discussion shows, for the characterization of the wave function φ_A, we can use any of the equivalent sets given by Eqs. (4.92b)–(4.92f).

We now write down the form of the one-electron function φ_A. Just as in nonrelativistic theory, we choose φ_A to be of the same form as the eigenfunctions of the one-electron part of the many-electron Hamiltonian. Our many-electron Hamiltonian is given by Eq. (4.85); the one-electron part of that operator is the Dirac Hamiltonian H_D. The main properties of the eigenfunctions of H_D are summarized in App. H. It is shown there that the eigenfunc-

TABLE 4.1

Spectroscopic notation	l	j	a	κ	$|\kappa|$
s	0	1/2	+1	−1	1
$\bar{\text{p}}$	1	1/2	−1	+1	1
p	1	3/2	+1	−2	2
$\bar{\text{d}}$	2	3/2	−1	+2	2
d	2	5/2	+1	−3	3
$\bar{\text{f}}$	3	5/2	−1	+3	3
f	3	7/2	+1	−4	4
$\bar{\text{g}}$	4	7/2	−1	+4	4
g	4	9/2	+1	−5	5
$\bar{\text{h}}$	5	9/2	−1	+5	5
h	5	11/2	+1	−6	6
$\bar{\text{i}}$	6	11/2	−1	+6	6
i	6	13/2	+1	−7	7
j	7	13/2	−1	+7	7

tions of H_D are of the form given by Eq. (H.27). Let us write down here that function, which is our starting point:

$$\psi = \begin{pmatrix} iu_{-\kappa m_j} f(r) \\ u_{\kappa m_j} g(r) \end{pmatrix}.$$
(4.100)

In this wave function, $u_{\kappa m_j}$ is given by Eq. (H.26),

$$u_{\kappa m_j} = \sum_{m_s} C\left(l\tfrac{1}{2}j; (m_j - m_s), m_s\right) Y_{l(m_j - m_s)} \eta_{m_s},$$
(4.101)

and the explanation of the symbols is given in the appendix. The radial functions g and f are the exact solutions of the Dirac equation, with g being the "large" and f the "small" component.

We now replace the Dirac functions f and g by arbitrary radial functions to be determined from the HFD equations. Let $P_{n\kappa}/r$ be the large and $Q_{n\kappa}/r$ the small component. Then we put our trial function in the form

$$\varphi_{n\kappa m_j} = \begin{pmatrix} i\dfrac{Q_{n\kappa}}{r} u_{-\kappa m_j} \\ \dfrac{P_{n\kappa}}{r} u_{\kappa m_j} \end{pmatrix}.$$
(4.102)

This is clearly a four-component function because $u_{\kappa m_j}$ is a two-component function, as we see from Eq. (H.19). In the previous formula, we have chosen the quantum number set given by Eq. (4.92d) for the characterization of the wave function. Thus, in Eq. (4.102), we have

$$A = \left(n_A, \kappa_A, m_{j_A}\right).$$
(4.103)

The derivation of the equations for $P_{n\kappa}$ and $Q_{n\kappa}$ is carried out exactly as in nonrelativistic theory. We evaluate the expectation value of the Hamiltonian, Eq. (4.89), with respect to determinantal functions containing one-electron functions of the type of φ_A, Eq. (4.102). Then, in the formula for the expectation value, the integrations with respect to the angular and spin parts are carried out. The results of these integrations are the appearance of certain constants very similar to the constants appearing in the corresponding nonrelativistic energy expression.

After the integration over the angular and spin parts, the expectation value of the energy is a functional of the radial functions. Variation of the energy with respect to these radial functions, under the subsidiary conditions of normalization and orthogonality, yields the HFD equations. The total energy of the atom is given by

$$E = \langle \Psi_T | H | \Psi_T \rangle, \qquad (4.104)$$

where Ψ_T is the normalized determinantal function given by Eq. (4.90). The complete Hamiltonian is given by Eq. (4.89). Here, Grant introduced the additional approximation of omitting the last term of the Hamiltonian, that is, omitting the retardation effects. Thus, in this formulation, the electron–electron interaction is represented by the electrostatic and magnetic terms.

For the energy given by Eq. (4.104), we obtain, similarly as in the nonrelativistic theory,

$$E = \sum_A I(A) + \sum_{A,B} [J(A, B) - K(A, B)], \qquad (4.105)$$

where

$$I(A) = \langle \varphi_A | H_D | \varphi_A \rangle, \qquad (4.106a)$$

$$J(A, B) = \langle \varphi_A(1) \varphi_B(2) | g_0(1, 2) | \varphi_A(1) \varphi_B(2) \rangle, \qquad (4.106b)$$

$$K(A, B) = \langle \varphi_A(1) \varphi_B(2) | g_0(1, 2) | \varphi_B(1) \varphi_A(2) \rangle. \qquad (4.106c)$$

In these formulas, $g_0(1, 2)$ is the operator given by Eq. (4.88a). The summation over A is for all quantum-number quartets in the atom and the summation over A and B is for all distinct pairs of quartets. In constructing the formula for the total energy, Eq. (4.105), we evaluate first the matrix components given by Eqs. (4.106a) to (4.106c). After the matrix components are obtained, we substitute them into Eq. (4.104) and carry out the summations.

For the matrix components of the one-electron operator, we obtain

$$I(A) = c \int_0^\infty \left\{ P_A \left[\frac{dQ_A}{dr} - \frac{\kappa_A}{r} Q_A + \left(c - \frac{Z}{cr} \right) P_A \right] \right.$$
$$\left. - Q_A \left[\frac{dP_A}{dr} + \frac{\kappa_A}{r} P_A + \left(c + \frac{Z}{cr} \right) Q_A \right] \right\} dr. \quad (4.107)$$

In this formula, as in all subsequent formulas, the index A attached to the radial functions P_A and Q_A means only the quantum numbers $(n_A \kappa_A)$ or the equivalent sets $(n_A l_A j_A)$ or $(n_A j_A a_A)$ and not the full set of four quantum numbers as indicated by Eq. (4.103). The same remark applies to integrals like $I(A)$ and others in which we have only the radial functions.

The evaluation of the matrix components given by Eqs. (4.106b) and (4.106c) is divided into the evaluation of the electrostatic part and the construction of the formula for the magnetic part. For the "direct" part of the electrostatic interaction, we get

$$\left\langle \varphi_A \varphi_B \left| \frac{1}{r_{12}} \right| \varphi_A \varphi_B \right\rangle = \sum_k a^k \left(j_A m_{j_A}, j_B m_{j_B} \right) F_k^E(A, B). \quad (4.108)$$

In this formula, the constant a^k is the result of the angular and spin integrations. The permitted values of k are

$$k = 0, 2, 4, \ldots, \quad \min(2j_A - 1, 2j_B - 1), \quad (4.109)$$

and a^k itself is given by

$$a^k \left(j_A m_{j_A}, j_B m_{j_B} \right) = d^k \left(j_A m_{j_A}, j_A m_{j_A} \right) d^k \left(j_B m_{j_B}, j_B m_{j_B} \right). \quad (4.110)$$

The d^k coefficients can be given in terms of the Clebsch–Gordan coefficients:

$$d^k \left(jm_j, j'm_j' \right) = (-1)^{m_j + 1/2} \frac{[(2j + 1)(2j' + 1)]^{1/2}}{2k + 1}$$
$$\times C\left(jj'k; \tfrac{1}{2}, -\tfrac{1}{2} \right) C\left(jj'k; -m_j, m_j' \right). \quad (4.111)$$

The radial integral in Eq. (4.108) is given by

$$F_k^E(A, B) = \int_0^\infty \left[P_A^2(1) + Q_A^2(1) \right] \left[P_B^2(2) + Q_B^2(2) \right] L_k(1, 2) \, dr_1 \, dr_2,$$
$$(4.112)$$

where L_k is the function defined by Eq. (2.72).

For the "exchange" part of the electrostatic integral, we get

$$\left\langle \varphi_A \varphi_B \left| \frac{1}{r_{12}} \right| \varphi_B \varphi_A \right\rangle = \sum_k b^k \left(j_A m_{j_A}, j_B m_{j_B} \right) G_k^E(A, B). \quad (4.113)$$

In this formula, the constant b_k is given by the relationship

$$b^k \left(j_A m_{j_A}, j_B m_{j_B} \right) = \left[d^k \left(j_A m_{j_A}, j_B m_{j_B} \right) \right]^2, \quad (4.114)$$

where d^k is given by Eq. (4.111). The range of k in Eq. (4.113) is determined by the relationships:

$$|j_A - j_B| \leq k \leq j_A + j_B, \quad (4.115a)$$

and

$$j_A + j_B + k = \begin{cases} \text{even if} & a_A \neq a_B, \\ \text{odd if} & a_A = a_B. \end{cases} \quad (4.115b)$$

The radial integral in Eq. (4.113) is given by

$$G_k^E(A, B) = \int_0^\infty [P_A(1)P_B(1) + Q_A(1)Q_B(1)]$$

$$\times [P_A(2)P_B(2) + Q_A(2)Q_B(2)]L_k(1,2)\, dr_1\, dr_2. \quad (4.116)$$

We come now to the evaluation of the magnetic interaction. We get for the "direct" part

$$\left\langle \varphi_A\varphi_B \left| \frac{\boldsymbol{\alpha}(1)\cdot\boldsymbol{\alpha}(2)}{r_{12}} \right| \varphi_A\varphi_B \right\rangle = \sum_k f^k(j_A m_{j_A}, j_B m_{j_B})F_k^M(A, B). \quad (4.117)$$

In this formula, the coefficients f^k are obtained from the angular and spin integrations. For these coefficients, Grant derived an explicit formula for which we refer the reader to Grant's papers. The radial integral in Eq. (4.117) is given by

$$F_k^M(A, B) = \int_0^\infty P_A(1)Q_A(1)P_B(2)Q_B(2)L_k(1,2)\, dr_1\, dr_2. \quad (4.118)$$

The range of k is given by

$$k = 1, 3, \ldots, \quad \min(2j_A, 2j_B). \quad (4.119)$$

For the exchange part of the magnetic interaction, we get

$$\left\langle \varphi_A\varphi_B \left| \frac{\boldsymbol{\alpha}(1)\cdot\boldsymbol{\alpha}(2)}{r_{12}} \right| \varphi_B\varphi_A \right\rangle$$

$$= \sum_{kJ\gamma} g^k(j_A m_{j_A}, j_B m_{j_B}, J\gamma)G_k^M(A, B; \gamma). \quad (4.120)$$

This formula is somewhat more complicated that the others because instead of one, we have three summation indices. The index γ can take only the values of -1, 0, and $+1$. For these values, we have three different radial integrals, which are defined as follows:

$$G_k^M(AB;1) = \int P_A(1)Q_B(1)P_A(2)Q_B(2)L_k(1,2)\,dr_1\,dr_2, \quad (4.121a)$$

$$G_k^M(AB;0) = \int P_A(1)Q_B(1)Q_A(2)P_B(2)L_k(1,2)\,dr_1\,dr_2 \quad (4.121b)$$

$$G_k^M(AB;-1) = \int Q_A(1)P_B(1)Q_A(2)P_B(2)L_k(1,2)\,dr_1\,dr_2. \quad (4.121c)$$

For the g_k coefficients, Grant has constructed explicit formulas for which we refer the reader to the original publications. Likewise, explicit rules are given for the range of the summations with respect to the indices k and J.

At this point, it is useful to compare the relativistic and nonrelativistic formulas. The relativistic formulas for the matrix components of the electrostatic interaction are given by Eqs. (4.108) and (4.113). As we see from Eqs. (C.24) and (C.26) from App. C and also from Eqs. (2.70b) and (2.70c), the relativistic formulas are closely analogous to the nonrelativistic. In fact, a comparison between Eqs. (4.112) and (2.71b) shows that in the nonrelativistic limit, that is, for $Q_A = Q_B = 0$, the radial integral $F_k^E(AB)$ becomes identical with $F_k(nl, n'l')$ if we put $A = (nl)$ and $B = (n'l')$. The same conclusion can be drawn from the comparison of the exchange integrals given by Eqs. (4.116) and (2.71c). There is also a strong similarity between the c^k coefficients, which occur in the nonrelativistic matrix components, and the d^k, which occur in the relativistic. In fact, for the c^k coefficients, Grant quoted the relationship:

$$(-1)^k c^k(lm_l, l'm_l') = (-1)^{m_l'} \frac{[(2l+1)(2l'+1)]^{1/2}}{2k+1} C(ll'k;0,0)$$

$$\times C(ll'k; -m_l, m_l'), \quad (4.122)$$

and the d^k coefficients satisfy the closely analogous expression, Eq. (4.111). On the right side of both equations, we have the Clebsch–Gordan coefficients, for which explicit formulas are available (see App. E). In fact, the d^k coefficients can be considered as the jj-coupling analogs of the c^k. We note that Grant has given tables of the numerical values of the d^k coefficients and also of the values of a^k and b^k. These tables are the counterparts of Slater's tables for the c^k coefficients (Slater II, 281). We summarize the main formulas for these coefficients in App. E.

Reviewing the formulas for the matrix components of the magnetic interaction, Eqs. (4.117) and (4.120), we observe that although these do not have nonrelativistic analogs, they have the same simple structure as the matrix components of the electrostatic interaction. It is clear from Grant's work that the construction of the angular coefficients of the magnetic interaction is more complicated that the construction of the angular coefficients of the electrostatic interaction; however, once these coefficients are computed, the matrix components become simple functionals of the radial orbitals similarly as in the case of the matrix components of the electrostatic interaction.

We now construct the energy expression for an atom with complete groups, that is, we substitute the matrix components into Eq. (4.105) and carry out the summations. In the nonrelativistic case, a group was defined as the electron states with common n and l. Here a group is defined as the electrons with the same (nlj), or, equivalently, the electrons with the same nja). The electron states within a group are labeled by m_j, which means that in each group, we have $2j + 1$ electron states. The group is complete if all states are filled.

First, we write down the expression for the first term of Eq. (4.105). Since $I(A)$, as we see from Eq. (4.107), does not depend on m_j, we simply get

$$\sum_A q(A)I(A), \tag{4.123}$$

where $A \equiv (nlj)$, and $q(A)$ is the occupation number of group A.

Next, we construct the double sums in Eq. (4.105). The double sum means a summation for all pairs of $(nljm_j)$ quartets. Let $A \equiv (n_A l_A j_A)$ and $B \equiv (n_B l_B j_B)$. Then the double sums will take the form

$$\sum_{AB} = \sum_A \sum_B \sum_{m_{j_A}} \sum_{m_{j_B}} = \sum_A \sum_{m_{j_A}} \sum_{m'_{j_A}} + \sum_A \sum_B \sum_{m_{j_A}} \sum_{m_{j_B}}, \quad (B \neq A) \tag{4.124}$$

where we have divided the summation into two terms. In the first term, we sum over all (m_{j_A}, m'_{j_A}) pairs within one group and then we sum up the result for all groups. In the second term, the summation is over all pairs of groups, that is, A and B denote different groups.

Let us start with the direct electrostatic integral in Eq. (4.108). Applying Eq. (4.124), we get

$$\sum_{A,B} \left\langle \varphi_A \varphi_B \left| \frac{1}{r_{12}} \right| \varphi_A \varphi_B \right\rangle = \sum_A \sum_{m_{j_A}} \sum_{m'_{j_A}} \sum_k a^k(j_A m_{j_A}, j_A m'_{j_A}) F_k^E(A, A)$$

$$+ \sum_A \sum_B \sum_{m_{j_A}} \sum_{m_{j_B}} \sum_k a^k(j_A m_{j_A}, j_B m_{j_B}) F_k^E(A, B).$$
$$(B \neq A)$$

$$\tag{4.125}$$

It is easy to show, on the basis of the properties of the angular coefficient, that

$$\sum_{m'_j} a^k(jm_j, j'm'_j) = q(j')\delta(k,0), \tag{4.126}$$

where the summation over m'_j is over a complete group and $q(j')$ is the occupation number of the complete, j' group. Substituting this into the second term of Eq. (4.125), we get

$$\sum_{\substack{A}} \sum_{\substack{B \\ (B \neq A)}} \sum_{m_{j_A}} q(B) F_0^E(A, B) = \sum_{\substack{A}} \sum_{\substack{B \\ (B \neq A)}} q(A) q(B) F_0^E(A, B). \tag{4.127}$$

When substituting Eq. (4.126) into the first term of Eq. (4.125), we must multiply by a factor of $\frac{1}{2}$ since the m_{j_A} and m'_{j_A} summations are independent; with the factor of $\frac{1}{2}$, we avoid counting any of the pairs twice. Thus, we obtain

$$\frac{1}{2} \sum_{A} \sum_{m_{j_A}} q(A) F_0^E(A, A) = \sum_{A} \frac{1}{2} [q(A)]^2 F_0^E(A, A). \tag{4.128}$$

This expression includes the electrostatic interaction of state A with itself.

Next, let us consider the exchange part of the electrostatic integral, Eq. (4.113). Applying Eq. (4.124), we get

$$\sum_{A,B} \left\langle \varphi_A \varphi_B \left| \frac{1}{r_{12}} \right| \varphi_B \varphi_A \right\rangle$$

$$= \frac{1}{2} \sum_{A} \sum_{m_{j_A}} \sum_{m'_{j_A}} \sum_{k} b^k(j_A m_{j_A}, j_A m'_{j_A}) G_k^E(A, A)$$

$$+ \sum_{A} \sum_{\substack{B \\ (B \neq A)}} \sum_{m_{j_A}} \sum_{m_{j_B}} \sum_{k} b^k(j_A m_{j_A}, j_B m_{j_B}) G_k^E(A, B). \tag{4.129}$$

where we have multiplied the first term by $\frac{1}{2}$ in order to avoid counting any $(m_{j_A} m'_{j_A})$ pairs twice.

For the b_k coefficients the following relationship is valid:

$$\sum_{m'_j} b^k(jm_j, j'm'_j) = \frac{1}{2} q(j') \Gamma_{jj'}^k, \tag{4.130}$$

where the $\Gamma_{jj'}^k$ coefficients are given in terms of the Clebsch–Gordan constants. We have the relationships

$$\Gamma_{jj'}^k = \frac{2}{2k+1}\left[C\left(jj'k;\frac{1}{2},-\frac{1}{2}\right)\right]^2 = \frac{2}{2j'+1}\left[C\left(jkj';\frac{1}{2},0\right)\right]^2$$

$$= \frac{2}{2j+1}\left[C\left(j'kj;\frac{1}{2},0\right)\right]^2. \tag{4.131}$$

By putting Eq. (4.130) into the formula given by Eq. (4.129), we obtain

$$\sum_{A,B}\left\langle\varphi_A\varphi_B\left|\frac{1}{r_{12}}\right|\varphi_B\varphi_A\right\rangle = \frac{1}{2}\sum_A\sum_k\frac{1}{2}[q(A)]^2\Gamma_{j_Aj_A}^k G_k^E(A,A)$$

$$+ \sum_A\sum_{\substack{B\\(B\neq A)}}\sum_k q(A)q(B)\frac{1}{2}\Gamma_{j_Aj_B}^k G_k^E(A,B). \tag{4.132}$$

In the first term, the exchange interaction of state A with itself is included. This compensates for the electrostatic self-interaction that is contained in Eq. (4.128). Indeed, for the $k=0$ term, we get

$$\Gamma_{jj'}^0 = \frac{2}{q(j')}\delta(jj'), \tag{4.133}$$

and so we can write Eq. (4.132) in the form

$$\sum_{A,B}\left\langle\varphi_A\varphi_B\left|\frac{1}{r_{12}}\right|\varphi_B\varphi_A\right\rangle = \frac{1}{2}\sum_A q(A)G_0^E(A,A)$$

$$+ \sum_A\sum_{k>0}\frac{1}{4}[q(A)]^2\Gamma_{j_Aj_A}^k G_k^E(A,A)$$

$$+ \sum_A\sum_{\substack{B\\(B\neq A)}}\sum_k\frac{1}{2}q(A)q(B)\Gamma_{j_Aj_B}^k G_k^E(A,B). \tag{4.134}$$

Let us denote by E^E that part of the total energy that is generated by the electrostatic interaction. On the basis of Eqs. (4.123), (4.125), (4.127), (4.128), and (4.134), we obtain

$$
\begin{aligned}
E^E = &\sum_A q(A)I(A) + \sum_A \frac{1}{2}[q(A)]^2 F_0^E(A, A) \\
&+ \sum_A \sum_B q(A)q(B)F_0^E(A, B) \\
&{}_{(B \neq A)} \\
&- \frac{1}{2} \sum_A q(A)G_0^E(A, A) - \sum_A \sum_{k>0} \frac{1}{4}[q(A)]^2 \Gamma_{j_A j_A}^k G_k^E(A, A) \\
&- \sum_A \sum_B \sum_k \frac{1}{2} q(A)q(B)\Gamma_{j_A j_B}^k G_k^E(A, B). \\
&{}_{(B \neq A)}
\end{aligned}
\tag{4.135}
$$

The second and fourth terms of this formula can be combined, because, as we see from Eqs. (4.112) and (4.116), we have the relationship

$$
G_k^E(A, A) = F_k^E(A, A). \tag{4.136}
$$

This is as it should be because the fourth term represents the self-exchange energy that compensates for the self-electrostatic energy contained in the second term. Combining these two terms we get

$$
\begin{aligned}
E^E = &\sum_A q(A)I(A) + \sum_A \tfrac{1}{2}q(A)[q(A) - 1]F_0^E(A, A) \\
&+ \sum_A \sum_B q(A)q(B)F_0^E(A, B) - \sum_A \sum_{k>0} \tfrac{1}{4}[q(A)]^2\Gamma_{j_A j_A}^k F_k^E(A, A) \\
&{}_{(B \neq A)} \\
&- \sum_A \sum_B \sum_k \tfrac{1}{2}q(A)q(B)\Gamma_{j_A j_B}^k G_k^E(A, B). \\
&{}_{(B \neq A)}
\end{aligned}
\tag{4.137}
$$

We come now to the magnetic interaction. The "direct" part of that interaction is given by Eq. (4.117). We have to place that expression into the summation described by Eq. (4.124). It is easy to show that for complete groups, the result will be zero. This is so because the f^k coefficients satisfy the relationship

$$
\sum_{m_j'} f^k(jm_j, j'm_j') = 0, \tag{4.138}
$$

if the m'_j summation is for a complete group. Thus, the "direct" part of the magnetic interaction will be zero.

We substitute the exchange integral into the summation given by Eq. (4.124) and obtain

$$\sum_{A,B} \left\langle \varphi_A \varphi_B \left| \frac{\alpha(1) \cdot \alpha(2)}{r_{12}} \right| \varphi_B \varphi_A \right\rangle$$

$$= \frac{1}{2} \sum_A \sum_{m_{j_A}} \sum_{m'_{j_A}} \sum_{kJ\gamma} g^k(j_A m_{j_A}, j_A m'_{j_A}; J\gamma) G_k^M(AA;\gamma)$$

$$+ \sum_A \sum_B \sum_{m_{j_A}} \sum_{m_{j_B}} \sum_{kJ\gamma} g^k(j_A m_{j_A}, j_B m_{j_B}; J\gamma) G_k^M(AB;\gamma). \quad (4.139)$$
$$\scriptstyle (B \neq A)$$

First, let us consider the second term. For this term, we have the relationship

$$\sum_{m_{j_B}} g^k(j_A m_{j_A}, j_B m_{j_B}; J\gamma) = \tfrac{1}{2} q(B) \Gamma_{j_A j_B}^k(a_A a_B; J\gamma). \quad (4.140)$$

The $\Gamma_{j_A j_B}^k$ coefficients in this formula should not be confused with the Γ's of Eq. (4.130). For these new Γ coefficients, Grant derived an explicit formula for which we refer the reader to the original publications. In evaluating the first term of Eq. (4.139), we observe from the formulas given by Eqs. (4.121a) to (4.121c) that the G_k^M radial integrals are the same for all three values of γ if $A \equiv B$, that is,

$$G_k^M(AA;1) = G_k^M(AA;0) = G_k^M(AA;-1). \quad (4.141)$$

This being the case, G_k^M does not depend on γ, and we can write

$$G_k^M(AA;\gamma) = G_k^M(A,A), \quad (4.142)$$

for all γ and we can carry out the summation over γ in the first term of Eq. (4.139). We obtain

$$\Gamma_{j_A j_A}^k(J) = \sum_\gamma \Gamma_{j_A j_A}^k(a_A a_A; J\gamma), \quad (4.143)$$

and again Grant derived an explicit expression for this constant. By substituting Eqs. (4.140), (4.142), and (4.143) into Eq. (4.139), we obtain

$$
\sum_{A,B} \left\langle \varphi_A \varphi_B \left| \frac{\boldsymbol{\alpha}(1) \cdot \boldsymbol{\alpha}(2)}{r_{12}} \right| \varphi_B \varphi_A \right\rangle
$$

$$
= \frac{1}{4} \sum_A \sum_{kJ} [q(A)]^2 \Gamma_{j_A j_A}^k (J) G_k^M (A, A)
$$

$$
+ \sum_A \sum_B \sum_{kJ\gamma} \frac{1}{2} q(A) q(B) \Gamma_{j_A j_B}^k (a_A a_B; J\gamma) G_k^M (AB; \gamma). \quad (4.144)
$$

Let us denote the magnetic part of the total energy by E^M. Since the direct term is zero, E^M will be identical with the expression in Eq. (4.144). Taking into account the fact that according to Eqs. (4.118), (4.121), and (4.142), we have the relationship

$$
F_k^M (AA) = G_k^M (AA), \quad (4.145)
$$

we obtain for E^M the expression

$$
E^M = \sum_A \sum_{kJ} \frac{1}{4} [q(A)]^2 \Gamma_{j_A j_A}^k (J) F_k^M (A, A)
$$

$$
+ \sum_A \sum_{\substack{B \\ (B \neq A)}} \sum_{kJ\gamma} \frac{q(A) q(B)}{2} \Gamma_{j_A j_B}^k (a_A a_B; J\gamma) G_k^M (AB; \gamma). \quad (4.146)
$$

The total energy of an atom with complete groups is

$$
E = E^E + E^M, \quad (4.147)
$$

where E^E is given by Eq. (4.137), and E^M by Eq. (4.146).

As in the case of the nonrelativistic HF theory, we can give a plausible physical interpretation to each term in the relativistic energy expression. Let us consider first E^E in Eq. (4.137). In that formula, the first term is the sum of one-electron energies. The second term is the intragroup electrostatic energy summed over all groups. The third is the electrostatic interaction between different groups. The fourth is the intragroup exchange energy and the fifth is the exchange interaction between different groups.

The magnetic energy given by Eq. (4.146) is of an equally simple structure. The expression does not contain any "direct" magnetic interaction; that interaction is zero for complete groups. The energies in E^M both come from exchange integrals and the first is the intragroup magnetic interaction and the second is the magnetic interaction between different groups.

It is extremely interesting to compare the relativistic and nonrelativistic HF energies for an atom with complete groups. The comparison is, of course, done only for E^E because the magnetic energy does not occur in the nonrelativistic model. For E^E, the relativistic energy is given by Eq. (4.137), which we write down here again because we want the two formulas side by side:

$$E^E = \sum_A q(A)I(A) + \sum_A \frac{1}{2}q(A)[q(A) - 1]F_0^E(AA)$$

$$+ \sum_A \sum_{\substack{B \\ (B \neq A)}} q(A)q(B)F_0^E(A, B) - \sum_A \sum_{k>0} \frac{1}{4}[q(A)]^2\Gamma_{j_A j_A}^k F_k^E(A, A)$$

$$- \sum_A \sum_{\substack{B \\ (B \neq A)}} \sum_k \frac{1}{2}q(A)q(B)\Gamma_{j_A j_B}^k G_k^E(A, B). \tag{4.148}$$

The nonrelativistic HF energy, in Slater's formulation, is given by Eq. (2.116). We have

$$E_F = \sum_{nl} q(nl)I(nl) + \sum_{(nl)} \frac{1}{2}q(nl)[q(nl) - 1]F_0(nl, nl)$$

$$+ \sum_{\substack{(nl)(n'l') \\ (n'l')>(nl)}} q(nl)q(n'l')F_0(nl, n'l')$$

$$- \sum_{(nl)} \sum_{k>0} \frac{1}{2}q(nl)c^k(l0, l0)F_k(nl, nl)$$

$$- \sum_{\substack{(nl)(n'l') \\ (n'l')>(nl)}} \sum_k 2[(2l + 1)(2l' + 1)]^{1/2}c^k(l0, l'0)G_k(nl, n'l'). \tag{4.149}$$

This formula is obtained from Eq. (2.116) by making some small changes in the notation. We have put $q(nl) = 2(2l + 1)$.

Also, we have extracted the $k = 0$ term from the fourth term of that equation and combined it with the first term, as in Eq. (2.117). Further, we multiplied the double summations with a factor of 2 so that they are now summations for distinct pairs, as in Eq. (4.148).

Comparing these two formulas, we see that they are not only similar, but, *mutatis mutandis*, they are identical. A few observations will show the accuracy of this statement. First, both energy expressions consist of five terms that have identical physical meanings. The dependence of E^E on the radial integrals is of the same structure as the dependence of E_F on the corresponding nonrelativistic radial integrals. We should recall that $F_k^E(AB)$ becomes identical with $F_k(nl, n'l')$ if we take the nonrelativistic limit, that is, if we put $Q_A = Q_B = 0$ and identify A with (nl) and B with $(n'l')$. Thus, we get the first three terms of E_F, that is, the terms representing the electrostatic interactions, from the first three terms of E^E by putting $Q_A = Q_B = 0$ and replacing A by (nl) and B by $(n'l')$.

The similarity between the terms representing the exchange interactions is also very strong. We get the radial integrals of the exchange terms in E_F from the corresponding radial integrals in E^E by again taking the nonrelativistic limit. The angular coefficients are different in the two formulas, reflecting the difference between the LS and jj couplings. Nevertheless, even there, we must observe that both c^k and Γ^k can be expressed, in a similar fashion, in terms of the Clebsch–Gordan coefficients. This is clear from Eqs. (4.122) and (4.131).

Summing up, it is accurate to say that the electrostatic part of the relativistic energy of an atom with complete groups is the straightforward generalization of the corresponding nonrelativistic energy expression.

B. The HFD Equations and Their Physical Meaning

We now proceed to the derivation of the HFD equations. We take the total energy given by Eq. (4.147) and vary it under the subsidiary conditions of normalization and orthogonality. First, in analogy to the function introduced by Eq. (2.77a), we define the following auxiliary functions:

$$\frac{1}{r} Y_k^E(AB|r) = \int_0^\infty [P_A(r')P_B(r') + Q_A(r')Q_B(r')] L_k(r,r')\, dr', \quad (4.150)$$

and

$$\frac{1}{r} Y_k^M(AB|r) = \int_0^\infty P_A(r')Q_B(r') L_k(r,r')\, dr'. \quad (4.151)$$

Using these functions, we can write our radial integrals in more convenient forms. We obtain

$$F_k^E(AB) = \int_0^\infty [P_A^2(r) + Q_A^2(r)] \frac{1}{r} Y_k^E(BB|r)\, dr$$

$$= \int_0^\infty [P_B^2(r) + Q_B^2(r)] \frac{1}{r} Y_k^E(AA|r)\, dr; \quad (4.152)$$

$$G_k^E(AB) = \int_0^\infty [P_A(r)P_B(r) + Q_A(r)Q_B(r)] \frac{1}{r} Y_k^E(AB|r)\, dr; \quad (4.153)$$

$$F_k^M(AB) = \int_0^\infty P_A(r)Q_A(r) \frac{1}{r} Y_k^M(BB|r)\, dr$$

$$= \int_0^\infty P_B(r)Q_B(r) \frac{1}{r} Y_k^M(AA|r)\, dr; \quad (4.154)$$

$$G_k^M(AB;1) = \int_0^\infty P_A(r)Q_B(r) \frac{1}{r} Y_k^M(AB|r)\, dr; \quad (4.155)$$

$$G_k^M(AB;0) = \int_0^\infty P_A(r) Q_B(r) \frac{1}{r} Y_k^M(BA|r)\, dr$$

$$= \int_0^\infty P_B(r) Q_A(r) \frac{1}{r} Y_k^M(AB|r)\, dr; \tag{4.156}$$

$$G_k^M(AB;-1) = \int_0^\infty P_B(r) Q_A(r) \frac{1}{r} Y_k^M(BA|r)\, dr. \tag{4.157}$$

The orthogonality and normalization are expressed by

$$\int_0^\infty \left[P_A(r) P_B(r) + Q_A(r) Q_B(r) \right] dr = \delta(A,B). \tag{4.158}$$

We vary the energy with respect to P_A and Q_A independently. Let δ_A denote the variation with respect to P_A and Q_A, that is, let us subject P_A to the small variation δP_A and Q_A to the small variation δQ_A. Then we get

$$\delta_A F_k^E(AB) = 2 \int_0^\infty \left[P_A(r)\delta P_A(r) + Q_A(r)\delta Q_A(r) \right] \frac{1}{r} Y_k^E(BB|r)\, dr,$$

and

$$\tag{4.159}$$

$$\delta_A F_k^E(AA) = 4 \int_0^\infty \left[P_A(r)\delta P_A(r) + Q_A(r)\delta Q_A(r) \right] \frac{1}{r} Y_k^E(AA|r)\, dr. \tag{4.160}$$

Similar expressions can be constructed for the other radial integrals. The subsidiary conditions, Eq. (4.158), can be taken into account by Lagrangian multipliers.

As a result of the variation principle, we obtain the HFD equations

$$\frac{dP_A}{dr} + \frac{\kappa_A}{r} P_A + \left[2c + \frac{1}{c}\left(\frac{Y^E(A|r)}{r} - \varepsilon_A \right) \right] Q_A$$

$$- \frac{Y^M(A|r)}{rc} P_A - W_Q(A|r) + V_P(A|r)$$

$$= \sum_{(B \neq A)} \frac{1}{c} \varepsilon_{AB}\, \delta(j_A j_B)\, \delta(a_A a_B)\, Q_B; \tag{4.161}$$

and

$$\frac{dQ_A}{dr} - \frac{\kappa_A}{r}Q_A - \frac{1}{c}\left(\frac{Y^E(A|r)}{r} - \varepsilon_A\right)P_A$$

$$+ \frac{Y^M(A|r)}{rc}Q_A + W_P(A|r) - V_Q(A|r)$$

$$= - \sum_{\substack{B \\ (B \neq A)}} \frac{1}{c}\varepsilon_{AB}\delta(j_Aj_B)\delta(a_Aa_B)P_B. \tag{4.162}$$

In these equations, ε_A is the orbital parameter and the ε_{AB} are the Lagrangian multipliers. The definitions of the functions occurring in the equations are as follows:

$$\frac{1}{r}Y^E(A|r) = \frac{Z}{r} - \sum_{A'}q(A')\frac{1}{r}Y_0^E(A'A'|r)$$

$$+ \sum_k \frac{1}{2}[q(A)]\Gamma_{j_Aj_A}^k\frac{1}{r}Y_k^E(AA|r); \tag{4.163a}$$

$$\frac{1}{r}Y^M(A|r) = \sum_{kJ}\frac{1}{4}q(A)\Gamma_{j_Aj_A}^k(J)\frac{1}{r}Y_k^M(AA|r); \tag{4.163b}$$

$$W_P(A|r) = -\frac{1}{rc}\sum_{\substack{B \\ (B \neq A)}}\sum_k\frac{q(B)}{2}\Gamma_{j_Aj_B}^kY_k^E(AB|r)P_B(r); \tag{4.164}$$

$$W_Q(A|r) = -\frac{1}{rc}\sum_{\substack{B \\ (B \neq k)}}\sum_k\frac{q(B)}{2}\Gamma_{j_Aj_B}^kY_k^E(AB|r)Q_B(r); \tag{4.165}$$

$$V_P(A|r) = -\frac{1}{rc}\sum_{\substack{B \\ B \neq A)}}\sum_{kJ}\frac{q(B)}{2}\left[\Gamma_{j_Aj_B}^k(a_Aa_B;J,-1)Y_k^M(BA|r)\right.$$

$$\left. + \frac{1}{2}\Gamma_{j_Aj_B}^k(a_Aa_B;J,0)Y_k^M(AB|r)\right]P_B(r); \tag{4.166}$$

$$V_Q(A|r) = -\frac{1}{rc}\sum_{\substack{B \\ (B \neq A)}}\sum_{kJ}\frac{q(B)}{2}$$

$$\times\left[\Gamma_{j_Aj_B}^k(a_Aa_B;J,1)Y_k^M(AB|r)\right.$$

$$\left. + \frac{1}{2}\Gamma_{j_Aj_B}^k(a_Aa_B;J,0)Y_k^M(BA|r)\right]Q_B(r). \tag{4.167}$$

The HFD equations, Eqs. (4.161) and (4.162), look complicated, but only because of the many indices of the angular coefficients and not because of the complicated structure of the radial functions. In fact, the equations are only slightly more complicated than the nonrelativistic equations, which, as we have seen in the preceding chapters, have a fairly straightforward mathematical structure.

In order to analyze the structure of the HFD equations and establish their physical meaning, let us omit the magnetic terms from the equations and let us write down the results. We obtain, by omitting all Y_k^M functions,

$$\frac{dP_A}{dr} + \frac{\kappa_A}{r}P_A + \left[2c + \frac{1}{c}\left(\frac{1}{r}Y^E(A|r) - \varepsilon_A\right)\right]Q_A - W_Q(A|r)$$

$$= \sum_{(B \neq A)} \frac{1}{c}\varepsilon_{AB}\delta(j_A j_B)\delta(a_A a_B)Q_B, \qquad (4.168)$$

$$\frac{dQ_A}{dr} - \frac{\kappa_A}{r}Q_A - \frac{1}{c}\left(\frac{1}{r}Y^E(A|r) - \varepsilon_A\right)P_A + W_P(A|r)$$

$$= -\sum_{(B \neq A)} \frac{1}{c}\varepsilon_{AB}\delta(j_A j_B)\delta(a_A a_B)P_B. \qquad (4.169)$$

For comparison, let us write down here the nonrelativistic HF equations. According to Eq. (2.88), we have

$$\left\{-\frac{1}{2}\frac{d^2}{dr^2} + \frac{l(l+1)}{2r^2} - \frac{Z}{r} + U(nl|r)\right\}P_{nl} = \varepsilon_{nl}P_{nl}, \qquad (4.170)$$

where the HF potential is given by

$$U(nl|r) = \sum_{(n'l')} q(n'l')\frac{1}{r}Y_0(n'l', n'l'|r) - \frac{1}{r}Y_0(nl, nl)$$

$$- \sum_{k>0} c^k(l0, l0)\frac{1}{r}Y_k(nl, nl|r)$$

$$- \sum_{\substack{n'l'k \\ (n'l' \neq nl)}} \left(\frac{2l'+1}{2l+1}\right)^{1/2} c^k(l0, l'0)\frac{1}{r}Y_k(nl, n'l'|r)\frac{P_{n'l'}(r)}{P_{nl}(r)}. \qquad (4.171)$$

This formula is obtained from Eq. (2.87) by using Eqs. (2.86), (2.100a), and (2.101a). Now let us consider the potential in Eq. (4.168), which is the equation for the large component P_A. The terms representing the potentials are

$$\frac{1}{cr}Y^E(A|r)Q_A - W_Q(A|r) = \frac{1}{c}\left[\frac{Z}{r} - \sum_{A'}q(A')\frac{1}{r}Y_0^E(A'A'|r)\right.$$

$$\left. + \sum_{k}\frac{q(A)}{2}\Gamma_{j_Aj_A}^k\frac{1}{r}Y_k^E(AA|r)\right]Q_A(r)$$

$$+ \frac{1}{c}\sum_{\substack{B \\ (B \neq A)}}\sum_{k}\frac{q(B)}{2}\Gamma_{j_Aj_B}^k\frac{1}{r}Y_k^E(AB|r)Q_B(r).$$

$$(4.172)$$

In order to show the similarity between the formulas in Eqs. (4.171) and (4.172), we make a few small changes in the notation of Eq. (4.172). Using Eq. (4.133), we detach the $k = 0$ term from the third term of Eq. (4.172). Also, we formally change the W_Q term into a local potential by dividing it by $Q_A(r)$. Thus, we get

$$\frac{1}{cr}Y^E(A|r)Q_A(r) - W_Q(A|r)$$

$$= \frac{1}{c}\left\{\frac{Z}{r} - \sum_{A'}q(A')\frac{1}{r}Y_0^E(A'A'|r) + \frac{1}{r}Y_0^E(AA|r)\right.$$

$$+ \sum_{k>0}\frac{q(A)}{2}\Gamma_{j_Aj_A}^k\frac{1}{r}Y_k^E(AA|r)$$

$$\left. + \sum_{\substack{B \\ (B \neq A)}}\sum_{k}\frac{q(B)}{2}\Gamma_{j_Aj_B}^k\frac{1}{r}Y_k^E(AB|r)\frac{Q_B(r)}{Q_A(r)}\right\}Q_A(r). \quad (4.173)$$

It is now evident that the sum of the nuclear potential and U in Eq. (4.170) have the same structure as the relativistic potential in Eq. (4.173). The physical meaning of the terms of the relativistic potential are the same as the physical meaning of the terms of the nonrelativistic. The only difference (apart from the trivial difference of -1, which is the result of the different derivations) is in the last exchange term in which we have Q_B and Q_A instead of P_B and P_A as in Eq. (4.171). (In this comparison, we put $B = (n'l')$ and $A = (nl)$.)

Let us introduce a symbol for the whole expression in Eq. (4.173). We write

$$
U_P(A|r) \equiv \frac{1}{c} \left\{ \frac{Z}{r} - \sum_{A'} q(A') \frac{1}{r} Y_0^E(AA'|r) + \frac{1}{r} Y_0^E(AA|r) \right.
$$

$$
+ \sum_{k>0} \frac{q(A)}{2} \Gamma_{j_A j_A}^k \frac{1}{r} Y_k^E(AA|r)
$$

$$
\left. + \sum_{\substack{B \\ (B \neq A)}} \sum_k \frac{q(B)}{2} \Gamma_{j_A j_B}^k \frac{1}{r} Y_k^E(AB|r) \frac{Q_B(r)}{Q_A(r)} \right\}. \quad (4.174)
$$

The P index indicates that this is the potential in the equation for the large component. Formally, it calls attention that we have Q_B and Q_A in the last exchange term. Naturally, the potential depends on A. Using this function, we can write Eq. (4.168) in the form

$$
\frac{dP_A}{dr} + \frac{\kappa_A}{r} P_A + \left[2c - \frac{\varepsilon_A}{c} + U_P(A|r) \right] Q_A(r)
$$

$$
= \sum_B \frac{1}{c} \varepsilon_{AB} \delta(j_A j_B) \delta(a_A a_B) Q_B. \quad (4.175)
$$

We can rewrite the Q equation similarly. Let

$$
U_Q(A|r) = \frac{1}{cr} Y^E(A|r) - W_P(A|r), \quad (4.176)
$$

and the reader can easily construct an expression corresponding to Eq. (4.174) from this formula. We obtain for Eq. (4.169),

$$
\frac{dQ_A}{dr} - \frac{\kappa_A}{r} Q_A - \left[-\frac{\varepsilon_A}{c} + U_Q(A|r) \right] P_A
$$

$$
= - \sum_{(B \neq A)} \frac{1}{c} \varepsilon_{AB} \delta(j_A j_B) \delta(a_A a_B) P_B. \quad (4.177)
$$

After the HFD equations have been solved, the total energy can be computed from Eq. (4.148) if the magnetic interactions are omitted and from Eqs. (4.146), (4.147), and (4.148) if the magnetic interactions are included. We show that if the orbital parameters are known, the energy expression can be put into a very simple form. We present this argument for the case when only the electrostatic part is considered. By multiplying Eq. (4.175) from the left by $-cQ_A$, Eq. (4.177) by cP_A, integrating, and taking into account the formula for $I(A)$, Eq. (4.107), we obtain

$$
q(A)I(A) = q(A)(c^2 - \varepsilon_A) - q(A)[q(A) - 1]F_0^E(A, A)
$$
$$
+ \frac{1}{2}[q(A)]^2 \sum_{k>0} \Gamma_{j_A j_A}^k F_k^E(A, A)
$$
$$
- \sum_{\substack{B \\ (B \neq A)}} \left[q(A)q(B)F_0^E(A, B) \right.
$$
$$
\left. - \sum_k \frac{q(A)q(B)}{2} \Gamma_{j_A j_B}^k G_k^E(A, B) \right]. \quad (4.178)
$$

By substituting this into the energy expression, Eq. (4.148), we obtain

$$
E^E = \sum_A \left\{ q(A)(c^2 - \varepsilon_A) - \frac{q(A)}{2}[q(A) - 1]F_0^E(A, A) \right.
$$
$$
\left. + \sum_{k>0} \frac{[q(A)]^2}{4} \Gamma_{j_A j_A}^k F_k^E(A, A) \right\}. \quad (4.179)
$$

From this formula, all integrals with $B \neq A$ are eliminated.

We now turn to the discussion of the physical meaning of the HFD equations. Let us consider first the equations derived in the absence of magnetic interactions, Eqs. (4.175) and (4.177). It will be useful to recall that the Hamiltonian underlying these equations is

$$
H = \sum_{k=1}^N H_D(k) + \frac{1}{2} \sum_{\substack{k,j=1 \\ (k \neq j)}}^N \frac{1}{r_{kj}}, \quad (4.180)
$$

where H_D is the Dirac one-electron Hamiltonian. The total wave function of the atom is given by the determinant in Eq. (4.90) and the one-electron wave functions are the four-component Dirac functions given by Eq. (4.102).

What is the most conspicuous feature of the HFD equations relative to the nonrelativistic HF equations? The most important feature is that we have obtained separate equations for each $l + \frac{1}{2}$ and $l - \frac{1}{2}$ pairs (more accurately, we get for each combination, a pair of equations for P_A and Q_A). In the nonrelativistic HF model, we get one equation for each (nl) group. That means that we get one energy parameter for each (nl) combination. In the relativistic model, we get separate equations and separate energy levels for each (nlj) or (nla) group. This is evident at once because our P_A and Q_A wave functions are actually abbreviations for $P_{n_A l_A j_A}$ and $Q_{n_A l_A j_A}$; also, we see the presence of κ_A in the equations. The presence of κ_A means that the energy parameter will depend on j_A and l_A because, as we have clarified it before, the absolute value of κ_A determines j_A and its sign determines the l_A from which j_A is generated. As we have seen, positive κ means $j = l - \frac{1}{2}$ and negative κ means $j = l + \frac{1}{2}$.

Thus, clearly, the structure of the HFD equations shows that

> The Hartree–Fock–Dirac equations have different eigenfunctions and different orbital parameters for $l, j = l + \frac{1}{2}$, and for $l, j = l - \frac{1}{2}$, that is, these equations contain automatically the most important relativistic effect, the spin–orbit coupling; thus, they provide the correct type of energy spectrum for the core orbitals in accordance with the empirical x-ray spectra.

Of course, only after solving the equations will it be possible to say whether the equations reproduce accurately the empirical x-ray spectra. We examine this question in the next chapter. But it is important to note that the spin–orbit interaction is built into the HFD formalism even if the electron–electron interaction is represented in the Hamiltonian only by the electrostatic term e^2/r_{12}. Thus, the most important feature of the relativistic HF model is the replacement of the nonrelativistic one-electron Hamiltonian operator with the Dirac relativistic one-electron operator, H_D. The replacement of the electrostatic interaction by the Breit interaction operator is important for conceptual as well as for computational reasons, but it is not the most important feature of the HFD model.

If we are looking at the equations in which the functions representing the magnetic interactions are present, Eqs. (4.161) and (4.162), then we realize that here we have the spin–other-orbit and spin–spin interactions built into the Hamiltonian that forms the background of these equations. Thus, these interactions are built into these equations. It is interesting that in the HFD model, these interactions are represented by potential-like terms. Finally, we recall that the retardation effects were omitted from the formalism. We also observe that the relativistic change in the electron mass at high velocities is built into the formalism in the presence as well as in the absence of the magnetic effects.

C. Incomplete Groups and Configuration Averages

As we have mentioned before, in relativistic Hartree–Fock theory, a group is defined as the set of electron states with common (nlj) or (nja). The maximum occupation number in a group is $2j + 1$; this is the number of possible m_j values, that is, the number of electron states with common j and different m_j. A group is complete if all states are filled and incomplete if the occupation number $q(nlj)$ is less than $2j + 1$.

In Sec. 3.3, we have seen that it is very useful to introduce the concept of the average energy of a configuration. We have shown that in LS coupling, the energy of any multiplet can be written as the sum of the average energy plus a few LS-dependent terms. In view of this, it is clear that the average energy of a configuration plays a central role in the nonrelativistic HF theory.

It was demonstrated by Grant that in the relativistic HF theory, it is possible to define an average energy for an incomplete group in a way that is completely analogous to the nonrelativistic theory. We now write down the average energy of an atom with one or several incomplete groups. In doing this, it will not be necessary to go through the arguments of Sec. 3.3 again. We could go through those arguments adjusting them to the (jj) coupling of the relativistic theory, which replaces the LS coupling that we have used in building up the method in Sec. 3.3. Such a lengthy procedure is not necessary, however, since we recall that for *complete groups*, the total relativistic energy of an atom generated by the electrostatic interaction is the close analog of the corresponding nonrelativistic energy expression. We have demonstrated the close correspondence between these two expressions when we made a comparison between the two energy formulas, given by Eqs. (4.148) and (4.149), at the end of Sec. 4.3/A.

We now show that the correspondence principle that connects those two equations can be used in a very simple fashion to set up the formulas for the relativistic average energy of a configuration. We assume that the relativistic formula for the average energy will be the analog of the nonrelativistic average-energy formula and we assume that the correspondence between them will be the same as between the relativistic formula for complete groups, Eq. (4.148), and the nonrelativistic, Eq. (4.149).

First, we show how the correspondence can be put into mathematical form. Let us consider, for complete groups, the following energy expression:

$$E = \sum_A q(A)I(A) + \sum_A \frac{q(A)}{2}[q(A) - 1]F_0^E(A, A)$$

$$+ \sum_A \sum_B q(A)q(B)F_0^E(A, B) - \sum_A \sum_{k>0} \frac{q(A)}{2}\Lambda(A, A, k)F_k^E(A, A)$$
$$\scriptstyle (B \neq A)$$

$$- \sum_A \sum_B \sum_k q(A)\Lambda(A, B, k)G_k^E(A, B). \tag{4.181}$$
$$\scriptstyle (B \neq A)$$

We state that this formula contains both Eqs. (4.148) and (4.149). We define the symbol $\Lambda(A, B, k)$ as follows:

$$\Lambda(A, B, k) = \tfrac{1}{2}q(B)\Gamma_{j_A j_B}^k, \tag{4.182}$$

for the relativistic case and

$$\Lambda(nl, n'l', k) = \sqrt{\frac{2l' + 1}{2l + 1}}\, c^k(l0, l'0), \tag{4.183}$$

for the nonrelativistic case. In the Λ symbol, we put $A = nl$, $B = n'l'$ for the nonrelativistic case.

Using these definitions, we see that

$$\Lambda(A, A, k) = \tfrac{1}{2} q(A) \Gamma^{k}_{j_A j_A}, \tag{4.184}$$

in the relativistic case and

$$\Lambda(nl, nl, k) = c^{k}(l0, l0), \tag{4.185}$$

in the nonrelativistic case.

Now let us consider the fourth term of Eq. (4.181). Using Eq. (4.184), we get for the coefficient in that term

$$\tfrac{1}{2} q(A) \Lambda(A, A, k) = \tfrac{1}{4} [q(A)]^2 \Gamma^{k}_{j_A j_A}, \tag{4.186}$$

which is what we have in the fourth term of Eq. (4.148). Likewise, we get, using Eq. (4.182), for the coefficient of the fifth term of Eq. (4.181),

$$q(A) \Lambda(A, B, k) = \frac{q(A) q(B)}{2} \Gamma^{k}_{j_A j_B}, \tag{4.187}$$

which is what we have in the fifth term of Eq. (4.148). Thus, it is shown that Eq. (4.181) reproduces the relativistic energy for complete groups, Eq. (4.148).

It is easy to see that Eq. (4.181) reproduces also the nonrelativistic energy, given by Eq. (4.149). Let us consider first the fourth term of our expression in Eq. (4.181). For the nonrelativistic case, we get, using Eq. (4.185),

$$\tfrac{1}{2} q(A) \Lambda(A, A, k) = \tfrac{1}{2} q(nl) \Lambda(nl, nl, k)$$

$$= \tfrac{1}{2} q(nl) c^{k}(l0, l0), \tag{4.188}$$

and this is the coefficient that we have in the fourth term of Eq. (4.149). For the fifth term of Eq. (4.181), we get, using Eq. (4.183),

$$q(A) \Lambda(A, B, k) = q(nl) \Lambda(nl, n'l', k)$$

$$= q(nl) \sqrt{\frac{2l' + 1}{2l + 1}} \, c^{k}(l0, l'0). \tag{4.189}$$

These expressions are all for complete groups, therefore, $q(nl) = 2(2l + 1)$. So we get

$$q(A)\Lambda(A, B, k) = 2(2l + 1)\sqrt{\frac{2l' + 1}{2l + 1}}\, c^k(l0, l'0)$$

$$= 2\sqrt{(2l + 1)(2l' + 1)}\, c^k(l0, l'0). \qquad (4.190)$$

This is the coefficient in the fifth term of Eq. (4.149). By using the relationships given by Eqs. (4.188) and (4.190), putting $q(A) = q(nl)$, $q(B) = q(n'l')$, and taking the nonrelativistic limit by putting $Q_A = Q_B = 0$, we indeed get the nonrelativistic formula, Eq. (4.149), from our general expression, Eq. (4.181). Thus, we see that the correspondence between the relativistic and nonrelativistic energies of atoms with complete groups is established by the relationships given by Eqs. (4.182) and (4.183).

It is easy now to set up the relativistic formula for the average energy. The nonrelativistic energy is given by Eq. (3.83):

$$E_{av.} = \sum_{(nl)} \left\{ q(nl)I(nl) + \frac{q(nl)}{2}[q(nl) - 1] \right.$$

$$\left. \times \left[F_0(nl, nl) - \sum_{k>0} \frac{c^k(l0, l0)}{4l + 1} F_k(nl, nl) \right] \right\}$$

$$+ \sum_{\substack{(nl)\ (n'l') \\ (nl)<(n'l')}} q(nl)q(n'l')$$

$$\times \left\{ F_0(nl, n'l') - \sum_{k} \frac{c^k(l0, l'0)}{2\sqrt{(2l + 1)(2l' + 1)}} G_k(nl, n'l') \right\}. \qquad (4.191)$$

The relativistic formula must be cpnstructed in such a way as to become identical with Eq. (4.191) in the nonrelativistic limit. We put $(nl) \rightarrow A$, $(n'l') \rightarrow B$, and replace the F_k and G_k integrals with their relativistic analogs. For the terms that contain the c^k coefficients, we use our correspondence principle embodied in Eqs. (4.182) and (4.183). In applying the correspondence principle, we must keep in mind that it was derived for complete groups.

In the third term of Eq. (4.191), we have $c^k(l0, l0)/(4l + 1)$. We transform this using Eqs. (4.185) and (4.184). Also, we recall that

$$4l + 1 = [q(nl)]_{comp.} - 1 = 2(2l + 1) - 1, \qquad (4.192)$$

thus, the relativistic analog of this will be

$$[q(A)]_{\text{comp.}} - 1 = (2j_A + 1) - 1 = 2j_A. \tag{4.193}$$

Thus, we put

$$\frac{c^k(l0, l0)}{4l + 1} \rightarrow \frac{1}{2}\frac{q(A)}{2j_A}\Gamma^k_{j_A j_A} = \frac{1}{2}\frac{2j_A + 1}{2j_A}\Gamma^k_{j_A j_A}, \tag{4.194}$$

where we have taken into account that in Eq. (4.184), $q(A)$ is for complete groups, that is, $q(A) = 2j_A + 1$. For the coefficient in the fifth term of Eq. (4.191), we obtain

$$\frac{c^k(l0, l0)}{2\sqrt{(2l + 1)(2l' + 1)}} \rightarrow \sqrt{\frac{2l + 1}{2l' + 1}}\frac{\Lambda(nl, n'l', k)}{2\sqrt{(2l + 1)(2l' + 1)}}$$

$$= \frac{\Lambda(nl, n'l', k)}{2(2l' + 1)} = \frac{\Lambda(A, B, k)}{q(B)}$$

$$= \frac{1}{2}\frac{q(B)}{q(B)}\Gamma^k_{j_A j_B} = \frac{1}{2}\Gamma^k_{j_A j_B}. \tag{4.195}$$

Thus, we obtain the relativistic average-energy formula:

$$E^E_{\text{av}} = \sum_A \left\{ q(A)I(A) + \frac{q(A)}{2}[q(A) - 1] \right.$$

$$\times \left[F^E_0(A, A) - \sum_{k>0}\frac{2j_A + 1}{4j_A}\Gamma^k_{j_A j_A}F^E_k(A, A) \right] \right\}$$

$$+ \sum_A \sum_B q(A)q(B)\left\{ F^E_0(A, B) - \sum_k\frac{1}{2}\Gamma^k_{j_A j_B}G^E_k(A, B) \right\}. \tag{4.196}$$
$$\scriptstyle (A<B)$$

It will be useful to have the formula for the energy expression of complete groups side by side with the average-energy formula; therefore, we write it here again:

$$E^E = \sum_A \left\{ q(A)I(A) + \frac{q(A)}{2}[q(A) - 1]F^E_0(A, A) \right.$$

$$\left. - \sum_{k>0}\frac{1}{4}[q(A)]^2\Gamma^k_{j_A j_A}F^E_k(A, A) \right\}$$

$$+ \sum_A \sum_B q(A)q(B)\left\{ F^E_0(A, B) - \sum_k\frac{1}{2}\Gamma^k_{j_A j_B}G^E_k(A, B) \right\}. \tag{4.197}$$
$$\scriptstyle (A<B)$$

It is easy to see that, exactly in the same way as in the nonrelativistic case, the average-energy formula becomes identical with the exact if we put $q(A) = 2j_A + 1$ and $q(B) = 2j_B + 1$ in the former. In fact, we see that the two formulas are formally identical except for the third term, which is the intragroup exchange energy. It should be noted that the formal equality does not mean actual identity because in Eq. (4.196), $q(A)$ and $q(B)$ are arbitrary within the restriction of $q_A \le 2j_A + 1$ and $q_B \le 2j_B + 1$, and in Eq. (4.197), the meanings of these symbols are

$$q_A = 2j_A + 1, \qquad q_B = 2j_B + 1. \tag{4.198}$$

Except for the third term, the identity of the two formulas is evident if, in the average-energy expression, we put the values given by Eq. (4.198). For the coefficient of the third term of Eq. (4.196), we get, with $q_A = 2j_A + 1$,

$$\frac{q(A)}{2}[q(A) - 1]\frac{2j_A + 1}{4j_A} = \frac{2j_A + 1}{2}[2j_A]\frac{2j_A + 1}{4j_A} = \frac{1}{4}[q(A)]^2, \tag{4.199}$$

which is what we have in the third term of Eq. (4.197).

In order to construct the average of the magnetic interaction energy, let us write down here the formula for complete groups. According to Eq. (4.146), we have

$$E^M = \sum_A \sum_{kJ} \frac{1}{4}[q(A)]^2 \Gamma^k_{j_A j_A}(J) F^M_k(A, A)$$
$$+ \sum_A \sum_B \sum_{kJ\gamma} \frac{q(A)q(B)}{2} \Gamma^k_{j_A j_B}(a_A a_B; J\gamma) G^M_k(A, B; \gamma). \tag{4.200}$$
$$(A < B)$$

In this formula, we have to put $2j_A + 1$ for $q(A)$ and $2j_B + 1$ for $q(B)$. Comparing this formula with the corresponding terms of Eq. (4.197), which are the third and the fifth terms, we see that, *mutatis mutandis*, they are identical. The changes that have to be made in Eq. (4.197) in order to get Eq. (4.200) are as follows:

$$\Gamma^k_{j_A j_A} \rightarrow \sum_J \Gamma^k_{j_A j_A}(J), \tag{4.201}$$

and

$$\Gamma^k_{j_A j_B} \rightarrow \sum_{J\gamma} \Gamma^k_{j_A j_B}(a_A a_B; J\gamma). \tag{4.202}$$

Thus, we obtain the formula for the average of the magnetic energy by simply making these changes in the third and fifth terms of Eq. (4.196). The result is

$$
E_{av}^M = \sum_{A,J,k} \frac{q(A)}{2} [q(A) - 1] \frac{2j_A + 1}{4j_A} \Gamma_{j_Aj_A}^k(J) F_k^M(A, A)
$$

$$
+ \sum_{\substack{A,B\ k,J,\gamma \\ (A<B)}} q(A)q(B) \tfrac{1}{2}\Gamma_{j_Aj_B}^k(a_Aa_B; J\gamma) G_k^M(A, B; \gamma). \quad (4.203)
$$

As we see immediately, this formula reproduces Eq. (4.200) if we apply it to complete groups. The identity of the two second terms is evident; we obtain the coefficient of the first term of Eq. (4.200) from the corresponding coefficient of Eq. (4.203) by using Eq. (4.199). Therefore, the relativistic energy of the average of an arbitrary configuration is given by the formula

$$
E_{av}^{Rel} = E_{av}^E + E_{av}^M, \quad (4.204)
$$

where the electrostatic part is given by Eq. (4.196) and the magnetic part by Eq. (4.203).

We note that for E_{av}^M, Grant has given an even simpler formula. This formula is

$$
E_{av}^M = \sum_{A,k} \frac{q(A)}{2} [q(A) - 1] \frac{2j_A + 1}{4j_A} \hat{\Gamma}_{j_Aj_A}^k F_k^M(A, A)
$$

$$
+ \sum_{\substack{(A<B)\ k,\gamma}} \frac{q(A)q(B)}{2} \hat{\Gamma}_{j_Aj_B}^k(a_Aa_B; \gamma) G_k^M(A, B; \gamma), \quad (4.205)
$$

which differs from the formula in Eq. (4.203) in that the structure of the angular coefficient is somewhat simpler. We note that the two types of coefficients are connected by the relationship,

$$
\hat{\Gamma}_{j_Aj_A}^k = \sum_{\gamma} \hat{\Gamma}_{j_Aj_A}^k(a_Aa_A; \gamma), \quad (4.206)
$$

and for the coefficients on the right side of this equation, we give explicit formulas at the end of App. E.

We write down now the HFD equations for the average energy of a configuration in the absence of magnetic interactions. For the large component P_A and for the small component Q_A, we obtain the equations

$$\frac{dP_A}{dr} + \frac{\kappa_A}{r} P_A - \left[2c + \frac{1}{c} \left(\frac{Y^E(A|r)}{r} - \varepsilon_A \right) \right] Q_A + W_Q(A|r)$$

$$= - \sum_{(B \neq A)} \frac{\varepsilon_{AB}}{c} q(B) \delta(\kappa_A, \kappa_B) Q_B, \qquad (4.207)$$

$$\frac{dQ_A}{dr} - \frac{\kappa_A}{r} Q_A + \frac{1}{c} \left(\frac{Y^E(A|r)}{r} - \varepsilon_A \right) P_A - W_P(A|r)$$

$$= \sum_{(B \neq A)} \frac{\varepsilon_{AB}}{c} q(B) \delta(\kappa_A, \kappa_B) P_B. \qquad (4.208)$$

In these equations, the zero level of the energy is defined in such a way that the electron at rest at infinity has zero energy. The functions W_P and W_Q are given by Eqs. (4.164) and (4.165). The potential function $(1/r)Y^E(A|r)$ is defined as follows

$$\frac{1}{r}Y^E(A|r) = \frac{Z}{r} - \sum_{A'} q(A') \frac{1}{r} Y_0^E(A'A'|r) + \frac{1}{r} Y_0(AA|r)$$

$$+ \sum_{k>0} [q(A) - 1] \frac{2j_A + 1}{4j_A} \Gamma_{j_A j_A}^k \frac{1}{r} Y_k^E(AA|r). \qquad (4.209)$$

As we see, by comparing this with Eq. (4.163a), the only difference between the potential function for an average configuration and the potential function for complete groups is in the term representing the intragroup exchange energy, which is the fourth term in our Eq. (4.209). We see also that we obtain the formula in Eq. (4.163a) by putting $q(A) = 2j_A + 1$ in Eq. (4.209); we must recall that in Eq. (4.163a), we have complete groups, that is, we have $q(A) = 2j_A + 1$.

We close this section by listing some information related to the relativistic HFD model.

1. Koopmans' theorem, which we discussed for the nonrelativistic HF model in Sec. 2.6, is valid again here. That is, the orbital parameter ε_A is equal to the removal energy of an electron from the state φ_A, assuming that the removal does not disturb the remaining orbitals.

2. For an atom with complete groups, the off-diagonal components of the Lagrangian can be eliminated by a unitary transformation exactly in the same way as in nonrelativistic theory.

3. Throughout the discussions in this chapter, we have omitted the retardation terms from the energy expressions. These are represented by the

operators given by Eq. (4.88b). The matrix elements of those operators were investigated by Kim[28] and also by Grant.[26] We summarize the results here. The direct matrix elements vanish, in the same way as for the magnetic interaction. The exchange part of the matrix components can be written in the form

$$\langle \varphi_A \varphi_B | g_1(1,2) | \varphi_B \varphi_A \rangle = \sum_{k,\gamma} h_k^\gamma(j_A m_A, j_B m_B) G_k^M(A,B,\gamma), \quad (4.210)$$

where G_k^M is defined by Eqs. (4.121a)–(4.121c). For the h_k^γ constants, Grant has given an explicit formula. Thus, it is interesting to note that the retardation term can be written in the same mathematical form as the term resulting from the magnetic interaction.

CHAPTER 5

A DISCUSSION
OF SOME REPRESENTATIVE
HARTREE – FOCK RESULTS

5.1. INTRODUCTION: THE RELATIONSHIP OF THE HF
MODEL TO EXACT QUANTUM MECHANICS

The scope of the chapter; relationship between the exact relativistic Schroedinger equation and the HFD model; relationship between the HFD model and the Breit – Pauli approximation; definition of relativistic and nonrelativistic correlation energies; a few words about the numerical methods used to obtain HF results.

5.2. TOTAL ENERGIES AND ELECTRON CONFIGURATIONS OF ATOMS

A. Comparison between the HFD and Breit – Pauli total energies of neutral atoms. B. The empirical values of the correlation energy for $N \leq 20$; construction of a semitheoretical correlation energy formula for all N. C. The total relativistic HF energies of atoms and ions between Li and U; relative size of relativistic energy and correlation energy; general statement about the accuracy of the relativistic HFD model; the nonrelativistic HF total energies for atoms between Li and U; estimates for the exact eigenvalue of the nonrelativistic Hamiltonian; general statement about the accuracy of the nonrelativistic HF model. D. Determination of the electron configurations of atoms with the HF model.

5.3. RADIAL DENSITIES AND RELATED QUANTITIES

A. General form of the radial densities of atoms across the periodic system; shell structure; the size of atoms versus Z; dependence of the density on increasing Z; general statement about the one-electron densities of electrons with a common principal quantum number; comparison between relativistic and nonrelativistic densities. B. The empirical and HF values of the diamagnetic susceptibilities for noble gases.

5.4. IONIZATION POTENTIALS AND MANY-VALENCE-ELECTRON ENERGIES

A. Comparison between measured and calculated ionization potentials; general statement about the connection between the error in the total energies and the error in ionization potentials; comparisons between theory and experiment for incomplete p, d, and f subshells. B. Comparison between empirical and theoretical values for the energies of n valence electrons; the dependence of correlation energy on the number of valence electrons; the accuracy of the HF model in calculating total energies and calculating valence-electron energies.

5.5. THE BINDING ENERGIES OF ATOMIC ELECTRONS

A. Comparison between the calculated relativistic energy levels and the empirical energy levels of electrons in the ground state; accuracy of the core-level calculations. B. Determination of the core / valence separation on the basis of the position of the d and f energy levels relative to the valence and core energies.

5.6. ANALYTIC HF CALCULATIONS

A. The need for analytic HF functions; early attempts to construct simple atomic orbitals; Slater's work introducing Slater-type functions; the general formulation of the Hartree – Fock – Roothaan analytic procedure; the HFR method for average of configurations and for complete groups; relativistic formulation; calculations of HFR functions.

5.7. SUMMARY

The summary of the conclusions reached by the review of selected Hartree – Fock calculations.

5.1. INTRODUCTION: THE RELATIONSHIP OF THE HF MODEL TO EXACT QUANTUM MECHANICS

Having developed the HF model in nonrelativistic as well as in relativistic form, we now have reached the point where we must ask two questions:

1. What kind of information can we obtain from this model concerning the electronic structure of atoms?
2. How accurate will be the information?

The answer to the first question is that the HF model provides an enormous amount of information on all aspects of the electronic structure of atoms. We discuss some of the information in this chapter. The answer to the second question cannot be given in simple terms. As we will show, the accuracy of the information depends on the quantity we are considering. For some quantities, the model provides highly accurate results; for others, the quality of the results ranges from fair to poor.

We want to define the scope of this chapter. It is not our intention to present a survey of the very large number of HF calculations. Such a survey is outside of the scope of this book. Our goal is to present and analyze the results of *some* representative HF calculations.

For this presentation, we have selected a number of physical quantities that are the most important for atomic structure. These quantities are the total energies and electron configurations of atoms; the radial densities, both total and one-electron, and some quantities related to the densities; the ionization potentials and many-valence-electron binding energies; and, finally, the binding energies of core electrons. HF calculations for these quantities constitute the bulk of the information provided by the HF model about the electronic structure of atoms.

Besides being important for atomic structure, the discussion of these calculations will enable us to draw general conclusions about the quality, that is, the accuracy of the HF model. We can summarize the purpose of this chapter by saying that our goal is to list and analyze a set of selected HF results that describe the most important atomic properties and at the same time give us an accurate idea about the quality of the model.

We note that the list of quantities to be discussed does not include the energy levels of the optical spectra. The reason for this omission is that here we concentrate on what might be called core-electron properties; the theory of valence electrons and the results obtained with that theory will be presented and discussed in the subsequent chapters.

We begin the evaluation of the HF model by establishing its place in quantum mechanics. Let us define H_{ex} as the *exact* Hamiltonian operator for the atom under consideration. We define this quantity in terms of Dirac's formulation of quantum mechanics.[29] Let $H_{ex}, A, B, C, \ldots$ be a complete set of commuting observables, that is, a set of commuting Hermitian operators for which there is only one joint eigenfunction belonging to any set of eigenvalues. The operators $A, B, C, \ldots$ represent angular momenta and spin and because we concentrate on the theory of the Hamiltonian in this book, these operators need not be specified here. About this complete set of observables, we assume that they are *exact*, that is, all eigenvalues, including the total energy of the system that is the eigenvalue of H_{ex}, will reproduce the results of the measurements *exactly*.

Let E_{ex} be the eigenvalue of H_{ex}, that is, let us define the Schroedinger equation

$$H_{ex}\psi_{ex} = E_{ex}\psi_{ex}. \tag{5.1}$$

At the present time, H_{ex} is not known although in Sec. 4.2/E, we have pointed out that both H_{ex} and E_{ex} can be meaningfully defined by means of quantum electrodynamics. In atomic structure calculations, we approximate H_{ex} and we have several choices for the approximation. First, let us approximate H_{ex} by the relativistic Breit Hamiltonian, which is given by Eq. (4.85). We have

$$H_B = \sum_{k=1}^{N} H_D(k) + \sum_{k,j=1}^{N} \{g_0(k,j) + g_1(k,j)\}, \qquad (5.2)$$

where H_D is the relativistic Dirac one-electron Hamiltonian and g_0 and g_1 are the Breit interaction operators. The operator g_0, given by Eq. (4.88a), represents the electrostatic and magnetic interactions, and g_1, given by Eq. (4.88b), represents the retardation effects.

Thus, our first approximation to the exact equation, Eq. (5.1), is

$$H_B \psi_B = E'_B \psi_B. \qquad (5.3)$$

Within the framework of the Breit theory, we can distinguish three stages of approximations. The first stage is the theory itself, that is, when we consider the whole operator given by Eq. (5.2). The second stage consists of the omission of the retardation effects, that is, this is an approximation in which we omit the last term of Eq. (5.2). The Hamiltonian in this approximation is

$$H_B = \sum_{k=1}^{N} H_D(k) + \sum_{k,j=1}^{N} g_0(k,j). \qquad (5.4)$$

The third stage consists of the omission of the magnetic interactions, that is, this is an approximation in which our Hamiltonian is

$$H_B = \sum_{k=1}^{N} H_D(k) + \sum_{k,j=1}^{N} \frac{1}{r_{kj}}. \qquad (5.5)$$

We can substitute into Eq. (5.3) either one of the three Hamiltonians, Eq. (5.2), (5.4), or (5.5).

The Hartree–Fock–Dirac (HFD) model, which we developed in Chapter 4, is an independent particle model resting on any one of these Hamiltonian operators. Here we must discuss whether it is meaningful to build a HF model on these Hamiltonians. We pointed out in Section 4.2/E that the exact eigenvalue of Eq. (5.3) cannot be meaningfully defined. Thus, the Hamiltonians given by Eqs. (5.2), (5.4), and (5.5) are meaningful *models*, but Eq. (5.3) has meaning only if the solution is given in terms of *first-order perturbation theory*. We have seen, however, in Chapter 3 that the HF model essentially consists of the minimization of the energy expression, which is obtained from first-order perturbation theory. This was clearly demonstrated in Sec. 3.2. Thus, by building up a HF model on any one of the Hamiltonians given by Eqs. (5.2), (5.4), and (5.5), we do not violate the rules laid down in Sec. 4.2/E concerning the Schroedinger equation, Eq. (5.3).

Let E_B be the total energy of an atom computed from the HFD model. We want to define the quality of this model in terms of the total energy of the atom. Since the HFD model is an approximation of the exact Schroedinger equation, Eq. (5.1), we can define the quality of the model by introducing the quantity

$$E_B^c = E_{ex} - E_B. \qquad (5.6)$$

E_B^c is the *relativistic correlation energy*. The concept of correlation energy was introduced into quantum mechanics by Wigner[30] and it means the interaction energy that is not taken into account by an independent particle model. This quantity plays an important role in atomic physics and is discussed in detail in subsequent parts of this work. Here it signifies only the deviation of E_B from E_{ex}.

We note that this definition of the correlation energy is not quite consistent. The logical thing to do would be to define the quality of the HFD model in terms of the Schroedinger equation, Eq. (5.3), that is, to define the correlation energy as the difference between the eigenvalue of that equation and E_B. This is not possible, however, because, as we have seen, the solution of Eq. (5.3) can be defined only up to the first order of perturbation theory. The use of E_{ex} can be justified by saying that because the Breit interaction operator represents the leading term of the exact relativistic interaction, calculations made by using this operator should be very good approximations of E_{ex}.

Another choice for an approximation of H_{ex} is the Breit–Pauli Hamiltonian operator, which we derived in Sec. 4.2/D. As we have seen in that section, the four-component Breit equation can be approximated by a two-component equation, which we have written down, for two electrons, in Eq. (4.73). We want to write this equation for an atom in the absence of external electric and magnetic fields. Let

$$H_P = H_0 + H_1, \qquad (5.7)$$

where H_0 is the nonrelativistic Hamiltonian,

$$H_0 = \sum_{j=1}^{N} \left(\frac{p_j^2}{2m} - \frac{Ze^2}{r_j} \right) + \sum_{j,k=1}^{N} \frac{e^2}{r_{jk}}. \qquad (5.8)$$

The operator H_1 is the relativistic part of the Hamiltonian. From Eq. (4.73), we obtain, in the absence of external electric and magnetic fields (but taking into account the electric field of the nucleus),

$$
H_1 = \sum_{j=1}^{N} \left\{ -\frac{p_j^4}{8m^3c^2} - \frac{ie\hbar}{4m^2c^2}(\mathscr{E}_j \cdot \mathbf{p}_j) - \frac{e\hbar}{4m^2c^2}[\mathscr{E}_j \cdot (\boldsymbol{\sigma}_j \times \mathbf{p}_j)] \right\}
$$

$$
+ \sum_{j<k=1}^{N} \left\{ -\frac{e^2(\mathbf{p}_j \cdot \mathbf{p}_k)}{m^2c^2 r_{jk}} + \frac{e^2}{2m^2c^2 r_{jk}} \right.
$$

$$
\times \left[(\mathbf{p}_j \cdot \mathbf{p}_k) - \frac{(\mathbf{r}_{kj} \cdot \mathbf{p}_j)(\mathbf{r}_{kj} \cdot \mathbf{p}_k)}{r_{jk}^2} \right]
$$

$$
+ \frac{e\hbar}{4m^2c^2} \frac{2e}{r_{jk}^3} \left[(\mathbf{r}_{jk} \times \mathbf{p}_k) \cdot \boldsymbol{\sigma}_j + (\mathbf{r}_{kj} \times \mathbf{p}_j) \cdot \boldsymbol{\sigma}_k \right]
$$

$$
+ \left(\frac{e\hbar}{2mc} \right)^2 \left[-\frac{8\pi}{3}(\boldsymbol{\sigma}_j \cdot \boldsymbol{\sigma}_k)\delta(\mathbf{r}_{jk}) \right.
$$

$$
\left. \left. + \frac{1}{r_{jk}^3} \left((\boldsymbol{\sigma}_j \cdot \boldsymbol{\sigma}_k) - \frac{3(\boldsymbol{\sigma}_j \cdot \mathbf{r}_{jk})(\boldsymbol{\sigma}_k \cdot \mathbf{r}_{jk})}{r_{jk}^2} \right) \right] \right\}. \tag{5.9}
$$

In the last two equations, we have switched back to cgs units in order to maintain the notation of Sec. 4.2.

We now form the Schroedinger equation with the Hamiltonian in Eq. (5.7). We obtain

$$
H_P \psi_P = (H_0 + H_1)\psi_P = E_P \psi_P, \tag{5.10}
$$

where E_P is the eigenvalue of H_P. Because the Hamiltonian H_P is a 2-component approximation of H_B, there is no conceptual problem with the definition of E_P.

We have discussed the physical meaning of the operator before, so let us recapitulate quickly the physical meaning of each term. We have in H_1 the relativistic velocity correction, the Darwin term, and the spin–orbit interaction. The fourth term is the magnetic interaction and the fifth represents the retardation effect. Next comes the spin–other-orbit interaction, and the last term is the spin–spin interaction energy.

It is easy to establish the connection with the 4-component Breit Hamiltonians in Eqs. (5.2), (5.4), and (5.5). The complete Breit operator in Eq. (5.2) corresponds to the complete operator H_1. In the Hamiltonian of Eq. (5.4), the retardation effects are omitted; therefore, this corresponds to the omission of the fifth term from H_1. Finally, the operator in Eq. (5.5) contains only the high-speed correction and the spin–orbit interaction. We obtain the corresponding operator from H_1 by omitting all but the first three terms.

Now we could build up a Hartree–Fock model based on operator H_P. This would differ from what we called the HFD model because the letter was based on the Breit Hamiltonian H_B. The way such a model is usually built up is first to omit the relativistic operator H_1 and formulate a nonrelativistic HF model with the nonrelativistic Hamiltonian H_0. This is the model that we have presented in Chapters 2 and 3. Let us denote the nonrelativistic HF energy by E_F. Then, the relativistic energy is computed by first-order perturbation theory in which the perturbing operator is H_1 and the wave functions that are used for the calculation of the first-order correction are the nonrelativistic HF wave functions.

It is clear that such a procedure does not correspond to the conventional use of perturbation theory for solving Eq. (5.10). The conventional procedure would be to omit H_1 from Eq. (5.10) and obtain the exact solutions of the unperturbed equation

$$H_0 \psi_0 = E_0 \psi_0. \tag{5.11}$$

By having obtained E_0, the relativistic correction would be computed from first-order perturbation theory, that is, the relativistic correction would be the expectation value of H_1 with respect to the solutions of Eq. (5.11). Let us denote this quantity by E'_R. Thus, we would approximate the eigenvalue of Eq. (5.10) by first-order perturbation theory, that is, by the formula

$$E_P = E_0 + E'_R. \tag{5.12}$$

Now let us introduce the nonrelativistic correlation energy with the definition

$$E_0^c = E_0 - E_F, \tag{5.13}$$

that is, let us define the correlation energy as the difference between the exact and HF energies in the nonrelativistic approximation. We need this quantity because we want to evaluate the HF model relative to the Breit–Pauli equation, Eq. (5.10); in order to approximate E_P we would need, according to Eq. (5.12), the exact solution of Eq. (5.11), E_0, which we definitely do not have.

By putting E_0 from Eq. (5.13) into Eq. (5.12), we get

$$E_P = E_F + E_0^c + E'_R. \tag{5.14}$$

This is still not a satisfactory formula for the approximation of E_P because E'_R is the relativistic correction computed using the eigenfunctions of Eq. (5.11). As we do not have these eigenfunctions, we introduce the approximation

$$E'_R = E_R. \tag{5.15}$$

In this equation, E_R is the expectation value of H_1 with respect to the HF wave function. This is a plausible approximation because it can be assumed that the expectation value computed with the HF wave function will be close to the expectation value computed with the eigenfunctions of Eq. (5.11).

By putting Eq. (5.15) into Eq. (5.14), we obtain

$$E_P = E_F + E_0^c + E_R. \tag{5.16}$$

Now if the Hamiltonian operator in Eq. (5.7) is an accurate approximation to H_{ex}, then we can also assume that E_P will be an accurate approximation of E_{ex}. Thus, we put

$$E_P = E_F + E_0^c + E_R = E_{ex}. \tag{5.17}$$

For comparison, we again write down Eq. (5.6):

$$E_{ex} = E_B + E_B^c. \tag{5.18}$$

As the concluding piece of this argument, we introduce the approximation

$$E_0^c = E_B^c = E_c, \tag{5.19}$$

that is, we assume that the nonrelativistic correlation energy closely approximates the relativistic. We are compelled to introduce this hypothesis for the following reasons. At the present time, we have a general theory of electron correlation from the Thomas–Fermi model. This is a nonrelativistic theory. A relativistic theory we do not have. On the other hand, we will calculate the empirical value of the correlation energy from Eq. (5.18) by equating E_{ex} with the empirical total energy of the atom and E_B with the energy obtained from the HFD model. The correlation energy obtained from that equation is obviously the relativistic. However, this is available only for atoms up to $Z = 20$, because the empirical total energy is available only up to $Z = 20$. For heavier atoms, we must use the nonrelativistic correlation energy. Hence, the necessity for the approximation of Eq. (5.19). (See Section 5.2/B, which follows).

We sum up. The most important quantity for the evaluation of the accuracy of the HF model is the total energy that is being minimized in the derivation of the equations for the HF orbitals. We have two models, both relativistic, for approximating the empirical total energy. The HFD model, which is based on the Breit Hamiltonian operator, yields the formula

$$E_{ex} = E_B + E_c, \tag{5.20}$$

and the perturbation theory, which is based on the Breit–Pauli Hamiltonian, yields

$$E_{ex} = E_F + E_c + E_R. \tag{5.21}$$

In both equations, E_c is the correlation energy. This quantity, which we introduced for both the relativistic and nonrelativistic models, is assumed to satisfy the relationship

$$E_c = E_0^c = E_B^c, \tag{5.22}$$

that is, it is assumed that the relativistic and nonrelativistic correlation energies are equal.

If we assume, as we should, that the two-component Hamiltonian, Eq. (5.7), is a good representation of the Breit Hamiltonian, Eq. (5.2), then we can assume that the two formulas, Eqs. (5.20) and (5.21), will yield almost the same value for E_{ex}. If that is the case, then we obtain the relationship

$$E_B = E_F + E_R, \tag{5.23}$$

that is, the total energy obtained from the HFD model should be well approximated by the sum of the nonrelativistic HF energy E_F and the relativistic correction E_R. We will see that this is indeed the case.

Finally, we note that the eigenvalue of the nonrelativistic Schroedinger equation, Eq. (5.11), is not a directly measurable quantity because the empirical total energy always contains the relativistic effects. We can relate this eigenvalue, however, to the empirical E_{ex} by using Eqs. (5.13) and (5.21). From those equations, we obtain

$$E_0 = E_{ex} - E_R. \tag{5.24}$$

We are now ready to look at the numerical results for the physical quantities that we have listed at the beginning of the section. Before we do that, however, this is probably a good place to say a few words about the computational labor that is needed to obtain numerical results from the HF formalism.

As we have outlined before, the purpose of this book is to concentrate on the *theory* of atomic structure. Nevertheless, because in this chapter we are discussing numerical results, it will be appropriate to establish a "point of contact" with the computational aspects of the theory. From this discussion, the reader who is studying atomic structure theory for the first time will get some idea about the nature of the computational work involved in obtaining the results discussed in this chapter. We will not go into a detailed discussion of the computational work. However, for readers who are interested in such work, we give some references that can serve as starting points for a study of the computational aspects of the theory.

In Sec. 2.1, we described the ideas of the numerical method, called the self-consistent field, that Hartree designed for the solution of his equations. Let us recapitulate the main features of that method.

In the preceding chapters, we developed the HF model in several forms. We developed the nonrelativistic model for atoms with complete groups, and for atoms with incomplete groups in LS coupling; also, we developed the method for the average of a configuration. The relativistic model was developed for atoms with complete groups and for configuration averages. In all these formulations, the formalism has the same basic mathematical structure. In all formulations, we have a set of coupled integro-differential equations for the one-electron wave functions and one-electron orbital parameters. Hartree has shown that the solutions of these equations can be obtained by iteration; we described the main steps of one of the iteration cycles in Sec. 2.1. In each cycle of iteration, the one-electron wave functions are obtained by numerical integration, that is, the wave functions (more accurately: the radial parts of the wave functions) are obtained in the form of numerical tables. In such a table, we have the numerical values of the wave functions at a set of appropriately chosen radius values.

As we described in Sec. 2.1, the iteration process is considered completed when the initial and final potentials and/or wave functions are identical to a certain predetermined accuracy. The results of a HF calculation are the set of numerical tables for the wave functions and the set of values for the orbital parameters. By having determined the wave functions, the expectation value of any operator, including the total energy of the atom (which is not equal to the sum of orbital parameters), can be computed. Thus, when we talk about results obtained with the HF model, we are talking about making a HF calculation for an atom and then computing the expectation values of various operators with the HF one-electron wave functions.

The self-consistent-field method is laborious. In each cycle of iteration one must solve a set of integro-differential equations with a fixed potential. In the nonrelativistic formulation, we have *one* equation for each (nl) group. In the relativistic formulation, we have *two* equations for each (nlj) group; one equation is for the large component and the other for the small. In addition, in the relativistic formulation, we have a much larger set of groups: the (nl) group of the nonrelativistic model splits into two groups, the (nl) and the $(n\bar{l})$, in the relativistic formulation. Thus, in general, we have in each cycle, a number of equations to solve with the number of equations being much larger in the relativistic model than in the nonrelativistic. Therefore, the computational labor is large because of two main reasons: first, we have a large number of equations to solve in each cycle, and, second, the number of cycles may be large.

We hope that even such a short description will suffice to make the point that in the years following the introduction of the method by Hartree, that is, from about 1930 to 1960, when high-speed computers were not yet available, HF calculations for atoms were considered as major pieces of research. The development of fast computers with large memories in the 1960s enabled scientists to develop effective computational programs for HF calculations; beginning around 1970, HF calculations started to be made on a large scale, that is, calculations, including relativistic calculations, were made for a large number of atoms and ions.

As the starting points for the reader interested in the numerical aspects of the theory, we mention two books. Developments concerning the numerical aspects of the theory were summarized, up to 1958, in Hartree's book [Hartree, 63]. The modern methods, that is, methods involving large computers, are described by Froese-Fischer in her book.[31] Many of the HF calculations discussed in this chapter were made using the program designed by Froese-Fischer or by using similar programs. In connection with relativistic calculations, we mention the review article by Desclaux.[32]

We note that one of the main features of the HF calculations is that the results, the one-electron wave functions, are given in the form of numerical tables. A procedure in which the one-electron wave functions are given in the form of analytic expressions was developed by Roothaan. This method is discussed in Sec. 5.6.

5.2. TOTAL ENERGIES AND ELECTRON CONFIGURATIONS OF ATOMS

A. Comparison of the Hartree – Fock – Dirac Model with the Breit – Pauli Procedure

In the preceding section, we introduced a hierarchy of approximations through which we can approach the exact total energy of an atom, E_{ex}. We have seen that there are basically two relativistic models that can be used. One was the Hartree–Fock–Dirac model; the other was the perturbation approach called the Breit–Pauli procedure. In Eq. (5.23), we formulated the conjecture that these two models provide very nearly equal total energies. Our first task is to test the validity of this conjecture.

Calculations of various atomic quantities with the Breit–Pauli approach have been carried out for a large number of atoms and ions by Fraga and collaborators. The results of these calculations are conveniently tabulated in a book (Fraga, Karkowski, and Saxena, hereafter referred to as FKS; see the References). We recall that in these calculations, the nonrelativistic HF equations are solved and the relativistic corrections are computed by first-order perturbation theory using the operator in Eq. (5.9). The resultant energy we denote by $\hat{E}_R$, that is, we put

$$\hat{E}_R = E_F + E_R, \tag{5.25}$$

where E_F is the nonrelativistic HF energy, and E_R is the relativistic correction.

The total energies of atoms and other atomic properties were calculated by Desclaux[33] using the HFD model. The basis for these calculations was the Hamiltonian given by Eq. (5.4), that is, the retardation terms were omitted. The HFD equations solved were the ones given in Eqs. (4.207) and (4.208), that is, the HFD equations were derived in the absence of magnetic interactions. The magnetic interactions were then computed from first-order perturbation theory with the HFD wave functions serving as the unperturbed solutions. We denote these total energies by E_B. It is clear from the description that these calculations are roughly equivalent to a Breit–Pauli procedure in which the retardation terms are omitted.

The statement given in the preceding paragraph, that the equations solved by Desclaux were Eqs. (4.207) and (4.208), must be qualified. The equations that were actually used by Desclaux were slightly modified in order to achieve a better correspondence with the LS coupling. It was argued that in many atoms, the *LS* coupling is dominant. The HFD equations are based on *jj* coupling, therefore, they may not be the best approximations in some cases. In order to take this problem into account, Desclaux introduced a generalized average energy, E_{av}^G, for incomplete groups. We explain this with the example of two *p* electrons.

In *LS* coupling, the p^2 configuration may produce the multiplets ^3P, ^1D, and ^1S. What is meant by the average of the p^2 configuration is the weighted average of the three multiplets. In the HFD model, a *p* electron may be associated with the $j = l - \frac{1}{2} = \frac{1}{2}$ or the $j = l + \frac{1}{2} = \frac{3}{2}$ values. Thus, a *p* state may become a $p_{1/2}$ or a $p_{3/2}$ state. Thus, what is in LS coupling *one p^2* configuration, in *jj* coupling may become one of the following configurations: $(p_{1/2}p_{1/2})$, $(p_{1/2}p_{3/2})$, and $(p_{3/2}p_{3/2})$. The assumption that underlies the averaging in Sec. 4.3/C is that we select one of these configurations as the actual configuration of the two electrons and average over the energies of the possible two-electron states that can be formed for that configuration. Let us call this average E_{av}. This is the average that is built into Eqs. (4.207)–(4.208).

Desclaux modified the equations by replacing E_{av} with the E_{av}^G, where the latter is defined for the p^2 configuration as

$$E_{av}^G(p^2) = \left[E_{av}\left(p_{1/2}^2\right) + 8E_{av}(p_{1/2}p_{3/2}) + 6E_{av}\left(p_{3/2}^2\right)\right]/15. \quad (5.26)$$

In this equation, E_{av} is the conventional average for that configuration. The factors are equal to the degeneracy of the states associated with the configuration. For example, the $(p_{1/2}p_{3/2})$ configuration may generate $J_1 = 1$ and $J_1 = 2$. The degeneracy is $\Sigma(i)(2J_i + 1) = 8$. The denominator of Eq. (5.26) is the sum of the degeneracies. Therefore, although E_{av} is the average over the states that can be generated by a configuration, E_{av}^G is the average over the (jj) configurations, which can be formed from a single LS configuration.

We summarize the results of calculations in Table 5.1. Listed are the total energies of neutral atoms from $Z = 2$ to $Z = 92$. From left to right, we have the name of the atom, the atomic number, the relativistic energy, $\hat{E}_R$, computed by Fraga, Karkowski, and Saxena [FKS, 119], the relativistic energy E_B computed by Desclaux,[33] and the nonrelativistic HF energy [FKS, 136]. In the next two columns, we have the relativistic corrections in percents for both calculations. We have

$$\Delta \hat{E}_R = 100 \frac{\hat{E}_R - E_F}{\hat{E}_R}, \tag{5.27}$$

and

$$\Delta E_B = 100 \frac{E_B - E_F}{E_B}. \tag{5.28}$$

Thus, $\Delta \hat{E}_R$ shows how large the percentage of the total energy is the relativistic correction in the Breit–Pauli procedure and ΔE_B shows the same quantity for the HFD approximation.[†]

The numbers bear out our assumption that the two approaches lead essentially to the same results. We have the following main points. For small Z values, both $\Delta \hat{E}_R$ and ΔE_B are negligible. Both become 0.1% for Ne, $Z = 10$. From then on, the numbers are follows:

	$\Delta \hat{E}_R$ (%)	ΔE_B (%)
$Z = 20$	0.4	0.4
$Z = 40$	1.50	1.55
$Z = 50$	2.27	2.40
$Z = 70$	4.16	4.70
$Z = 92$	6.70	8.37

Thus, the difference between $\Delta \hat{E}_R$ and ΔE_B is negligible up to about $Z = 40$, small between $Z = 40$ and $Z = 50$, and becomes somewhat larger for values greater than $Z = 70$. However, the *difference* is still less than 1% at $Z = 70$. The difference becomes greater than 1% at about $Z = 82$ (lead).

Therefore, we conclude that the Breit–Pauli approach and the HFD model are leading essentially to the same total energies for neutral atoms; the conjecture that we formulated in Eq. (5.23) is proved to a high degree of accuracy. The difference is negligible up to $Z = 40$ and small even for heavy

[†]In this table, as well as in all subsequent tables, we have the absolute values of the energies. We should keep in mind that the total energies of atoms are always negative. The relativistic correction and the correlation energy are also negative; the magnetic energy is positive and the retardation effects are negative.

TABLE 5.1. The Total Energies of Neutral Atoms Calculated with the Relativistic Breit–Pauli Method by Fraga, Karkowski, and Saxena and with the Hartree–Fock–Dirac Formalism by Desclaux

Atom	Z	$\hat{E}_R$	E_B	E_F	$\Delta \hat{E}_R$	ΔE_B
He	2	2.86137	2.86175	2.86169	−0.01	0.002
Li	3	7.43269	7.43327	7.43273	−0.0005	0.007
Be	4	14.57434	14.5752	14.57303	0.009	0.01
B	5	24.53397	24.5350	24.52906	0.02	0.02
C	6	37.70068	37.6732	37.68866	0.03	−0.04
N	7	54.42602	54.3229	54.40098	0.05	−0.14
O	8	74.85626	74.8172	74.80947	0.06	0.01
F	9	99.48894	99.4897	99.40944	0.08	0.08
Ne	10	128.6732	128.674	128.5472	0.10	0.10
Na	11	162.0526	162.053	161.8591	0.12	0.12
Mg	12	199.9006	199.901	199.6145	0.14	0.14
Al	13	242.2862	242.286	241.8768	0.17	0.17
Si	14	289.4238	289.403	288.8544	0.2	0.19
P	15	341.4911	341.420	340.7188	0.23	0.21
S	16	398.5331	398.503	397.5050	0.26	0.25
Cl	17	460.8253	460.821	459.4822	0.29	0.29
Ar	18	528.5436	528.540	526.8178	0.33	0.325
K	19	601.3544	601.352	599.1648	0.36	0.36
Ca	20	679.4827	679.502	676.7580	0.40	0.40
Sc	21	763.1308	763.133	759.7359	0.44	0.445
Ti	22	852.5600	852.531	848.4059	0.49	0.48
V	23	947.9192	947.852	942.8846	0.53	0.52
Cr	24	1,049.358	1,049.21	1,043.310	0.58	0.56
Mn	25	1,157.073	1,156.87	1,149.866	0.62	0.60
Fe	26	1,270.974	1,270.88	1,262.444	0.67	0.66
Co	27	1,391.444	1,391.42	1,381.415	0.72	0.72
Ni	28	1,518.588	1,518.64	1,506.872	0.77	0.77
Cu	29	1,652.560	1,652.71	1,638.951	0.82	0.83
Zn	30	1,793.569	1,793.78	1,777.849	0.88	0.89
Ga	31	1,941.370	1,941.63	1,923.271	0.93	0.94
Ge	32	2,096.103	2,096.42	2,075.460	0.99	1.00
As	33	2,257.906	2,258.28	2,234.236	1.05	1.06
Se	34	2,426.779	2,427.30	2,399.866	1.10	1.10
Br	35	2,682.924	2,603.59	2,572.443	1.17	1.20
Kr	36	2,786.450	2,787.28	2,752.054	1.23	1.26
Rb	37	2,977.057	2,978.07	2,938.358	1.3	1.33
Sr	38	3,174.949	3,176.18	3,131.546	1.37	1.40
Y	39	3,380.203	3,381.68	3,331.685	1.43	1.48
Zr	40	3,593.070	3,594.81	3,538.995	1.50	1.55
Nb	41	3,813.653	3,815.67	3,753.554	1.57	1.63
Mo	42	4,042.061	4,044.45	3,975.444	1.65	1.71
Tc	43	4,278.442	4,281.19	4,202.790	1.72	1.78
Ru	44	4,522.737	4,526.11	4,441.489	1.80	1.87
Rh	45	4,775.224	4,779.23	4,685.802	1.87	1.95
Pd	46	5,035.984	5,040.71	4,937.783	1.95	2.04

TABLE 5.1. (*Continued*)

Atom	Z	$\hat{E}_R$	E_B	E_F	$\Delta\hat{E}_R$	ΔE_B
Ag	47	5,305.132	5,310.66	5,197.519	2.03	2.13
Cd	48	5,582.824	5,589.05	5,465.137	2.11	2.22
In	49	5,868.688	5,875.84	5,740.172	2.19	2.31
Sn	50	6,163.025	6,171.21	6,022.934	2.27	2.40
Sb	51	6,465.924	6,475.24	6,313.490	2.36	2.50
Te	52	6,777.393	6,788.06	6,611.785	2.44	2.60
I	53	7,097.617	7,109.76	6,917.986	2.53	2.70
Xe	54	7,426.674	7,440.46	7,232.141	2.62	2.80
Cs	55	7,764.312	7,779.91	7,553.935	2.71	2.90
Ba	56	8,110.739	8,128.34	7,883.547	2.80	3.01
La	57	8,465.959	8,485.87	8,221.068	2.89	3.12
Ce	58	8,830.527	8,852.82	8,566.928	2.98	3.22
Pr	59	9,204.545	9,229.40	8,921.188	3.07	3.34
Nd	60	9,588.095	9,615.86	9,283.888	3.17	3.45
Pm	61	9,981.295	10,012.3	9,655.106	3.27	3.57
Sm	62	10,384.30	10,481.8	10,034.96	3.36	3.68
Eu	63	10,797.24	10,835.5	10,423.55	3.46	3.80
Gd	64	11,219.98	11,262.6	10,820.63	3.56	3.92
Tb	65	11,652.88	11,700.3	11,226.58	3.66	4.05
Dy	66	12,096.04	12,148.7	11,641.45	3.76	4.17
Ho	67	12,549.56	12,607.80	12,065.30	3.86	4.30
Er	68	13,013.53	13,078.0	12,498.16	3.96	4.43
Tm	69	13,488.13	13,559.3	12,940.18	4.06	4.56
Yb	70	13,973.51	14,051.9	13,391.46	4.16	4.70
Lu	71	14,469.75	14,555.9	13,851.82	4.27	4.84
Hf	72	14,976.72	15,071.3	14,321.25	4.38	4.98
Ta	73	15,494.51	15,598.3	14,799.79	4.48	5.12
W	74	16,023.28	16,136.9	15,287.55	4.59	5.26
Re	75	16,563.07	16,687.4	15,784.54	4.70	5.41
Os	76	17,113.89	17,249.9	16,290.65	4.81	5.56
Ir	77	17,675.99	17,824.6	16,806.11	4.92	5.71
Pt	78	18,249.44	18,411.7	17,330.94	5.03	5.87
Au	79	18,834.34	19,011.3	17,865.22	5.14	6.03
Hg	80	19,430.84	19,623.5	18,409.01	5.29	6.19
Tl	81	20,038.65	20,248.3	18,961.83	5.37	6.35
Pb	82	20,658.07	20,886.0	19,524.02	5.49	6.52
Bi	83	21,289.14	21,536.7	20,095.60	5.61	6.69
Po	84	21,931.95	22,200.7	20,676.52	5.72	6.86
At	85	22,586.61	22,878.2	21,266.88	5.84	7.04
Rn	86	23,253.25	23,569.1	21,866.77	5.96	7.22
Fr	87	23,931.70	24,237.8	22,475.88	6.08	7.27
Ra	88	24,622.10	24,992.3	23,094.32	6.20	7.59
Ac	89	25,324.40	25,724.9	23,722.18	6.33	7.78
Th	90	26,039.12	26,471.9	24,359.78	6.45	7.98
Pa	91	26,766.43	27,233.7	25,007.23	6.57	8.17
U	92	27,506.37	28,010.5	25,664.48	6.70	8.37

atoms. Maximum deviation is of the order of 1% of the total energy. Therefore, it is accurate to say that both models give the total relativistic HF energy within an accuracy of 1%, and at a much higher accuracy for low Z.

Because the two different models lead approximately to the same results, a question now arises: Which model should be used in the evaluation of the HF theory? In this chapter, we will use the results of the two models interchangeably, that is, we will consider the two models as being of equal accuracy. The Breit–Pauli approach will be used in a large number of cases, not because it is more accurate than the HFD model, but because results obtained with this model are available in a larger number of cases and are tabulated in a convenient fashion in the book of Fraga, Karkowski, and Saxena.

B. Determination of Correlation Energies

We turn now to the concept of the correlation energies, which we defined for the relativistic case by Eq. (5.6) and for the nonrelativistic by Eq. (5.13). As we have seen, this quantity characterizes the deviation of the HF total energies from the eigenvalues of the corresponding Hamiltonian operators.

For the purposes of this discussion, we would like to have a general theory of the electron correlation that would enable us to calculate accurately the correlation energy for any atom or ion. Such a theory was developed by Gombas[34] using the Thomas–Fermi model. According to that theory, the correlation energy is given by the formula

$$E_c = -\alpha N \quad \text{(a.u.)}, \tag{5.29}$$

where N is the number of electrons in the atom. For the constant α, Gombas derived the value

$$\alpha = 0.056. \tag{5.30}$$

This is what we have called the nonrelativistic correlation energy.

How accurate is the formula in Eq. (5.29)? In order to answer this question, we must define the empirical value of the correlation energy. We can use the relationship given by Eq. (5.6) for this purpose. According to that, we have

$$E_B^c = E_{ex} - E_B, \tag{5.31}$$

where E_{ex} is the empirical value of the total energy, and E_B is the energy obtained from the HFD model. E_B^c is the relativistic correlation energy, and we can compute this quantity if E_{ex} and E_B are available.

Here we are confronted by two difficulties. First, the empirical total energy of atoms is known only up to $Z = 20$. Second, the quantity computed from this equation is the relativistic correlation energy, whereas Eq. (5.29) gives the nonrelativistic. Thus, the two equations, Eq. (5.29) and Eq. (5.31), do not, a priori, describe the same quantity and, in addition, the correlation energy computed from Eq. (5.31) is available only up to $Z = 20$.

The best procedure for handling this problem appears to be the following. First, we will use Eq. (5.31) to define the empirical value of the relativistic correlation energy for atoms up to $Z = 20$. Then, in order to avoid insurmountable difficulties, we will adopt the approximation that we have already written down in Eq. (5.22), according to which the relativistic and nonrelativistic correlation energies are equal. Thus, we will assume that the correlation energy computed from Eq. (5.31) is the same as the correlation energy computed from Eq. (5.29).

Comparison of the values obtained from Eq. (5.29) with the empirical values shows that Eq. (5.29) gives poor values with the α determined from the Thomas–Fermi model. It was shown by Szasz, Berrios-Pagan, and McGinn,[35] however, that the formula in Eq. (5.29) describes accurately the correlation energy if α is adjusted properly to the empirical values. Thus, we have a situation not uncommon for the Thomas–Fermi model, in which the model provides the accurate functional dependence of a quantity, but gives a poor approximation for the values of the parameters.

Therefore, we determine the correlation energy for all atoms and ions by assuming that Eq. (5.29) is valid. We determine the value of α by adjusting the E_c curve (in reality, a straight line) to the empirical. Figure 5.1 shows the situation for neutral atoms. The abscissa is the number of electrons up to $N = 20$. The ordinate is $|E_c|$. The solid line connects the points that designate the empirical values of $|E_c|$. The dashed line is from our formula, Eq. (5.29), and we adjusted α in such a way that the straight line gives a good approximation to the empirical points. We have carried out this procedure for neutral atoms and for singly, doubly, and triply ionized atoms, that is, for $q = 0, 1, 2, 3$, where $q = Z - N$. The results for α are as follows:

$q = Z - N$	α
0	0.03197
1	0.03175
2	0.03133
3	0.03072

In these calculations, which are the extensions of the earlier work,[35] the total energies are the sum of all ionization potentials, that is,

$$E_{ex} = - \sum_{i=1}^{N} I_i, \tag{5.32}$$

where I_1 is the first ionization potential of the neutral atom, I_2 is the second ionization potential, and so on. The summation is for all electrons in the

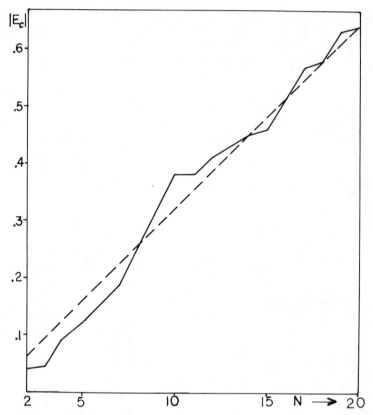

Fig. 5.1. Correlation energies of neutral atoms from $Z = 2$ to $Z = 20$. The full zigzag line connects the empirical values. The dashed line is given by Eq. (5.29) with $\alpha = 0.03197$.

atom, that is, the total energy is equal to the energy needed to remove all electrons. The values of the I_k's are taken from the list given by Moore[36] and it is only up to $N = 20$ that all ionization potentials are available, that is, we can construct E_{ex} only up to $N = 20$.

Table 5.2 contains the data underlying the diagram in Fig. 5.1. In this table, which is for neutral atoms, that is, for $q = 0$, we have, from left to right, the atom; the atomic number, which is also the number of electrons; the empirical total energy, E_x; and the relativistic total energy, $\hat{E}_R$. The empirical total energy is computed from Eq. (5.32), that is, this quantity is equal to the sum of all ionization potentials. The relativistic energy is obtained from the formula of Eq. (5.25) and we used the data supplied by Fraga, Karkowski, and Saxena [FKS, 119]. These relativistic energies are based on the Breit–Pauli procedure, whereas in Eq. (5.31), we had the total energy computed with the HFD model; however, as we have stated in Sec.

TABLE 5.2. Comparison of the Empirical Values of the Correlation Energy with the Values Obtained from Eq. (5.29)

Atom	Z	$E_{ex.}$	$\hat{E}_R$	$E_C(\text{exp.})$	E_C	ΔE_C
He	2	2.9034546	2.861376	0.04207	0.06394	−51.9
Li	3	7.4781698	7.432698	0.04547	0.09591	−110.9
Be	4	14.668835	14.57434	0.09449	0.12788	−35.3
B	5	24.658838	24.53397	0.12487	0.15985	−28.0
C	6	37.85667	37.70068	0.15598	0.19182	−22.9
N	7	54.61334	54.42602	0.18732	0.22379	−19.5
O	8	75.11187	74.85626	0.25560	0.25576	−0.06
F	9	99.80871	99.48894	0.31977	0.28773	10.0
Ne	10	129.05355	128.6732	0.38035	0.31970	15.9
Na	11	162.43238	162.0526	0.37977	0.35167	7.4
Mg	12	200.31455	199.9006	0.41395	0.38364	7.3
Al	13	242.71827	242.2862	0.43206	0.41561	3.8
Si	14	289.87552	289.4238	0.45172	0.44758	0.9
P	15	341.95549	341.4911	0.46439	0.47955	−3.2
S	16	399.04572	398.5331	0.51262	0.51152	0.2
Cl	17	461.39243	460.8253	0.56713	0.54349	4.2
Ar	18	529.12576	528.5436	0.58216	0.57546	1.1
K	19	601.98295	601.3544	0.62855	0.60743	3.4
Ca	20	680.12183	679.4827	0.63913	0.63940	−0.04

5.2/A, we will always assume that the two models are interchangeable and we can substitute $\hat{E}_R$ for E_B. The fifth column contains the empirical value of E_c, which we denote by $E_c(\text{exp})$, and the sixth column is the correlation energy computed from Eq. (5.29) with the adjusted α value. The last column gives the deviation

$$\Delta E_c = 100 \frac{E_c(\text{exp}) - E_c}{E_c(\text{exp})}, \tag{5.33}$$

in percents. As we see, the deviation of the analytic formula from the empirical values is fairly large for small Z's, where the fluctuations in the empirical curve are large. Beginning with Na ($Z = 11$), however, we have satisfactorily small deviations.

Having determined the α values, we now have a formula for the correlation energy for all N values with $q = 0, 1, 2$, and 3. We have seen that these curves are constructed in such a way that they give a good approximation to the empirical value for $N \leq 20$. For heavier atoms, the accuracy of this formula is not known. Our attitude will be that the formula is probably very accurate for all N. We base this assumption on two facts. The first fact is that the Thomas–Fermi model generally describes accurately the (N, Z) dependence of the physical quantities, that is, we can assume that the N dependence of E_c is correctly described by Eq. (5.29). The second fact is that the calculations of Mann and Waber[37] for the lanthanide atoms led to the formula given by Eq. (5.29). This is discussed in the next section.

C. The Total Energies of Atoms and Ions

Having constructed semitheoretical formulas for the correlation energy in the preceding section, we are now ready to look at the total energies of atoms and ions. The results will be presented for neutral atoms ($q = 0$). First, we present the results obtained with relativistic calculations, and after that, we discuss the nonrelativistic results.

Table 5.3 contains the relativistic results for neutral atoms. Because the meaning of the table is different for $N \leq 20$ and for $N > 20$, we divided it by a line after $N = 20$. First, we discuss the data for $N \leq 20$.

Going from left to right in the table, we have the atom, the atomic number, and the empirical values of the total energy computed from Eq. (5.32). Next comes the relativistic total energy, computed, as all relativistic energies in this section with the exception of the lanthanide calculations of Mann and Waber, with the Breit–Pauli procedure by Fraga, Karkowski, and Saxena [FKS, 119]. The next column is the empirical correlation energy computed from the formula

$$E_c = E_{ex} - \hat{E}_R,\qquad(5.34a)$$

where E_{ex} is the empirical total energy, and $\hat{E}_R$ is the relativistic total energy. The next column gives the correlation energy in percents, that is,

$$\Delta E_c = 100\frac{E_{ex} - \hat{E}_R}{E_{ex}}.\qquad(5.34b)$$

The next column is the relativistic correction, that is,

$$E_R = \hat{E}_R - E_F,\qquad(5.35)$$

where E_F is the nonrelativistic HF energy. We had E_F in Table 5.1 and did not put it into this table. The values of E_F are taken from the tables of Fraga, Karkowski, and Saxena [FKS, 139]. The last column gives the relativistic correction in percents, that is,

$$\Delta E_R = 100\frac{\hat{E}_R - E_F}{\hat{E}_R}.\qquad(5.36)$$

TABLE 5.3. The Results of the Relativistic HF Calculations for the Total Energies of Neutral Atoms from He to U

Atom	Z	$E_{ex.}$	$\hat{E}_R$	E_C	ΔE_C	E_R	ΔE_R
He	2	2.9034546	2.861376	0.0420786	1.45	−0.000321	−0.01
Li	3	7.4781698	7.432698	0.045471	0.61	0.000031	−0.0004
Be	4	14.668835	14.57434	0.094494	0.64	0.000131	0.0009
B	5	24.658838	24.53397	0.124867	0.51	0.00049	0.02
C	6	37.85667	37.70068	0.155989	0.41	0.012020	0.03
N	7	54.61334	54.42602	0.187319	0.34	0.025039	0.05
O	8	75.11187	74.85626	0.255609	0.34	0.046790	0.06
F	9	99.808710	99.48894	0.319770	0.32	0.079500	0.08
Ne	10	129.05355	128.6732	0.380349	0.29	0.125999	0.10
Na	11	162.43238	162.0526	0.379779	0.23	0.193500	0.12
Mg	12	200.31455	199.9006	0.413950	0.21	0.286100	0.14
Al	13	242.71827	242.2862	0.432069	0.18	0.409400	0.17
Si	14	289.87552	289.4238	0.451719	0.15	0.569400	0.20
P	15	341.95549	341.4911	0.464389	0.14	0.772300	0.23
S	16	399.04572	398.5331	0.512620	0.13	1.02809	0.26
Cl	17	461.39243	460.8253	0.567130	0.12	1.34309	0.29
Ar	18	529.12576	528.5436	0.582159	0.11	1.72580	0.32
K	19	601.98295	601.3544	0.628550	0.10	2.18959	0.36
Ca	20	680.12183	679.4827	0.639130	0.09	2.72470	0.40
Sc	21	763.80217	763.1308	0.67137	0.088	3.39490	0.44
Ti	22	853.26334	852.5600	0.70334	0.082	4.15410	0.49
V	23	948.65451	947.9192	0.73531	0.077	5.03459	0.53
Cr	24	1,050.1252	1,049.358	0.76728	0.073	6.04799	0.58
Mn	25	1,157.8722	1,157.073	0.79925	0.069	7.20699	0.62
Fe	26	1,271.8052	1,270.974	0.83122	0.065	8.53000	0.67
Co	27	1,392.3071	1,391.440	0.86319	0.061	10.0290	0.72
Ni	28	1,519.4831	1,518.588	0.89516	0.059	11.7159	0.77
Cu	29	1,653.4871	1,652.560	0.92713	0.056	13.6090	0.82
Zn	30	1,794.5281	1,793.569	0.95910	0.053	15.7199	0.88
Ga	31	1,942.3610	1,941.370	0.99107	0.051	18.0990	0.93
Ge	32	2,097.1260	2,096.103	1.02304	0.049	20.7430	0.99
As	33	2,258.9610	2,257.906	1.05501	0.047	23.6700	1.04
Se	34	2,427.8659	2,426.779	1.08698	0.044	26.9130	1.11
Br	35	2,604.0429	2,602.924	1.11895	0.043	30.4810	1.17
Kr	36	2,787.6009	2,786.450	1.15092	0.041	34.3959	1.23
Rb	37	2,978.2398	2,977.057	1.18289	0.039	38.6989	1.30
Sr	38	3,176.1638	3,174.949	1.21486	0.038	43.4030	1.37
Y	39	3,381.4498	3,380.203	1.24683	0.037	48.5179	1.43
Zr	40	3,594.3488	3,593.070	1.27880	0.036	54.0749	1.50
Nb	41	3,814.9637	3,813.653	1.31077	0.034	60.0989	1.57
Mo	42	4,043.4037	4,042.061	1.34274	0.033	66.6169	1.64
Tc	43	4,279.8167	4,278.442	1.37471	0.032	73.6520	1.72
Ru	44	4,524.1436	4,522.737	1.40668	0.031	81.2479	1.79
Rh	45	4,776.6626	4,775.224	1.43865	0.030	89.4220	1.87
Pd	46	5,037.4546	5,035.984	1.47062	0.029	98.2009	1.95

TABLE 5.3. (*Continued*)

Atom	Z	$E_{ex.}$	$\hat{E}_R$	E_C	ΔE_C	E_R	ΔE_R
Ag	47	5,306.6345	5,305.132	1.50259	0.028	107.6130	2.03
Cd	48	5,584.3585	5,582.824	1.53456	0.027	117.6870	2.11
In	49	5,870.2545	5,868.688	1.56653	0.027	128.5160	2.20
Sn	50	6,164.6235	6,163.025	1.59850	0.026	140.0910	2.27
Sb	51	6,467.5544	6,465.924	1.63047	0.025	152.434	2.36
Te	52	6,779.0554	6,777.393	1.66244	0.024	165.608	2.44
I	53	7,099.3114	7,097.617	1.69441	0.024	179.631	2.53
Xe	54	7,428.4003	7,426.674	1.72638	0.023	194.533	2.62
Cs	55	7,766.0703	7,764.312	1.75835	0.023	210.377	2.71
Ba	56	8,112.5293	8,110.739	1.79032	0.022	227.192	2.80
La	57	8,467.7812	8,465.959	1.82229	0.021	244.891	2.89
Ce	58	8,832.3812	8,830.527	1.85426	0.021	263.599	2.98
Pr	59	9,206.4312	9,204.545	1.88623	0.020	283.357	3.08
Nd	60	9,590.0132	9,588.095	1.91820	0.020	304.207	3.17
Pm	61	9,983.2451	9,981.295	1.95017	0.019	326.189	3.27
Sm	62	10,386.282	10,384.30	1.98214	0.019	349.340	3.36
Eu	63	10,799.254	10,797.24	2.01411	0.019	373.690	3.46
Gd	64	11,222.026	11,219.98	2.04608	0.018	399.350	3.56
Tb	65	11,654.958	11,652.88	2.07805	0.018	426.299	3.66
Dy	66	12,098.15	12,096.04	2.11002	0.017	454.590	3.76
Ho	67	12,551.702	12,549.56	2.14199	0.017	484.260	3.86
Er	68	13,015.704	13,013.53	2.17396	0.017	515.370	3.96
Tm	69	13,490.335	13,488.13	2.20593	0.016	547.950	4.06
Yb	70	13,975.747	13,973.51	2.23790	0.016	582.050	4.16
Lu	71	14,472.019	14,469.75	2.26987	0.016	617.930	4.27
Hf	72	14,979.021	14,976.72	2.30184	0.015	655.470	4.38
Ta	73	15,496.843	15,494.51	2.33381	0.015	694.720	4.48
W	74	16,025.645	16,023.28	2.36578	0.015	735.730	4.59
Re	75	16,565.467	16,563.07	2.39775	0.014	778.530	4.70
Os	76	17,116.319	17,113.89	2.42972	0.014	823.240	4.81
Ir	77	17,678.451	17,675.99	2.46169	0.014	869.880	4.92
Pt	78	18,251.933	18,249.44	2.49366	0.014	918.500	5.03
Au	79	18,836.865	18,834.34	2.52563	0.013	969.120	5.14
Hg	80	19,433.397	19,430.84	2.55760	0.013	1,021.83	5.26
Tl	81	20,041.239	20,038.65	2.58957	0.013	1,076.82	5.37
Pb	82	20,660.691	20,658.07	2.62154	0.013	1,134.05	5.50
Bi	83	21,291.793	21,289.14	2.65351	0.012	1,193.54	5.61
Po	84	21,934.635	21,931.95	2.68548	0.012	1,255.43	5.72
At	85	22,589.327	22,586.61	2.71745	0.012	1,319.73	5.84
Rn	86	23,255.999	23,253.25	2.74942	0.012	1,386.48	5.96
Fr	87	23,934.481	23,931.70	2.78139	0.012	1,455.82	6.08
Ra	88	24,624.913	24,622.10	2.81336	0.011	1,527.78	6.20
Ac	89	25,327.245	25,324.40	2.82533	0.011	1,602.22	6.33
Th	90	26,041.997	26,039.12	2.87730	0.011	1,679.34	6.45
Pa	91	26,769.339	26,766.43	2.90927	0.011	1,759.20	6.57
U	92	27,509.311	27,506.37	2.94124	0.011	1,841.89	6.70

Let us look at the columns containing the correlation energy and the relativistic energy in percents. The correlation energy is the largest for He and decreases steadily as we go to higher N values. With the relativistic energy, it is the opposite: it is very small for He and increases steadily with increasing N. Thus, the accuracy of the relativistic HF calculations increases with increasing N because this accuracy can be measured by the decline in correlation energy.

Looking at the percent values of the correlation energy, we see that it ranges from 0.6% for Li to 0.1% for Ca. Because in this range we have the empirical total energies for comparison, these values for the correlation energies give us an accurate gauge for the accuracy of the relativistic HF theory. We have seen also that for this range, the results of the HFD calculations and the results of the Breit–Pauli calculations are practically identical. Thus, we can say that, for neutral atoms: In the range $N = 3$ to $N = 20$, the relativistic HF theory reproduces the empirical values of total energies with high accuracy.

We turn now to the atoms with $N > 20$. The meanings of the columns are the same as before with the exception of the experimental energy and the correlation energy. For these atoms, we do not know the empirical binding energy, and, consequently, we do not know the empirical correlation energy. Thus, for this range, the column headed by E_c contains the correlation energy computed from our formula, Eq. (5.29), with the α adjusted to the empirical values in the range $N \leqq 20$. In the column headed by E_{ex}, we have now

$$E_{ex} = \hat{E}_R + E_c, \tag{5.37}$$

that is, we *estimate* the empirical energies by using our semitheoretical formula for the correlation energy.

Let us now take a look at the table in its entirety. The correlation energy declines from 0.6% for Li ($Z = 3$) to 0.01% for U ($Z = 92$). The relativistic energy increases with increasing N from negligible for Li to 6.7% for U. For almost all N values, except for very small N, the relativistic energy is much larger than the correlation energy. Thus, the accuracy of the relativistic HF model depends almost entirely on how accurately the relativistic correction is given by the model.

As we see from the table, we have constructed estimates for the total energies of atoms with $N > 20$. These estimates are given by the formula

$$E_{ex} = E_F + E_R + E_c, \tag{5.38}$$

where E_F is the nonrelativistic HF energy, E_R is the relativistic correction, and E_c is computed from Eq. (5.29) with α adjusted to the empirical values for $N \leq 20$. The crucial question, defining the quality of the relativistic HF model, is now: How accurate are these estimates?

We are able to give a clear-cut answer. In Eq. (5.38), the nonrelativistic HF energy, E_F, is an arbitrary but uniquely defined reference point. The accuracy of E_{ex} will depend on the accuracy of E_R and E_c. We have seen in Sec. 5.2/A that the uncertainty in the relativistic total energy, that is, in $\hat{E}_R = E_F + E_R$, can be reliably estimated as not more than about 1% with much less for most atoms. We do not have a similar figure for the correlation energy. We have presented arguments in favor of the accuracy of Eq. (5.29), but we have no figures to estimate the error in E_c. Fortunately, for a judgement on the accuracy of Eq. (5.38), we do not really need a very accurate formula for E_c. We see from the table that for most atoms, E_R is much larger than E_c, in many cases larger by a magnitude. Thus, even if our formula, Eq. (5.29), would be in error by a factor of 2, this would not appreciably upset the argument, which boils down to the relativistic energy being much more important than the correlation energy for most atoms.

The preceding, of course, is not valid for $N \leq 20$, but in that range, we did not use Eq. (5.29). For $N \leq 20$, we have the empirical total energies and with that the empirical correlation energies.

Having concluded that in the $N > 20$ range, the relativistic energy is more important than the correlation energy, we can say that the estimates for the total energy are probably in error that is not greater than 1% and in fact much less for most atoms. This 1% is our estimate for the maximum difference between the two available major relativistic calculations that we have analyzed in Sec. 5.2/A. Thus, we conclude that: The relativistic HF model reproduces the total energies of neutral atoms with high accuracy. The maximum error in the calculated values is estimated at 1%; it is probably much less for many atoms.

With the data supplied by Fraga, Karkowski, and Saxena, it is easy to draw up similar tables for atoms with any degree of ionization. For $N > 20$, Eq. (5.29) can be used for the correlation energies. We have, in fact, constructed such tables for $q = 1$, 2, and 3, with the α values mentioned in Sec. 5.2/B. There is no need, however, to include those tables here because the conclusions that can be drawn from them are exactly the same as the conclusions drawn from Table 5.3.

Thus, now we can register the first major piece of information about the structure of atoms that we have obtained from the HF model. We have obtained reliable estimates for the total energies of neutral atoms between $Z = 2$ and $Z = 92$ and we have obtained similar results for positive ions with $q = 1$, 2, and 3 (tables not shown).

We turn now to the examination of nonrelativistic results. These results cannot be directly compared with empirical data because the empirical data always contain the relativistic effects. The significance of the nonrelativistic results is twofold. First, we need the nonrelativistic HF energies, E_F, as the reference point in Eq. (5.38). Second, the optical spectra, that is, the spectra generated by valence electrons, are in most cases dominated by LS coupling. Therefore, for the theory of valence electrons, which will be presented in

Part Two, the nonrelativistic HF model is very important. Thus, it is important to examine the results obtained with this model for the total energies of atoms.

We present our results for neutral atoms in Table 5.4. In this table, we have, from left to right, the atom; the atomic number; the nonrelativistic HF energy, E_F; and the correlation energy, E_c. For atoms with $N \leq 20$, the correlation energy is the empirical; for atoms with $N > 20$, the correlation energy is computed from Eq. (5.29). The next column in the table contains

$$E_0 = E_F + E_c, \tag{5.39}$$

that is, this quantity is our *estimate* for the eigenvalue of the exact nonrelativistic Hamiltonian given by Eq. (5.8). The quantity E_0 should be a good approximation to the solution of Eq. (5.11). The last column gives the correlation energy in percent, that is,

$$\Delta E_c = 100 \frac{E_0 - E_F}{E_0}. \tag{5.40}$$

The results lend themselves to a clear interpretation. Let us take a look at Table 5.4. We omit the He atom from the argument since that is in a class by itself. From Li to U, the correlation energy increases steadily as prescribed by the linear dependence on N, Eq. (5.29). Nevertheless, the percentage of the total energy that makes up the correlation energy decreases strongly as we go from Li to U. In fact, the correlation energy is 0.6% for Li ($Z = 3$), 0.1% for Ca ($Z = 20$), 0.04% for Zr ($Z = 40$), 0.02% for Yb ($Z = 70$), and 0.01% for U ($Z = 92$). Thus, the correlation energy is less than 0.1% for atoms heavier than Ca ($Z = 20$) and less than 0.04% for atom heavier than Zr ($Z = 40$).

Now we ask the question: How accurate are our estimates for E_0? Obviously, the accuracy depends on the accuracy of Eq. (5.29) for E_c. As we have seen, we do not have a reliable estimate for the accuracy of the correlation-energy formula. Fortunately, the smallness of the correlation energy relative to the total energy enables us to make a rough calculation as follows. In a "bad-case scenario" in which the formula for E_c would be in error by a factor of 2, we would have the previous percentages roughly doubled. That would mean that our results for E_0 would still be accurate within 0.2% for atoms heavier than calcium and ($Z = 20$). This can be considered high accuracy. In fact, our formula for the correlation energy is probably not as inaccurate as that and the error in E_0 is not more than 0.10–0.15% for atoms heavier than calcium. Thus: The nonrelativistic HF model reproduces the total energies of neutral atoms with high accuracy. The error in the results for E_0 is estimated to be not more than 0.2% for atoms heavier than calcium and probably smaller for many atoms.

TABLE 5.4. The Results of the Nonrelativistic HF Calculations for the Total Energies of Neutral Atoms from He to U

Atom	Z	E_F	E_C	E_0	ΔE_C
He	2	2.861697	0.042078	2.90377	1.449
Li	3	7.432729	0.045471	7.47820	0.608
Be	4	14.57303	0.094495	14.66752	0.644
B	5	24.52906	0.124868	24.65393	0.506
C	6	37.68866	0.155989	37.84465	0.412
N	7	54.40098	0.187319	54.58830	0.343
O	8	74.80947	0.255609	75.06508	0.340
F	9	99.40944	0.319770	99.72921	0.321
Ne	10	128.5472	0.380349	128.92755	0.295
Na	11	161.8591	0.379779	162.23888	0.234
Mg	12	199.6145	0.413950	200.02845	0.207
Al	13	241.8768	0.432069	242.30887	0.178
Si	14	288.8544	0.451719	289.30612	0.156
P	15	340.7188	0.464389	341.18319	0.136
S	16	397.5050	0.512619	398.01762	0.129
Cl	17	459.4822	0.567130	460.04933	0.123
A	18	526.8178	0.582159	527.39996	0.110
K	19	599.1648	0.628550	599.79335	0.105
Ca	20	676.7580	0.639130	677.39713	0.0943
Sc	21	759.7359	0.67137	760.40727	0.0882
Ti	22	848.4059	0.70334	849.10924	0.0828
V	23	942.8846	0.73531	943.61991	0.0779
Cr	24	1,043.310	0.76728	1,044.0772	0.0735
Mn	25	1,149.866	0.79925	1,150.6652	0.0694
Fe	26	1,262.444	0.83122	1,263.2752	0.0658
Co	27	1,381.415	0.86319	1,382.2781	0.0624
Ni	28	1,506.872	0.89516	1,507.7671	0.0594
Cu	29	1,638.951	0.92713	1,639.8781	0.0565
Zn	30	1,777.849	0.95910	1,778.8081	0.0539
Ga	31	1,923.271	0.99107	1,924.2620	0.0515
Ge	32	2,075.36	1.02304	2,076.3830	0.0493
As	33	2,234.236	1.05501	2,235.2910	0.0472
Se	34	2,399.866	1.08698	2,400.9529	0.0453
Br	35	2,572.443	1.11895	2,573.5619	0.0434
Kr	36	2,752.054	1.15092	2,753.2049	0.0418
Rb	37	2,938.358	1.18289	2,939.5408	0.0402
Sr	38	3,131.546	1.21486	3,132.7608	0.0388
Y	39	3,331.685	1.24683	3,332.9318	0.0374
Zr	40	3,538.995	1.27880	3,540.2738	0.0361
Nb	41	3,753.554	1.31077	3,754.8647	0.0349
Mo	42	3,975.444	1.34274	3,976.7867	0.0337
Tc	43	4,204.790	1.37471	4,206.1647	0.0327
Ru	44	4,441.489	1.40668	4,442.8956	0.0317
Rh	45	4,685.802	1.43865	4,687.2406	0.0307
Pd	46	4,937.783	1.47062	4,939.2536	0.0298

TABLE 5.4. (*Continued*)

Atom	Z	E_F	E_C	E_0	ΔE_C
Ag	47	5,197.519	1.50259	5,199.0215	0.0289
Cd	48	5,465.137	1.53456	5,466.6715	0.0281
In	49	5,740.172	1.56653	5,741.7385	0.0273
Sn	50	6,022.934	1.59850	6,024.5325	0.0265
Sb	51	6,313.490	1.63047	6,315.1204	0.0258
Te	52	6,611.785	1.66244	6,613.4474	0.0251
I	53	6,917.986	1.69441	6,919.6804	0.0244
Xe	54	7,232.141	1.72638	7,233.8673	0.0238
Cs	55	7,553.935	1.75835	7,555.6933	0.0233
Ba	56	7,883.547	1.79032	7,885.3373	0.0227
La	57	8,221.068	1.82229	8,222.8902	0.0221
Ce	58	8,566.928	1.85426	8,568.7822	0.0216
Pr	59	8,921.188	1.88623	8,923.0742	0.0211
Nd	60	9,283.888	1.91820	9,285.8062	0.0206
Pm	61	9,655.106	1.95017	9,657.0561	0.0202
Sm	62	10,034.96	1.98214	10,036.942	0.0197
Eu	63	10,423.55	2.01411	10,425.564	0.0193
Gd	64	10,820.63	2.04608	10,822.676	0.0189
Tb	65	11,226.58	2.07805	11,228.658	0.0185
Dy	66	11,641.45	2.11002	11,643.56	0.0181
Ho	67	12,065.30	2.14199	12,067.442	0.0177
Er	68	12,498.16	2.17396	12,500.334	0.0174
Tm	69	12,940.18	2.20593	12,942.385	0.0170
Yb	70	13,391.46	2.23790	13,393.697	0.0167
Lu	71	13,851.82	2.26987	13,854.089	0.0164
Hf	72	14,321.25	2.30184	14,323.551	0.0161
Ta	73	14,799.79	2.33381	14,802.123	0.0158
W	74	15,287.55	2.36578	15,289.915	0.0154
Re	75	15,784.54	2.39775	15,786.937	0.0152
Os	76	16,290.65	2.42972	16,293.079	0.0149
Ir	77	16,806.11	2.46169	16,808.571	0.0146
Pt	78	17,330.94	2.49366	17,333.433	0.0144
Au	79	17,865.22	2.52563	17,867.745	0.0141
Hg	80	18,409.01	2.55760	18,411.567	0.0139
Tl	81	18,961.83	2.58957	18,964.419	0.0136
Pb	82	19,524.02	2.62154	19,526.641	0.0134
Bi	83	20,095.6	2.65351	20,098.253	0.0132
Po	84	20,676.52	2.68548	20,679.205	0.0130
At	85	21,266.88	2.71745	21,269.597	0.0128
Rn	86	21,866.77	2.74942	21,869.519	0.0126
Fr	87	22,475.88	2.78139	22,478.661	0.0124
Ra	88	23,094.32	2.81336	23,097.133	0.0121
Ac	89	23,722.18	2.84533	23,725.025	0.0120
Th	90	24,359.78	2.87730	24,362.657	0.0118
Pa	91	25,007.23	2.90927	25,010.139	0.0116
U	92	25,664.48	2.94124	25,667.421	0.0114

With the data supplied by FKS, it is easy to draw up similar tables for positive ions. We have constructed tables for atoms with $q = 1, 2$, and 3. There is no need to include these tables because the conclusions that can be drawn from them are the same as the conclusions drawn from the table for neutral atoms.

Finally, we would like to discuss in this section the calculations of Mann and Waber for lanthanide atoms. Mann and Waber carried out relativistic HFD calculations for the lanthanide atoms, that is, for atoms between La ($Z = 57$) and Yb ($Z = 70$). The equations that were solved were Eqs. (4.168) and (4.169). These are the HFD equations without magnetic and retardation terms, that is, this is an approach similar to the method used by Desclaux, and the underlying Hamiltonian is given by Eq. (5.5). In addition, Mann and Waber computed the magnetic and retardation energies using first-order perturbation theory. An interesting feature is the calculation of correlation energy using a formula that is different from ours, but, like ours, was derived from the Thomas–Fermi model. It was shown that the computed correlation can be approximated by the formula

$$E_c^{MW} = -0.0425 \, N \ \text{(a.u.)}. \tag{5.41}$$

There are two points to be noted here. First, Mann and Waber were able to approximate the computed values of the correlation energy by the same formula that was derived by Gombas. Second, the α constant that they obtained is different from Gombas' value and also different from ours. We have for (neutral atoms):

Gombas:	$\alpha = 0.0560$
Mann and Waber:	$\alpha = 0.0425$
This work:	$\alpha = 0.03197$

The α value obtained by Mann and Waber is between the other two values and deviates from ours by about 30%. This is evidence showing the reliability of our correlation energy formula, Eq. (5.29).

Another interesting feature of these calculations is that the HFD energy, the magnetic energy, the retardation energy, and the correlation energy are listed separately; therefore, we get a clear picture about their relative magnitudes. Here are the results for Yb ($Z = 70$):

$E(\text{HFD}) = -14067.765$ a.u. (100%)
$E(\text{Mag.}) = +15.394225$ a.u. (0.109%)
$E(\text{Ret.}) = -1.4961634$ a.u. (0.011%)
$E(\text{Corr.}) = -3.0406588$ a.u. (0.022%)
$E(\text{Total}) = -14056.908$ a.u.

In this table, E(Total) means the sum of the four preceding energies. The percents are calculated by dividing the absolute values by the absolute value of E(HFD). The figures show that if we consider the magnetic effects, the retardation effects, and the correlation effects as perturbations, then the magnetic effects are the most important, with the correlation being the second and the retardation effects, the smallest, the third.

D. Electron Configurations

We introduced the concept of electron configuration in a formal way in Sec. 3.1. We have seen that in LS coupling, a configuration looks like this:

$$(n_1 l_1)^{q(n_1 l_1)} (n_2 l_2)^{q(n_2 l_2)} \cdots (n_N l_N)^{q(n_N l_N)},$$

where $q(nl)$ is the occupation number of the group with the quantum numbers (nl). In table 3.1, we listed the configurations of all neutral atoms with the occupation numbers tabulated for each atom. This table will be referred to as the list of *experimental* electron configurations.

How is such a table constructed? We have taken the data from the Atomic Energy Level Tables [Atomic Energy Level Tables, AET, I, II, III, IV; see References]. These tables contain the data about the optical spectra that are generated by valence electrons. A detailed discussion of these spectra is given in Sec. 6.1. Here it will be sufficient to outline how the electron configurations are derived from these spectra.

First, in most cases, the number of electrons in an atom can be determined from the position of the atom in the periodic table, that is, from chemical data. Most of the electrons will be in complete groups whose size we know from the Pauli exclusion principle: a complete group can hold a maximum of $2(2l + 1)$ electrons, that is, an (ns) group will hold 2, an (np) group will hold 6, and so on. The only part of the configuration that must be determined from the spectra is the configuration of valence electrons. This is a fairly complicated task, but it has been carried out for most atoms and the results are listed in the tables.

We note that it is somewhat of a misnomer to designate these configurations as experimental. It is true that the tables contain the empirical values of the optical-energy levels. But it is also true that the classification of spectral lines by configurations is a largely theoretical task. The theory used here is the quantum theory of LS coupling, based on a HF model, or, more precisely, on an *orbital* approximation. However, strictly speaking, there are no orbitals in an atom and the exact eigenfunction for N electrons cannot be written in the form of a HF wave function. Therefore, the configurations are experimental in the sense that they are derived using experimental data. The configurations listed in the Atomic Energy Level Tables are those from which the complex structure of the empirical spectra can be derived.

In the relativistic HF theory, we characterize an electron state by the quantum numbers (nlj) or (nja). Thus, an electron configuration will have the form

$$(n_1 l_1 j_1)^{q(n_1 l_1 j_1)} (n_2 l_2 j_2)^{q(n_2 l_2 j_2)} \cdots,$$

where $q(nlj)$ is the occupation number of the group (nlj). Another notation is to write l for $j = l + \frac{1}{2}$ and $\bar{l}$ for $j = \bar{l} - \frac{1}{2}$. In this notation, the configuration has the form

$$(n_1 l_1)^{q(n_1 l_1)} (n_1 \bar{l}_1)^{q(n_1 \bar{l}_1)} \cdots.$$

In the relativistic model, the number of electrons in a group is $2j + 1$, that is, $2l + 2$ for l and $2\bar{l}$ for $\bar{l}$. We have the following occupation numbers for complete groups: $q = 2$ for s, $q = 2$ for $\bar{p}$, $q = 4$ for p, $q = 4$ for $\bar{d}$, $q = 6$ for d, and so on. The lowest energy part of a large atom will have the configuration

$$(1s)^2 (2s)^2 (2\bar{p})^2 (2p)^4 (3s)^2 (3\bar{p})^2 (3p)^4 (3\bar{d})^4 (3d)^6 \cdots,$$

and so on. It is evident, of course, that the relativistic configuration determines the nonrelativistic because the former determines the number of electrons with (nl). For a complete group in LS coupling, from the relativistic model, we get

$$q_{LS}(nl) = q_{jj}(n\bar{l}) + q_{jj}(nl) = 2\bar{l} + 2l + 2 = 2(2l + 1), \qquad \text{for } \bar{l} = l,$$

which is the correct result. Thus, it is evident that if we want to have a consistent relativistic model, then this model must produce the same (nl) configuration as the nonrelativistic model; more accurately, both models must produce the same configuration that is derived from the optical spectra.

We want to elucidate this point. Let us suppose that the empirical spectrum leads to the $(np)^3$ configuration in (LS) coupling. Then, in (jj) coupling, we must have $(n\bar{p})^2(np)$ or $(n\bar{p})(np)^2$; we cannot have $(n\bar{p})^2(n + 1, s)$, because this last would not, in the nonrelativistic limit, reduce to $(np)^3$. The relativistic model cannot lead to a configuration that is not compatible with the optical spectra.

We are now coming to the main point of this section, which is that the HF model can be used to determine the electron configuration. The method is simple: HF calculations must be carried out for close-lying configurations. The configuration with the lowest total energy will be the "correct" one. If the model is realistic, the theoretically determined configuration must be identical with the empirical.

Calculations to test the electron configurations of atoms were carried out by Larson and Waber[38] and by Fricke, Greiner, and Waber[39] using the nonrelativistic HF theory. In all but a few cases, the results were in agreement with empirical data. Exceptions were the atoms Cr, Nb, Tc, Pd, Re, and Os. Calculations with the relativistic HF model were carried out by Mali and Hussonnois.[40] Because the relativistic model gives the total energies of neutral atoms more accurately than the nonrelativistic, we want to look at the relativistic calculations in more detail.

Mali and Hussonnois carried out HFD calculations using the equations for complete groups, Eqs. (4.168) and (4.169). For incomplete groups, the same equations were used with the difference that the occupation numbers were adjusted to the actual occupation numbers of the incomplete groups. Using the orbitals obtained this way, Mali and Hussonnois also calculated the average energy of the configuration using Eq. (4.196). Calculations were done for several close-lying configurations of all neutral atoms. The results agreed with the empirical data with some exceptions. The atoms for which the configurations obtained were not identical with the experimental are listed here in such a way that on the left we have the configuration obtained with the HFD model and on the right we have the experimental configuration:

Cr	$(\text{Ar})(3\bar{d})^4(4s)^2;$	$(\text{Ar})(3d)^5(4s);$
Cu	$(\text{Ar})(3\bar{d})^4(3d)^5(4s)^2;$	$(\text{Ar})(3d)^{10}(4s);$
Nb	$(\text{Kr})(4\bar{d})^3(5s)^2;$	$(\text{Kr})(4d)^4(5s);$
Tc	$(\text{Kr})(4\bar{d})^4(4d)^2(5s);$	$(\text{Kr})(4d)^5(5s)^2;$
Pd	$(\text{Kr})(4\bar{d})^4(4d)^5(5s);$	$(\text{Kr})(4d)^{10}.$

In addition, there were systematic discrepancies between the theoretical and experimental data for the lanthanide atoms.

We note here that similar discrepancies were reported by Mann and Waber in the calculations carried out for lanthanides. In these calculations, which we mentioned before, Mann and Waber obtained, for configurations different from the experimental, energies that were deeper than the energies obtained for the empirical configurations. However, Mann and Waber disregarded these results and considered the results obtained for the empirical configurations as the correct ones.

Here are some numbers for a typical case in which the theoretical configuration is different from the experimental. We list the average energies obtained[40] for three different configurations of the Yb atom ($Z = 70$):

$(4\bar{f})^6(4f)^7(5\bar{d})(6s)^2$	$E = -14,069.4203 \text{ a.u.}$
$(4\bar{f})^6(4f)^8(6s)^2$	$E = -14,069.3615 \text{ a.u.}$
$(4\bar{f})^6(4f)^6(5\bar{d})^2(6s)^2$	$E = -14,069.2335 \text{ a.u.}$

The experimental configuration is $(4f)^{14}(6s)^2$. As we see, the HFD model gives as the correct configuration $(4f)^{13}(5d)(6s)^2$, which is different from the experimental.

We want to analyze the nature of the problem that we encounter in a calculation of this type. The correctness of the answer depends on the correct configuration yielding the lowest energy. Let us take a look at the size of the energy differences. We get for them $\Delta_1 = 0.0588$ and $\Delta_2 = 0.128$, respectively. At the same time, the ionization potential, that is, the binding energy of the outermost electron of the atom is $I = 0.23$ a.u. Also, we have seen at the end of Sec. 5.2/C that the magnetic energy is $E(\text{Mag.}) = 15.39$ a.u., the retardation energy is $E(\text{Ret.}) = -1.496$ a.u., and the correlation energy is $E(\text{Corr.}) = -3.041$ a.u. In the calculations of Mali and Hussonnois, the magnetic, retardation and correlation effects were omitted. We observe also, comparing the previous numbers with the numbers quoted at the end of Sec. 5.2/C, that there is a difference between the HFD result of Mann and Waber, which is $E(\text{HFD}) = -14067.765$ a.u., and the result of Mali and Hussonnois, which is $E(\text{HFD}) = -14069.4203$ a.u. The difference is $\Delta E = 1.6553$ a.u.

Therefore, as we see from these numbers, the HFD model gives slightly different energies as we move an electron from the $(4f)$ orbital into the $(5\bar{d})$, that is, we do get an energy minimum at one of the close-lying configurations as we move some of the outer electrons from one incomplete group to another. But the energy difference that results from such a change is smaller than the energy needed to remove an electron from the atom and it is also much smaller than the magnetic, retardation, and correlation effects. Also, the difference between the two HFD calculations that we have mentioned here is larger than the difference between the configurations. (In the two calculations, slightly different equations (potentials) were used.)

In the preceding sections, we have recorded with satisfaction the result that the total energies of atoms are estimated by the relativistic model with a 1% accuracy and probably with a higher accuracy for many atoms. Now we see, however, that if we would like to use the HF model for an absolutely reliable determination of electron configurations, then an accuracy much higher than 1% would be needed.

This does not mean that the results of the configuration-determining calculations are meaningless. It is a fortunate circumstance that what we need here is only an energy *difference* and we do not even need the size of that difference; the correctness of the result is determined by the sign of the difference. Thus, even if the energy values quoted here have a built-in inaccuracy that is greater than the difference between configurations, we still may get the correct result because the built-in errors may be of the same size for all the configurations considered. In other words, the computed energies may be in error, but the errors may shift the computed energies all in one direction. If that is the case, then the result for a *difference* between the energies may still come out correctly. In fact, it is clear that this is what is happening, because the calculations have yielded the correct configuration in most of the cases.

Summing up, we can say the following. Calculations for the determination of electron configurations of neutral atoms have given the correct results in the majority of cases. For a small number of atoms, the configurations obtained were different from the experimental; systematic differences occurred for the lanthanides. An analysis of the results shows that such calculations are taxing the accuracy of the HF model to the utmost because of the smallness of the energy difference between configurations. In order to obtain a reliable result for the difference between two configurations, the total energies of both configurations must be computed with very high accuracy.

5.3. RADIAL DENSITIES AND RELATED QUANTITIES

A. The General Form of Radial Densities and the Size of Atoms

We proceed now to an examination of the radial densities provided by the HF model. The total radial density for the nonrelativistic model was defined by Eqs. (2.29), (2.30), and (2.31). This quantity is the amount of electron charge that can be found between the radii r and $r + dr$ and it is, after the total energy, the most important quantity obtained from the HF model.

There are numerous highly accurate tables available giving the values of HF densities. The most important is the work of Froese-Fischer,[41] who published the radial densities of all neutral atoms from He to Rn, which were obtained from nonrelativistic average-of-configuration calculations. Another

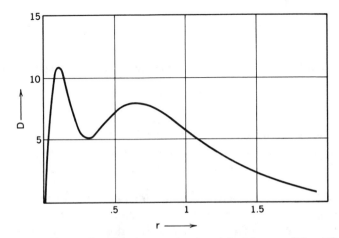

Fig. 5.2. The total radial electron density of the Ne atom ($Z = 10$).

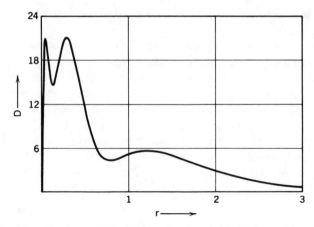

Fig. 5.3. The total radial electron density of the Ar atom ($Z = 18$).

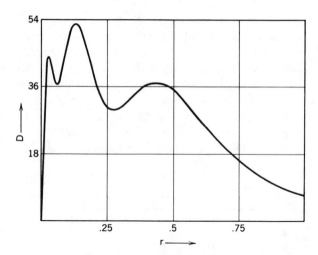

Fig. 5.4. The total radial electron density of the Kr atom ($Z = 36$).

important work is the paper presented by Clementi. For the range up to $Z = 30$, Clementi calculated[42] analytic HF orbitals for neutral atoms as well as for many positive and negative ions. (For analytic HF orbitals, see Sec. 5.6.)

In Figs. 5.2 to 5.5, we have the total radial densities of the neutral atoms of noble gases, that is, the densities of Ne ($Z = 10$), Ar ($Z = 18$), Kr ($Z = 36$), and Xe ($Z = 54$). These densities, along with the radial density of mercury, which we presented in Fig. 2.1, form a representative group cover-

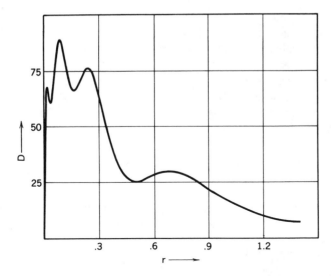

Fig. 5.5. The total radial electron density of the Xe atom ($Z = 54$).

ing the whole periodic system. These densities are from the tables of Froese-Fischer. Looking at these diagrams, we see that they show the characteristic maxima and minima that we discussed in Sec. 2.1 in connection with Hartree's first results. We called these diagrams the "images" of atoms. The HF calculations show that all atoms have the same type of image, that is, all atoms show the characteristic maxima and minima in their densities. The maxima signify the presence of *shells* in the electronic structure of atoms, where, as we stated before, a shell is defined as the assembly of electrons with the same principal quantum numbers. All atoms have a shell structure, that is, all atoms have a set of radii around which the electron densities are concentrated.

We now look more closely at the densities. A conspicuous feature is that the highest value of the electron density, which is generally reached in the second shell (the L shell) increases strongly as we go from lower to higher Z values. (The highest density is in the K shell in light atoms.) Although the highest density is about 10 atomic units for Ne, it is about 150 for Hg. On the other hand, while the inner density increases, the size of the atoms is not a linear function of Z; especially, it cannot be said that the atoms are getting larger with increasing Z.

The size of an atom can be defined only with a certain degree of arbitrariness because the one-electron wave functions and the one-electron densities decline exponentially for large r. We adopted the definition of Fraga et al. [FKS, 46], according to which the size of the atom is equal to the radius, where the maximum is in the outermost one-electron density. Figure 5.6 shows the size of the neutral atoms from hydrogen ($Z = 1$) to uranium ($Z = 92$). (The values of R are given in Å units.)

This figure gives a great deal of highly interesting information. In the figure, the values given by Fraga et al. versus Z are plotted. The points giving these values are connected by straight lines. These lines, of course, do not have a direct physical meaning, but they give the trend of the R versus Z connection.

First, we see that R fluctuates with Z, that is, as we said, R does not generally increase (or decrease) with Z. Instead, R is a function of the electron configuration. There are distinct minima in R reached at the atoms He, Ne, A, Kr, Xe, and Rn. These atoms have complete outermost groups; He has $(2s)^2$ and the others have $(np)^6$ configurations in their outermost shell. The reader should note that the compactness of these atoms is coming from complete *groups* not from complete *shells*; in the Ar atom, for example, we have the $(3s)^2(3p)^6$ configuration that would be a complete shell only if we would add the $(3d)^{10}$ group.

There are also distinct maxima in the radii reached at atoms Li, Na, K, Rb, Cs, and Fr. These are the alkali atoms in which an (ns) electron is added to the $(n-1, s)^2(n-1, p)^6$ configuration. Here the Pauli exclusion principle pushes the outermost electron to the area well outside of the filled groups and the strong increase in the radius of these atoms is because the added electron has a principal quantum number larger than the principal quantum number of the filled groups. Thus, this increase is a shell-structure effect. The increase in the radius enables the classification of the added electron as the *valence electron*, that is, the electron that is easily detachable to form compounds with other atoms.

We also have some smaller secondary maxima at the atoms Al, Ga, In, and Tl. In these atoms, we have one p electron added to the closed s group. Clearly, the addition of this electron, even though here the added p electron and the s electrons have the same principal quantum number, means that the added electron is pushed, because of the Pauli principle, to an area outside of the area of the filled s group. Thus, this is a group (as opposed to shell) effect.

It is also very interesting to look at the "curves" between the maxima and the minima, that is, between the atoms Li and Ne, Na and A, K and Kr, and so on. We see that here the size of the atoms decline and in fact, in those regions of Z where an incomplete p group or d group is being filled up, the decline of R could be approximated with a monotonic R-versus-Z function. For the range from the K atom to Kr, we would need two $R(Z)$ functions; one for the Z range, where the d group is filled up and another in which the p group is completed. Similar is the situation in the Rb to Xe range and in the Cs to Rn area. The point is that although overall the radii of atoms fluctuate, in the areas of Z between a maximum and a minimum R, the size

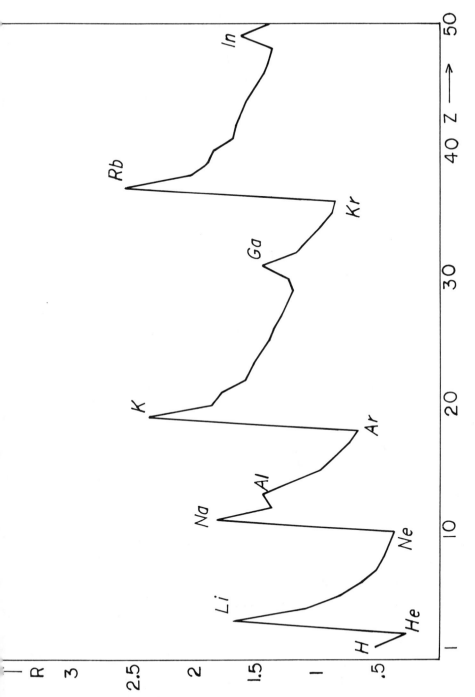

Fig. 5.6. Atomic radii versus atomic number Z.

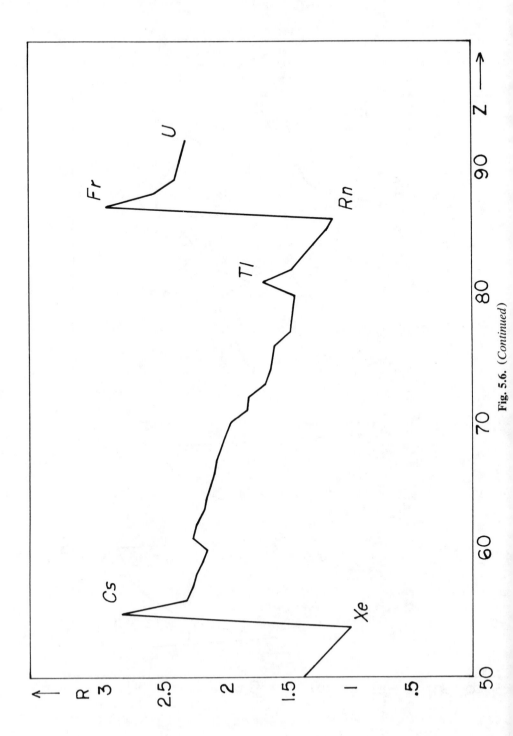

Fig. 5.6. (*Continued*)

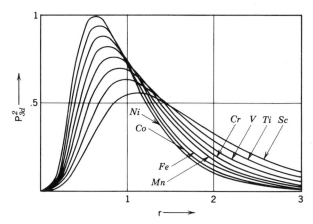

Fig. 5.7. The one-electron radial densities of the $3d$ electrons in the neutral atom series Sc to Ni.

of the atoms could be represented, in good approximation, with a continuous and monotonic $R(Z)$ function.

There are many situations in which the properties of molecules and solids depend strongly on the electronic structure of the atoms from which these compounds are formed. A typical example is ferromagnetism. It is well known that in ferromagnetic compounds, the quantity $x = R/2r_{3d}$ plays an important role. Here R is the internuclear separation in the crystal and r_{3d} is the radius of the $3d$ electrons in the free atom. Ferromagnetism occurs in compounds for which $x > 1.5$. This condition is fulfilled for Fe, Co, and Ni, but not for Mn. From the point of view of electron densities, the increase in the size of the densities of $3d$ electrons is a crucial factor here.

In order to have some idea of how an incomplete group is built up, we present, in Fig. 5.7, the one-electron radial densities of the $3d$ electron for the atoms Sc ($Z = 21$) to Ni ($Z = 28$). The densities are from the work of Froese-Fisher.[41]

In this diagram, we see that the one-electron radial densities contract as we go from lower to higher Z values. For the range of atoms shown in the diagram, the contraction of the maxima is about 0.7 a.u. We have constructed similar diagrams for the one-electron p densities between Al ($Z = 13$) and Ar ($Z = 18$) and also for the sequence of one-electron f densities for the atoms between Ce ($Z = 58$) and Dy ($Z = 66$). These diagrams are not shown. The interesting feature here is that all three sequences show contraction with decreases with increasing azimuthal quantum number. The approximate range of contraction is 1.5 a.u. for the p sequence, it is 0.7 a.u. for the d sequence, and 0.2 a.u. for the f sequence. In fact, the f electron densities are so closely packed that it would be difficult to have a presentable diagram in the size of a book.

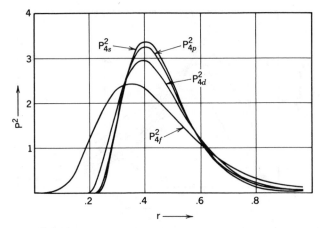

Fig. 5.8. The one-electron radial densities of the $4s$, $4p$, $4d$, and $4f$ electrons in the Pb^{4+} atom ($Z = 82$).

For the atoms Mn–Fe–Co–Ni, which we mentioned in connection with ferromagnetism, the change in the maximum of the $3d$ density is only about 0.125 a.u. Thus, we see that very small changes in the electronic structure of the outer electrons can bring about crucial changes in the physical properties of compounds that are built from these atoms. It is also demonstrated that the HF model gives an accurate description of these changes in the electronic structure.

Another feature of the electronic structure is demonstrated in Fig. 5.8. For this demonstration, we have selected a heavy atom, Pb^{4+}, which possesses a filled N shell, that is, it possesses a filled shell with the principal quantum number, $n = 4$. This is the largest shell in the periodic system because the shells with $n > 4$ do not fill up. Figure 5.8 shows the one-electron densities of the $4s$, $4p$, $4d$, and $4f$ electrons in this atom.[†] As we see, the main maxima of these densities are, in a very good approximation, at the same radius. (The $4f$ density, which does not have any nodes because of the $n - l - 1 = 4 - 3 - 1 = 0$ rule, deviates a little from the common position of the maximum.) This diagram explains the existence of the shell structure. The shell structure, which is visible on the total density, is created by the one-electron densities, which have the same n but different l, having their main maxima at the same place in a very good approximation. If this would not be the case, if these densities would cover a wider range, then the total density would not show the characteristic maxima and minima that we

[†]From the work of Saxena, Ref. 43.

described as shell structure. Thus, we may formulate the rule:

> The HF model shows that the radial densities of electrons with the same principal quantum number have their main maxima at the same radius in a very good approximation. This is what generates the *shell structure.*[‡]

It is noteworthy that when a valence electron is added to a complete $(np)^6$ group, it has the principal quantum number $(n + 1)$, and that is the reason why the radial density of this electron is pushed so far out from the region occupied by the $(np)^6$ group. On the other hand, when an (np) electron is added to an $(ns)^2$ group, the increase in the size of the atom is small, as we have seen from Fig. 5.6.

We note that the HF equations of the (nl) orbitals with the same n but different l do not show that the maxima will be located in the same position. It can be shown, however, by formulating the HF model with pseudopotentials,[†] that the total potential, when it contains a pseudopotential, is very nearly independent of l for electrons with common n but different l. From this, the previous rule follows immediately.

Finally, we want to compare the nonrelativistic and relativistic densities. Such a comparison was carried out for H-like atoms by Burke and Grant.[44] They have shown for an *H*-like atom with $Z = 80$ (mercury) that the relativistic one-electron radial electron densities are more compressed than the nonrelativistic, that is, generally the relativistic electron density has a higher maximum than the nonrelativistic and the relativistic maximum is closer to the nucleus than the nonrelativistic. They have also shown that the density with $j = l - \frac{1}{2}$ is more inward than the density with $j = l + \frac{1}{2}$.

Using a simplified relativistic HF model, Kahn, Hay, and Cowan have shown[45] (see Chapter 9) that in a heavy atom like U $(Z = 92)$, the relativistic s and p densities are more compressed toward the nucleus than the nonrelativistic densities, whereas for the d and f electrons, the opposite is true. Thus, there appears to be a difference between the densities in a H-like atom and the densities in a heavy atom.

B. Diamagnetic Susceptibilities

The comparison between the measured and calculated values of the diamagnetic susceptibility of an atom can serve as a guide for the quality of the calculated electron density. The theoretical formula for the diamagnetic

---

[‡]The rule is valid only if the shell is completely filled. The situation is quite different if the d or f group is incomplete. See Sec. 5.5/B.
[†]This will be presented in the second volume.

susceptibility is

$$\chi = -N_A \frac{e^2}{6mc^2} \overline{r^2},$$

where $\overline{r^2}$ is given by

$$\overline{r^2} = \int \psi^* r^2 \psi \, dv.$$

By substituting the HF total wave function into this formula, we obtain a calculated value for χ (N_A is the Avogadro number).

Unfortunately, most of the measured data are given for molecules and crystals, whereas the previous formula is for free atoms. Direct comparison between theory and experiment can be made for noble gases. Here are some data: the calculated HF values are from Fraga et al. [FKS, 353]; the empirical values are in parentheses and both values are in 10^{-6} cm^3/gram-atom units.

He	$\chi = -1.88$	(-1.88);
Ne	$\chi = -7.42$	(-6.74);
A	$\chi = -20.62$	(-19.6);
Kr	$\chi = -31.31$	(-28.8);
Xe	$\chi = -49.62$	(-43.9).

The agreement with experiment is quite good. We must remember, however, that this is a test for the density only in the outer regions, that is, in those regions where r^2 is large. The comparison does not say anything about the detailed structure of the density in the inner part of the atom.

5.4. IONIZATION POTENTIALS AND MANY-VALENCE-ELECTRON ENERGIES

A. Ionization Potentials

Our next task is to examine the ionization potentials yielded by the HF model. We introduce some definitions first.

The ionization potential is defined as the energy necessary to remove one electron from the outermost group of an atom. Let E_{exp} be the empirical total energy of the atom and let E_{exp}^+ be the energy of the atom with the outermost electron removed. Then the ionization potential is the positive quantity

$$I_{exp} = |E_{exp} - E_{exp}^+|. \tag{5.42}$$

If the atom is the neutral species and E_{exp}^+ is the singly positive ion, then we talk about the first ionization potential; if E_{exp} is the energy of the first positive ion and E_{exp}^+ is the energy of the doubly positive ion, then we talk about the second ionization potential; and so on.

Now let I_{rel} be the ionization potential calculated with the relativistic HF model, that is, let

$$I_{rel} = |E_{rel} - E_{rel}^+| \qquad (5.43)$$

where E_{rel} is the calculated total relativistic energy of the atom and E_{rel}^+ is the calculated relativistic energy of the ion. Here, again, we can define the first, second, and so on, ionization potentials.

We can calculate the potentials also from the nonrelativistic model. Let I_{HF} be the calculated nonrelativistic ionization potential, that is, let

$$I_{HF} = |E_{HF} - E_{HF}^+|, \qquad (5.44)$$

where E_{HF} and E_{HF}^+ are the calculated nonrelativistic energies for the atom and the ion, respectively. Here, again, we can define the first, second, and so on, ionization potentials.

In Table 5.5, we have the first ionization potentials for all atoms from Li to U. In the first column, we have the chemical symbol of the atom; in the second, we have Z; and the third column contains I_{exp} (in electron-volts). The experimental values are from the compilation of Moore.[36] The next column contains the calculated I_{rel} values that were tabulated by Fraga et al. [FKS, 259]. The next column contains Δ_{rel}, which is defined as follows:

$$\Delta_{rel} = 100 \frac{I_{exp} - I_{rel}}{I_{exp}}. \qquad (5.45)$$

This quantity gives the deviation of the relativistic ionization potential from the empirical in percent. This quantity is positive, in general, because generally I_{rel} is smaller than I_{exp}; there are cases, however, when Δ_{rel} is negative, which means that I_{rel} comes out larger than I_{exp}.

In the column next to the last, we have I_{HF}, which is taken from Fraga's tables [FKS, 259]. The last column contains Δ_{HF} defined as

$$\Delta_{HF} = 100 \frac{I_{exp} - I_{HF}}{I_{exp}}, \qquad (5.46)$$

which means that this quantity gives the deviation of I_{HF} from I_{exp} in percent.

We compare the results of the relativistic calculations with the empirical data in Fig. 5.9. In this diagram, we have the ionization potentials as functions of Z. We have connected the points that indicate the empirical potentials by a full line and the points that indicate the calculated relativistic

TABLE 5.5. The Calculated and Empirical Values of the First Ionization Potentials of Atoms between Li and U

Atom	Z	I_{exp}	I_{rel}	Δ_{rel}	I_{HF}	Δ_{HF}
Li	3	5.392	5.39	0.0	5.39	0.0
Be	4	9.322	8.04	13.7	8.04	13.7
B	5	8.298	7.93	4.43	7.93	4.43
C	6	11.260	10.8	4.08	10.8	4.08
N	7	14.534	13.9	4.36	14.0	3.67
O	8	13.618	11.9	12.6	11.9	12.6
F	9	17.422	15.7	9.88	15.7	9.88
Ne	10	21.564	19.8	8.18	19.8	8.18
Na	11	5.139	4.96	3.48	4.95	3.67
Mg	12	7.646	6.61	13.5	6.60	13.6
Al	13	5.986	5.50	8.11	5.50	8.11
Si	14	8.151	7.63	6.39	7.65	6.14
P	15	10.486	9.99	4.73	10.0	4.63
S	16	10.360	9.03	12.8	9.03	12.8
Cl	17	12.967	11.8	8.99	11.8	8.99
Ar	18	15.759	14.7	6.72	14.8	6.08
K	19	4.341	4.02	7.39	4.00	7.85
Ca	20	6.113	5.14	15.9	5.12	16.2
Sc	21	6.540	6.31	3.51	6.14	6.11
Ti	22	6.820	6.17	9.53	5.95	12.7
V	23	6.740	6.08	9.79	5.81	13.8
Cr	24	6.766	4.98	26.4	4.64	31.4
Mn	25	7.435	9.65	− 29.8	9.38	− 26.2
Fe	26	7.870	8.31	− 5.59	7.96	− 1.14
Co	27	7.860	8.24	− 4.83	7.79	0.89
Ni	28	7.635	8.16	− 6.87	7.62	0.19
Cu	29	7.726	6.71	13.15	6.04	21.8
Zn	30	9.394	7.78	17.20	7.64	18.7
Ga	31	5.999	5.63	6.15	5.61	6.48
Ge	32	7.899	7.44	5.81	7.50	5.05
As	33	9.810	9.30	5.19	9.52	2.96
Se	34	9.752	8.39	13.9	8.33	14.6
Br	35	11.814	10.7	9.43	10.8	8.58
Kr	36	13.999	12.9	7.85	13.2	5.71
RB	37	4.177	3.78	9.50	3.72	10.9
Sr	38	5.695	4.73	16.9	4.66	18.2
Y	39	6.38	6.00	5.96	5.49	13.9
Zr	40	6.84	5.61	17.9	4.97	27.3
Nb	41	6.88	5.18	24.4	4.41	35.9
Mo	42	7.099	3.91	44.9	2.95	58.4
Tc	43	7.28	6.91	5.08	6.12	15.9
Ru	44	7.37	5.48	25.6	4.52	38.7
RH	45	7.46	4.86	34.8	3.72	50.1
Pd	46	8.34	4.23	49.3	2.88	65.5
Ag	47	7.576	2.53	66.6	0.939	87.6

TABLE 5.5. (*Continued*)

Atom	Z	I_{exp}	I_{rel}	Δ_{rel}	I_{HF}	Δ_{HF}
Cd	48	8.993	7.31	18.7	6.94	22.8
In	49	5.786	5.25	9.26	5.17	10.6
Sn	50	7.344	6.76	7.95	6.87	6.45
Sb	51	8.641	8.26	4.41	8.69	−0.567
Te	52	9.009	7.74	14.1	7.59	15.7
I	53	10.451	9.61	8.07	9.69	7.28
Xe	54	12.13	11.3	6.84	11.7	3.54
Cs	55	3.894	3.43	11.9	3.33	14.5
Ba	56	5.212	4.21	19.22	4.09	21.5
La	57	5.57	11.3	−102.6	8.70	−55.9
Ce	58	5.47	10.4	−90.12	7.50	−37.1
Pr	59	5.42	11.0	−102.9	7.83	−44.4
Nd	60	5.49	11.2	−104.0	7.81	−42.2
Pm	61	5.55	10.6	−90.99	6.98	−25.7
Sm	62	5.63	9.58	−70.16	5.52	1.95
Eu	63	5.67	15.5	−173.4	12.2	−115.2
Gd	64	6.14	14.5	−136.1	10.9	−77.52
Tb	65	5.85	13.6	−132.5	9.63	−64.6
Dy	66	5.93	14.3	−141.1	10.1	−70.3
Ho	67	6.02	15.3	−154.1	10.9	−81.1
ER	68	6.10	14.6	−139.3	9.66	−58.4
Tm	69	6.18	13.6	−120.1	8.19	−32.5
Yb	70	6.254	4.97	20.5	4.68	25.2
LU	71	5.426	8.11	−49.5	6.39	−17.7
Hf	72	7.00	7.36	−5.14	5.28	24.6
Ta	73	7.89	7.27	7.86	4.73	40.0
W	74	7.98	5.96	25.3	2.91	63.5
Re	75	7.88	8.39	−6.47	5.99	23.9
Os	76	8.70	6.97	19.8	4.11	52.7
Ir	77	9.10	6.32	30.55	2.97	67.4
Pt	78	9.00	6.14	31.7	2.26	74.9
Au	79	9.225	5.49	40.5	1.03	88.8
Hg	80	10.437	7.89	24.4	6.78	35.0
Tl	81	6.108	5.22	14.5	4.93	19.3
Pb	82	7.416	6.76	8.84	6.99	5.74
Bi	83	7.289	7.24	0.67	8.30	−13.9
Po	84	8.42	8.26	1.90	7.81	7.24
At	85		8.79		8.90	
Rn	86	10.748	9.41	12.44	10.3	4.17
Fr	87		3.40		3.16	
Ra	88	5.279	3.95	25.2	3.67	30.5
Ac	89	6.90	10.8	−56.5	5.93	14.0
Th	90		10.1		4.57	
Pa	91		10.3		4.30	
U	92		10.1		3.48	

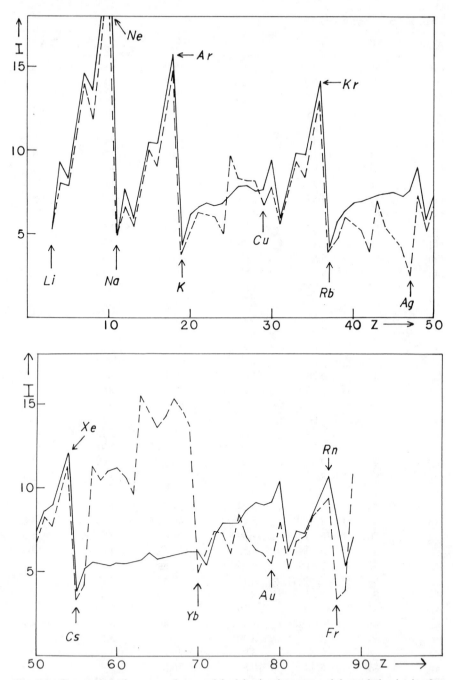

Fig. 5.9. Comparison between the empirical ionization potentials and the ionization potentials calculated with the relativistic HF model. Ionization potentials versus Z for neutral atoms. The solid line connects the empirical points; the dashed line connects the HF values.

values by a dashed line. These lines, of course, do not have physical meaning because only the measured and calculated points are physically meaningful. Nevertheless, as we will see presently, the lines, which show the trend of the measured and calculated values, are essential for the evaluation of the results.

Now let us take a look at the table. The deviations, Δ_{rel}, are ranging from about 3% to approximately 150%, where this last is the worst case. Most of the deviations are a few percents, that is, they are below 10%. Our first observation is that the calculated results for ionization potentials do not approximate the empirical values nearly as well as was the case with the total energies.

It is not difficult to see why this is the case. By definition, the ionization potential is a small number that is the difference of two large numbers. The calculated value of this small number will be very sensitive to errors in the calculated values of the two large numbers. We can easily analyze the situation by a simple calculation.

Let E and E^+ be the exact and $\hat{E}$ and $\hat{E}^+$ be the approximate total energies of the atom and of the ion, respectively. Let us assume that the approximate values deviate from the exact as follows:

$$\hat{E} = E(1 + x), \tag{5.47a}$$

$$\hat{E}^+ = E^+(1 + y). \tag{5.47b}$$

The meaning of x and y can be seen easily because we get from the previous formulas that the errors in percent in the values of E and E^+ are

$$\Delta(E) = -100x, \tag{5.48a}$$

and

$$\Delta(E^+) = -100y. \tag{5.48b}$$

Indeed, we have

$$\Delta(E) = -100x = 100\frac{E - \hat{E}}{E}. \tag{5.49}$$

The error in the ionization potential will be

$$\Delta(I) = 100\frac{I - \hat{I}}{I}, \tag{5.50}$$

where I and $\hat{I}$ are the exact and approximate values, respectively. Because we are not interested in the sign of $\Delta(I)$ in this argument, we can calculate

$\Delta(I)$ by omitting the absolute value signs from I and $\hat{I}$ and we get

$$\Delta(I) = 100 \frac{-Ex + E^+y}{E - E^+}. \tag{5.51}$$

Let us assume that

$$\alpha = y - x, \qquad y = \alpha + x, \tag{5.52}$$

and we get from Eq. (5.51)

$$\Delta(I) = -100x + 100 \frac{E^+}{E - E^+}\alpha. \tag{5.53}$$

The first term here is the error in E, as we see from Eq. (5.49). The second term gives that part of the error in I that arises from the fact that E and E^+ are calculated with different accuracy. This term is zero if $\alpha = 0$, that is, this term will be zero if $x = y$, where x gives the error in E, and y gives the error in E^+.

An example will elucidate Eq. (5.53). For the O atom and O^+ ion, we have the following empirical energies:

$$E = -75.11187 \text{ a.u. (O atom)};$$
$$E^+ = -74.611393 \text{ a.u. } (O^+ \text{ ion}).$$

These energies are approximated by the relativistic HF values. The energies and the errors relative to the exact values are given as

$$\hat{E} = -74.85626 \text{ a.u.} \qquad \Delta(E) = 0.34\%;$$
$$\hat{E}^+ = -74.41909 \text{ a.u.} \qquad \Delta(E^+) = 0.26\%.$$

Let us now calculate $\Delta(E)$ from Eq. (5.49). We get

$$-100x = 0.34\%.$$

This is the first term of Eq. (5.53). For the second term, we obtain, using E, E^+, and Eq. (5.52),

$$100 \frac{E^+}{E - E^+}\alpha = 12.31\%.$$

The total error in the ionization potential is then, according to Eq. (5.53),

$$0.34\% + 12.31\% = 12.65\%.$$

This is the value that we have in Table 5.5 (the slight difference is due to different rounding off in the two calculations).

Thus, we see that the total energies of the neutral O atom and positive O^+ ion were reproduced by the relativistic HF model with errors of 0.34% and 0.26%, respectively. The error in the calculated value of the ionization potential is, however, 12.6%! This large error is not coming from large errors in the calculated values of the total energies, but from the *difference* in the errors of the atom and of the positive ion.

Therefore, we are able to formulate the following general rules:

1. If a calculation reproduces the energies of the atom and of the ion with the same accuracy, then the calculated value of the ionization potential will show the same error.

2. If a calculation gives different errors in the energies of the atom and of the ion, the difference in the errors will show up magnified in the calculation of the ionization potential. This magnification may be very large.

Returning to Table 5.5, we want to look closer at the relativistic values. We observe that some of the Δ_{rel} values came our negative, which means that in Eq. (5.50), $\hat{I}$ is larger in absolute value than I (in the table, I_{rel} is larger that I_{exp}). This is, of course, yet another manifestation of the phenomenon that we have just described about the connection between the accuracies of the atomic total energies and the accuracy of the ionization potential. The percentages came out negative because the calculated values of the ionization potentials came out very inaccurately in these cases.

Now let us examine Fig. 5.9, which shows the empirical ionization potentials and the results of the relativistic HF calculations. We see that the empirical ionization potential is strongly dependent on the electron configuration. The "curve" I_{exp} versus Z (in reality a zigzag line) has minima at the atoms Li, Na, K, Rb, and Cs, that is, at the alkali atoms, which have the (ns) electron outside of closed $(n-1, p)^6$ groups. It is easy to remove this electron, hence, the minima in the I_{exp}. The empirical values show maxima at the atoms Ne, A, Kr, Xe, and Rn; these are the atoms with the compact $(np)^6$ groups, which are difficult to break up. Comparing the calculated values with the empirical, we see that the calculated values reproduce the empirical very accurately between Li and K, between Cu and Kr, between Ag and Xe, and between Au and Rn. As we see from the electron-configuration table, Table 3.1, these are the atoms with incomplete p groups. In these ranges, not only the reproduction of the ionization potentials is excellent, but the zigzag line representing the calculated values runs parallel to the line representing the empirical values. This means that the changes that are taking place in the electronic structure of these atoms, as the p group is being filled up, are accurately reproduced by the relativistic HF model.

We see from the diagram also that the calculated values do not reproduce the empirical "trend" very well between K and Cu, between Rb and Ag, and between Yb and Au. These are the atoms in which an incomplete d group is being filled up. Thus, in these ranges of Z, the results for the ionization potentials are poor. Finally, the calculated values are completely unrealistic in the range between Cs and Yb; these are the atoms in which an incomplete f group is being filled up.

We conclude that the relativistic HF model reproduces the ionization potentials of atoms with incomplete p groups very accurately, reproduces the ionization potentials of atoms with incomplete d groups poorly, and does not give meaningful results for atoms with incomplete f groups.

Comparing the relativistic and nonrelativistic results, we observe that the errors in the two sets of numbers are almost identical for the light atoms up to Ca. For greater Z, the relativistic values are generally better, although there are cases in which the nonrelativistic calculation yielded a better result. The fact that we do not always get better results is due to the circumstance that the relativistic Hamiltonian underlying the relativistic HF model is itself only a *model* of the unknown, exact, relativistic, Hamiltonian operator.[†]

B. Many-Valence-Electron Energies

This section is a precursor of later chapters in the sense that here we discuss valence-electron properties as opposed to the properties of the whole atom. First, let us introduce the quantities that we need in this discussion.

Let I_k be the kth ionization potential of the atom. In the context of the previous section, I_1 is the ionization potential of the neutral atom, I_2 is the ionization potential of the singly positive ion, and so on. Let the empirical total energy of the n valence electrons be

$$E_n^{\text{exp}} = -(I_1 + I_2 + \cdots + I_n) \qquad (n = 1, 2, \cdots), \qquad (5.54)$$

where I_k is the empirical ionization potential. For $n = 1$, the energy of the valence electron is equal to the negative of the first ionization potential.

Let the calculated relativistic energy of the n valence electrons be

$$E_n^{\text{rel}} = -(I_1^{\text{rel}} + I_2^{\text{rel}} + \cdots + I_n^{\text{rel}}), \qquad (5.55)$$

where I_k^{rel} is the kth calculated relativistic ionization potential. Likewise, let the nonrelativistic total energy of the n valence electrons be

$$E_n^{HF} = -(I_1^{HF} + I_2^{HF} + \cdots + I_n^{HF}), \qquad (5.56)$$

[†]In connection with this problem, see also the observations concerning the connection between LS and jj coupling, Sec. 5.2/A.

where I_k^{HF} is the kth calculated nonrelativistic ionization potential of the atom.

Let us define the correlation energy of the n valence electrons as

$$E_n^c = E_n^{\text{exp}} - E_n^{\text{rel}}, \tag{5.57}$$

where it is clear that this is the relativistic correlation energy. In accordance with our discussion of Sec. 5.2/B, we assume that the relativistic and nonrelativistic correlation energies are equal and we refer to this quantity simply as the correlation energy.

As we have seen, the exact nonrelativistic energy of a system is not observable because the empirical data always contain the relativistic effects. We can estimate, however, the eigenvalue of the nonrelativistic Hamiltonian operator of the valence electrons by the formula

$$E_n^e = E_n^{HF} + E_n^c, \tag{5.58}$$

which is the same approximation as in Eq. (5.39). In Eq. (5.39), the relationship was written down for the whole atom, whereas here it is written down for n valence electrons.

Let Δ_c be the size of the correlation energy in percent, that is, let

$$\Delta_c = 100 \frac{E_n^{\text{exp}} - E_n^{\text{rel}}}{E_n^{\text{exp}}}. \tag{5.59}$$

This quantity relates the correlation energy to the empirical total energy. A similar quantity is Δ_e, where

$$\Delta_e = 100 \frac{E_n^c}{E_n^e} = 100 \frac{E_n^{\text{exp}} - E_n^{\text{rel}}}{E_n^{HF} + E_n^c}, \tag{5.60}$$

which shows that Δ_e relates the correlation energy to the nonrelativistic total energy, estimated by Eq. (5.58). In other words, Δ_e shows how large the correlation energy is, in percent, relative to the nonrelativistic total energy.

In Tables 5.6 and 5.7, we have the results provided by the HF model. The headings of the tables are self-explanatory in terms of the quantities that we have defined. The empirical ionization potentials are taken from Moore's tables[36]; we have in our tables the data for those atoms for which the empirical data are available. We have designated as valence electrons those outer electrons that are conventionally designated as valence electrons. For Cu and Zn, we have two sets of data: in one case, the 10 d electrons plus the 1 or 2 s electrons are considered to be valence electrons, so the number of valence electrons is $n = 11$ and $n = 12$. In the other case, the d electrons

TABLE 5.6. The Empirical and the Calculated Relativistic n-Valence-Electron Energies of Some Selected Atoms between Li and Ra

Atom	Z	n	E_n^{exp}	E_n^{rel}	E_n^{corr}	Δ_{corr}
Li	3	1	5.392	5.39	0.003	0.04
Be	4	2	27.533	26.14	1.393	5.1
B	5	3	71.382	69.13	2.252	3.1
C	6	4	148.022	145.0	3.022	2.0
N	7	5	266.943	263.1	3.843	1.4
O	8	6	433.092	427.3	5.792	1.3
F	9	7	658.82	651.1	7.720	1.2
Ne	10	8	953.586	944.6	8.986	0.9
Na	11	1	5.139	4.96	0.179	3.5
Mg	12	2	22.681	21.31	1.371	6.0
Al	13	3	53.261	51.20	2.061	3.9
Si	14	4	103.129	100.03	3.099	3.0
P	15	5	176.784	173.09	3.694	2.1
S	16	6	276.549	270.83	5.719	2.1
Cl	17	7	408.87	401.40	7.470	1.8
Ar	18	8	577.74	568.00	9.740	1.7
K	19	1	4.341	4.02	0.321	7.4
Ca	20	2	17.984	15.94	2.044	11.4
Sc	21	3	44.1	41.41	2.690	6.1
Ti	22	4	91.157	87.27	3.887	4.2
V	23	5	162.637	157.48	5.157	3.2
Cr	24	6	263.186	255.98	7.206	2.7
Mn	25	7	394.612	387.55	7.062	1.8
Fe	26	8	559.561	549.91	9.651	1.7
Co	27	9	763.35	753.34	10.010	1.3
Ni	28	10	1011.873	997.66	14.213	1.4
Cu	29	11	1304.948	1290.31	14.638	1.1
Zn	30	12	1650.88	1634.68	16.200	0.98
Cu	29	1	7.726	6.71	1.016	13.4
Zn	30	2	27.358	24.68	2.778	10.1
Ga	31	3	57.219	54.13	3.089	5.4
Ge	32	4	103.763	99.34	4.423	4.2
As	33	5	169.554	164.2	5.354	3.1
Se	34	6	254.706	248.29	6.416	2.5
Br	35	7	368.214	357.3	10.914	2.96
Kr	36	8	508.008	493.4	14.608	2.9
Rb	37	1	4.117	3.78	0.397	9.5
Sr	38	2	16.725	15.13	1.595	9.5
Y	39	3	39.14	36.10	3.040	7.8
Zr	40	4	77.3	73.11	4.190	5.4
Nb	41	5	135.09	128.68	6.410	4.7
Mo	42	6	226.009	205.21	20.799	9.2
Ag	47	1	7.576	2.53	5.046	66.6
Cd	48	2	25.901	22.81	3.091	11.9

TABLE 5.6. (*Continued*)

Atom	Z	n	E_n^{exp}	E_n^{rel}	E_n^{corr}	Δ_{corr}
In	49	3	52.685	48.85	3.835	7.3
Sn	50	4	93.212	88.36	4.852	5.2
Sb	51	5	150.671	143.36	7.311	4.8
Te	52	6	222.429	213.94	8.489	3.8
Cs	55	1	3.894	3.43	0.464	11.9
Ba	56	2	15.216	13.65	1.566	10.3
Au	79	1	9.225	5.49	3.735	40.5
Hg	80	2	29.193	24.49	4.703	16.1
Tl	81	3	56.366	49.62	6.746	11.9
Pb	82	4	96.701	88.76	7.945	8.2
Ra	88	2	15.426	12.82	2.606	16.9

are viewed as core electrons, so in this case, only the s electrons are counted as valence electrons, which means that $n = 1$ and $n = 2$.

The data show a clear-cut pattern. Let us look at the sequences of atoms between Li and Ne, between Na and A, between K and Zn (considered as a 12-electron system), between Cu and Kr (Cu considered as a 1-electron system), between Rb and Mo, between Ag and Te, and between Au and Bi. In all these sequences, with the exception of the Rb → Mo sequence, we have the incomplete p groups filling up. In the Rb → Mo sequence, we have a d group being filled up. As we see from Table 5.6, the correlation energy peaks at $n = 1$ or $n = 2$, usually, at $n = 2$, and then declines steadily as we go to larger n values. This means that the HF model gives a relatively poor result for one or two valence electrons. The model becomes steadily more accurate as the number of valence electrons increases. In the Li → Ne, Na → A, K → Zn, and Cu → Kr sequences, we find the smallest correlation energy in the atoms with the largest n. These are the atoms Ne, A, Zn, and Kr, that is, atoms with completed groups. In the Rb → Mo set, the situation is more complex because, although the correlation energy diminishes between Rb and Nb ($n = 5$), it becomes suddenly larger in Mo ($n = 6$). In the Ag → Te and in the Au → Bi sequence, we again have monotonic decline in the correlation energy (the Au → Bi sequence is actually only an Au → Pb sequence because we do not have the theoretical values for Bi; we have included Bi because we have the empirical value for the total energy).

We note that for light atoms, that is, for the atoms between Li and A, the correlation energy ranges from about 1% to about 6%. This is much larger than the corresponding percentages for the total correlation energy of all electrons in the neutral atoms of this Z range. The reason for this is that apparently the HF model reproduces the energy of a few valence electrons less accurately than the energy of the neutral atom with its larger number of electrons. We also observe that the percentages become larger as we go to

**TABLE 5.7. The Estimated Values of the Exact Nonrelativistic
n-Valence-Electron Energies of Some Selected Atoms
between Li and Ra**

Atom	Z	n	E_n^{HF}	E_n^{corr}	E_n^e	Δ_e
Li	3	1	5.39	0.003	5.393	0.04
Be	4	2	26.14	1.393	27.533	5.0
B	5	3	69.13	2.252	71.382	3.1
C	6	4	144.9	3.022	147.922	2.0
N	7	5	263.1	3.843	266.943	1.4
O	8	6	427.3	5.792	433.092	1.3
F	9	7	651.2	7.720	658.920	1.2
Ne	10	8	944.7	8.986	953.686	0.9
Na	11	1	4.96	0.179	5.139	3.5
Mg	12	2	21.3	1.371	22.671	6.0
AL	13	3	51.0	2.061	53.061	3.9
Si	14	4	99.95	3.099	103.049	3.0
P	15	5	172.7	3.694	176.394	2.1
S	16	6	270.53	5.719	276.249	2.1
Cl	17	7	400.1	7.470	407.570	1.8
Ar	18	8	567.0	9.740	576.740	1.7
K	19	1	4.00	0.321	4.321	7.4
Ca	20	2	16.42	2.044	18.464	11.1
Sc	21	3	41.44	2.690	44.130	6.1
Ti	22	4	87.45	3.887	91.337	4.2
V	23	5	158.11	5.157	163.267	3.1
Cr	24	6	257.04	7.206	264.246	2.7
Mn	25	7	389.78	7.062	396.842	1.8
Fe	26	8	552.06	9.651	561.711	1.7
Co	27	9	755.09	10.010	765.100	1.3
Ni	28	10	1002.22	14.213	1016.433	1.4
Cu	29	11	1296.14	14.638	1310.773	1.1
Zn	30	12	1641.74	16.200	1657.940	0.97
Cu	29	1	6.04	1.016	7.056	14.4
Zn	30	2	24.14	2.778	26.918	10.3
Ga	31	3	53.31	3.089	56.399	5.5
Ge	32	4	98.2	4.423	102.623	4.3
As	33	5	162.72	5.354	168.074	3.2
Se	34	6	246.13	6.416	252.546	2.5
Br	35	7	353.9	10.914	364.814	3.0
Kr	36	8	489.1	14.608	503.708	2.9
Rb	37	1	3.72	0.397	4.117	9.6
Sr	38	2	14.96	1.595	16.555	9.6
Y	39	3	36.09	3.04	39.13	7.8
Zr	40	4	73.37	4.19	77.56	5.4
Nb	41	5	129.51	6.41	135.92	4.7
Mo	42	6	206.85	20.799	227.649	9.1
Ag	47	1	0.939	5.046	5.985	84.3

TABLE 5.7. (*Continued*)

Atom	Z	n	E_n^{HF}	E_n^{corr}	E_n^e	Δ_e
Cd	48	2	21.74	3.091	24.831	12.4
In	49	3	47.07	3.835	50.905	7.5
Sn	50	4	85.87	4.852	90.722	5.3
Sb	51	5	140.29	7.311	147.601	4.9
Te	52	6	209.49	8.439	217.979	3.9
Cs	55	1	3.33	0.464	3.794	12.2
Ba	56	2	13.26	1.566	14.826	10.5
Au	79	1	1.03	3.735	4.765	78.4
Hg	80	2	21.48	4.703	26.183	17.9
Tl	81	3	44.83	6.746	51.576	13.1
Pb	82	4	81.89	7.945	89.835	8.8
Ra	88	2	11.89	2.606	14.496	17.9

higher Z values despite the fact that we are dealing with the same small number of valence electrons in the low and high Z ranges. The results are especially poor for Ag ($Z = 47, n = 1$) and for Au ($Z = 79, n = 1$). The Cu atom, when considered as a 1-electron system, is also conspicuous with its high correlation energy, 13%.

The significance of Table 5.7 is that here we have accurate estimates for the nonrelativistic exact n-electron energies, E_n^e. Looking at the percentages, Δ_e, we see that up to $Z = 18$, the percentages are practically the same as the percentages in Table 5.6. This means that, in this range, the estimate for the nonrelativistic energy is practically the same as the empirical total energy, that is, the relativistic correction is very small. Looking at the higher Z values, we observe that in some cases, the nonrelativistic energy is deeper than the empirical. This is a discrepancy that is probably caused by inaccuracies in the calculated relativistic energies that are used to obtain the correlation energies, which, in turn, are used in calculating the estimates for E_n^e. Another reason for the discrepancy might be that the assumption, according to which the relativistic and nonrelativistic correlation energies are equal, may not hold very well in some cases.

5.5. THE BINDING ENERGIES OF ATOMIC ELECTRONS

A. *X*-Ray Energy Levels

In this section, we discuss the energy levels of the atomic electrons in the ground state of the atom. Although the majority of these levels are in the *x*-ray region, the discussion includes the valence-electron energies in the ground state, which are in the optical region. When atomic energy levels are mentioned, one usually thinks of the optical spectra of atoms that are generated by the excited states of valence electrons; in this section, "atomic energy levels" means the energy levels of all electrons, core and valence, in the ground state. The discussion of the optical spectra, that is, the discussion of what are usually called atomic energy levels, begins in Sec. 6.1.

A systematic study of atomic core energies was first carried out by Slater[46] for neutral atoms up to $Z = 41$ (Nb). A comprehensive critical evaluation of the x-ray measurements was done by Bearden,[47] and on the basis of this work, energy level tables for the core electrons were set up by Bearden and Burr.[48] The data of Bearden and Burr were transformed into data for free atoms by Lotz.[49] Our experimental values for the energy levels of electrons in free atoms are taken from the tables of Lotz.

We want to compare the experimental data with the energy levels obtained from the HF model. An accurate comparison between theory and experiment is possible only with the relativistic model because the nonrelativistic model does not give the spin–orbit splitting, which is the conspicuous feature of the x-ray energy levels. In fact, this was one of the reasons why the relativistic model had to be formulated.

We now compare the orbital-energy levels, obtained from the relativistic HF model, with the empirical data in Lotz's tables. We must recall that such a comparison involves Koopmans' theorem. As we have seen in Sec. 2.6, the binding energy of an electron in the atom is equal to the orbital-energy parameter of the HF model if we assume that Koopmans' theorem is valid, that is, if we assume that the removal of an electron leaves the orbitals of all other electrons unchanged. By comparing the orbital parameters with Lotz's data, we are assuming the validity of Koopmans' theorem.

In Table 5.8, we have the experimental and theoretical values of the electron energy levels for Ne ($Z = 10$), S($Z = 16$), Fe ($Z = 26$), Kr ($Z = 36$), Ag ($Z = 47$), Cs ($Z = 55$), Dy ($Z = 66$), and Hg ($Z = 80$). The layout is self-explanatory. The quantity denoted by Δ is the difference between the calculated and empirical values in percents. The calculated values are taken from the work of Desclaux.[33] All energies are in electron-volts.

From the table, we see that the agreement between theory and experiment is reasonable good. In the majority of cases, the deviation is a few percents. Looking more closely, we see that in all cases, the agreement is very good for the innermost electrons, fair for the electrons with medium energies, and poor for the outer subshells. More accurately, the agreement is quite good for the outermost group, which contains the valence electrons, but poor for the subshells immediately below the valence subshell. It is generally valid that the quantity Δ increases as we move from the innermost shell up to the shells with higher principal quantum numbers. In fact, Δ takes a jump when we move from the shell with n to the shell with $n + 1$. Thus, for example, in the Hg atom, Δ is less than 1% for the K and L shells, it is between 1% and 2% for the M shell, jumps to between 1% and 20% for the N shell, and is

TABLE 5.8. Comparison of the Calculated Energy Levels of Atomic Electrons with the Empirical Data (All Energies in Electron-Volts)

State	Calc.	Obs.	Δ	Calc.	Obs.	Δ	Calc.	Obs.	Δ
	Ne ($Z = 10$)			S ($Z = 16$)			Fe ($Z = 26$)		
1s+	892.7	870.1	2.6	2,511	2476	1.4	7,177	7,117	0.8
2s+	52.66	48.47	8.6	246	232	6.2	883	851	3.8
2p−	23.20	21.66	7.1	183	170	7.5	757	726	4.3
2p+	23.07	21.56	7.0	181	168	8.0	745	713	4.5
3s+				24.14	20.20	19.5	116	98	18
3p−				11.64	10.36	12.4	76	61	25
3p+				11.55	—	—	75	59	27
3d−							16	9	80
3d+							16	—	—
4s+							7.174	7.870	−8.8
	Kr ($Z = 36$)			Ag ($Z = 47$)			Cs ($Z = 55$)		
1s+	14,409	14,327	0.6	25,656	25,520	0.5	36,175	35,987	0.5
2s+	1,961	1,927	1.7	3,862	3,812	1.3	5,777	5,717	1.0
2p−	1,765	1,731	1.9	3,578	3,530	1.3	5,420	5,362	1.1
2p+	1,710	1,678	1.9	3,402	3,357	1.3	5,066	5,014	1.0
3s+	305	292	4.6	745	724	2.9	1,245	1,220	2.1
3p−	234	222	5.6	630	608	3.7	1,095	1,068	2.5
3p+	226	214	5.7	598	577	3.7	1,025	1,000	2.6
3d−	102.7	95.0	8.2	398	379	4.9	765	742	3.1
3d+	101.4	93.8	8.1	391	373	4.9	750	728	3.1
4s+	32.31	27.51	17.4	116	101	15	254	233	8.9
4p−	14.73	14.67	0.4	78	69	13	197	174	13.5
4p+	13.99	14.00	−0.06	73	63	16	183	164	11.7
4d−				14	11	30	90	81	11
4d+				14	10	36	87	79	11
5s+				6.451	7.58	−15	26	25	42
5p−							20	14	41
5p+							18	12.3	46
6s+							3.49	3.89	−10.3
	Dy ($Z = 66$)			Hg ($Z = 80$)					
1s+	54,092	53,792	0.5	83,627	83,108	0.6			
2s+	9,129	9,050	0.9	14,968	14,845	0.8			
2p−	8,661	8,585	0.9	14,332	14,214	0.8			
2p+	7,859	7,794	0.8	12,381	12,288	0.8			
3s+	2,084	2,048	1.8	3,621	3,567	1.5			
3p−	1,880	1,846	1.9	3,336	3,283	1.6			
3p+	1,711	1,678	1.9	2,898	2,852	1.6			
3d−	1,365	1,335	2.2	2,432	2,390	1.8			
3d+	1,328	1,298	2.3	2,340	2,300	1.7			
4s+	437	416	5.1	833	806	3.4			
4p−	355	331	7.3	711	683	4.0			
4p+	317	297	6.7	603	579	4.2			
4d−	181	164	10.2	402	382	5.4			
4d+	174	157	10.6	382	363	5.3			
4f−	13.5	6	125	122	107	13.7			

TABLE 5.8. (*Continued*)

State	Calc.	Obs.	Δ	Calc.	Obs.	Δ	Calc.	Obs.	Δ
$4f+$	12.5	—	—	117	103	13.8			
$5s+$	60	50	20	138	125	11.0			
$5p-$	35	33	6.5	96	85	13.2			
$5p+$	30	28	7.8	77	68	13.7			
$5d-$	—	—	—	18	14	26.3			
$5d+$	—	—	—	16	12	30.3			
$6s+$	5.10	5.93	-13.9	8.92	10.4	-14.2			

between 10% and 30% for the O shell. Thus, the agreement between theory and experiment becomes poorer as we move from the inner electrons to the outer electrons.

It is noteworthy that the calculated values are always below the empirical except in the case of the valence electrons. We note also that although the agreement between theory and experiment is reasonable in most cases, it is nowhere as good as in the case of the total energies of atoms.

B. Determination of the Core / Valence Separation

We have seen in Sec. 5.2/D that the HF model can be used to determine the electron configurations of atoms and that the configurations obtained from the theory are, in most cases, identical with the configurations derived from the optical spectra. We now show that the relativistic HF model can be used to establish another important atomic property, which is the core/valence separation.

In Sec. 6.1, we will discuss the optical spectra of atoms and we will see that the valence electrons of an atom are the electrons that generate the optical spectrum. We now show that the binding energies can also be used to define the orbitals that can be classified as valence orbitals. We discuss here those atoms in which the incomplete d and f subshells are being built up.

In Fig. 5.10, we plot the binding energies of the atoms in the series extending from Sc ($Z = 21$) to Zn ($Z = 30$). The binding energies are from the relativistic HF calculations of Desclaux.[33] In the upper part of the diagram, we have the orbital energies of the ($4s^+$) and ($3d^+$) electrons. The energy scale on the left side of the diagram refers to these energy levels. On the lower part of the diagram, we have the energy levels of the ($3p +$) electrons. The energy scale on the right side of the diagram refers to these levels. Note the difference in magnitude between the two scales.

In this diagram, we have the iron series atoms, that is, the atoms in which the ($3d$) shell is built up. Question: Are the ($3d$) electrons valence or core electrons? We answer this question on the basis of the position of the ($3d$)

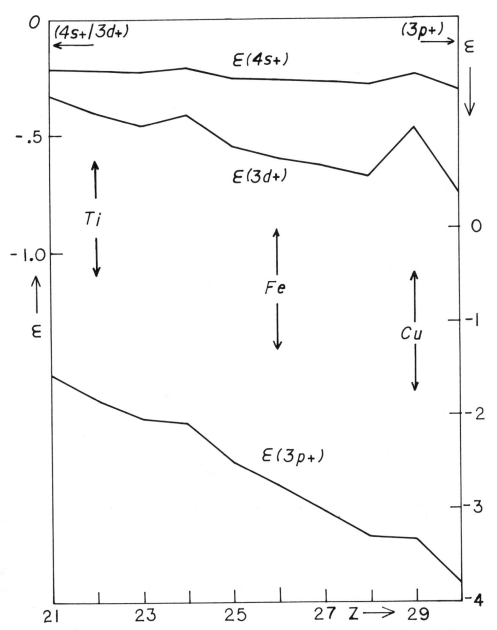

Fig. 5.10. The relative position of the $(4s+)$, $(3d+)$, and $(3p+)$ energy levels in the iron series atoms between Sc and Zn.

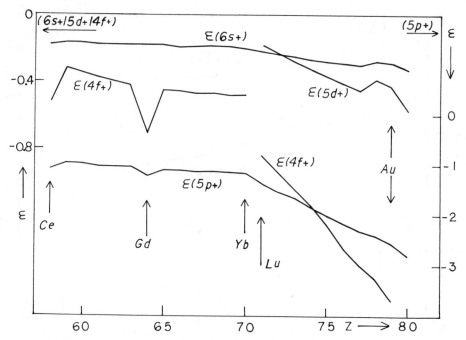

Fig. 5.11. The relative position of the $(6s+)$, $(5d+)$, $(4f+)$, and $(5p+)$ energy levels in the atoms between Ce ($Z = 58$) and Hg ($Z = 80$).

levels. It is obvious that the electrons in the outermost shell, the $(4s)$ electrons, are valence electrons. We can also assume that the $(3p)$ electrons in their compact $(3p)^6$ group are core electrons. Thus, the $(3d)$ electrons are valence electrons if their energies are close to the $(4s)$ energies and they are core electrons if their energies are close to the $(3p)$ levels.

The diagram provides a clear answer. Throughout the series, the $(3d)$ levels are quite close to the $(4s)$ levels; in fact, the $(4s)$ levels are fairly constant at about -0.2 a.u. and the $(3d)$ levels are about -0.5 a.u. On the other hand, the energies of the $(3p)$ orbitals are ranging from about -2.0 a.u. to -4.0 a.u. So, as we see from the diagram, the distance between the $(3d)$ and the $(4s)$ is about 0.3 a.u. and the distance between the $(3d)$ and the $(3p)$ is about 2.5 a.u.; the latter is about 8 times larger than the former. Thus, it is clear that if we accept this criterion as a valid definition, we must classify the $(3d)$ electrons as valence electrons. We note that the argument is valid for the $(3d)^{10}$ configuration of the Cu atom as well.

We have drawn up a similar diagram for the series Y ($Z = 39$) to Ag ($Z = 47$), but it is so similar to Fig. 5.10 that it does not need to be presented. Instead, we look now at Fig. 5.11, in which we have data for the atoms from Ce ($Z = 58$) to Hg ($Z = 80$). In the upper part of the diagram, we have first the $(6s+)$ levels for all atoms. These are remarkably constant for the whole range at about -0.2 a.u. Also in the upper part, we have the $(4f+)$ levels for the rare earths, that is, from Ce $(Z = 58)$ to Yb $(Z = 70)$. On the right side of the upper part, we have the $(5d+)$ levels from Lu $(Z = 71)$ to Hg $(Z = 80)$. The energy scale on the left side of the diagram refers to the $(6s+)$ and $(5d+)$ levels plus the $(4f+)$ levels of the rare earths.

In the lower part of the diagram, we have the first $(5p+)$ levels for the whole range and also the $(4f)$ levels for the Lu–Hg range. These energies are defined in terms of the scale on the right side of the diagram.

We ask the question again: Are the $(4f+)$ electrons of the rare earth atoms and $(5d+)$ electrons of the third long period valence electrons or core electrons? Assuming that the position of the energy levels is a valid criterion, we must give the answer that these electrons are valence electrons. The position of the $(5d+)$ levels for Lu $(Z = 71)$ and for Hf $(Z = 72)$ are especially interesting: they are actually above the $(6s+)$ levels, although only by a very small amount for Hf. These are the only atoms for which the (nd) level is higher than the $(n+1, s)$. It is also very interesting to compare the $(4f+)$ electrons in the rare earth range with the position of the $(4f+)$ levels in the Lu–Hg series. In the Lanthanide series, the $(4f+)$ levels are fairly constant and are clearly in the valence range. After the $(4f)^{14}$ configuration is completed at Yb $(Z = 70)$, the $(4f)$ levels becomes more negative as we proceed from Yb to the Hg $(Z = 80)$. Here in this range, the $(4f)$ electrons rapidly become core electrons, as we see from the fact that their energy levels are very close to the $(5p)$ levels and actually sink below the $(5p)$ levels at $Z = 74$ (W). Thus, we see clearly that for the range $Z = 58$ to 70, the $(4f)$ electrons are valence electrons, and from $Z = 71$ to $Z = 80$, they are core electrons. (Note that the two set of $(4f)$ levels are on different scales.)

Summarizing the preceding discussion, we can say that the relativistic HF model can be used successfully to identify the valence electrons in any atom, that is, it can be used to establish the core/valence separation. The conclusions reached here are in agreement with the conclusions reached on the basis of optical spectra, which we will discuss in Sec. 6.1. The interesting feature of this discussion is that this is a purely theoretical argument based on the HF model and it forms the counterpart of the experimental determination derived from the optical spectra.

5.6. ANALYTIC HF CALCULATIONS

A. Introduction

In discussing the machinery of actual HF calculations, we have pointed out that the exact solutions of the HF equations, the one-electron functions, are obtained in these calculations as numerical tables, that is, a set of numbers that gives the value of the wave function at a selected set of radii. This method of presentation had been introduced by Hartree in his very first calculations and has been the standard procedure ever since. It is clear that the applications of such numerical functions are difficult in the problems of molecular quantum mechanics. Therefore, a number of methods have been developed for the calculation of HF orbitals, or good approximations to them, in analytic form.

Analytic HF calculations do not belong to the main line of this book because their significance lies in applications rather than in the study of the atomic structure. Nevertheless, they are very important in molecular quantum mechanics and also important in the method of configuration interaction (CI), which will be presented in the second volume of this work. Thus, in keeping with our policy of establishing *points of contact* with adjacent fields, in this case with molecular quantum mechanics, we discuss here analytic methods in some detail.

B. Calculations of Simple One-Electron Wave Functions

In the early years of quantum mechanics, when fast computers were not yet available, the question arose whether it is possible to construct atomic wave functions that are built from simple combinations of elementary functions. Attempting to answer this question, Fock and Petrashen[50] made some simple calculations for light atoms. We present some of their work here.

Because the calculations were done for atoms with complete groups, the starting point is the energy expression for complete groups, Eq. (2.116). In order to calculate the integrals $I(nl)$, $F_k(nl, n'l')$, and $G_k(nl, n'l')$, which occur in the energy expression, Fock and Petrashen used the following trial functions for the one-electron wave functions of the $1s$, $2s$, and $2p$ electrons:

$$P(1s|r) = Are^{-\alpha r}, \tag{5.61a}$$

$$P(2s|r) = Br\left[1 - \tfrac{1}{3}(\alpha + \beta)r\right]e^{-\beta r}, \tag{5.61b}$$

$$P(2p|r) = Cr^2e^{-\gamma r}. \tag{5.61c}$$

The quantities are in atomic units. A, B, and C are normalization constants for which one obtains

$$A = (4\alpha^3)^{1/2}, \tag{5.62a}$$

$$B = \left[\frac{12\beta^5}{\alpha^2 - \alpha\beta + \beta^2}\right]^{1/2}, \tag{5.62b}$$

$$C = 2(\gamma^5/3)^{1/2}. \tag{5.62c}$$

The $2s$ wave function, Eq. (5.61b), is constructed in such a way as to be orthogonal to the $1s$ wave function at any values of α and β. The real numbers α, β, and γ are variational parameters that were determined from the energy minimum principle.

By putting the trial functions given before into the energy expression, Eq. (2.116), one obtains that expression as a function of the α, β, and γ parameters. Fock and Petrashen found that the minimum energy and the corresponding parameter values for the Na^+ ion were

$$E = -4358.8 \text{ eV};$$
$$\alpha = 10.68, \qquad \beta = 4.22, \qquad \gamma = 3.49.$$

Comparing this with the nonrelativistic HF value given in Table 5.8, we have

$$E = -4397.62 \text{ eV},$$

and the comparison shows that the calculations of Fock and Petrashen reproduce the exact HF value with a 0.88% accuracy.

It is interesting that trial functions as simple as these can reproduce the HF energy with a less than 1% error. However, this result, and similar results for other atoms, show only that the total energy can be reproduced with a high accuracy with simple expressions for the wave functions. The results do not mean that other quantities, like the total radial density, come out accurately from such calculations. These results are just the demonstration of a theorem of quantum mechanics according to which a poor trial function can give very good results for the total energy in a variational calculation.[†] But the results of Fock and Petrashen also demonstrated that meaningful results can be obtain with simple wave functions, at least for light atoms.

Numerous such calculations for light atoms are available. We mention the early work of Morse, Young, and Haurwitz,[51] which contains calculations for several multiplets of the atoms from He to Mg. The reader might also look at the discussion of such calculations in Slater's book [Slater I, 348].

C. Introduction of the Slater-Type Orbitals (STOs)

In an early paper, Slater introduced a method that proved to have far-reaching influence on atomic and molecular quantum mechanics. We call this the method of Slater-type orbitals, STOs.

According to this work, the one-electron wave functions of an atom can well be approximated by the expression

$$P(nl|r) = Ar^{n^*} \exp[-(Z - \gamma)r/n^*]. \tag{5.63}$$

The most conspicuous feature of this method is that Slater approximated the atomic one-electron functions by nodeless functions, that is, by functions

[†]A clear demonstration of this is given in the book by Shiff, Ref. 20, in Problem 5 on p. 295.

that do not have points where they are zero except at $r = 0$ and $r = \infty$. Such wave functions are called *Slater-type orbitals, or STOs*.

For the determination of the parameters, Slater set up semiempirical rules. The one-electron wave functions given by Eq. (5.63) are solutions of a Schroedinger equation with the following potential:

$$U(r) = -\frac{Z - \gamma}{r} + \frac{n^*(n^* - 1)}{2r^2}, \tag{5.64}$$

and the energy, associated with the STOs, is

$$E = -\frac{(Z - \gamma)^2}{2(n^*)^2}. \tag{5.65}$$

Slater demanded that the parameters γ and n^* be determined in such a way that the expectation values of some typical quantities, like ionization potentials, x-ray levels, and diamagnetic susceptibilities, should be good approximations to the empirical when computed with the STOs. By this method, Slater arrived at the following rules.

The effective principal quantum number n^* should be related to the real one as follows: $n = 1$, $n^* = 1$; $n = 2$, $n^* = 2$; $n = 3$, $n^* = 3$; $n = 4$, $n^* = 3.7$; $n = 5$, $n^* = 4$; and $n = 6$, $n^* = 4.2$. In order to determine γ, we arrange the electrons into the groups

$$(1s), (2s, 2p), (3s, 3p), (3d), (4s, 4p),$$
$$(4d, 4f), (5s, 5p), \ldots,$$

that is, we put the (ns) and (np) electrons in one group and the (nd) and (nf) into another. Then, for γ, we postulate the following:

1. The electrons in any selected group contribute to γ the value of 0.35 except in the $(1s)$ group, where the contribution is 0.3.
2. The electrons outside of a selected group do not contribute to the γ of the selected group.
3. In the case of a selected (ns, np) group, the electrons in an inner group with principal quantum number $(n - 1)$ contribute to γ the value of 0.85. The electrons of the inner groups with principal quantum number less than $(n - 1)$ contribute 1.0. In the case of an (fd) group, all inner electrons contribute to γ the value of 1.0.

Slater has shown that by following these rules, one gets good approximations for the quantities mentioned before, that is, for ionization potentials, x-ray energies, and diamagnetic susceptibilities.

Slater's work was, and is, interesting per se, but the real significance of this work lies outside of the scope of the original formulation. Slater's ideas found application in two areas, unrelated to the original work:

1. In molecular quantum mechanics, the use of STOs proved to be very useful. Also, in configuration interaction (CI) studies of atoms, the use of STOs proved very effective. Thus, Slater introduced a simple function that is particularly adapted to certain types of atomic and molecular studies.

2. The work of Slater has shown that the energy levels of atomic electrons and the total radial electron densities of atoms can well be approximated by *modeless* one-electron wave functions that are not orthogonalized to each other. In a later part of this book, we will see that such STOs can be considered as *pseudoorbitals*. As we show in subsequent parts of this work, it is possible to formulate the HF method in terms of pseudoorbitals and such a formulation is exactly equivalent to the conventional formulation.[53] Thus, Slater's work is the precursor of the Hartree–Fock–Pseudopotential[53] method.

D. The Hartree – Fock – Roothaan Expansion Method

The approaches that we have discussed in the previous sections are ad hoc methods. A systematic method for the construction of analytic HF orbitals was developed by Roothaan.[54] Since this method has its roots in molecular quantum mechanics, in presenting this method, we must start with a short digression in that direction.

The direct generalization of the HF model in molecular physics is the so-called Hund–Mulliken, or molecular orbital (MO), method. In this method, just like in the HF model of atoms, each electron is assigned a one-electron wave function that is called a molecular orbital (MO). The salient feature of this function is that it extends to the whole area of the molecule.

Although the basic idea of this model was introduced by Hund and Mulliken, the mathematical framework of the theory is the work of Roothaan. Roothaan postulated that the molecular orbitals should be approximated by linear combinations of atomic orbitals (LCAO-MO). Already at this step, we have arrived at the main motivation for the introduction of analytic HF functions. The question is this: If we approximate the MOs by atomic orbitals, what kind of atomic orbitals should we use? Because the molecules do not possess the central symmetry of the atoms, the HF orbitals, which are given in the form of numerical tables, would be very difficult to use in the construction of molecular orbitals. This is the argument for the construction of atomic orbitals in analytic form, that is, for the construction of analytic HF functions.

We present here that part of Roothaan's work that deals with the construction of atomic one-electron HF wave functions. We do not discuss the molecular part of Roothaan's method, which is the main part of his effort. Thus, the reader must be aware that what follows here is only part of Roothaan's work.

The goal is, therefore, to put the HF one-electron functions into analytic form. Now, as we have seen in the preceding parts of this work, the HF model can be developed in relativistic as well as in nonrelativistic form; it has different forms for atoms with complete groups, for atoms with one or more incomplete groups, and for the average of an atomic configuration. The Hartree–Fock–Roothaan (HFR) method can be applied to any of these formulations. Because of its central position in atomic theory, we demonstrate the application of the HFR method for the average of a configuration. We discuss also the case of atoms with complete groups.

We have derived the HF equations for the average-of-configuration model by varying the energy expression given by Eq. (3.96) under the subsidiary conditions given by Eq. (3.98). The results were the HF equations given by Eq. (3.100). The energy minimum principle expressed by Eq. (3.99) is equivalent to the following set of conditions:

$$\langle \delta P_{nl} | H_{av} - \varepsilon_{nl} | P_{nl} \rangle - \sum_{\substack{n' \\ (n' \neq n)}} \lambda_{nl,n'l} \langle \delta P_{nl} | P_{n'l} \rangle = 0 \qquad \text{(for all } (nl)), \quad (5.66)$$

where the H_{av} operator is given in Eq. (3.119b), that is,

$$H_{av} = -\frac{1}{2} \frac{d^2}{dr^2} + \frac{l(l+1)}{2r^2} - \frac{Z}{r} + U_{av}, \qquad (5.67)$$

and the potential U_{av} is given by Eq. (3.116). The HF equations in Eq. (3.100) are obtained from the conditions given by Eq. (5.66).

Now let us assume that the HF one-electron wave function can be put in the analytic form

$$P_{nl}(r) = \sum_{n''p''} C_{n''p''}(nl) \chi_{n''p''}^{l}(r), \qquad (5.68)$$

where the function $\chi_{n''l''}^{l}(r)$ is a normalized Slater-type orbital (STO), that is, it has the form

$$\chi_{n''l''}^{l}(r) = A_{n''p''}^{l} r^{n''} e^{-(\xi_{n''p''}^{l})r}. \qquad (5.69)$$

In this STO, n'' is an effective principal quantum number that should not be confused with the real quantum number n of P_{nl}. The parameter p'' is just a serial number permitting us to have several different χ functions with the same effective principal quantum number. l is the real azimuthal quantum number of P_{nl}. The normalization constant is

$$A_{n''p''}^{l} = \left[\frac{\left(2\xi_{n''p''}^{l}\right)^{2n''+1}}{2n''} \right]^{1/2}. \qquad (5.70)$$

As we see from Eq. (5.69), the zeta parameter, and, consequently also the χ function, is independent of the real principal quantum number n.

Thus, the HF orbital is expressed as an expansion of the analytic functions χ, which we call basis functions. The expansion coefficients, $C_{n''p''}(nl)$, are forming a different set of each (nl). We have one set of basis functions for each l. There will be a set of ξ's and a set of basis functions for the s orbitals, for the p orbitals, for the d orbitals, and so on. There will be a separate set of expansion coefficients for the $1s$ orbital, for the $2s$ orbital, for the $3s$ orbital, and so on.

Next, we substitute the expansion given by Eq. (5.68) into the variational condition, Eq. (5.66). By fixed basis functions, the variation indicated by δP_{nl} means the variation with respect to all coefficients in the expansion of P_{nl}, Eq. (5.68). Before we carry out the variation, a small change is necessary in Eq. (5.66). We put the Lagrangian multipliers in such a form that shows the explicit dependence on P_{nl}. This is accomplished by multiplying Eq. (3.100) from the left by P_{nl} and integrating over r. Using Eq. (5.67), we obtain

$$\lambda_{nl,n'l} = \langle P_{n'l}|H_{av}|P_{nl}\rangle \qquad (n' \neq n). \tag{5.71}$$

Now we substitute this into Eq. (5.66) and change the summation index n', which is a dummy variable, to n'''. Then from Eq. (5.66), we get

$$\langle \delta P_{nl}|H_{av} - \varepsilon_{nl}|P_{nl}\rangle$$
$$- \sum_{n'''} \langle \delta P_{nl}|P_{n'''l}\rangle\langle P_{n'''l}|H_{av}|P_{nl}\rangle = 0 \qquad \text{(for each } (nl)). \tag{5.72}$$

Next, we substitute the expansion, Eq. (5.68), into this formula via the δP_{nl} and the P_{nl}. We obtain

$$\delta P_{nl} = \delta\left[\sum_{n'p'}C_{n'p'}(nl)\chi^l_{n'p'}(r)\right], \tag{5.73a}$$

$$P_{nl} = \sum_{n''p''}C_{n''p''}(nl)\chi^l_{n''p''}(r), \tag{5.73b}$$

and

$$\langle \delta P_{nl}|H_{av} - \varepsilon_{nl} - \Omega(nl)|P_{nl}\rangle$$
$$= \sum_{n'p'}\sum_{n''p''}[\delta C_{n'p'}(nl)]C_{n''p''}(nl)$$
$$\times \langle \chi^l_{n'p'}|H_{av} - \varepsilon_{nl} - \Omega(nl)H_{av}|\chi^l_{n''p''}\rangle = 0. \tag{5.74}$$

We have one such equation for each (nl) group. The Ω operator is shorthand for the following expression:

$$\Omega(nl) = \sum_{n'''}|P_{n'''l}\rangle\langle P_{n'''l}| \qquad (n''' \neq n). \tag{5.75}$$

Now we carry out the variation in Eq. (5.74) and we recall that each coefficient belonging to the expansion of P_{nl} must be varied independently. We obtain

$$\sum_{n''p''} C_{n''p''}(nl) \left[M^{(nl)}(n'p'|n''p'') - \varepsilon_{nl} S^{(l)}(n'p'|n''p'') \right] = 0$$

$$\text{(for each } (n'p')). \quad (5.76)$$

In this equation, we introduced the following notations:

$$M^{(nl)}(n'p'|n''p'') \equiv \langle \chi^l_{n'p'} | H_{av} - \Omega(nl) H_{av} | \chi^l_{n''p''} \rangle, \quad (5.77a)$$

$$S^{(l)}(n'p'|n''p'') \equiv \langle \chi^l_{n'p'} | \chi^l_{n''p''} \rangle. \quad (5.77b)$$

The physical meaning of Eq. (5.76) is simple. The indices $(n'p')$ and $(n''p'')$ are the summation indices in the HFR expansion of the orbital P_{nl}. Both the C coefficients and the M-matrix components depend on the real quantum numbers (nl) of the group. The matrix components of the overlap between the basis functions, the S-matrix components, depend on the azimuthal quantum number l. We have one equation like Eq. (5.76) for each pair of expansion coefficients $(n'p')$; thus, in Eq. (5.76), we have a set of linear homogeneous algebraic equations for the determination of the expansion coefficients, $C_{n'p'}(nl)$. The solutions of this set of equations will provide the expansion of the HF orbital, in terms of basis functions, as given by Eq. (5.68). In order to get all HF functions in the expanded analytic form, we have to solve an equation system like Eq. (5.76) for each (nl) group.

The condition for the existence of a set of solutions of Eq. (5.76) is that the *secular equation* should be zero:

$$\det \left[M^{(nl)}(n'p'|n''p'') - \varepsilon_{nl} S^{(l)}(n'p'|n''p'') \right] = 0. \quad (5.78)$$

In the Hartree–Fock–Roothaan method, the calculation of HF orbitals is carried out as follows. First, the basis functions of the expansion, Eq. (5.68), are selected. These basic functions are a separate set for each l occurring in the atom. Then the matrix components, M and S, are constructed for the secular equations. These will be different for each (nl) group. In order to construct these matrix components, an initial set of C coefficients must also be chosen because the M components depend on the density matrices of the HF orbitals via the potential U_{av} in Eq. (5.67) and via the operator Ω. After the M and S matrices are constructed for each (nl), the secular equation, Eq. (5.78), must be solved for each (nl). From this step, we get the orbital parameter ε_{nl}. With this ε_{nl}, we must go back to the system of equations given by Eq. (5.76) from which we obtain the expansion coefficients with the exception of one; this one can be used to normalize the orbital P_{nl}. By having obtained the orbital parameters and the expansion coefficients for each (nl), the calculation of a set of solutions is completed.

It is evident that this procedure, as described here, is only one cycle in the iteration process that is needed to get the Roothaan-type solutions of the HF equations. In the previous procedure, we started with an initial set of C coefficients for the construction of the matrix in the secular equation. The final set of C coefficients, which we obtain from the procedure outlined here, in general, will be different from the initial set. The same is true for the orbital parameters. Thus, we must repeat the outlined procedure with the new set of coefficients serving as starting values. The iteration process must be continued until the final and initial coefficients and orbital parameters become identical.

We want to describe clearly the difference, from the mathematical point of view, between the numerical and analytic HF procedures. In the numerical procedure, we have one integro-differential equation for each (nl) group. The potential (operator) in these equations depends on all HF orbitals; therefore, the equations are coupled. The solutions are obtained by an iteration procedure that we have called the "self-consistent field."

In the analytic HF method, we get a set of linear homogeneous algebraic equations plus one secular equation (determinantal equation) for each (nl) group. The matrix components of the linear equations and the matrix of the secular equation depend on all HF orbitals, that is, on the expansion coefficients of all orbitals. The solutions are again obtained by an iteration procedure.

Thus, from the mathematical point of view, the main difference between the two methods is that in the numerical HF method, we have a set of integro-differential equations; in the analytic procedure, we have a set of linear homogeneous algebraic equations plus a set of determinantal equations. In both cases, the solutions are obtained by iteration.

Using the language of Dirac's formulation of the quantum mechanics, we can say that Roothaan transformed the HF equations into a matrix representation defined by the basis functions. Depending on the basis functions, many different representations are possible.

We want to discuss shortly the case of complete groups. If we have an atom with complete groups, then the HF equations are given by Eq. (2.97b). Using Eq. (2.98) for the potential operator, we can write the HF equations in the form

$$H^{(l)}P_{nl} = \varepsilon_{nl}P_{nl}, \tag{5.79}$$

where

$$H^{(l)} = \frac{1}{2}\frac{d^2}{dr^2} + \frac{l(l+1)}{2r^2} - \frac{Z}{r} + U^{(l)}(r). \tag{5.80}$$

The operator U^l is defined by Eq. (2.98). An inspection of that formula reveals that this operator depends only on l and not on n. Thus, the whole Hamiltonian in this formulation depends only on l.

A further simplification here is that we do not have any Lagrangian multipliers; those can be eliminated by a unitary transformation.

Transforming the problem into the Roothaan representation, we again get the secular equation given by Eq. (5.76), but now M will be simpler and will not depend on n. We obtain

$$\sum_{n''p''} C_{n''p''}(nl) \left[M^{(l)}(n'p'|n''p'') - \varepsilon_{nl} S^{(l)}(n'p'|n''p'') \right] = 0$$

$$\text{(for each } (n'p')), \quad (5.81)$$

where, instead of Eq. (5.77a), we have now

$$M^{(l)}(n'p'|n''p'') \equiv \langle \chi^l_{n'p'} | H^{(l)} | \chi^l_{n''p''} \rangle, \qquad (5.82)$$

and for the S matrix, we again have Eq. (5.77b). The secular equation becomes

$$\det \left[M^{(l)}(n'p'|n''p'') - \varepsilon_{nl} S^{(l)}(n'p'|n''p'') \right] = 0. \qquad (5.83)$$

Thus, in the case of complete groups, the matrix of the linear equations and the matrix of the secular equation will depend on n only through the orbital parameter ε_{nl}. This means that the different orbital parameters for a given l will be the eigenvalues of the same secular equation. Thus, for example, the ε_{1s} will be the lowest eigenvalue of the secular equation for $l = 0$, ε_{2s} the next higher eigenvalues, and so on.

We note that, on the basis of the arguments presented here, it is easy to construct the HFR equations for the case of atoms with one or more incomplete groups. All one has to do is to insert the appropriate HF Hamiltonian operator into the M matrix of the secular equation and to do this in such a way that the Lagrangian multipliers (if there are any) are included in the substitution.

As we have seen in the preceding chapters, the HF model can be formulated relativistically as well as nonrelativistically. The relativistic Hartree–Fock–Roothaan procedure was formulated by Kim.[28] In this formulation, Kim used the Hamiltonian given by Eq. (5.5) in which we have the relativistic Dirac one-electron operators H_D and the ordinary electrostatic interaction terms. The one-electron wave functions were the 4-component Dirac-type functions given by Eq. (4.102). The radial functions in that expression were expanded in terms of STOs. The relativistic interaction between the electrons was represented by the Breit operator and was treated as a first-order perturbation. Kim formulated the theory for atoms with complete groups. Calculations were presented for the atoms He, Be, and Ne.

Large-scale calculations with the nonrelativistic HFR method were carried out by Clementi.[42] Clementi constructed computer programs for the large-scale production of atomic HFR wave functions. Calculations were first made[42] for a large number of atoms and ions up to $Z = 30$ (zinc). Later Clementi and Roetti[56] extended these calculations up to $Z = 54$ (xenon). A significant feature of Clementi's calculations was that besides the calculations of the expansion coefficients, he also carried out the optimization of the orbital exponents. By this, we mean the repetition of the full self-consistent calculation of the expansion coefficients with different sets of orbital ξ parameters. (See Eq. (5.69)). By minimizing the energy with respect to the orbital exponents, Clementi achieved very good results with relatively short expansions for the HF orbitals.

5.7. SUMMARY

At the beginning of the chapter, we formulated the question: How accurate is the HF model? The review of the calculated values of various quantities provided a fairly clear answer to this question. We summarize our conclusions.

The calculated quantities can be divided into two main groups. First, we reviewed the results for calculated energies, that is, for the total energies of atoms as well as for the binding energies of atomic electrons. Second, we reviewed the results for radial densities that are, of course, based on the results for one-electron HF wave functions. We also considered various quantities that are related to either of these two groups.

In connection with the total relativistic energies of atoms, we have established that calculations based on the HFD model and calculations based on the Breit–Pauli perturbation procedure lead to the same results in very good approximation. The difference between the two is never greater than 1% of the total energy and it is much less for many atoms. Because the experimental total energies are known only up to Ca ($Z = 20$), we have used the comparison between the two relativistic methods to estimate the maximum error with which a relativistic calculation reproduces the total energy of an atom and established the maximum error as 1%.

Next, we used the relativistic calculations to construct a semitheoretical formula for the correlation energy. Using this formula, we established that the correlation energy is important for light atoms and declines as we go to higher Z; the exact opposite is the case with the relativistic correction, which is small for light atoms and increases rapidly with increasing Z.

We have also reviewed the calculations of total energy with the nonrelativistic HF model and established that the maximum error in these calculations, relative to the exact eigenvalue of the nonrelativistic Schroedinger equation, is 0.2% of the total energy and much less for many atoms.

We have reviewed, in connection with the total energies, the calculation of ionization potentials and many-valence-electron energies. It is established that relativistic HF calculations give excellent results for the ionization energies in case of incomplete p subshells, fair results for incomplete d

subshells, and poor results for f subshells. The calculations of many-valence-electron energies show that the relative size of the correlation energy is large for one or two valence electrons and declines as the number of valence electrons increases. It is established that the accuracy of the HF model is not as high for the total energy of a few valence electrons as for the total energy of the whole atom with a large number of electrons.

Still in connection with total energy calculations, we have discussed the theoretical determination of electron configurations, that is, the theoretical determination of the occupation numbers of electron states. We established that the HF model in most cases yields the correct occupation numbers. We have seen, however, that there are discrepancies in a number of cases. It is established that a theoretical determination of occupation numbers requires the calculation of the total energy with an accuracy much higher than what is possible with the HF model.

A comparison between the calculated and empirical binding energies of electrons (x-ray levels for core electrons and optical levels for the ground state of the valence electrons) shows that the relativistic HF model yields good results for the innermost electrons, but the quality of the results declines as we move from the inner to the outer shells. It is established that the core/valence separation can be determined by using the orbital parameters of the relativistic HF model; identifying the valence electrons this way, we obtain the same results as were obtained from the analysis of the optical spectra.

Turning to the second group of quantities, the radial densities and quantities related to them, we observed that HF model gives an accurate picture of the electron distribution inside of an atom by providing the one-electron wave functions. By providing accurate one-electron wave functions, the HF model produces a quantity that is vitally important in many molecular and solid-state calculations, besides providing the explanation of such fundamental properties as the shell structure of atoms. Unfortunately, a direct comparison between theory and experiment is not possible in the case of densities or, more accurately, we have not found references to an existing method with which such comparison could be made. Indirect comparison is provided by the calculation of diamagnetic susceptibilities, where the agreement between theory and experiment is quite good in the case of noble gases. We have used the densities to determine the sizes of the atoms and found that the atomic sizes, as functions of Z, lead to the same conclusions about the structure of atoms as the conclusions reached on the basis of chemical data.

Finally, we noted that the HF model can be transformed into matrix representation and the one-electron wave functions can be determined as linear combinations of simple analytic functions. These are important in molecular calculations as well as in configuration interaction studies.

PART TWO

VALENCE-ELECTRON THEORIES

CHAPTER 6

HARTREE – FOCK THEORY FOR VALENCE ELECTRONS

6.1. INTRODUCTION: AN OVERVIEW OF ATOMIC SPECTRA

Description of the basic structure of atomic spectra; atoms with one, two, and more valence electrons; identification of terms by configuration and by L, S, and J; the Rydberg-type structure of the series; special features of the rare-earth and noble-gas spectra.

6.2. LS COUPLING FOR SEVERAL INCOMPLETE GROUPS

Description of the general method for the construction of the diagonal as well as nondiagonal matrix components of the Hamiltonian; explicit formulas for the $(l^q l')$-type configurations; list of references for the tables of coefficients.

6.3. HF THEORY FOR ONE VALENCE ELECTRON: "FROZEN-CORE" APPROXIMATION

Formulation of the HF model for an atom with the core orbitals fixed (frozen); comparison with the fully varied HF model; advantages of the frozen core; the dependence of the HF potential on the valence-electron orbital; Koopmans' theorem in the frozen-core approximation.

6.4. HF THEORY FOR SEVERAL VALENCE ELECTRONS: "EXTENDED FROZEN-CORE" APPROXIMATION

Conventional formulation of the HF model for an atom with the $(l^q l')$ configuration; separation of the core and nonrunning valence contributions in the HF potential; introduction of the "extended frozen core" and the simplifications achieved by this step; Koopmans' theorem for extended frozen core; summary of the properties of the HF model with the frozen-core approximation.

6.5. RADIATION TRANSITION PROBABILITIES

Survey of Einstein's formula for the transition probabilities; derivation of the formula for dipole transitions. A. Transition probabilities, oscillator strengths, selection rules, and summation rules for one-electron atoms. B. Construction of the matrix components of dipole moment for atoms with several valence electrons; formulas in the *LS* coupling and in the *LSJM* coupling; transitions involving equivalent electrons; fractional parentage for transitions involving equivalent electrons.

6.6. SPIN – ORBIT COUPLING

The Hamiltonian operator with spin – orbit coupling; construction of the matrix component for atoms with one and with several valence electrons; transitions in the presence of spin – orbit interaction.

6.7. ZEEMAN EFFECT

The Hamiltonian with spin – orbit coupling and external magnetic field; the matrix components of the magnetic interaction; the Lande *g* factor; the *g* factor for states that are linear combinations of different (*LS*) terms.

6.8. SUMMARY AND DISCUSSION

Summary of the main steps in the empirical classification of spectra; the hierarchy of approximations leading to the theoretical description of the spectra; the necessity for the introduction of correlation effects into the theory.

6.1. INTRODUCTION: AN OVERVIEW OF ATOMIC SPECTRA

A very large amount of information about the electronic structure of atoms comes from the analysis of the atomic spectra. The bulk of information on the spectra is contained in the four-volume work entitled *Atomic Energy Levels*, published by the National Bureau of Standards (Atomic Energy Level Tables, AET, I, II, III, IV). The first three volumes by Moore contain the data on atoms and ions of the elements from H to Ac ($Z = 89$). The fourth volume by Martin, Zalubas, and Hagan contains the data on the rare-earth elements, that is, on atoms and ions from La ($Z = 57$) to Lu ($Z = 71$). Our goal in this section is to take a look at this work and summarize those features of the empirical material that are most important for the quantum theory of the electronic structure. At this point, the reader is expected to review the introduction to the tables and to be familiar with their layout and the notation used.

The tables contain the measured energy levels of the valence electrons of atoms and ions. The levels are given in wave-number units (cm^{-1}) and the arrangement is such that the ground state, that is, the lowest state of the atom is given as the zero level. The higher levels are given by their positive distance from the ground state. Let ε_k be a negative energy level, as we have used this concept in this work, and let E_k be the same level in the tables. Let

E_0 be the quantity called the "series limit" in the tables. Then

$$\varepsilon_k = -(E_0 - E_k). \tag{6.1}$$

If we denote the ground state by ε_0, then

$$\varepsilon_0 = -E_0. \tag{6.2}$$

Next, we clarify the question of units. As Hartree pointed out, using atomic units in atomic calculations is not only consistent, but frees the results from the uncertainties in the measured values of fundamental constants. For this reason, we have used, wherever possible, atomic units in our formulas and will continue to do so. On the other hand, in the tables, the energy levels are given, with very high "spectroscopic accuracy," in cm^{-1} units. The connection between the atomic units and the cm^{-1} units is as follows.

The Planck formula for the energy of radiation is

$$E = h\nu, \tag{6.3a}$$

with ν being the frequency. If k is the wave number and c is the speed of light, then

$$E = (hc)k. \tag{6.3b}$$

Using the recent values for h and c, we get for the conversion factor (hc)

$$h = 6.6260755 \times 10^{-34} \text{ J} \cdot \text{s};$$
$$c = 299{,}792{,}458 \text{ m/s};$$
$$hc = 1986.4475 \times 10^{-26} \text{ J} \cdot \text{cm}. \tag{6.3c}$$

Using the number in Eq. (6.3b), we get the energy in joules if k is in cm^{-1}. We get the energy in eV units if we use the conversion factor

$$1 \text{ eV} = 1.602177 \times 10^{-19} \text{ J}, \tag{6.3d}$$

with which we obtain

$$E \text{ (eV)} = 1.23984 \times 10^{-4} k \text{ (cm}^{-1}). \tag{6.3e}$$

In order to get the energy in atomic units, we quote the formula

$$1 \text{ a.u.} = \frac{e^2}{a_0} = 2(R_\infty)(hc), \tag{6.3f}$$

where a_0 is the smallest Bohr radius, and R_∞ is the Rydberg constant for infinite nuclear mass. Using

$$R_\infty = 1.097373153 \times 10^7 \text{ m}^{-1}, \tag{6.3g}$$

and using for (hc) the value given before, we obtain

$$1 \text{ a.u.} = 43.597481 \times 10^{-19} \text{ J}, \tag{6.3h}$$

and

$$1 \text{ J} = 2.2937106 \times 10^{17} \text{ a.u.} \tag{6.3i}$$

Using this number in Eq. (6.3c), we obtain

$$E \text{ (a.u.)} = 4.5563356 \times 10^{-6} k \text{ (cm}^{-1}). \tag{6.3j}$$

This formula gives the energy in atomic units if k is given in cm^{-1}. We have the correspondence relationships:

$$1 \text{ cm}^{-1} = 4.5563356 \times 10^{-6} \text{ a.u.} \tag{6.3k}$$

$$1 \text{ cm}^{-1} = 1.23984 \times 10^{-4} \text{ eV.} \tag{6.3l}$$

The inverse relationships are

$$1 \text{ a.u.} = 219,474.6 \text{ cm}^{-1}, \tag{6.3m}$$

and

$$1 \text{ eV} = 8065.56 \text{ cm}^{-1}. \tag{6.3n}$$

From these formulas, we recover the familiar conversion factor:

$$1 \text{ a.u.} = 27.2112 \text{ eV.} \tag{6.3o}$$

Thus, the situation is this. The theory can be built up consistently using atomic units and the experimental values can be given consistently in cm^{-1} units. The conversion factors will depend on the accuracy of the measured values of the constants h, c, R_∞, and e. The numbers used here are from *The 1986 Adjustment of the Fundamental Physical Constants* by Cohen and Taylor, CODATA, Bulletin No. 63 (Pergamon Press).

We note that the conversion factors used in the various publications differ widely. For example, in the compilation of ionization potentials, which we used in Sec. 5.4/A, Moore used the values

$$1 \text{ eV} = 8065.73 \text{ cm}^{-1},$$
$$1 \text{ cm}^{-1} = 1.23981 \times 10^{-4} \text{ eV.}$$

The value used by Martin, Zalubas, and Hagan in the lanthanide volume of the AET is

$$1 \text{ eV} = 8065.48 \text{ cm}^{-1}.$$

Now we turn to the discussion of the tables. The first impression that we receive from the tables is that atoms have a very large number of energy levels, which, at first sight, appear to be arranged in complex random patterns. Analysis of the spectra, however, leads very quickly to some general rules that can be summarized as follows.

1. *All spectra are generated by valence electrons.* Conversely, one can *define* valence electrons as the electrons generating the spectra. In Table 6.1, we summarize the configurations of the valence electrons for all neutral atoms from H ($Z = 1$) to U ($Z = 92$). Thus, our first conclusion is that the spectra are generated by a small number of valence electrons and the other electrons, which we call *inner* or *core* electrons, do not directly participate in the formation of the spectra.

From here on, we divide the discussion into two parts, first, considering all atoms and ions with the exception of the lanthanides and noble gases and then reviewing these as a separate group. Let A be an atom or ion and let A^+ be the positive ion of A, that is, let A^+ be the atom A with the outermost electron removed. A review of the tables show that most energy levels have the following structure:

$$[\text{Configuration of } A^+] \, [\text{running electron}] \qquad (6.4)$$

This symbolic representation means that an energy level of atom A is constructed in such a way that we take first a configuration of the positive ion A^+ and add to this configuration one electron, the "running" electron, in the state (nl). Thus, most spectral lines are generated by the transition of a single electron, the running electron, from one state to another.

We note that in the symbolic representation, the "configuration of A^+" is not necessarily the ground state of atom A^+. Here we are talking about *a* configuration of A^+ that may be the ground state or it may be an excited state of the atom A^+. The spectrum of atom A may have many series if A^+ has numerous low-lying configurations.

We now look at some examples. The example of the atoms OV and OVI is discussed in Sec. 5.1 of the introduction of Vol. I of AET. As we see there, the $(1s)^2(2s)^2 \, ^2S$ configuration of OVI generates the following levels of OV:

$$\left[(1s)^2(2s)^2 \, ^2S\right](n, l) \qquad (6.5a)$$

where (nl) is $3s$ for the $^{1,3}S$ state of OV, is $2p$ for the $^{1,3}P$ state, and is $3d$ for the $^{1,3}D$ state. Thus, here atom A^+ is the OVI, and by adding one electron to the state $[(1s)^2(2s)^2 \, ^2S]$, we obtain several levels of OV.

Similarly, if we consider the $[(1s)^2(2p)]^2P$ state, then, by adding one electron, we can generate the states

$$\left[(1s)^2(2p) \, ^2P\right](nl), \qquad (6.5b)$$

where we get for $(nl) = 3s$, the $^{1,3}P$ states, for $(nl) = 3p$, the $^{1,3}S$, $^{1,3}P$, and $^{1,3}D$ states, and for $(nl) = 3d$, the $^{1,3}P$, $^{1,3}D$, and $^{1,3}F$ states.

TABLE 6.1. The Valence-Electron Configurations of the Neutral Atoms from H to U

Atom	Z	Valence configuration
H	1	$(1s)$
He	2	$(1s)^2$
Li	3	$(2s)$
Be	4	$(2s)^2$
B	5	$(2s)^2(2p)$
C	6	$(2s)^2(2p)^2$
N	7	$(2s)^2(2p)^3$
O	8	$(2s)^2(2p)^4$
F	9	$(2s)^2(2p)^5$
Ne	10	$(2p)^6$
Na	11	$(3s)$
Mg	12	$(3s)^2$
Al	13	$(3s)^2(3p)$
Si	14	$(3s)^2(3p)^2$
P	15	$(3s)^2(3p)^3$
S	16	$(3s)^2(3p)^4$
Cl	17	$(3s)^2(3p)^5$
Ar	18	$(3p)^6$
K	19	$(4s)$
Ca	20	$(4s)^2$
Sc	21	$(3d)(4s)^2$
Ti	22	$(3d)^2(4s)^2$
V	23	$(3d)^3(4s)^2$
Cr	24	$(3d)^5(4s)$
Mn	25	$(3d)^5(4s)^2$
Fe	26	$(3d)^6(4s)^2$
Co	27	$(3d)^7(4s)^2$
Ni	28	$(3d)^8(4s)^2$
Cu	29	$(3d)^{10}(4s)$
Zn	30	$(3d)^{10}(4s)^2$
Ga	31	$(4s)^2(4p)$
Ge	32	$(4s)^2(4p)^2$
As	33	$(4s)^2(4p)^3$
Se	34	$(4s)^2(4p)^4$
Br	35	$(4s)^2(4p)^5$
Kr	36	$(4p)^6$
Rb	37	$(5s)$
Sr	38	$(5s)^2$
Y	39	$(4d)(5s)^2$
Zr	40	$(4d)^2(5s)^2$
Nb	41	$(4d)^4(5s)$
Mo	42	$(4d)^5(5s)$
Tc	43	$(4d)^5(5s)^2$
Ru	44	$(4d)^7(5s)$
Rh	45	$(4d)^8(5s)$

TABLE 6.1. (*Continued*)

Atom	Z	Valence configuration
Pd	46	$(4d)^{10}$
Ag	47	$(4d)^{10}(5s)$
Cd	48	$(4d)^{10}(5s)^2$
In	49	$(5s)^2(5p)$
Sn	50	$(5s)^2(5p)^2$
Sb	51	$(5s)^2(5p)^3$
Te	52	$(5s)^2(5p)^4$
I	53	$(5s)^2(5p)^5$
Xe	54	$(5p)^6$
Cs	55	$(6s)$
Ba	56	$(6s)^2$
La	57	$(5d)(6s)^2$
Ce	58	$(4f)(5d)(6s)^2$
Pr	59	$(4f)^3(6s)^2$
Nd	60	$(4f)^4(6s)^2$
Mp	61	$(4f)^5(6s)^2$
Sm	62	$(4f)^6(6s)^2$
Eu	63	$(4f)^7(6s)^2$
Gd	64	$(4f)^7(5d)(6s)^2$
Tb	65	$(4f)^9(6s)^2$
Dy	66	$(4f)^{10}(6s)^2$
Ho	67	$(4f)^{11}(6s)^2$
Er	68	$(4f)^{12}(6s)^2$
Tm	69	$(4f)^{13}(6s)^2$
Yb	70	$(4f)^{14}(6s)^2$
Lu	71	$(5d)(6s)^2$
Hf	72	$(5d)^2(6s)^2$
Ta	73	$(5d)^3(6s)^2$
W	74	$(5d)^4(6s)^2$
Re	75	$(5d)^5(6s)^2$
Os	76	$(5d)^6(6s)^2$
Ir	77	$(5d)^7(6s)^2$
Pt	78	$(5d)^9(6s)$
Au	79	$(5d)^{10}(6s)$
Hg	80	$(5d)^{10}(6s)$
Tl	81	$(6s)^2(6p)$
Pb	82	$(6s)^2(6p)^2$
Bi	83	$(6s)^2(6p)^3$
Po	84	$(6s)^2(6p)^4$
At	85	$(6s)^2(6p)^5$
Rn	86	$(6p)^6$
Fr	87	$(7s)$
Ra	88	$(7s)^2$
Ac	89	$(6d)(7s)^2$
Th	90	$(6d)^2(7s)^2$
Pa	91	$(5f)^2(6d)(7s)^2$
U	92	$(5f)^3(6d)(7s)^2$

We now turn to a more complex atom, S ($Z = 16$). On page 181 of Vol. I, we find the energy levels of the neutral atom. The observed terms are summarized on page 183. According to that summary, the observed levels are of the form

$$(3s)^2(3p)^4 \quad \text{(ground state)};$$
$$(3s)(3p)^5;$$

$$\left.\begin{array}{l} \left[(3s)^2(3p)^{3\,4}S\right](nx); \\[2pt] \left[(3s)^2(3p)^{3\,2}D\right](nx); \\[2pt] \left[(3s)^2(3p)^{3\,2}P\right](nx); \end{array}\right\} \qquad \begin{array}{ll} (ns), & n \geq 4; \\ (np), & n \geq 4; \\ (nd), & n \geq 3. \end{array} \qquad (6.6a)$$

In the table on page 184, which contains the energy levels of the S^+ ion, the first three levels are

$$(3s)^2(3p)^{3\,4}S;$$
$$(3s)^2(3p)^{3\,2}D;$$
$$(3s)^2(3p)^{3\,2}P. \qquad (6.6b)$$

As we see, these are the "parent" configurations of the last three levels of the S atom listed before. From these configurations, the $(3s)^2(3p)^4$ can also be generated by adding a $(3p)$ electron in the appropriate coupling. The fourth level in the S^+ spectrum is $(3s)(3p)^4$, which is the parent configuration of the $(3s)(3p)^5$ level of the neutral S atom.

The presence of an incomplete d group among the valence electrons gives rise to spectra of even greater complexity. Let us consider the Ti atom ($Z = 22$). The ground-state valence configuration is $(3d)^2(4s)^2$. On page 278 of Vol. I, we have the list of observed terms, a part of which is as follows:

$$\left.\begin{array}{l} \left[(3d)^2(4s)\,^4F\right](nx); \\[4pt] \left[(3d)^{3\,4}F\right](nx); \\[4pt] \left[(3d)^2(4s)\,^2F\right](nx); \\[8pt] \left[(3d)^2(4s)\,^2D\right](nx); \\[4pt] \left[(3d)^{3\,2}G\right](nx); \\[4pt] \left[(3d)^{3\,4}P\right](nx); \\[4pt] \left[(3d)^{3\,2}P\right](nx); \\[4pt] \left[(3d)^2(4s)\,^4P\right](nx); \end{array}\right\} \qquad (6.7)$$

$$\vdots \qquad .$$

The parent configurations are in brackets. Now looking at the spectrum of the Ti^+ atom on page 279, we find the first 5 lines as follows:

$$\left. \begin{array}{l} \left[(3d)^{2\,3}F(4s)\right]\,^4F; \\ \left[(3d)^3\right]\,^4F; \\ \left[(3d)^{2\,3}F(4s)\right]\,^2F; \\ \left[(3d)^{2\,1}D(4s)\right]\,^2D; \\ \left[(3d)^3\right]\,^2G; \end{array} \right\} \qquad (6.8)$$

$$\vdots \qquad \cdot$$

Comparing these configurations with the configurations in square brackets of the first five lines of the S spectrum, we see that they are identical. Clearly, the S and S^+ spectra obey the simple "build-up principle" formulated before. And, of course, looking at the spectrum of the S^{++} atom on page 282, we see that the first two lines are $(3d)^{2\,3}F$ and $(3d)^{2\,1}D$, which are the parent configurations of the S^+ lines listed before. And so on.

Summing up, we see that the general rule formulated before is demonstrated in these special cases. We can make another important observation by looking at these lists of levels. The energy levels are identified primarily by the configuration of the valence electrons. The coupling of the electrons is also important, however; different couplings lead to different energies. Thus, the following two lines,

$$\begin{array}{l} \left[(3d)^2(4s)\,^4F\right](nx); \\ \left[(3d)^2(4s)\,^2D\right](nx); \end{array} \qquad (6.9)$$

have the same parent *configuration*, but different *couplings*, and, therefore, they generate two different series of energy levels.

Finally, last but not least, the tables show that the spectra exhibit a fine structure. By this, we mean that often a level that is identified by configuration and coupling splits into a number of closely spaced lines that are identified by the different total (or inner) quantum number J. For example, on page 274, where we have the spectrum of the neutral Ti atom, we see around the middle the level $[(3d)^2(4s)\,^4F](4p)\,^5F$. Here the parent configuration consists of the two $(3d)$ electrons and the $(4s)$ electron coupled to a ^4F, and this group is then coupled to a $(4p)$ state to give a ^5F. Here we have, instead of a single line, a group of five close-lying levels that form the fine structure. The levels in the group are identified by the J value, which runs from 1 to 5. We have such fine structure wherever the coupling of L and S permit different values for J.

Next, we want to look more closely at the group of lines called *series*. Let us consider first an atom with a single valence electron. The spectrum of K ($Z = 19$) is on page 227. Here the group below the valence electron is $(3p)^6\,^1S$ and this group does not participate in the formation of the spectrum. Thus, the various series are characterized by the (nl) of the valence electron. The fine structure is a doublet because for any l, except zero, we have

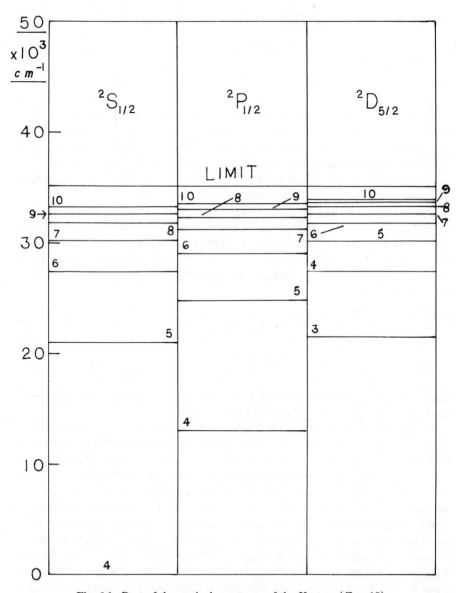

Fig. 6.1. Part of the optical spectrum of the K atom ($Z = 19$).

$j = l \pm \frac{1}{2}$. In Fig. 6.1, we have the ^2S, $^2P_{1/2}$ and $^2D_{5/2}$ series plotted in such a way that the numbers next to the lines are the principal quantum numbers of the running electron. The ground state, $(4s)\,^2S$, is on the lower left side and the limit is indicated as a horizontal line across.

This diagram shows something very interesting. If we would plot all these lines as they are listed in the table, that is, if we would plot them by their increasing distance from the ground state, they would appear to be of random structure. Separating the *series* by grouping together the lines with common L (which is also the common l because $L = l$) reveals that each series has a Rydberglike structure. By this, we do not mean that they can be described by a Rydberg-type mathematical formula. What we mean is that each series is Rydberglike with the distance between the lines diminishing as we go toward higher n values, similarly as in the case of the H spectrum.

Next, we look at the spectrum of Ca $(Z = 20)$, which has two valence electrons. The spectrum is on page 242 of the tables. We have selected the three series

$$\left[(4s)\,^2S\right](ns)\,^1S;$$

$$\left[(4s)\,^2S\right](np)\,^1P; \qquad (6.10)$$

$$\left[(4s)\,^2S\right](nd)\,^3D_1.$$

In Fig. 6.2, we have the first few lines of the three series and we observe that their structure is exactly like the structure of the K lines, that is, the series are Rydberglike in the sense as we defined this concept before.

Finally, in Fig. 6.3, we have a part of the spectrum of the S atom $(Z = 16)$. In the tables, this is on page 181. We plotted the first few lines of the series

$$\left.\begin{array}{l} \left[(3s)^2(3p)^3\,^4S\right](np)\,^3P_2; \\ \left[(3s)^2(3p)^3\,^4S\right](ns)\,^3S_1; \\ \left[(3s)^2(3p)^3\,^4S\right](nd)\,^5D_4. \end{array}\right\} \qquad (6.11)$$

The parent configurations are the same here. The principal quantum numbers are indicated in the figure. The ground state, which is $(3p)^4\,^3P_2$, is not shown in the figure; note the scale on the left side of the diagram. As we see, the general structure of these series is exactly the same as in the case of the K and Ca atoms despite the fact that the number of valence electrons is different in the three cases.

We now summarize the salient features of the spectra. This summary refers to all atoms and ions that are in the AET with the exception of the Lanthanides and the noble gases. The energy levels of atoms can be identified by the electron configuration and by the coupling of the valence

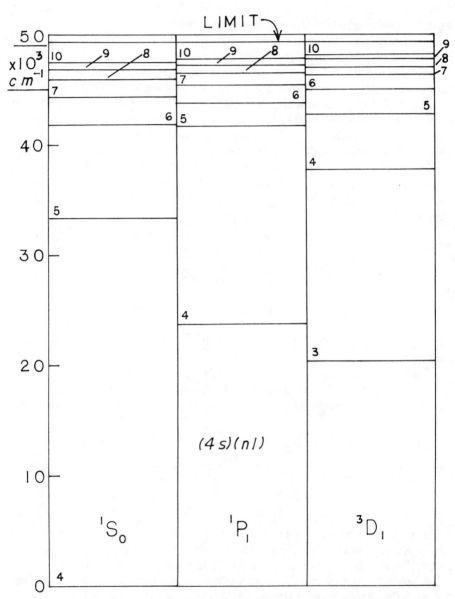

Fig. 6.2. Part of the optical spectrum of the Ca atom ($Z = 20$).

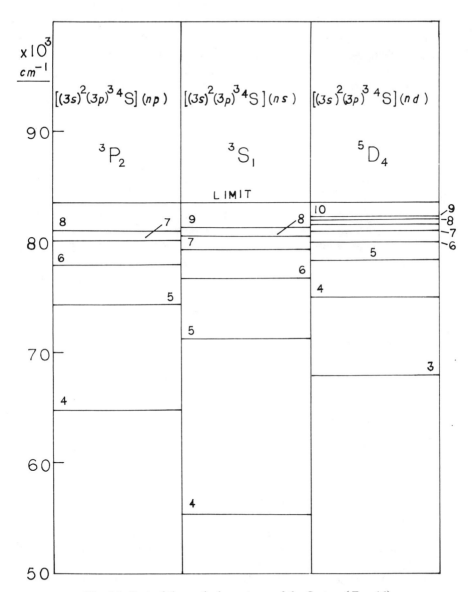

Fig. 6.3. Part of the optical spectrum of the S atom ($Z = 16$).

electrons; the fine structure requires the total quantum number J for identification. The series of lines, that is, energy levels, belonging to the same parent configuration and coupling and having the same final L, S, and J, but having different principal quantum numbers n for the running electron, show a Rydberglike structure, where this concept was defined before. The LS coupling is sufficient for the identification of the energy levels.

We turn now to the lanthanides and the noble gases. As we see from Vol. IV of the AET, the spectra of the rare-earth atoms are much more complex than the spectra of the other atoms and ions. There are several reasons for this.

The rare-earth elements comprise the group from La ($Z = 57$) to Yb ($Z = 70$). As we see from Table 6.1, these atoms have at least one incomplete group, the $(4f)^q$ group, among their valence electrons; some have, in addition, a $(5d)$ electron. Due to the high number of valence electrons and their high l values, it is often impossible to describe the levels of these atoms by a single configuration. In many cases, the energy levels must be represented by a mixture of configurations. The wave function describing such a level is the linear combination of the wave functions describing the various participating configurations. In the tables, the squared expansion coefficients are given as percentages. (The sum of the expansion coefficient squares is, of course, equal to 1, that is, equal to 100%.)

A second complicating feature is that LS coupling does not work in many cases. In order to describe these spectra, there are five different coupling schemes that are needed. These are

1. LS coupling
2. $J_1 j$ or $J_1 J_2$ coupling
3. jj coupling
4. $J_1 l$ or $J_1 L_2$ coupling
5. LS_1 coupling

The first, LS coupling, is the familiar scheme that we used in the description of the nonrelativistic HF model. In the $(J_1 j)$ notation, J_1 is the total angular momentum of a group of electrons and j is the total angular momentum of a single electron ($j = l \pm \frac{1}{2}$). In this scheme, these are coupled according to quantum mechanical rules. In the $(J_1 J_2)$ notation, both symbols refer to groups of electrons. The (jj) coupling is the same as in the relativistic HF model. The $(J_1 l)$ notation means coupling of the total quantum number of a group of electrons to the azimuthal quantum number of a single electron. In the $(J_1 L_2)$ scheme, J_1 means the total angular momentum of a group, and L_2 is the total azimuthal quantum number of another group; in this scheme, these two are coupled. In the LS_1 scheme, L is the total azimuthal quantum number of one group, and S_1 is the total spin of another.

There is nothing intrinsically difficult about these various coupling schemes, but they require different term symbols and they are certainly making the classification of energy levels much more complicated than in the case of a simple LS coupling.

Another feature of the rare-earth spectra is that the spin–orbit interaction is more important here than in the simpler atoms. There are cases in which different energy levels are characterized by the same J value but by different Lande g factors. (See, for example, page 31 of Vol. IV of the tables.) The g factors are obtained from measuring the splitting of single levels into a group of levels in a magnetic field (the Zeeman effect). Different g factors are indications of different configurations. In the LS coupling scheme, the g factor is related to L, S, and J by the familiar formula. If the term cannot be described by a single configuration, then the g factor must be constructed using the g factors of the participating configurations from which the final wave function is formed by linear combination. This is lucidly described in Sec. 2.4 of Vol. IV of the tables and need not be repeated here. (See also Sec. 6.7.)

The noble gases form a special group. In these atoms, the spectrum is formed by breaking up the tightly bound $(np)^6$ valence subshell. The LS coupling does not hold in these cases. What is being used here is the scheme that we called $(J_1 l)$ coupling before. Here J_1 is the total angular momentum of the $(np)^5$ electrons and l is the azimuthal quantum number of the running electron. For the $(np)^5$ configuration, we have only the ^2P term, so the possible J_1 values are $J_1 = \frac{1}{2}, \frac{3}{2}$. For a given l, the coupling of J_1 and l provides the following K values (the usual notation for this quantum number):

$$K = |J_1 - l|, \ldots (J_1 + l). \tag{6.12a}$$

The total angular momentum of the system is obtained by coupling the spin of the running electron with the K values. Thus, we obtain

$$J = \left| K - \tfrac{1}{2} \right| \cdots \left(K + \tfrac{1}{2} \right). \tag{6.12b}$$

The energy level is designated by the K value in square brackets, like $[K]$, and by the J value. Thus, for example, if $l = 1$, then the possible K values are $\frac{1}{2}$ and $\frac{3}{2}$ for $J_1 = \frac{1}{2}$, and $\frac{1}{2}, \frac{3}{2}$, and $\frac{5}{2}$ for $J_1 = \frac{3}{2}$. The possible J values are

$$
\begin{aligned}
K &= \tfrac{1}{2}, & J &= 0, 1; \\
K &= \tfrac{3}{2}, & J &= 1, 2; \\
K &= \tfrac{5}{2}, & J &= 2, 3.
\end{aligned}
\tag{6.13}
$$

The two series with $J_1 = \frac{1}{2}$ and $J_1 = \frac{3}{2}$ have two different series limits, the former being the ground state of the positive ion with $J = \frac{1}{2}$, and the latter the ground state of the ion with $J = \frac{3}{2}$.

We now summarize the properties of the rare-earth atoms and ions and the properties of the noble gases that set them apart from other atoms. The rare-earth atoms require special treatment because (1) some of the energy levels cannot be represented by a single configuration, and (2) in many cases, the LS coupling is not suitable for the description of the spectrum. The first problem is treated by mixing the configurations, which also results in a special mixing formula for the g factors. The second problem is treated by the introduction of four new coupling schemes in addition to the LS coupling. Special attention must be paid to the spin–orbit interaction, which is more important here than in the simpler atoms. The noble gases are also different from the other atoms because, here, instead of using LS coupling, we must use what was described in connection with the rare-earth spectra as the $(J_1 l)$ coupling. When this scheme is used, the spectra of the noble-gas atoms are not excessively complicated.

6.2. LS COUPLING FOR SEVERAL INCOMPLETE GROUPS

Discussing the electron configurations of atoms in Sec. 3.1, we have established that in the ground state, the electrons of most atoms are either in complete groups or occupy configurations with only one incomplete group. We have set up the LS coupling scheme for such cases in Chapter 3.

Reviewing the spectra of atoms in the Atomic Energy Level Tables, we are compelled to observe that in contrast to the ground state, the excited states of most atoms consist of more than one incomplete group of electrons. As a prototype of configurations with more than one incomplete groups, we discuss here the following configuration:

$$\left[(n'' l'')^{q\,(2S_1+1)} L_1 \right] (nl)^{\,(2S+1)} L. \tag{6.14}$$

Here the parent configuration consists of the incomplete group of electrons with the quantum numbers $(n'' l'')$. The occupation number of these is q. They are coupled to a state with the quantum numbers $(L_1 S_1)$. The running electron has the quantum numbers (nl) and the parent group is coupled to the running electron to form a state with the quantum numbers (LS). Thus, we have two incomplete groups, one of which consists of a single electron.

What we need is the correct total wave function and total energy expression for the atom. In Sec. 3.1, we have outlined how energy expression can be constructed in the LS coupling. We have seen that the problem boils down to the construction of the secular equation whose matrix components are those of the Hamiltonian with respect to the total wave function. We need the diagonal as well as the off-diagonal matrix components. If the diagonal-sum rule can be used, only the diagonal components are needed. In the general case, however, we cannot rely on the diagonal-sum rule. As the azimuthal quantum number of the group $(n''l'')$ becomes larger (beginning with $l'' = 2$, d electrons), we are encountering more than one multiplet with the same $(L_1 S_1)$. In such a case, the diagonal-sum rule cannot be used and so we must construct the whole secular determinant, that is, we need the diagonal and the off-diagonal matrix components.

Let $\psi(\alpha L_1 S_1, LS)$ denote the wave function of the state given by Eq. (6.14). This wave function belongs to the configuration $(n''l'')^q(nl)$, which is not explicitly indicated. We want a general formula for the matrix component

$$\langle \psi(\alpha L_1 S_1, LS)|H|\psi(\alpha' L_1' S_1', LS)\rangle. \qquad (6.15)$$

In our formulas, α is a parameter needed for the characterization of the wave function in addition to the angular momentum and spin quantum numbers. We need the diagonal as well as the off-diagonal components of this type.

We have already seen matrix components of this type. For atoms with one incomplete group, the expression given by Eqs. (3.84) and (3.89) is a special case of Eq. (6.15). It will be useful to establish the connection between the two expressions explicitly. What we have in Eq. (3.84) is the diagonal case of Eq. (6.15), and we can write

$$\langle \alpha LS|H|\alpha LS\rangle = E_{\text{av}} + \sum_{k>0}^{2l''} f_k(\alpha LS) F_k(n''l'', n''l''), \qquad (6.16)$$

where we have omitted parameters $L_1 S_1$, but kept L because for some $(n''l'')^q$ configurations, there is more than one wave function with the same (LS).

A general remark about the structure of the matrix components given by Eq. (6.15). We have seen in Sec. 3.3 that the HF energy for atoms with two incomplete groups, one of which contains only one electron, was given by Eqs. (3.94) and (3.95). In our present notation, we can write

$$\langle \alpha L_1 S_1 LS|H|\alpha L_1 S_1 LS\rangle$$

$$= E_{\text{av}} + \sum_{k>0}^{2l''} f_k F_k(n''l'', n''l'') + \sum_{k=0}^{l_m} g_k F_k(nl, n''l'')$$

$$+ \sum_{k=0}^{l+l''} h_k G_k(nl, n''l''), \qquad (6.17)$$

where (nl) is the group with one electron. We recall that l_m is the lesser of $(2l)$ and $(2l'')$. In order to make the notation more accurate, we write

$$f_k = f_k(\alpha L_1 S_1 LS; \alpha L_1 S_1 LS); \tag{6.18a}$$

$$g_k = g_k(\alpha L_1 S_1 LS; \alpha L_1 S_1 LS); \tag{6.18b}$$

$$h_k = h_k(\alpha L_1 S_1 LS; \alpha L_1 S_1 LS). \tag{6.18c}$$

The structure of Eq. (6.17) is interesting and permits a simple physical interpretation. This is the HF energy of an atom with two incomplete groups in the $(n''l'')^q(nl)$ configuration. For this configuration, we can have numerous wave functions with different (LS) combinations. Each, or some, of the (LS) combinations can have different $(L_1 S_1)$ parents and there might be several α parameters for some of the $(L_1 S_1)$ combinations. All these parameters signify angular momentum and spin properties and indicate the angular momentum and spin operators that are diagonalized by the wave function that underlies the matrix component given by Eq. (6.17). The interesting fact is that all these angular-momentum properties are compressed into parameters f_k, g_k, and h_k. These parameters are fixed once parameters α, L_1, S_1, L, and S are fixed. On the other hand, the matrix component given by Eq. (6.17) can be used for any set of radial orbitals because the radial orbitals occur only in the F_k and G_k integrals, which are not affected by the "angular" coefficients f_k, g_k, and h_k.

We can put these observations into the following form. The HF model is defined as an approximation in which certain angular-momentum and spin operators are diagonalized by wave functions, which, at the same time, minimize the expectation value of the Hamiltonian. This division, into angular-momentum properties on the one hand and expectation value of the Hamiltonian on the other hand, is clearly visible in Eq. (6.17). The angular-momentum properties are determined by coefficients f_k, g_k and h_k. The minimization of the Hamiltonian will take place by varying the energy given by Eq. (6.17) and this variation will affect only the F_k and G_k integrals. The angular-momentum parameters will enter the variation process as fixed constants. Thus, we can summarize this digression into the physical meaning of Eq. (6.17) as follows:

1. The angular-momentum properties of the multiplet under consideration are given by parameters f_k, g_k, and h_k.
2. The minimization of the energy will be carried out by determining the correct values of the F_k and G_k integrals.
3. The "angular-momentum" and "energy" quantities are clearly separated in Eq. (6.17).

Next we turn to the off-diagonal matrix components given by Eq. (6.15). We state that these components will have the same general structure as the diagonal components. This can be seen easily by inspecting the derivations leading to Eqs. (3.84), (3.89), and (3.94). In those derivations, we argued on the basis of *configurations*. The angular-momentum properties did not enter the derivations explicitly because we did not actually determine coefficients f_k, g_k, and h_k. We determined the *types* of coefficients that occur and the limits of the summations.

Looking at Eq. (6.15), we see that the off-diagonal components given by that expression are off-diagonal with respect to the angular-momentum and spin parameters. The two wave functions in Eq. (6.15) are of the same configuration. From this it follows that these off-diagonal components will have the same general form as the diagonal components for the same configuration. Thus, we obtain, for the off-diagonal matrix component of Eq. (6.15),

$$
\langle \psi(\alpha L_1 S_1 LS)|H|\psi(\alpha' L_1' S_1', LS)\rangle = E_{\mathrm{av}}
$$

$$
+ \sum_{k>0}^{2l''} f_k(\alpha L_1 S_1 LS; \alpha' L_1' S_1' LS) F_k(n''l'', n''l'')
$$

$$
+ \sum_{k=0}^{l_m} g_k(\alpha L_1 S_1 LS; \alpha' L_1' S_1' LS) F_k(nl, n''l'')
$$

$$
+ \sum_{k=0}^{l+l''} h_k(\alpha L_1 S_1 LS; \alpha' L_1' S_1' LS) G_k(nl, n''l'').
$$

$$
(6.19)
$$

A general method for the calculation of the coefficients that occur in Eqs. (6.17) and (6.19) was developed by Racah.[57] Here we give explicit expressions for the $(n''l'')^q(nl)$ configuration. Using these expressions, one can construct the coefficients for any set of angular and spin parameters. We have the general formulas for g_k and h_k [Ref. 58, p. 348, and Slater II, 140, 146]:

$$
g_k(\alpha L_1 S_1 LS; \alpha' L_1' S_1' LS)
$$

$$
= (l''\|C^{(k)}\|l'')(l\|C^{(k)}\|l)
$$

$$
\times \delta(S_1 S_1')(-1)^{L_1'+l+L}\begin{Bmatrix} L_1 & l & L \\ l & L_1' & k \end{Bmatrix}
$$

$$
\times \langle (l'')^q \alpha L_1 S_1 \| U^{(k)} \| (l'')^q \alpha' L_1' S_1' \rangle,
$$

$$
(6.20)
$$

and

$$h_k(\alpha L_1 S_1 LS; \alpha' L_1' S_1' LS)$$

$$= \frac{1}{2} \left[(l'' \| C^{(k)} \| l) \right]^2 \left\{ \delta_{\beta\beta'} \frac{q(n''l'')}{(2l''+1)(2l+1)} \right.$$

$$- \sum_{r=1}^{r_m} (-1)^r (2r+1) \begin{Bmatrix} l'' & l'' & r \\ l & l & k \end{Bmatrix}$$

$$\times \left[\delta(S_1 S_1')(-1)^{L_1+l+L} \begin{Bmatrix} L_1 & l & L \\ l & L_1' & r \end{Bmatrix} \right.$$

$$\times \langle (l'')^q \alpha L_1 S_1 \| U^{(r)} \| (l'')^q \alpha' L_1' S_1' \rangle$$

$$+ 4(-1)^{L_1'+S_1'+l+1/2+L+S} \begin{Bmatrix} L_1 & l & L \\ l & L_1' & r \end{Bmatrix} \begin{Bmatrix} S_1 & \frac{1}{2} & S \\ \frac{1}{2} & S_1' & 1 \end{Bmatrix}$$

$$\left. \left. \times \left(\tfrac{3}{2} \right)^{1/2} \langle (l'')^q \alpha L_1 S_1 \| V^{(r1)} \| (l'')^q \alpha' L_1' S_1' \rangle \right] \right\}. \tag{6.21}$$

In these formulas, $(l_1 \| C^{(k)} \| l_2)$ is the matrix component of the spherical harmonies, which is related to our c^k coefficients, which occur frequently in the formulas of the HF method, as follows:

$$(l_1 \| C^{(k)} \| l_2) = \left[(2l_1 + 1)(2l_2 + 1) \right]^{1/2} c^k(l_1 0, l_2 0). \tag{6.21a}$$

The quantities with $U^{(k)}$ and $V^{(r1)}$ are Racah coefficients and $\{ \cdots \}$ is the $6j$ symbol. Parameter β indicates a full set of angular parameters, that is, we have

$$\delta_{\beta\beta'} = \delta(\alpha\alpha') \delta(L_1 L_1') \delta(S_1 S_1'), \tag{6.22}$$

and r_m is the lesser of $(2l)$ and $(2l'')$.

Coefficient f_k in Eq. (6.19) is obtainable from Eq. (6.20) by putting $l \equiv l''$ and $L_1' = L_1$. Thus, f_k is diagonal in all parameters except α and depends only on parameters $(\alpha\alpha' L_1 S_1)$; it is essentially the same as in the case of an $(n''l'')^q$ configuration without the running electron.

The f_k, g_k, and h_k coefficients can easily be constructed from the formulas given by Eqs. (6.20) and (6.21). The necessary constants are tabulated in Ref. 58 as follows:

$C^{(k)}$ coefficients:	p. 677
$6j$ symbols:	p. 646, 147
$U^{(k)}$ and $V^{(r1)}$ coefficients:	p. 679

An even more convenient way to obtain the f_k, g_k, and h_k coefficients is to use Slater's tables. Slater tabulated the numerical values of many coefficients for a number of configurations. Some of the configurations covered by Slater's tables are more complex than the example for which we have written down Eq. (6.19). Unfortunately, Slater's tables are restricted in other respects; for example, he does not give any coefficients for configurations containing f electrons.[†] Here is a list of references to Slater's tables:

Configuration	Reference
$(p)^q$	Slater II, 287;
$(p)^q l$	Slater II, 287–292;
$(l)^q s$	Slater II, 292;
$(l)^q l' s$	Slater II, 293;
$(d)^q$	Slater II, 294–295;
$(d)^q l$	Slater II, 296–321.

As we see from this list, the $p^q l$, $l^q s$, and $d^q l$ configurations are of the same type for which we have written down the general formula, Eq. (6.19). The configuration $l^q l' s$ is slightly more complex and is not included in Eq. (6.19). On the other hand, from the formulas given in Eqs. (6-19)–(6.21), we can construct the coefficients for any set of quantum numbers, including those not included in Slater's tables.

This concludes the discussion of the *LS* coupling method for configurations of the type given by Eq. (6.14). This type represents a large part of the spectra with the exception of lanthanides and noble gases. The important feature of this presentation was not so much that we have given explicit formulas for the construction of the matrix components. Rather, what we want to emphasize is the *method* by which Eq. (6.19) was constructed.

In Sec. 3.3, we have shown how the general form for the matrix components can be established. The derivation of the explicit expressions for the f_k, g_k, and h_k coefficients belongs to the angular-momentum and spin theory and is outside of the scope of this work in which we concentrate on the theory of the Hamiltonian operator.

The angular-momentum and spin coefficients can be constructed similarly for other configurations not represented by Eq. (6.14). The reader interested in the more complex cases of *LS* coupling, as well as in the other coupling schemes necessary for the lanthanides and for the noble gases, is referred to the book by Cowan,[58] which is heavily slanted in the direction of angular-momentum and spin formalism. This book is recommended for being the most recent treatment of the subject as well as for its extensive tables of various coefficients.

[†]The f_k coefficients for the $(nf)^q$ configurations are given by Nielson and Koster, Ref. 14.

Before closing the section, we emphasize again that the formulas in this section determine the angular-momentum and spin properties but do not say anything about the radial orbitals occurring in the F_k and G_k integrals. The great strength of Slater's and Racah's methods lies in that they determine the angular-momentum and spin properties for an arbitrary set of radial orbitals, leaving the latter to be determined from the energy minimum principle.

6.3. HF THEORY FOR ONE VALENCE ELECTRON: "FROZEN-CORE" APPROXIMATION

We now proceed to the development of HF theory for atoms with one valence electron. Let us first establish how we would treat such atoms if we would be working with the exact Schroedinger equation. The Hamiltonian for an atom with $N + 1$ electrons is given by

$$H = \sum_{i=1}^{N+1} f_i + \frac{1}{2} \sum_{i,j=1}^{N+1} \frac{1}{r_{ij}}, \qquad (6.23)$$

where $f = t + g$, with t being the kinetic-energy operator, and g the nuclear potential. The Schroedinger equation is

$$H\psi_k = E_k\psi_k, \qquad (6.24)$$

where the lowest energy level of this equation corresponds to the ground state and the higher levels to the excited states. The wave functions belonging to different states will be orthogonal to each other as well as to the ground state.

If we want to formulate a HF model for this problem, we must remember that the HF model is a variational approximation. Thus, we must recall how excited states are treated in variation theory. As is known, the variational calculation of the ground state always provides an upper limit to the exact eigenvalue. Let the sequence of the eigenvalues of Eq. (6.24) be $E_0 < E_1 < E_2 \cdots < E_k$, with E_0 being the ground state. Let the corresponding exact eigenfunctions be $\psi_0, \psi_1, \ldots, \psi_k$. Let ψ be a variational wave function with which we want to approximate E_{k+1}. We can expand ψ in terms of the complete set of eigenfunction:

$$\psi = \sum_{i=0}^{\infty} c_i\psi_i, \qquad (6.25)$$

where

$$c_i = \langle \psi_i|\psi \rangle. \qquad (6.26)$$

Now, if we orthogonalize ψ to $\psi_0, \psi_1, \ldots, \psi_k$, then

$$c_i = 0 \qquad (i = 0, 1, \ldots, k), \tag{6.27}$$

and

$$\psi = \sum_{i=k+1}^{\infty} c_i \psi_i. \tag{6.28}$$

For the expectation value of the Hamiltonian, we get

$$\langle \psi | H | \psi \rangle = \sum_{i=k+1}^{\infty} E_i |c_i|^2, \tag{6.29}$$

and if we replace all the eigenvalues by the lowest in the set, E_{k+1}, in this expression we obtain

$$\langle \psi | H | \psi \rangle \geq E_{k+1} \sum_{i=k+1}^{\infty} |c_i|^2, \tag{6.30}$$

from which it follows that

$$\langle \psi | H | \psi \rangle \geq E_{k+1}, \tag{6.31}$$

where we have used the closure relationship, taking into account Eq. (6.27).

Thus, in the variation theory, the approximate energy will be an upper limit if the trial function is orthogonalized to all lower-lying excited-state wave functions. The trouble with this prescription is that if we calculate all levels with the HF approximation, then we do not have the exact solutions of Eq. (6.24), to which we should orthogonalize the HF wave functions in order to ensure that the calculations will yield upper limits.

The next best procedure to the orthogonalization of the trial function to all lower-lying exact solutions is the orthogonalization of the HF wave function to all HF functions representing lower-lying energy levels. This is not a very simple procedure because, in the case of excited states, the HF orbitals must be calculated separately for each excited state, that is, we have to take into account the orthogonality requirement by introducing Lagrangian multipliers with respect to lower-lying excited states. Thus, the HF equation for an excited state will depend on the orbitals of lower-lying excited states.

We now show that these difficulties can be eliminated by introducing the so-called "frozen-core" approximation. This approximation was first introduced by Fock[5] for atoms with one valence electron, and in the form that we call the "extended frozen-core" approximation by Cohen and Kelly[59] for atoms with several valence electrons.

We start to build up the $(N + 1)$-electron atom by considering first the positive ion with N electrons. Let $\varphi_1 \cdots \varphi_N$ be spin-orbitals for the N electrons and let the HF wave function be the single determinant:

$$\psi = (N!)^{-1/2} \det[\varphi_1 \varphi_2 \cdots \varphi_N]. \tag{6.32}$$

As we have shown in Sec. 2.2, the orbitals can be chosen to be orthogonal and normalized without loss of generality. Therefore, we have

$$\langle \varphi_i | \varphi_k \rangle = \delta_{ik} \qquad (i, k = 1, \ldots, N). \tag{6.33}$$

The exact Hamiltonian for the N-electron system is

$$H = \sum_{i=1}^{N} f_i + \frac{1}{2} \sum_{i,j=1}^{N} \frac{1}{r_{ij}}, \tag{6.34}$$

where f is the same as in Eq. (6.23). The expectation value of H with respect to the ψ of Eq. (6.32) is given by

$$E_C = \langle \psi | H | \psi \rangle. \tag{6.35}$$

Application of the variation principle to this energy expression under the subsidiary conditions given by Eq. (6.33) yields the HF equations. Because we are dealing with a single-determinantal wave function, the off-diagonal Lagrangian multipliers can be eliminated by a unitary transformation and we obtain the HF equation in the form

$$H_F \varphi_k = \varepsilon_k \varphi_k \qquad (k = 1, \ldots, N), \tag{6.36}$$

where

$$H_F = f + U, \tag{6.37a}$$

and

$$U = \sum_{j=1}^{N} U_j, \tag{6.37b}$$

with

$$U_j(1)F(1) = \int \frac{1}{r_{12}} \left[|\varphi_j(2)|^2 F(1) - \varphi_j(1)\varphi_j(2)F(2) \right] dq_2. \tag{6.37c}$$

In the rest of the derivation, we consider the HF equations as solved, that is, we assume that the orbitals $\varphi_1 \cdots \varphi_N$ and orbital parameters $\varepsilon_1 \cdots \varepsilon_N$ are available.

Next, we add the valence electron to the N-electron atom and assume that the addition of it to the atom has no effect on the N core orbitals, that is, we say that the core orbitals are "frozen." The total wave function of the $(N + 1)$-electron system is the determinant

$$\psi = [(N + 1)!]^{-1/2} \det[\varphi_1 \cdots \varphi_N \varphi], \qquad (6.38)$$

where we denoted the valence electron orbital by φ. Because this is still a single-determinantal wave function, we can again assume that all orbitals are orthogonal without loss of generality. That is, in addition to Eq. (6.33), we can write

$$\langle \varphi | \varphi_i \rangle = 0 \qquad (i = 1, \ldots, N), \qquad (6.39a)$$

and

$$\langle \varphi | \varphi \rangle = 1. \qquad (6.39b)$$

The exact Hamiltonian of the $(N + 1)$-electron system is given by Eq. (6.23). The expectation value of that Hamiltonian with respect to the determinant given by Eq. (6.38) is the total energy of the system. We obtain

$$E_T = \langle \psi | H | \psi \rangle. \qquad (6.40)$$

Let us introduce the spin-dependent density matrix

$$\rho(qq') = \sum_{i=1}^{N+1} \varphi_i(q) \varphi_i^*(q'). \qquad (6.41)$$

The total density becomes with this notation

$$\rho(q) = \rho(qq) = \sum_{i=1}^{N+1} |\varphi_i(q)|^2. \qquad (6.42)$$

Using these relationships, we can write Eq. (6.40) in the form

$$\begin{aligned} E_T = \sum_{i=1}^{N+1} \langle \varphi_i | f | \varphi_i \rangle &+ \frac{1}{2} \int \frac{\rho(q)\rho(q')}{|\mathbf{r} - \mathbf{r'}|} \, dq \, dq' \\ &- \frac{1}{2} \int \frac{\rho(qq')\rho^*(qq') \, dq \, dq'}{|\mathbf{r} - \mathbf{r'}|}, \end{aligned} \qquad (6.43)$$

which is, or course, the same type of expression as in Eq. (2.42). Now let

$$\rho(qq') = \rho_0(qq') + \varphi(q)\varphi^*(q'), \qquad (6.44)$$

where

$$\rho_0(qq') = \sum_{i=1}^{N} \varphi_i(q)\varphi_i^*(q'), \qquad (6.45a)$$

is the density matrix of the electron core. The total electron density of the core is given by

$$\rho_0(q) = \rho_0(qq). \qquad (6.45b)$$

Using Eq. (6.44), we obtain the total energy E_T in the form

$$E_T = E_C + E_v, \qquad (6.46a)$$

where the core energy E_C is given by

$$E_C = \sum_{i=1}^{N} \langle \varphi_i|f|\varphi_i \rangle + \frac{1}{2} \int \frac{\rho_0(q)\rho_0(q')\,dq\,dq'}{|\mathbf{r} - \mathbf{r}'|}$$
$$- \frac{1}{2} \int \frac{\rho_0(qq')\rho_0^*(qq')}{|\mathbf{r} - \mathbf{r}'|}\,dq\,dq', \qquad (6.46b)$$

and the energy of the valence electron is

$$E_v = \langle \varphi|f|\varphi \rangle + \int \frac{\rho_0(q')|\varphi(q)|^2}{|\mathbf{r} - \mathbf{r}'|}\,dq\,dq'$$
$$- \int \frac{\rho_0(qq')\varphi^*(q)\varphi(q')}{|\mathbf{r} - \mathbf{r}'|}\,dq\,dq'. \qquad (6.46c)$$

If we would write out Eq. (6.35) in detail, using the density matrix for the core, we would see immediately that the core energy in Eq. (6.46b) is identical with the expression in Eq. (6.35). Thus, the core energy does not change by the addition of the valence electron.

Next, we determine the valence orbital φ. We vary E_T, given by Eq. (6.46a), with respect to φ^* under the subsidiary conditions given by Eqs. (6.39a) and (6.39b). In the variation, we keep the core orbitals, $\varphi_1 \cdots \varphi_N$, fixed; that is the meaning of the "frozen-core." In this way we obtain the HF equation:

$$H_F\varphi = \varepsilon\varphi + \sum_{i=1}^{N} \lambda_i\varphi_i, \qquad (6.47)$$

where H_F is the same as in Eq. (6.36), and the Lagrangian multipliers are given by

$$\lambda_i = \langle \varphi_i | H_F | \varphi \rangle. \tag{6.48}$$

Because the core orbitals satisfy Eq. (6.36) and H_F is Hermitian, we have

$$\lambda_i = \varepsilon_i \langle \varphi_i | \varphi \rangle = 0, \tag{6.49}$$

and so the HF equation for the valence orbital becomes

$$H_F \varphi = \varepsilon \varphi. \tag{6.50}$$

Inspecting the derivation leading to this equation, we see that it is valid regardless of whether we are talking about the ground state or about an excited state. Thus, denoting the ground- and excited-state orbital parameters by $\varepsilon_v, v = 0, 1, \ldots, k$, and the corresponding orbitals by φ_v, we can write Eq. (6.50) in the form

$$H_F \varphi_v = \varepsilon_v \varphi_v \qquad (v = 0, 1, \ldots, k, \ldots). \tag{6.51}$$

Using Eq. (6.51), we can write the valence-electron energy, given by Eq. (6.46c), in the form

$$E_v = \langle \varphi_v | H_F | \varphi_v \rangle = \varepsilon_v. \tag{6.52}$$

The total energy of the atom is given by Eq. (6.46a). Using Eq. (6.52), we obtain for the total energy in the excited state v[†]

$$E_T(v) = E_C + \varepsilon_v. \tag{6.53}$$

The orbital parameter of the HF equations, therefore, can be written in the form

$$\varepsilon_v = E_T(v) - E_C. \tag{6.54}$$

Thus, the limit of the negative orbital energy, the zero level, is reached when

$$E_T = E_C, \tag{6.55}$$

that is, we get $\varepsilon_v = 0$ when the electron is removed and the total energy is equal to the core energy. In Eqs. (6.1) and (6.2), we have defined the notation of the Atomic Energy Level Tables. The ε_k symbol in Eq. (6.1) is the empirical equivalent to the ε_v of Eq. (6.54), that is, ε_k is the total energy of

[†]From this equation, it is clear that for two excited states v and v',

$$E_T(v) - E_T(v') = \varepsilon_v - \varepsilon_{v'},$$

that is, Koopmans' theorem is valid exactly in the frozen-core approximation.

the atom reduced by the core energy. For the ground state, we have

$$\varepsilon_v \text{ (ground state)} \approx \varepsilon_0 = -E_0, \qquad (6.56)$$

where ε_0 and E_0 are defined by Eq. (6.2).

Our result, embodied in Eq. (6.51), is that by introducing the frozen-core approximation, we have indeed eliminated the difficulties connected to the need of orthogonalizing the excited-state orbitals to the lower-lying states. It is clear from Eq. (6.51) that the wave functions belonging to different excited states will be orthogonal because they are the eigenfunctions of a joint Hamiltonian operator. Thus, with the core orbitals fixed, each excited state has become independent of other excited states.

It is very useful to review what the procedure would be without the frozen-core approximation. A fully consistent application of the HF model to excited states would consider each state as an $(N + 1)$-electron problem. The exact Hamiltonian is given by Eq. (6.23). The total wave function for each excited state would be the single determinant

$$\psi_T(v) = [(N + 1)!]^{-1/2} \det [\varphi_1 \cdots \varphi_N \varphi_v], \qquad (6.57)$$

and we can assume that for each excited state, the orbitals are orthogonal to each other and normalized. We would obtain the HF equations for the state v by varying the expectation value of the Hamiltonian with respect to $\psi_T(v)$. We would obtain HF equations for the core orbitals as well and we would get a separate set of equations for each excited state. Let us consider what would be the equation for the valence orbital.

The equation for the valence orbital φ_v would be the conventional HF equation plus Lagrangian multipliers ensuring orthogonality to the lower-lying excited states. Thus, we would get

$$H_F^{(v)} \varphi_v = \varepsilon_v \varphi_v + \sum_{i=1}^{N} \lambda_{vi} \varphi_i^{(v)} + \sum_{v'=0}^{v-1} \lambda_{vv'} \varphi_{v'}. \qquad (6.58)$$

In this equation, the Lagrangian multipliers λ_{vi} ensure the orthogonality with respect to the core orbitals and the multipliers $\lambda_{vv'}$ take into account the orthogonality with respect to the lower-lying excited states.[†] The Hamiltonian operator is the same as in Eq. (6.36). The superscript v indicates that the operator depends on the core orbitals belonging to φ_v, that is, on the core orbitals that must be made fully self-consistent with φ_v. In a consistent application of the HF model, the core orbitals must be made self-consistent with each φ_v; that is what we indicated with the superscript v on $H_F^{(v)}$ and

---

[†]In reality, the orthogonality to lower-lying excited states should be expressed in terms of the $(N + 1)$-electron determinants. It is easy to see that in the case of one running electron, the orthogonality between two valence orbitals is equivalent to the orthogonality between the corresponding $(N + 1)$-electron determinants.

also on the core orbital $\varphi_i^{(v)}$. Multiplying the equation from the left by $\varphi_k^{(v)*}$ $(k = 1, \ldots, N)$, and integrating, we get

$$\lambda_{vk} = \langle \varphi_k^{(v)} | H_F^{(v)} | \varphi_v \rangle \qquad (k = 1, \ldots, N). \tag{6.59}$$

Now let

$$\Omega^{(v)} = \sum_{i=1}^{N} |\varphi_i^{(v)}\rangle\langle \varphi_i^{(v)}|, \tag{6.60a}$$

and

$$\Pi^{(v)} = 1 - \Omega^{(v)}. \tag{6.60b}$$

Using these projection operators, we can write Eq. (6.58) in the form

$$\hat{H}_F^{(v)}\varphi_v = \varepsilon_v\varphi_v + \sum_{v'=0}^{v-1} \lambda_{vv'}\varphi_{v'}, \tag{6.61}$$

where

$$\hat{H}_F^{(v)} = \Pi^{(v)}H_F^{(v)}. \tag{6.62}$$

Now let $\bar{v}$ be another low-lying excited state. The HF equation for the valence orbital will be

$$\hat{H}_F^{(\bar{v})}\varphi_{\bar{v}} = \varepsilon_{\bar{v}}\varphi_{\bar{v}} + \sum_{v'=0}^{\bar{v}-1} \lambda_{vv'}\varphi_{v'}, \tag{6.63}$$

where the Hamiltonian in this equation has the same functional dependence on the core orbitals as the Hamiltonian in Eq. (6.62) except that the core orbitals are now self-consistent with $\varphi_{\bar{v}}$.

We now show that the Lagrangian multipliers vanish if the core orbitals are the same in two equations, Eqs. (6.61) and (6.63). Multiply Eq. (6.61) from the left by $\varphi_{\bar{v}}^*$ and integrate. We get

$$\lambda_{v\bar{v}} = \langle \varphi_{\bar{v}} | \hat{H}_F^{(v)} | \varphi_v \rangle. \tag{6.64}$$

Multiply Eq. (6.63) from the left by φ_v^* and integrate. Then we get

$$\langle \varphi_v | \hat{H}_F^{(\bar{v})} | \varphi_{\bar{v}} \rangle = 0. \tag{6.65}$$

Let us now assume that the core orbitals are the same for all excited states, that is, let us introduce the "frozen core." Then

$$\varphi_i^{(v)} \equiv \varphi_i^{(\bar{v})} \qquad (i = 1, \ldots, N), \tag{6.66}$$

for all excited states. Also,

$$\hat{H}_F^{(v)} \equiv \hat{H}_F^{(\bar{v})}. \tag{6.67}$$

Using this relationship and taking into account the form of the projection operator, Eq. (6.60a), we get for Eq. (6.65)

$$
\begin{aligned}
\langle \varphi_v | \hat{H}_F^{(\bar{v})} | \varphi_{\bar{v}} \rangle &= \langle \varphi_v | \hat{H}_F^{(v)} | \varphi_{\bar{v}} \rangle \\
&= \langle \Pi^{(v)} \varphi_v | H_F^{(v)} | \varphi_{\bar{v}} \rangle = \langle \varphi_v | H_F^{(v)} | \varphi_{\bar{v}} \rangle \\
&= 0.
\end{aligned} \tag{6.68}
$$

Using Eq. (6.60a), we obtain for Eq. (6.64)

$$\lambda_{v\bar{v}} = \langle \varphi_{\bar{v}} | \hat{H}_F^{(v)} | \varphi_v \rangle = \langle \varphi_{\bar{v}} | H_F^{(v)} | \varphi_v \rangle. \tag{6.69}$$

Comparing Eqs. (6.68) and (6.69), we conclude that

$$\lambda_{v\bar{v}} = 0, \tag{6.70}$$

where the last step requires the observation that $H_F^{(v)}$ is Hermitian. Substituting this result into Eq. (6.58), we obtain an equation in which we have only the Lagrangian multipliers with respect to the core orbitals. Those multipliers can be eliminated by a unitary transformation and we arrive back at Eq. (6.51).

This derivation demonstrated vividly how much simpler the HF model is with the "frozen core" than without it. In order to realize this, on need only to compare Eq. (6.51) with Eq. (6.58) and reflect on the amount of computational labor behind these equations. In the full formulation, represented by Eq. (6.58), we have a separate set of core orbitals for each excited state and the equation for one of the valence orbitals depends on all lower-lying excited states. These complications are absent from the frozen-core formulation.

Is the frozen-core approximation justifiable? Fock's original argument was the the valence orbital does not have much effect on the core distribution. This statement was contradicted by Hartree, who, in one of his early large-scale calculations, in the calculations for the Hg atom, found that the two $(6s)$ electrons of the atom have a considerable effect on the $(5d)^{10}$ electron distribution. From the large number of HF calculations, we could certainly select some cases in which the "frozen-core" approximation can be shown to be very accurate for some or all of the computed quantities. Likewise, we certainly could select cases in which the approximation would prove to be less accurate. The accuracy or inaccuracy of the approximation would also depend on the quantity that would be used for the comparison. Orbital-energy parameters are more likely to come out accurately than the radial electron densities, especially the "loose" density of an outer group. In our opinion, an argument that justifies a model by comparing it case by case to a more exact approximation is not very helpful because in order to do such a comparison, we have to calculate the more exact approximation and if we do that, then what is the model for? A model should be justified by plausible, general, mathematical, and physical arguments. If the result obtained with the model is not satisfactory, then a new and better approximation must be constructed and justified again by general mathematical, and physical arguments.

We adopt the position that the frozen-core approximation is primarily introduced for mathematical convenience, that is, it makes the treatment of excited states simpler conceptually and computationally. About the nature of this approximation, we can say, qualitatively, that the effect of the valence electrons is generally restricted to the electrons immediately below the valence shell. The quality of the approximation relative to the fully varied HF model will depend on how large the effect of the valence orbitals will be on the electrons immediately below the valence shell.

There is another argument in favor of the frozen core. As we will see, in discussing some numerical results (Sec. 7.2/D), the HF model yields poor results for the optical spectra of many-electron atoms. The reason for this is that for the optical spectrum, the core-valence correlation effects are important. In order to get accurate results, these effects must be taken into account. Thus, from the point of view of accuracy, the crucial question is not whether we consider the fully varied HF model or the frozen core. Rather, the crucial question will be whether we have incorporated the core-valence correlation effects into the theory or not. The point is that these correlation effects can be incorporated much more easily if we start from the frozen-core approximation.

Finally, we now clarify the angular-momentum and spin conditions and the form of the total potential in the theory of one-valence-electron atoms. A single-determinantal wave function, like ψ in Eq. (6.32), yields for complete groups the quantum numbers $L = S = M_L = M_S = 0$. Now if the orbital φ, in Eq. (6.38), is a spin orbital of the central-field type with the quantum numbers $(nlm_l m_s)$ then the $(N + 1)$-electron wave function of Eq. (6.38) will have the total quantum numbers $L = l$, $S = \frac{1}{2}$, $M_L = m_l$, and $M_S = m_s$. Thus, the levels of the spectrum generated by the valence orbital will be characterized by the (nl) of the excited state and will always be doublets with $L = l$. Thus, every excited state will be characterized by the notation

$$(nl)\,^2L, \tag{6.71}$$

where $L = l$.

We write down now the explicit formula for the total potential. For atoms with complete groups, the potential was given by Eq. (2.98). Our equation is Eq. (6.51); the potential in H_F is the HF potential for complete groups.

Changing Eq. (2.98) slightly, by writing it in the form of a local potential, we obtain

$$
U(r) = \sum_{\substack{n_j l_j \\ (\text{core})}} 2(2l_j + 1) \frac{1}{r} Y_0(n_j l_j n_j l_j | r)
$$

$$
- \sum_{\substack{n_j l_j \\ (\text{core})}} \sum_{k} \sqrt{\frac{2l_j + 1}{2l_v + 1}} \, c^k(l_v 0, l_j 0) \frac{1}{r} Y_k(n_j l_j n_v l_v | r) \frac{P_{n_j l_j}}{P_{n_v l_v}}, \quad (6.72)
$$

where the summations are over the complete groups of the core, and $(n_v l_v)$ are the valence-electron orbital parameters. The HF equation becomes (see Eq. (2.97b)),

$$
\left[-\frac{1}{2} \frac{d^2}{dr^2} + \frac{l_v(l_v + 1)}{2r^2} - \frac{Z}{r} + U(r) \right] P_{n_v l_v} = \varepsilon_{n_v l_v} P_{n_v l_v}. \quad (6.73)
$$

This equation is valid in the frozen-core approximation, for any excited state. It is interesting to observe that although the total potential U is a local potential in this equation, it is not the same local potential for all excited states. We see clearly in Eq. (6.72) that the last term, the exchange interaction, is a different function for each $(n_v l_v)$. Thus, we can now make the following statements:

1. In the frozen-core approximation, the Hamiltonian operator is the same for all excited states.
2. At the same time, the total *potential*, which can be formally written in local form, is a different function for each $(n_v l_v)$.
3. If we define the *effective potential* as

$$
U_{\text{eff}} = \frac{l_v(l_v + 1)}{2r^2} - \frac{Z}{r} + V(r) - A(n_v l_v | r), \quad (6.74)
$$

where V is the first term of Eq. (6.72), and A is the second, then, we see again that U_{eff} is different for each $(n_v l_v)$. We observe however, that, if we omit the exchange term A, then the effective potential becomes the joint potential for all excited states with the same l_v. That is, the potential depends on n_v only through the exchange term A.

6.4. HF THEORY FOR SEVERAL VALENCE ELECTRONS: "EXTENDED FROZEN-CORE" APPROXIMATION

We proceed now to build up the HF model for atoms with several valence electrons. The theory, which is a straightforward generalization of the model for one valence electron, is complicated by the fact that for several valence electrons, we have a number of incomplete groups and the wave function of the atom, in most cases, cannot be written in the form of a single determinant.

We build up the theory in two steps. First, we formulate the HF model in its conventional form, that is, in the same form in which we have formulated it for the ground state of atoms with several incomplete groups. In that formulation, the only novel feature is the orthogonality of the excited states relative to the lower-lying wave functions. As the second step, we introduce the frozen-core approximation and show that the formalism can be significantly simplified by this step.

Let us consider the case of an atom with two incomplete groups outside closed subshells. The valence electrons are the electrons in the incomplete groups. Let one of these groups be characterized by the quantum numbers $(n''l'')$ and the occupation number $q(n''l'')$. We assume that the other group consists of one valence electron with the quantum numbers (nl). This electron is the "running" electron.

We have formulated the HF equations for such a case in Sec. 3.3. In that formulation, it was tacitly assumed that the equations were valid for the ground state. Let us write down the equation for the running electron. The HF potential for that electron was given by Eq. (3.113). We have to substitute the HF potential into an equation of the type of Eq. (3.47). Equation (3.47) was written down for one incomplete group, but we would get the same equation for several groups except for the different potential. Thus, for the ground state of the running electron, we obtain

$$\left[-\frac{1}{2}\frac{d^2}{dr^2} + \frac{l(l+1)}{2r^2} - \frac{Z}{r} + U \right] P_{nl} = \varepsilon_{nl} P_{nl} + \sum_{n'} \lambda_{nn'} P_{n'l}, \quad (6.75)$$

were

$$U = U_{\text{av}} + \sum_{k=0}^{l_m} g_k \frac{1}{r} Y_k(n''l'', n''l''|r) + \sum_{k=0}^{l+l'} h_k \frac{1}{r} Y_k(nl, n''l''|r) \frac{P_{n''l''}}{P_{nl}}. \quad (6.76)$$

In Eq. (6.75), the Lagrangian multipliers ensure the orthogonality of the wave function to the core orbitals. U_{av} is the average potential of the configuration and it is given by the potential in Eq. (3.100). The g_k and h_k coefficients determine the angular-momentum and spin properties. We have discussed these coefficients in Sec. 6.2.

If we want to generalize Eq. (6.75) to excited states, the only change that we have to introduce is the application of Lagrangian multipliers, ensuring the orthogonality of an excited state to all lower-lying excited states. Let the Lagrangian multiplier ensuring that orthogonality be $\Gamma_{nn'}$. Also, let us change the notation by replacing (nl) with $(n_v l_v)$, where these quantum numbers now indicate an arbitrary valence-electron state. Then we obtain, by introducing the Γ's into Eq. (6.75),

$$
\left[-\frac{1}{2}\frac{d^2}{dr^2} + \frac{l_v(l_v + 1)}{2r^2} - \frac{Z}{r} + U \right] P_{n_v l_v}
$$

$$
= \varepsilon_{n_v l_v} P_{n_v l_v} + \sum_{n'} \lambda_{n_v n'} P_{n' l_v} + \sum_{n'_v = n_0}^{n_v - 1} \Gamma_{n_v n'_v} P_{n'_v l_v}. \tag{6.77}
$$

In this equation, the summation over n' is for all core orbitals with l_v. The summation over n'_v is for those excited states that have the azimuthal quantum number l_v and lie below the state $(n_v l_v)$; that is, the quantum number n_0 is the lowest principal quantum for which there is an excited state with l_v. The potential U is given by Eq. (6.76), and we get, with the new notation,

$$
U = U_{av} + \sum_{k=0}^{l_m} g_k \frac{1}{r} Y_k(n''l'', n''l''|r)
$$

$$
+ \sum_{k=0}^{l_v + l''} h_k \frac{1}{r} Y_k(n_v l_v, n''l'') \frac{P_{n''l''}}{P_{n_v l_v}}. \tag{6.78}
$$

These two equations, Eqs. (6.77) and (6.78), define the HF model for the running electron in the conventional formulation. As we see, the equation for $P_{n_v l_v}$ depends not only on all other orbitals in the atom, core and valence, but also on the orbitals of all lower-lying excited states. Because the core orbitals have to be made self-consistent with each excited-state orbital, we have a separate set of core orbitals for each such state. It is evident that in this form, the model is, conceptually as well as computationally, considerably more complicated than for the ground state.

It should be noted that the HF equations for the "nonrunning" valence electrons and for the core orbitals can be set up easily. The starting point for the formulation of these equations is Eq. (3.94), which gives the total energy of an atom with two incomplete groups.

Before proceeding to the frozen-core approximation, we want to write the HF potential in a form that will be useful later. We want to write the potential in Eq. (6.78) in such a form that the core and valence contributions are clearly separated. By valence contribution, we mean the electrostatic and exchange potentials generated by the "nonrunning" valence electrons, that is, by the electrons in the incomplete group characterized by $(n''l'')$. First, let us write down the general formula for U_{av}, which is the potential in Eq. (3.100):

$$
\begin{aligned}
U_{av} = &\sum_{\substack{(n'l') \\ \text{(all groups)}}} q(n'l')\frac{1}{r}Y_0(n'l', n'l'|r) \\
&- \frac{1}{r}Y_0(nl, nl|r) - \frac{q(nl)-1}{4l+1}\sum_{k>0}c^k(l0,l0)\frac{1}{r}Y_k(nl, nl|r) \\
&- \sum_{\substack{(n'l') \\ (n'l \neq nl)}} \frac{q(n'l')}{2(2l'+1)}\sqrt{\frac{2l'+1}{2l+1}}\sum_k c^k(l0,l'0)\frac{1}{r}Y_k(nl, n'l'|r)\frac{P_{n'l'}}{P_{nl}}.
\end{aligned}
$$

$$(6.79)$$

We separate the core from the valence contributions in such a way that in the first and fourth terms, we separate the $(n''l'')$ terms from the core contributions. Also, in the first term, we omit the term with (nl), which enables us to omit the second term. In the third term, $q(nl) = 1$, so that term is zero. In this way, we obtain the full core contribution. In order to get the full valence contribution, we must add to the potential obtained from Eq. (6.79) the terms arising from the g_k and h_k expressions in Eq. (6.78). These last represent electrostatic and exchange potentials generated by the "nonrunning" valence electrons.

The total potential, core plus valence, will be the expression

$$U = U_C + U_v, \tag{6.80}$$

where

$$
\begin{aligned}
U_C = &\sum_{\substack{(n'l') \\ \text{(core)}}} q(n'l')\frac{1}{r}Y_0(n'l', n'l'|r) \\
= &\sum_{\substack{(n'l') \\ \text{(core)}}} \frac{q(n'l')}{2(2l'+1)}\sqrt{\frac{2l'+1}{2l_v+1}}\sum_k c^k(l_v0, l'0)\frac{1}{r}Y_k(n_vl_v, n'l'|r)\frac{P_{n'l'}}{P_{n_vl_v}}, \tag{6.81}
\end{aligned}
$$

and

$$U_v = q(n''l'')\frac{1}{r}Y_0(n''l'', n''l''|r)$$

$$-\frac{q(n''l'')}{2(2l''+1)}\sqrt{\frac{2l''+1}{2l_v+1}}\sum_k c^k(l_v0, l''0)\frac{1}{r}$$

$$\times Y_k(n_v l_v, n''l''|r)\frac{P_{n''l''}}{P_{n_v l_v}} + \sum_{k=0}^{l_m} g_k \frac{1}{r}Y_k(n''l'', n''l''|r)$$

$$+ \sum_{k=0}^{l_v+l''} h_k \frac{1}{r}Y_k(n_v l_v, n''l''|r)\frac{P_{n''l''}}{P_{n_v l_v}}. \tag{6.82}$$

Next we show that both U_C and U_v can be written as linear operators that depend only on l_v, besides, of course, depending on the core and nonrunning valence orbitals. Inspecting Eqs. (6.81) and (6.82), we see that the terms that represent electrostatic interactions do not depend on the parameters of the running electron. The terms representing exchange interactions depend on both n_v and l_v. It is easy to see, however, that we can write U_C in the operator form

$$U_C(r)P_{n_v l_v}(r) = \int_0^\infty K_C(n'l', l_v|rr')P_{n_v l_v}(r')\, dr'. \tag{6.83}$$

The kernel K_C of this operator depends on all core orbitals, which are indicated by $(n'l')$. Besides that, K_C depends only on l_v, which becomes evident if we compare Eqs. (6.83) and (6.81). The comparison yields

$$K_C(n'l', l_v|rr') = \sum_{\substack{(n'l') \\ (\text{core})}} q(n'l')\frac{1}{r}Y_0(n'l', n'l'|r)\,\delta(r-r')$$

$$-\sum_{\substack{(n'l') \\ (\text{core})}} \frac{q(n'l')}{2(2l'+1)}\sqrt{\frac{2l'+1}{2l_v+1}}\sum_{k=0}^{l_v+l'} c^k(l_v0, l'0)$$

$$\times \frac{1}{r}P_{n'l'}(r)P_{n'l'}(r')L_k(rr'), \tag{6.84}$$

where L_k is given by Eq. (2.72). Similarly, we can write

$$U_v(r)P_{n_v l_v}(r) = \int_0^\infty K_v(n''l'', l_v|rr')P_{n_v l_v}(r')\, dr', \tag{6.85}$$

where

$$K_v(n''l'', l_v|rr')$$

$$= q(n''l'') \frac{1}{r} Y_0(n''l'', n''l''|r) \delta(r - r')$$

$$+ \sum_{k=0}^{l_m} g_k \frac{1}{r} Y_k(n''l'', n''l''|r) \delta(r - r')$$

$$- \frac{q(n''l'')}{2(2l''+1)} \sqrt{\frac{2l''+1}{2l_v+1}} \sum_{k=0}^{l_v+l''} c^k(l_v 0, l''0) \frac{1}{r} P_{n''l''}(r) P_{n''l''}(r') L_k(rr')$$

$$+ \sum_{k=0}^{l_v+l''} h_k \frac{1}{r} P_{n''l''}(r) P_{n''l''}(r') L_k(rr'). \tag{6.86}$$

In Eqs. (6.84) and (6.86), we incorporated the local potentials in the kernels by making use of the following property of the Dirac function:

$$\int_0^\infty \delta(r - r') f(r') \, dr' = f(r). \tag{6.87}$$

Having written both U_C and U_v as linear operators, we are now able to define a linear operator for the expression in square brackets of Eq. (6.77). We can write

$$H_F(n'l', n''l'', l_v) \equiv -\frac{1}{2} \frac{d^2}{dr^2} + \frac{l_v(l_v + 1)}{2r^2} - \frac{Z}{r}$$

$$+ U_C(n'l', l_v) + U_v(n''l'', l_v), \tag{6.88}$$

and it is clear on the basis of the properties of U_C and U_v that H_F is a linear operator depending on the core orbitals indicated by $(n'l')$, on the nonrunning valence orbitals denoted by $(n''l'')$, and on the running-electron parameter l_v. Using Eq. (6.88), we can write Eq. (6.77) in the form

$$H_F(n'l', n''l'', l_v) P_{n_v l_v} = \varepsilon_{n_v l_v} P_{n_v l_v} + \sum_{\substack{n' \\ (\text{core})}} \lambda_{n_v n'} P_{n'l_v} + \sum_{n'_v = n_0}^{n_v - 1} \Gamma_{n_v n'_v} P_{n'_v l_v}. \tag{6.89}$$

Next, we incorporate the core-valence Lagrangians into the Hamiltonian operator. Let

$$\Omega(n'l_v) = \sum_{\substack{n' \\ (\text{core})}} |P_{n'l_v}\rangle\langle P_{n'l_v}|, \tag{6.90}$$

and

$$\Pi(n'l_v) = 1 - \Omega(n'l_v), \tag{6.91}$$

where the summation over n' is for the core orbitals with l_v. We multiply Eq. (6.89) from the left by the core orbital $P_{n'l_c}$ and integrate. The Γ term will not yield anything because the orbitals in that term are all valence orbitals, which are orthogonal to the core orbitals. We get

$$\lambda_{n_c n'} = \langle P_{n'l_c} | H_F(n'l', n''l'', l_v) | P_{n_c l_c} \rangle, \tag{6.92}$$

and using Eqs. (6.90) and (6.91), we can write Eq. (6.89) in the form

$$\tilde{H}_F(n'l', n''l'', l_v) P_{n_c l_c} = \varepsilon_{n_c l_c} P_{n_c l_c} + \sum_{n'_c = n_0}^{n_v - 1} \Gamma_{n_c n'_c} P_{n'_c l_c}, \tag{6.93}$$

where

$$\tilde{H}_F(n'l', n''l'', l_v) = \Pi(n'l_v) H_F(n'l', n''l'', l_v). \tag{6.94}$$

We now introduce the frozen-core approximation in the form suggested by Cohen and Kelly.[59] In contrast to Fock, who suggested that the orbitals in the atomic core should be kept fixed for all valence electron states, we demand that not only the core orbitals, but also the orbitals of the nonrunning valence electrons be kept fixed, that is, we demand that *all* orbitals except the wave function of the running electron be kept frozen. We can formulate this demand by saying that the core states $P_{n'l'}$ and the nonrunning valence orbitals $P_{n''l''}$ be the same for all excited states of the running electron. We call this the "extended frozen-core" approximation.

Let us establish what will be the effect of this approximation on the HF equations. In order to do that, let us first consider the conventional equation for an excited state with the quantum numbers $\tilde{n}_v$ and l_v, with $\tilde{n}_v$ being less than n_v, that is, let us consider an excited state below n_v, but with the same azimuthal quantum number. We can write down the HF equation for such an orbital on the basis of Eq. (6.93). It will be

$$\tilde{H}_F\left(\tilde{n}', \tilde{l}', \tilde{n}''\tilde{l}'', l_v\right) P_{\tilde{n}_v l_v} = \varepsilon_{\tilde{n}_v l_v} P_{\tilde{n}_v l_v} + \sum_{n''_v = n_0}^{\tilde{n}_v - 1} \Gamma_{\tilde{n}_c n''_c} P_{n''_c l_c}. \tag{6.95}$$

In this equation, we have $(\tilde{n}', \tilde{l}')$ and $(\tilde{n}''\tilde{l}'')$. These quantum numbers indicate the core orbitals and the nonrunning valence orbitals, respectively. The reason why we have a set of orbitals in Eq. (6.95) different from the orbitals in Eq. (6.93) is that, as we have pointed out before, the core orbitals and the nonrunning valence orbitals must be made self-consistent with the valence orbital $P_{n_v l_v}$. Because in the two equations we have different orbitals for the running electron, we must have different sets of core orbitals and different orbitals for the nonrunning valence electrons. Thus, the set indicated by $(n'l'n''l'')$ is self-consistent with $P_{n_v l_v}$ and the set $(\tilde{n}'\tilde{l}'\tilde{n}''\tilde{l}'')$ is self-consistent with the orbital $P_{\tilde{n}_v l_v}$.

We now show that in the extended frozen-core approximation, the Hamiltonian operators in Eqs. (3.95) and (3.93) will become identical, as a result of which the Lagrangian multipliers ensuring orthogonality between the different excited states will become zero. Actually, the statement about the Hamiltonians becoming identical is self-evident because the extended frozen-core approximation is defined in such a way as to achieve this. The proof about the Lagrangians is simple. Let us multiply Eq. (6.93) from the left by $P_{\tilde{n}_v l_v}$ and integrate. We obtain

$$\Gamma_{n_v \tilde{n}_v} = \langle P_{\tilde{n}_v l_v} | \tilde{H}_F(n'l', n''l'', l_v) | P_{n_v l_v} \rangle, \tag{6.96}$$

where we recall that $\tilde{n}_v < n_v$; therefore, $\tilde{n}_v$ is among the values to which the summation in Eq. (6.93) is extended. Multiply Eq. (6.95) from the left by $P_{n_v l_v}$ and integrate. We obtain, in view of $n_v > \tilde{n}_v$,

$$\langle P_{n_v l_v} | \tilde{H}_F\left(\tilde{n}'\tilde{l}', \tilde{n}''\tilde{l}'', l_v\right) | P_{\tilde{n}_v l_v} \rangle = 0. \tag{6.97}$$

The introduction of the extended frozen core means that we can put

$$P_{n'l'} \equiv \tilde{P}_{n'l'}, \tag{6.98a}$$

and

$$P_{n''l''} \equiv \tilde{P}_{n''l''}. \tag{6.98b}$$

Using these relationships and taking into account Eqs. (6.94), (6.90), and (6.91), we get for the matrix component in Eq. (6.97)

$$\begin{aligned}
\langle P_{n_v l_v} | \tilde{H}_F & \left(\tilde{n}'\tilde{l}', \tilde{n}''\tilde{l}'', l_v\right) | P_{\tilde{n}_v l_v} \rangle \\
&= \langle P_{n_v l_v} | \tilde{H}_F(n'l', n'', l_v) | P_{\tilde{n}_v l_v} \rangle \\
&= \langle \Pi(n'l_v) P_{n_v l_v} | H_F(n'l', n''l'', l_v) | P_{\tilde{n}_v l_v} \rangle \\
&= \langle P_{n_v l_v} | H_F(n'l', n''l'', l_v) | P_{\tilde{n}_v l_v} \rangle = 0. \tag{6.99}
\end{aligned}$$

In the third line, we used the Hermitian character of the operator Π, and in the fourth, we made use of the relationship

$$
\begin{aligned}
\Pi(n'l_v)P_{n_v l_v} &= \left(1 - \Omega(n'l_v)\right)P_{n_v l_v} \\
&= P_{n_v l_v} - \sum_{\substack{n' \\ (\text{core})}} |P_{n'l_v}\rangle\langle P_{n'l_v}|P_{n_v l_v}\rangle \\
&= P_{n_v l_v},
\end{aligned}
\tag{6.100}
$$

which is valid because all orbitals of the running electron are orthogonal to the core orbitals. By using the properties of the projection operators in Eq. (6.96), we obtain

$$
\begin{aligned}
\Gamma_{n_v \tilde{n}_v} &= \langle \Pi(n'l_v)P_{\tilde{n}_v l_v}|H_F(n'l', n''l'', l_v)|P_{n_v l_v}\rangle \\
&= \langle P_{\tilde{n}_v l_v}|H_F(n'l', n''l'', l_v)|P_{n_v l_v}\rangle.
\end{aligned}
\tag{6.101}
$$

Comparing Eqs. (6.99) and (6.101) and recalling that H_F is Hermitian, we obtain

$$
\Gamma_{n_v \tilde{n}_v} = 0,
\tag{6.102}
$$

which is the statement we wanted to prove.

Therefore, our conclusion is that in the extended frozen-core approximation, the HF equation for the running electron of an atom with two incomplete groups is given by Eqs. (6.93) and (6.94) with the Lagrangians omitted:

$$
\tilde{H}_F(n'l', n''l'', l_v)P_{n_v l_v} = \varepsilon_{n_v l_v}P_{n_v l_v},
\tag{6.103}
$$

with

$$
\tilde{H}_F(n'l', n''l'', l_v) = \Pi(n'l_v)H_F(n'l', n''l'', l_v).
\tag{6.104}
$$

These equations are valid for any excited states. It is interesting to note that here we cannot say that two orbitals belonging to two different excited states will be orthogonal because they are the eigenfunctions of the same operator. Such a statement can only be made if the operator is Hermitian and H_F is, as we see from Eq. (6.104), not Hermitian. Nevertheless, two orbitals of different excited states *are* orthogonal. Let us write down the two equations:

$$
\tilde{H}_F P_{n_v l_v} = \varepsilon_{n_v l_v}P_{n_v l_v},
\tag{6.105}
$$

$$
\tilde{H}_F P_{n'_v l_v} = \varepsilon_{n'_v l_v}P_{n'_v l_v}.
\tag{6.106}
$$

Multiply the first equation from the left by $P_{n'_v l_v}$, the second by $P_{n_v l_v}$, and integrate. Subtract the second equation from the first. Then we get

$$\langle n_v l_v | \tilde{H}_F | n'_v l_v \rangle - \langle n'_v l_v | \tilde{H}_F | n_v l_v \rangle = (\varepsilon_{n_v l_v} - \varepsilon_{n'_v l_v}) \langle n'_v l_v | n_v l_v \rangle. \quad (6.107)$$

By using Eq. (6.100) in both matrix components, we obtain

$$\langle n_v l_v | H_F | n'_v l_v \rangle - \langle n'_v l_v | H_F | n_v l_v \rangle = (\varepsilon_{n_v l_v} - \varepsilon_{n'_v l_v}) \langle n'_v l_v | n_v l_v \rangle, \quad (6.108)$$

and now we can say that the left side is zero because H_F is Hermitian, and from that, it follows

$$\int P_{n'_v l_v} P_{n_v l_v} \, dr = 0. \quad (6.109)$$

Thus, two orbitals of the same l_v will be automatically orthogonal if they are solutions of Eq. (6.103). Because two orbitals of different l_v will be orthogonal because of their angular part, we can say that all excited-state wave functions will be orthogonal.

Next we derive the connection between $\varepsilon_{n_v l_v}$ and the total energy of the atom. From Eqs. (6.103) and (6.100), we obtain

$$\varepsilon_{n_v l_v} = \langle P_{n_v l_v} | H_F | P_{n_v l_v} \rangle. \quad (6.110)$$

Taking into account Eqs. (6.88), (6.83), (6.84), (6.85), and (6.86), we get

$$\varepsilon_{n_v l_v} = I(n_v l_v) + \langle P_{n_v l_v} | U_C | P_{n_v l_v} \rangle + \langle P_{n_v l_v} | U_v | P_{n_v l_v} \rangle, \quad (6.111)$$

where

$$\langle P_{n_v l_v} | U_C | P_{n_v l_v} \rangle = \sum_{\substack{(n'l') \\ (\text{core})}} q(n'l') F_0(n_v l_v, n'l') - \sum_{\substack{(n'l') \\ (\text{core})}} \frac{q(n'l')}{2(2l'+1)}$$

$$\times \sqrt{\frac{2l'+1}{2l_v+1}} \sum_k c^k(l_v 0, l'0) G_k(n_v l_v, n'l'), \quad (6.112)$$

and

$$\langle P_{n_v l_v} | U_v | P_{n_v l_v} \rangle$$
$$= q(n''l'') F_0(n_v l_v, n''l'')$$
$$+ \sum_{k=0}^{l_m} g_k F_k(n_v l_v, n''l'') + \sum_{k=0}^{l_v+l''} h_k G_k(n_v l_v, n''l'')$$
$$- \frac{n(n''l'')}{2(2l''+1)} \sqrt{\frac{2l''+1}{2l_v+1}} \sum_{k=0}^{l_v+l''} c^k(l_v 0, l''0) G_k(n_v l_v, n''l''). \quad (6.113)$$

The total energy of the atom was given by Eqs. (3.94) and (3.95):

$$E_{HF} = E_{av} + \sum_{k>0}^{2l''} f_k F_k(n''l'', n''l'')$$

$$+ \sum_{k=0}^{l_m} g_k F_k(n_v l_v, n''l'') + \sum_{k=0}^{l+l''} h_k G_k(n_v l_v, n''l''). \quad (6.114)$$

Now let us recall that because the atomic core consists of complete groups, the average energy gives the exact expression for the energy. The same is true for the core-valence interaction and for the intragroup electrostatic energy of the incomplete groups. If we inspect Eq. (6.112), we realize that that expression is the interaction of the running electron with the core. Likewise, Eq. (6.113) contains the formula for the interactions of the running electron with the valence electrons in the other incomplete group.

Let us now write E_{av} into this form:

$$E_{av} = E(\text{core}) + E(n_v l_v | \text{core}) + E(n''l'' | \text{core})$$

$$+ E(n_v l_v) + E(n''l'') + E(n_v l_v | n''l''). \quad (6.115)$$

Here $E(\text{core})$ is the total energy of the core, $E(n_v l_v | \text{core})$ is the interaction between the running electron and the core, and so on. Substituting this into (6.114), we obtain

$$E_{HF}(n_v l_v) = E(\text{core}) + E(n_v l_v | \text{core})$$

$$+ E(n''l'' | \text{core}) + E(n_v l_v) + E(n''l'')$$

$$+ E(n_v l_v | n''l'') + \sum_{k>0}^{2l''} f_k F_k(n''l'', n''l'')$$

$$+ \sum_{k=0}^{l_m} g_k F_k(n_v l_v, n''l'') + \sum_{k=0}^{l+l''} h_k G_k(n_v l_v, n''l''). \quad (6.116)$$

It is easy to see that Eq. (6.111) can be written in the following form:

$$\varepsilon_{n_v l_v} = I(n_v l_v) + E(n_v l_v | \text{core})$$

$$+ E(n_v l_v | n''l'') + \sum_{k=0}^{l_m} g_k F_k(n_v l_v, n''l'')$$

$$+ \sum_{k=0}^{l_v + l''} h_k G_k(n_v l_v, n''l''), \quad (6.117)$$

where we separated the interaction between the running and nonrunning valence electrons into that parts that is contained in E_{av} and the g_k and h_k terms that are the corrections to E_{av}.

Let us now subtract Eq. (6.117) from Eq. (6.116). We obtain

$$E_{HF}(n_v l_v) - \varepsilon_{n_v l_v} = E(\text{core}) + E(n''l''|\text{core})$$

$$+ E(n''l'') + \sum_{k>0}^{2l''} f_k F_k(n''l'', n''l''). \quad (6.118)$$

This expression has a transparently clear physical interpretation. On the right side, first, we have the core energy. The next term is the interaction between the nonrunning valence electrons and the core; this is the exact expression because the average energy gives the exact expression for this. The last two terms are the complete intragroup energy of the nonrunning valence electrons. $E(n''l'')$ is that part of this energy that is contained in E_{av} and the f_k term is the correction to E_{av}.

Let us call the core plus the nonrunning valence electrons the "extended core." Then we can write Eq. (6.118) in the form

$$E_{HF}(n_v l_v) - \varepsilon_{n_v l_v} = E(\text{extended core}), \quad (6.119)$$

or

$$E_{HF}(n_v l_v) = \varepsilon_{n_v l_v} + E(\text{extended core}). \quad (6.120)$$

Now let us consider the difference between the total energies of two excited states. This is the energy that is emitted in form of a photon in an optical transition. In constructing this difference, we recall that E (extended core) does not in any way depend on the excited-state parameters $(n_v l_v)$. We obtain

$$\Delta = E_{HF}(n_v l_v) - E_{HF}(n'_v l'_v)$$

$$= \varepsilon_{n_v l_v} - \varepsilon_{n'_v l'_v}. \quad (6.121)$$

We obtained an interesting result that can be formulated as follows:

In the framework of the HF model, when the extended frozen-core approximation is introduced, Koopmans' theorem is valid exactly, that is, the difference of total energies is equal to the difference of the orbital parameters.

Therefore, in this approximation, we can write

$$\nu = \frac{\varepsilon_{n_v l_v} - \varepsilon_{n'_v l'_v}}{h}, \quad (6.122)$$

where ν is the frequency of the emitted light, h is Planck's constant, and, of course, this equation is still subjected to the selection rules (see Sec. 6.5).

We quickly clarify the last sentence. From angular momentum and spin theory, it is known that one-electron transitions between multiplets are permitted only if the running electron in the atom makes the transition with

$$\Delta l = \pm 1, \tag{6.123}$$

and the total angular-momentum and spin quantum numbers are changing according to the rule

$$\Delta L = 0, \pm 1, \tag{6.124a}$$

$$\Delta S = 0. \tag{6.124b}$$

These equations are derived under the assumption of LS coupling, that is, under the assumption that we are working with total wave functions diagonalizing L^2 and S^2. An inspection of the derivation leading to Eq. (6.121) shows that the (LS) quantum numbers may be different for the states $(n_v l_v)$ and $(n'_v l'_v)$. The (LS) quantum numbers are in the g_k and h_k coefficients of the formula for $\varepsilon_{n_v l_v}$, which is given by Eq. (6.117). In deriving Eq. (6.121), we have assumed that the total intragroup energy of the nonrunning valence electrons was the same in the states $(n_v l_v)$ and $(n'_v l'_v)$. The intragroup energy of the nonrunning valence electrons is given by the last two terms of Eq. (6.118). In those terms, we have the f_k coefficients. The configuration of the nonrunning valence electrons is $(n''l'')^q$. For such a configuration, the f_k coefficients depend only on the $(L_1 S_1)$ of the parent multiplet. The f_k's do not depend on how the running electron is coupled to the parent configuration, that is, they do not depend on the total quantum numbers (LS). Thus, the f_k coefficients and along with them the total intragroup energy of the nonrunning valence electrons will be the same in the two excited states even if the total angular momentum is different. Thus, the derivation leading to Eq. (6.121) is indeed in full agreement with the selection rules.

We summarize the results of this section and of the preceding section. We have formulated the HF model for atoms with one valence electron and for atoms with several valence electrons. Among the atoms with several valence electrons, we have considered those that have two incomplete groups, one of which consisted of the running electron. Besides formulating the HF model in its conventional form, we also introduced the approximations that we called "frozen-core" and "extended frozen-core" approximations. It was shown that the model becomes much simpler, conceptually and computationally, if these approximation are introduced. In the simplified form, the model exhibited the following properties:

1. All states of the running electron were orthogonal to each other without any Lagrangian multipliers.
2. The wave functions of the running electron were the solutions of HF equations in which the Hamiltonian operator was Hermitian for one

valence electron and non-Hermitian for several valence electrons. In both cases, the radial valence orbitals were solutions of equations in which the Hamiltonian operators (and the nonlocal potentials) were the same for orbitals with the same azimuthal quantum member.

3. In the frozen-core and extended frozen-core approximations, Koopmans' theorem was proved to be valid exactly.

In connection with the model for several valence electrons, we would like to make some additional observations. As we noted in Sec. 6.1, there are many atomic spectra that can be characterized as one-electron spectra despite the large number of core electrons and despite the presence of other nonrunning valence electrons. Among these spectra, there are many that have the configuration $(n''l'')^q(nl)$, that is, they are of the type of configuration for which we have formulated the HF model here. Thus, the model developed in this section covers a large number of energy levels in the optical spectra.

In addition to this, it is clear, from the discussions of Secs. 3.3 and 6.2 and from the presentation in this section, that the HF model can easily be developed, with or without the frozen-core approximation, for multiplets generated by more than two incomplete groups, that is, for multiplets with electron configurations more complicated than the case treated in this section. Thus, on the basis of the techniques developed in this and in the preceding sections, the HF model can be formulated for the large majority of multiplets occurring in the atomic spectra.

6.5. RADIATION TRANSITION PROBABILITIES

When we reviewed the Atomic Energy Level Tables in Sec. 6.1, we were looking at energy levels that were produced by transitions from one state of the atom to another. In this section, we discuss the connection between such radiation-producing or absorbing transitions and the electronic structure of atoms.

Let a number of atoms be placed in electromagnetic radiation with the spectral energy density $\rho(\nu)$, where ν is the frequency of radiation. Under the influence of the radiation stimulated emission and absorption will take place; in addition, spontaneous emission will occur even in absence of radiation. The probability per unit time that an atom originally in state n will undergo a transition to state m is given by

$$R_{n \to m} = B_{n \to m} \rho(\nu_{mn}),$$

(6.125)

where

$$\nu_{mn} = \frac{1}{h}(E_m - E_n),$$

(6.126)

and we assumed that $n < m$, that is, $E_n < E_m$. This is the probability of stimulated absorption. The probability per unit time that an atom in state m will undergo a transition to state n is given by

$$R_{m \to n} = A_{m \to n} + B_{m \to n} \rho(\nu_{mn}). \tag{6.127}$$

In these formulas, $B_{n \to m}$ and $B_{m \to n}$ are the Einstein coefficients of stimulated absorption and emission, respectively; $A_{m \to n}$ is the coefficient of spontaneous emission. Using classical Boltzman statistics, Einstein derived that

$$B_{m \to n} = B_{n \to m}, \tag{6.128}$$

and

$$A_{m \to n} = \frac{16\pi^2 \hbar \nu_{mn}^3}{c^3} B_{m \to n}. \tag{6.129}$$

It is obvious that the coefficients will depend on the electronic structure of atoms undergoing the transitions. In this section, our goal is to derive formulas for $B_{m \to n}$ in terms of the atomic wave functions. From Einstein's formulas, we see that the emission and absorption processes are fully described by the A and B coefficients; we gain understanding of these processes by investigating the dependence of the coefficients on the electronic structure.

Let the atom be described by the Schroedinger equation:

$$H\Phi = i\hbar \frac{\partial \Phi}{\partial t}, \tag{6.130}$$

where the Hamiltonian is given by

$$H = H_0 + H', \tag{6.131}$$

and H_0 is the nonrelativistic exact Hamiltonian. H' is the operator describing the electromagnetic field and we assume that this operator can be treated as a perturbation.

We construct H' as follows. In Sec. 4.2/A, we have discussed the form of the Hamiltonian operator in the presence of an external field. For a one-electron atom, we have written down Eq. (4.8), which reads

$$H = \frac{1}{2m}\left(\mathbf{p} + \frac{e}{c}\mathbf{A}\right)^2 - eV. \tag{6.132}$$

Here $\mathbf{p}$ is the momentum of the electron, and $\mathbf{A}$ and V are the vector and scalar potentials, respectively, of the electromagnetic field. From Eq. (4.11), we get

$$H = \frac{p^2}{2m} + \frac{e}{mc}\mathbf{A} \cdot \mathbf{p} + \frac{e^2}{2mc^2}A^2 - eV. \tag{6.133}$$

If we work with electromagnetic fields that are weak relative to the field of the nucleus, we can omit A^2 and we get

$$H = \frac{p^2}{2m} - eV + \frac{e}{mc}\mathbf{A} \cdot \mathbf{p}. \tag{6.134}$$

Assuming that the electromagnetic radiation consists of a light wave that can be described by the vector potential, we can identify V with the internal potential of the atom and we can put for the unperturbed Hamiltonian

$$H_0 = \frac{p^2}{2m} - eV, \tag{6.135}$$

and then we get for the perturbation

$$H' = \frac{e}{mc}\mathbf{A} \cdot \mathbf{p}. \tag{6.136}$$

For an N-electron atom, we get

$$H' = \sum_{j=1}^{N} \frac{e}{mc}\mathbf{A}_j \cdot \mathbf{p}_j. \tag{6.137}$$

In Appendix J, we show that by using Eq. (6.137), we obtain from the time-dependent Schroedinger equation the approximate formula

$$B_{n \to m} = \frac{3\pi}{3\hbar^2}|R_{mn}|^2, \tag{6.138}$$

where

$$|R_{mn}|^2 = |X_{mn}|^2 + |Y_{mn}|^2 + |Z_{mn}|^2. \tag{6.139}$$

In this formula, we have for X

$$X = -e \sum_{j=1}^{N} x_j, \qquad (6.140)$$

where x_j is the x-coordinate of the jth electron. Also

$$X_{mn} = \int \psi_m^{0*} X \psi_n^0 \, dq, \qquad (6.141)$$

where ψ_m^0 is the exact eigenfunction of the unperturbed Hamiltonian, H_0. As we see, X_{mn} is the matrix component of the electric dipole moment with respect to the unperturbed eigenfunctions of the Hamiltonian, H_0.

We note that the basic approximation made in deriving the expression of Eq. (6.138) is that the electromagnetic field is constant within the atoms. If this approximation is not made, then the expression we get for the transition probability contains, besides the electric dipole term, also the magnetic dipole and electric quadrupole radiation terms. It is easy to see, however, that these terms will be small relative to the electric dipole radiation. Thus, the additional terms will be significant only in the case when the dipole radiation term vanishes because of the symmetry properties of the atomic wave functions.

We now establish the selection rules for dipole radiation and derive the formulas for the transition probabilities for some special cases. The selection rules are selecting the transitions that are permitted by the symmetry properties of the atomic wave functions; dipole radiation is not possible if the transition is ruled out by the symmetry properties.

The most general selection rule, which is valid under all circumstances, is that transitions are permitted only between an even and an odd state. We have seen that the transition probability is zero if the matrix component of the dipole radiation is zero. The electric dipole moment for one electron is $(-e\mathbf{r})$. This is an odd function because upon inversion of the coordinates, $\mathbf{r}$ goes into $(-\mathbf{r})$. Now if the product of the wave functions, $\psi_m^{0*} \psi_n^0$, is an even function, then the matrix component will be zero because the integrand will be an odd function in the integral giving the matrix component of the dipole moment. If $(-e\mathbf{r})$ is an odd function, $\psi_m^{0*} \psi_n^0$ must also be an odd function so that the product $(-e\mathbf{r}\psi_m^{0*} \psi_n^0)$ will be an even function. Thus, from the two eigenfunctions, ψ_m^0 and ψ_n^0, one must be odd and the other even. This rule is generally valid because it depends only on the symmetry and not on the specific form of the wave functions.

We now formulate the selection rules in terms of the quantum numbers that characterize the transition. We treat two special cases: First, we discuss atoms with one valence electron and then we proceed to the more general case of atoms with several valence electrons.

A. Atoms with One Valence Electron

In the HF model, an atom, in which there is a core with complete groups and a single valence electron, is characterized by a single-determinantal wave function; the wave function of the valence electron has the same angular part as in the H atom. Thus, for transitions involving the valence electron, the selection rules will be the same as for the H atom. Let n, l, m_l, and m_s be the quantum numbers of the valence electron in the initial state and let n', l', m'_l, and m'_s be the quantum numbers in the final state. Then, similarly to the H atom, we have

$$\Delta l = l - l' = \pm 1; \tag{6.142}$$

$$\Delta m_l = m_l - m'_l = 0, \pm 1. \tag{6.143}$$

Because the electric dipole moment does not depend on the electron spin, the spin quantum number does not change, that is, $\Delta m_s = 0$. The core having complete groups, we have for the total angular momentum and spin of the atom, L and S, respectively,

$$L = l, \qquad \Delta L = l - l'; \tag{6.144}$$

$$S = \tfrac{1}{2}, \qquad \Delta S = 0. \tag{6.145}$$

Next, we want a formula for the B coefficient in terms of the HF wave functions. We approximate ψ_m^0 and ψ_n^0 by the HF determinantal wave functions. Working with the frozen-core approximation, our core orbitals will be the same in the initial and final state. Let P_{nl} and $P_{n'l'}$ be the radial parts of the HF wave functions of the valence electron in the initial and final states. We want the matrix components of coordinates x, y, and x with respect to the initial and final wave functions. It is convenient to determine the matrix components of the combinations $(x \pm iy)$ and z. According to Slater [Slater II, 224], we get

$$\langle n, l, m_l | x + iy | n', l + 1, m_l - 1 \rangle$$
$$= \sqrt{(l - m_l + 1)(l - m_l + 2)}\, A(n, l; n', l + 1); \tag{6.146}$$

$$\langle n, l, m_l | x + iy | n', l - 1, m_l - 1 \rangle$$
$$= -\sqrt{(l + m_l - 1)(l + m_l)}\, A(n, l; n', l - 1). \tag{6.147}$$

From these formulas, we see that the matrix components of $(x + iy)$ are zero unless $m'_l = m_l - 1$ and, of course, $l' = l + 1$. Further, we get

$$\langle n, l, m_l | x - iy | n', l + 1, m_l + 1 \rangle$$
$$= -\sqrt{(l + m_l + 1)(l + m_l + 2)}\, A(n, l; n', l + 1); \quad (6.148)$$

$$\langle n, l, m_l | x - iy | n', l - 1, m_l + 1 \rangle$$
$$= \sqrt{(l - m_l - 1)(l - m_l)}\, A(n, l; n', l - 1); \quad (6.149)$$

and we see that here we must have $m'_l = m_l + 1$. Finally,

$$\langle n, l, m_l | z | n', l + 1, m_l \rangle$$
$$= \sqrt{(l + 1)^2 - m_l^2}\, A(n, l; n', l + 1); \quad (6.150)$$

$$\langle n, l, m_l | z | n', l - 1, m_l \rangle$$
$$= \sqrt{(l^2 - m_l^2)}\, A(n, l; n', l - 1); \quad (6.151)$$

and we see that in these matrix components, we must have $m'_l = m_l$. The quantities denoted by A are defined by the formulas

$$A(n, l; n', l + 1) = \frac{1}{\sqrt{(2l + 1)(2l + 3)}} \int_0^\infty r P_{nl}(r) P_{n'l+1}(r)\, dr; \quad (6.152)$$

$$A(n, l; n', l - 1) = \frac{1}{\sqrt{(2l - 1)(2l + 1)}} \int_0^\infty r P_{nl}(r) P_{n'l-1}(r)\, dr. \quad (6.153)$$

We see from these formulas that the transition probabilities will depend on the initial and final radial orbitals of the valence electron. In order to obtain a formula for the transition $(n, l) \rightarrow (n', l \pm 1)$, we must sum over the permitted m_l values. We see that transition from a state with m_l will be permitted into the states with $m_l, (m_l + 1)$, and $(m_l - 1)$. Because these transitions will have the same frequency—the energy levels depend only on n and l—we must sum over the permitted m_l transitions.

We introduce here an oscillator strength for the transition $(nlm_l) \rightarrow (n'l'm'_l)$ with the definition

$$f(n, l, m_l; n', l', m'_l) = \frac{E_{n'l'} - E_{nl}}{3} \{ |\langle n, l, m_l | x | n', l', m'_l \rangle|^2$$
$$+ |\langle n, l, m_l | y | n', l', m'_l \rangle|^2$$
$$+ |\langle n, l, m_l | z | n', l', m'_l \rangle|^2 \}. \quad (6.154)$$

For these oscillator strengths, we have the Kuhn–Thomas sum rule,[61] which reads

$$\sum_{n'l'm'_l} f(n, l, m_l; n'l'm'_l) = 1, \tag{6.155}$$

where the summation is for all states permitted by the selection rules. Constructing the matrix components of x, y, and z from the expressions given by Eqs. (6.146)–(6.151), we obtain

$$\sum_{m'_l} f(n, l, m_l; n', l + 1, m'_l)$$

$$= \frac{(l+1)(2l+3)}{3} (E_{n'l+1} - E_{nl})[A(n, l; n', l + 1)]^2, \tag{6.156}$$

and

$$\sum_{m'_l} f(n, l, m_l; n', l - 1, m'_l)$$

$$= \frac{l(2l-1)}{3} (E_{n'l-1} - E_{nl})[A(n, l; n', l - 1)]^2. \tag{6.157}$$

Using these results, we can rewrite the Kuhn–Thomas summation rule as follows:

$$\sum_{n'l'm'_l} f(n, l, m_l; n', l', m'_l) = \sum_{n'm'_l} f(n, l, m_l; n', l + 1, m'_l)$$

$$+ \sum_{n'm'_l} f(n, l, m_l; n', l - 1, m'_l) = 1, \tag{6.158}$$

where, of course, we can substitute the expressions in Eqs. (6.156) and (6.157) into the first and second terms on the left side of this relationship.

Using these equations, Wigner derived the sum rules:[62]

$$(2l+1) \sum_{n'} (E_{n'l+1} - E_{nl})[A(n, l; n', l + 1)]^2$$

$$= -(2l+1) \sum_{n'} (E_{n'l-1} - E_{nl})[A(n, l; n', l - 1)]^2 = 1. \tag{6.159}$$

B. Atoms with Several Valence Electrons

We now turn to atoms that have q valence electrons in an arbitrary configuration. We ask the question: What kind of transitions are permitted for such an atom? Let C_1 and C_2 be two configurations that differ by *one* orbital and $C_1 C_3$ be two configurations that differ by *two* orbitals. Transitions between two states are possible if the matrix component of the dipole moment is not zero. The dipole moment is a one-electron operator. By straightforward application of the methods presented in App. A, it is easy to show that the matrix component between C_1 and C_2 will be nonzero and the matrix component between C_1 and C_3 will be zero. It is the property of Slater determinants formed from orthonormal spin-orbitals that the matrix component of a one-electron operator with respect to two Slater determinants is zero if the two determinants differ by more than one spin-orbital.

The reader will observe that in the previous statement, we compared two configurations, that is, we compared two sets of (nl) values assigned to the valence electrons. The statement does not say anything about the angular-momentum and spin quantum numbers of the two configurations. Likewise, the statement concerned only electric dipole radiation and did not say anything about magnetic dipole and electric quadripole radiations. But we are able to conclude that

> Electric dipole radiation is possible only between configurations differing by one orbital.

We turn now to the angular-momentum and spin properties of the two configurations. Let us assume that we work with LS coupling and the energy levels are characterized by wave functions that diagonalize L^2, S^2, L_z, and S_z, that is, the total angular momentum, the total spin and their z-components. Let L and S be the quantum numbers in the initial state and L' and S' the numbers in the final state. Let the quantum numbers of the running electron be (nl) in the initial state and $(n'l')$ in the final state. Then the selection rules are as follows:

$$\Delta L = L - L' = 0, \pm 1; \qquad (6.160)$$

$$\Delta S = S - S' = 0; \qquad (6.161)$$

$$\Delta l = l - l' = \pm 1. \qquad (6.162)$$

Thus, we see that transitions are possible only between terms of the same multiplicity and we also see that the selection rule for the running electron is the same as for the one-electron atoms.

Next, we want a formula for the transition probabilities in terms of the radial orbitals of the valence electrons. It was shown by Slater that, using the theorems of Dirac and of Güttinger and Pauli, concerning angular-momentum and spin operators, one can construct explicit formulas for the transition probabilities. The general method for the construction of the dipole matrix components is, of course, Racah's method. We present both Slater's formulas and Racah's expression.

First, Slater constructed the matrix components of $(X \pm iY)$ and Z with respect to wave functions diagonalizing L^2, S^2, L_z, and L_s. Here are the formulas:

$$\langle L, M_L | X \pm iY | L, M_L \mp 1 \rangle$$

$$= A(L, L)\sqrt{(L \pm M_L)(L \mp M_L + 1)} ; \qquad (6.163)$$

$$\langle L, M_L | Z | L, M_L \rangle = A(L, L) M_L; \qquad (6.164)$$

$$\langle L, M_L | X \pm iY | L + 1, M_L \mp 1 \rangle$$

$$= \pm A(L, L + 1)\sqrt{(L \mp M_L + 1)(L \mp M_L + 2)} ; \quad (6.165)$$

$$\langle L, M_L | Z | L + 1, M_L \rangle = A(L, L + 1)\sqrt{(L + 1)^2 - M_L^2} ; \quad (6.166)$$

$$\langle L, M_L | X \pm iY | L - 1, M_L \mp 1 \rangle$$

$$= \mp A(L, L - 1)\sqrt{(L \pm M_L)(L \pm M_L - 1)} ; \qquad (6.167)$$

$$\langle L, M_L | Z | L - 1, M_L \rangle = A(L, L - 1)\sqrt{L^2 - M_L^2} . \qquad (6.168)$$

In these formulas, X, Y, and Z are equal to $\Sigma_i x_i$, $\Sigma_i y_i$, and $\Sigma_i z_i$, where the summations are over the valence electrons. The formulas reflect the selection rule given by Eq. (6.160). In addition, we have the selection rule given by Eq. (6.161) and we have, for M_L and M_S,

$$M_L' = M_L, M_L \pm 1; \qquad (6.169)$$

$$M_S' = M_S. \qquad (6.170)$$

The previous formulas contain the quantities denoted by $A(L, L')$. We observe here that these quantities must depend on the (nl) of the running electron as well as on the quantum numbers of the parent configuration. Let us assume that in the initial state, that is, in the state in which the running electron has the quantum numbers (nl), the remaining $q - 1$ electrons are coupled to an $(L_1 S_1)$ state by LS coupling. Likewise, let us assume that in the final state, in which the valence electron has the quantum numbers $(n'l')$, the remaining $q - 1$ electrons are again coupled to an $(L_1 S_1)$ state. Then we can write down the $A(L, L')$ coefficients for the two possible transitions, that is, for $l' = l + 1$ and for $l' = l - 1$. Here are the formulas:

$l \rightarrow l + 1$ transitions:

$$A(L, L) = -A(n, l; n', l + 1)$$

$$\times \frac{\sqrt{(L + l - L_1 + 1)(L - l + L_1)(l + L_1 - L + 1)(l + L_1 + L + 2)}}{2L(L + 1)};$$

$$(6.171)$$

$$A(L, L + 1) = A(n, l; n', l + 1)$$

$$\times \sqrt{\frac{(L + l - L_1 + 1)(L + l - L_1 + 2)(l + L_1 + L + 3)(l + L_1 + L + 2)}{4(L + 1)^2 \left[4(L + 1)^2 - 1 \right]}};$$

$$(6.172)$$

$$A(L, L - 1) = -A(n, l; n', l + 1)$$

$$\times \sqrt{\frac{(L - l + L_1 - 1)(L - l + L_1)(l + L_1 - L + 2)(l + L_1 - L + 1)}{4L^2(4L^2 - 1)}};$$

$$(6.173)$$

$l \rightarrow l - 1$ transitions:

$$A(L, L) = -A(n, l; n', l - 1)$$

$$\times \sqrt{\frac{(L + l - L_1)(L - l + L_1 + 1)(l + L_1 - L)(l + L_1 + L + 1)}{2L(L + 1)}};$$

$$(6.174)$$

$$A(L, L + 1) = -A(n, l; n', l - 1)$$

$$\times \sqrt{\frac{(L - l + L_1 + 1)(L - l + L_1 + 2)(l + L_1 - L)(l + L_1 - L - 1)}{4(L + 1)^2 \left[4(L + 1)^2 - 1 \right]}};$$

$$(6.175)$$

$$A(L, L - 1) = A(n, l; n', l - 1)$$

$$\times \sqrt{\frac{(L + l - L_1 - 1)(L + l - L_1)(l + L_1 + L + 1)(l + L_1 + L)}{4L^2(4L^2 - 1)}}.$$

$$(6.176)$$

In these formulas, the quantities denoted by $A(n, l; n', l')$ are given by Eqs. (6.152)–(6.153).

These results, of course, can also be derived from Racah's formalism. We define the tensor operators

$$T_1^{(1)} = -\frac{1}{\sqrt{2}}(X + iY),$$ (6.177)

$$T_0^{(1)} = Z,$$ (6.178)

$$T_{-1}^{(1)} = \frac{1}{\sqrt{2}}(X - iY).$$ (6.179)

The matrix components of these operators in terms of valence-electron wave functions that diagonalize L^2, S^2, L_z, and S_z are

$$\langle L, M_l | T_q^{(1)} | L', M_L' \rangle = (-1)^{M_L - L' - 1} \begin{pmatrix} L & L' & 1 \\ -M_L & M_L' & q \end{pmatrix} (L \| T^{(1)} \| L'),$$ (6.180)

where, of course, L' can be L, $L + 1$, and $L - 1$, and M_L' can be M_L, $M_L + 1$, and $M_L - 1$, and we have

$$(L \| T^{(1)} \| L) = \sqrt{L(L + 1)(2L + 1)}\, A(L, L);$$ (6.181)

$$(L \| T^{(1)} \| L + 1) = -\sqrt{(L + 1)(2L + 1)(2L + 3)}\, A(L, L + 1);$$ (6.182)

$$(L \| T^{(1)} \| L - 1) = \sqrt{L(2L - 1)(2L + 1)}\, A(L, L - 1).$$ (6.183)

In these formulas, the quantities denoted by $A(L, L')$ are given, for the L' values that are permitted by the selection rules, by Eqs. (6.171)–(6.176). In Eq. (6.180), we also have the Wigner $3j$ symbol. Equation (6.180) is equivalent to the set of formulas given by Eqs. (6.163)–(6.168).

Considering the spectra of atoms with several valence electrons, we mentioned in Sec. 6.1 that the spectra are showing a fine structure that, in the Atomic Energy Level Tables, appear as closely spaced lines with the same (LS) but different J values. The J value is the inner quantum number and it is related to the operator of the total angular momentum, $\mathbf{J} = \mathbf{L} + \mathbf{S}$. We must expand our discussion of transition probabilities to include this feature of the spectra. In the LS coupling scheme, the first step is to construct valence-electron wave functions that diagonalize L^2, S^2, L_z, and S_z. From these wave functions, it is possible to construct linear combinations that diagonalize L^2, S^2, J^2, and J_z, where J_z is the operator of the z-component of $\mathbf{J}$. Thus, here the good quantum numbers will be LSJ and M (but not M_L and M_S). Using Racah's method, one can easily construct formulas for the matrix components of the operators given by Eqs. (6.177)–(6.179). First, we quote the selection rules for the case when L, S, J, and M are good quantum numbers. We have

$$\Delta L = L - L' = 0, \pm 1; \qquad (6.184)$$

$$\Delta J = J - J' = 0, \pm 1; \qquad (6.185)$$

$$\Delta S = 0. \qquad (6.186)$$

In addition, we have again the rule that only one valence electron can make the transition and in that transition $\Delta l = \pm 1$.

We now quote the formulas for the matrix components:

$$\langle L, S, M | T_q^{(1)} | L', S', M' \rangle$$

$$= (-1)^{J+M} (LSJ \| T^{(1)} \| L'SJ') V(JJ'1; -MM'q)$$

$$= (-1)^{M-J'-1} \begin{pmatrix} J & J' & 1 \\ -M & M' & q \end{pmatrix} (LSJ \| T^{(1)} \| L'SJ'), \qquad (6.187)$$

where V is the Racah coefficient, which we have converted into a $3j$ symbol. For the quantities $(\cdots \| T^{(1)} \| \cdots)$, we have

$$(LSJ \| T^{(1)} \| L'SJ') = (-1)^{S+1+L+J'} \begin{Bmatrix} L & S & J \\ J' & L' & 1 \end{Bmatrix}$$

$$\times \sqrt{(2J+1)(2J'+1)} \, (L \| T^{(1)} \| L'), \qquad (6.188)$$

where $\{ \cdots \}$ is the $6j$ symbol. By putting Eq. (6.188) into Eq. (6.187), we get

$$\langle LJM \| T_q^{(1)} | L'J'M' \rangle$$

$$= (-1)^{M+S+L} \begin{pmatrix} J & J' & 1 \\ -M & M' & q \end{pmatrix} \begin{Bmatrix} L & J & S \\ J' & L' & 1 \end{Bmatrix}$$

$$\times \sqrt{(2J+1)(2J'+1)} \, (L \| T^{(1)} \| L'), \qquad (6.189)$$

where the $(L \| T^{(1)} \| L')$ are given by Eqs. (6.181)–(6.183) for the L' values permitted by the selection rules. We note that for M', we have the selection rule

$$\Delta M = M - M' = 0, \pm 1. \qquad (6.190)$$

The matrix components of the operators $T_q^{(1)}$ can be constructed from Eq. (6.189) using the relationship contained in Eqs. (6.181)–(6.183); those relationships in turn depend on the formulas given in Eqs. (6.171)–(6.176).

The elegancy and compactness of Racah's theory is demonstrated by showing that the rather lengthy formulas given by Eqs. (6.163)–(6.168) and by

Eqs. (6.171)–(6.176) can be restated in the following formula:

$$\langle lLM_L | T_q^{(1)} | l'L'M_L' \rangle$$

$$= (-1)^{l + L_1 + M_L} \begin{pmatrix} L & L' & 1 \\ -M_L & M_L' & q \end{pmatrix} \begin{Bmatrix} l & L & L_1 \\ L' & l' & 1 \end{Bmatrix}$$

$$\times \sqrt{(2L + 1)(2L' + 1)} \, (l\|T^{(1)}\|l'), \qquad (6.191)$$

where

$$(l\|T^{(1)}\|l) = 0; \qquad (6.192)$$

$$(l\|T^{(1)}\|l + 1) = -\sqrt{(l + 1)(2l + 1)(2l + 3)} \, A(n, l; n', l + 1); \quad (6.193)$$

$$(l\|T^{(1)}\|l - 1) = \sqrt{l(2l - 1)(2l + 1)} \, A(n, l; n', l - 1). \qquad (6.194)$$

Here the A's are the same as in Eqs. (6.152) and (6.153). In Eq. (6.191), the $(\cdots)$ is the $3j$ symbol and the $\{\cdots\}$ is the $6j$ symbol.

Next, we consider transitions in which the running electron has the same azimuthal quantum number in the initial or final state as the stationary valence electrons, that is, we consider transitions in which one of the states is such that the running electron and the stationary valence electrons are equivalent. These transitions are of the type

$$l^q \to l^{q-1}l', \qquad (6.195)$$

that is, the running electron makes the transition $l \to l'$. Let us denote the possible (LS) values of the l^q configuration by (LS), $(L'S')$, $(L''S'')$, and so on. Thus, the configuration has the following multiplets:

$$l^{q\,(2S+1)}L; \qquad l^{q\,(2S'+1)}L'; \qquad l^{q\,(2S''+1)}L''; \qquad (6.196)$$

and so on. Also, let the possible multiplets formed by the l^{q-1} configuration, which is the parent configuration of the $l^{q-1}l'$ be (L_1S_1), $(L_1'S_1')$, $(L_1''S_1'')$, and so on. Each of these parent multiplets can be coupled to the running electron and the resulting states will be denoted in the following way:

$$\left[l^{q-1\,(2S_1+1)}L_1 \right] l'^{\,(2S+1)}L, \qquad (6.197)$$

where the notation is the same as in Eq. (6.14).

What kind of transitions will there be between the states described by Eqs. (6.196) and (6.197)? The answer to this question was provided by Menzel and Goldberg[65] whose ideas were expanded into a full-scale theory by Racah.[57] Menzel and Goldberg suggested that the parent multiplet should not change in the transition. This implies that the multiplets of l^q, Eq. (6.196), should be decomposed into the form similar to those in Eq. (6.197), that is, the multiplets of l^q should be written in the form

$$
\left.
\begin{array}{l}
\left[l^{q-1\,(2S_1+1)}L_1 \right] l^{(2S+1)}L, \\[2mm]
\left[l^{q-1\,(2S_1'+1)}L_1' \right] l^{(2S'+1)}L', \\[2mm]
\left[l^{q-1\,(2S_1''+1)}L_1'' \right] l^{(2S''+1)}L'',
\end{array}
\right\}
\tag{6.198}
$$

where each of the (LS), (L', S'), and so on, pairs, are standing for several possible combinations. Among the different parents, there will be some that will lead to the same final (LS) values. Let us write down the following symbolic equation:

$$
l^{q\,(2S+1)}L
$$

$$
= \sqrt{n}\, a(\alpha SL;\, \alpha_1 S_1 L_1) \left[l^{q-1\,(2S_1+1)}L_1 \right] l^{(2S+1)}L
$$

$$
+ \sqrt{n}\, a(\alpha SL;\, \alpha_1' S_1' L_1') \left[l^{q-1\,(2S_1'+1)}L_1' \right] l^{(2S+1)}L
$$

$$
+ \sqrt{n}\, a(\alpha SL;\, \alpha_1'' S_1'' L_1'') \left[l^{q-1\,(2S_1''+1)}L_1'' \right] l^{(2S+1)}L
$$

$$
+ \cdots + \cdots.
\tag{6.199}
$$

On the left side of this symbolic equation, we have one of the possible LS combinations arising from the l^q configuration. On the right side, we have the same (LS) levels but now they are generated by different parents. The coefficient, $a(\alpha SL;\, \alpha_1 S_1 L_1)$, is the coefficient of fractional parentage that assigns different weights to the different parents. In $a(\alpha SL;\, \alpha_1 S_1 L_1)$, the symbols mean that this is the coefficient of the parent multiplet $(\alpha_1 S_1 L_1)$ generating the final (αSL). The α and α_1 symbolize whatever parameters are needed to identify the state besides $(L_1 S_1)$ and (LS). Number n is a normalization constant.

The meaning of the previous symbolic equation is simply that the $[l^{q\,(2S+1)}L]$ multiplet can be generated from several different parents with different weight factors.

Returning to our question about the possible transitions between l^q and $l^{q-1}\,l'$, we now state that the transition must satisfy the following relationships:

$$
\left[l^{q-1\,(2S_1+1)}l_1 \right] l^{(2S+1)}L \;\rightarrow\; \left[l^{q-1\,(2S_1+1)}L_1 \right] l'^{\,(2S'+1)}L'.
\tag{6.200}
$$

Thus, we see that in the transition, the parent quantum numbers do not change; the quantum numbers that change are subjected to the selection rules

$$\begin{cases} \Delta l = l' - l = \pm 1; \\ \Delta L = L' - L = 0, \pm 1; \\ \Delta S = S' - S = 0, \end{cases} \qquad (6.201)$$

which are the same as in the case of non-equivalent electrons. For the matrix components of the transition, we get

$$\langle l^q \alpha SLM_L | T_q^{(1)} | (l^{q-1} \alpha_1 S_1 L_1) l' SL'M_L' \rangle$$

$$= (-1)^{M_L - L' - 1} \begin{pmatrix} L & L' & 1 \\ -M_L & M_L' & q \end{pmatrix} \sqrt{n}\, a(\alpha SL; \alpha_1 S_1 L_1)$$

$$\times (lL_1 L \| T^{(1)} \| l' L_1 L'), \qquad (6.202)$$

where the matrix $T^{(1)}$ is given by Eq. (6.188). Changing the notation in Eq. (6.188), we get

$$(lL_1 L \| T^{(1)} \| l' L_1 L')$$

$$= (-1)^{L_1 + 1 + l + L'} \begin{Bmatrix} l & L & L_1 \\ L' & l' & 1 \end{Bmatrix} \sqrt{(2L + 1)(2L' + 1)}$$

$$\times (l \| T^{(1)} \| l'), \qquad (6.203)$$

where the $(l \| T^{(1)} \| l')$ coefficients are given by Eqs. (6.192)–(6.194). By putting Eq. (6.203) into Eq. (6.202), we get

$$\langle l^q \alpha SLM_L | T_q^{(1)} | (l^{q-1} \alpha_1 S_1 L_1) l' SL'M_L' \rangle$$

$$= (-1)^{M_L + L_1 + l} \begin{pmatrix} L & L' & 1 \\ -M_L & M_L' & q \end{pmatrix} \begin{Bmatrix} l & L & L_1 \\ L' & l' & 1 \end{Bmatrix}$$

$$\times \sqrt{(2L + 1)(2L' + 1)}\, (l \| T^{(1)} \| l')$$

$$\times \sqrt{n}\, a(\alpha SL; \alpha_1 S_1 L_1). \qquad (6.204)$$

As we see, the structure of this matrix component is the same as for transitions between non-equivalent electrons except that in the last term, we have the coefficient of fractional parentage. Otherwise, the symbols are the same as before, $(\cdots)$ is the $3j$ symbol, $\{ \cdots \}$ is the $6j$ symbol, and $(l \| T^{(1)} \| l')$ is given by Eqs. (6.192)–(6.194).

We note that explicit formulas for the matrix components and for coefficients of fractional parentage can be found in the book by Condon and Odabasi,[66] page 254. There are also tables of numerical values for a large number of configurations to be found there.

Summarizing this section, we see the basic ideas and main formulas for a theory of radiative transitions in the HF approximation. The discussion here has been restricted to (LSJ) coupling. Having absorbed the basic ideas, the reader will be able to set up the matrix components of transitions for any configuration. We observe that we have given summation rules only for the case of one-electron atoms; however, it is clear that knowing the matrix components of the operator $T_q^{(1)}$, one is able to calculate the matrix components of X, Y, and Z, and the oscillator strengths. This procedure can be carried out for any number of valence electrons in any configuration.

There is one point that must be emphasized. Regardless of whether we are considering atoms with one or with several valence electrons, the transition probabilities will depend on the wave function of the running electron in the initial and final states of the transition. In the HF model, these wave functions are the solutions of the HF equations, which we discussed in Secs. 6.3 and 6.4. Thus, the calculated results will contain the approximation consisting of the replacement of the exact nonrelativistic eigenfunctions ψ_m^0 and ψ_n^0 by the corresponding HF wave functions. We must keep in mind that the results of this section are valid only in the nonrelativistic HF model with (LSJ) coupling.

6.6. SPIN – ORBIT COUPLING

The most convenient starting point for the investigation of spin–orbit coupling is the Breit–Pauli Hamiltonian operator given by Eq. (5.7). The Hamiltonian has the form

$$H_p = H_0 + H' \tag{6.205}$$

where H_0 is the nonrelativistic Hamiltonian:

$$H_0 = \sum_{j=1}^{N} (t_j + g_j) + \frac{1}{2} \sum_{j,k=1}^{N} \frac{1}{r_{jk}}. \tag{6.206}$$

In this formula, we introduced atomic units and denoted the kinetic-energy operator by t and the nuclear potential by g. The relativistic part of the Hamiltonian, H', is given by Eq. (5.9).

A detailed discussion of the spin-dependent effects that are included in Eq. (5.9) is given in Sec. 9.4, in the framework of the effective Hamiltonian theory. It is shown there that for an atom with n valence electrons outside complete groups, the operator representing the spin-dependent effects can be brought to the form of Eq. (9.62), which we write down here:

$$H_s = \sum_{i=1}^{n} f_i(r_i)\mathbf{s}_i \cdot \mathbf{L}_i + \sum_{\substack{i,j=1 \\ i \neq j}}^{n} g_{s0}(ij) + \sum_{\substack{i,j=1 \\ (i \neq j)}}^{n} h_{s0}(ij). \qquad (6.207)$$

The precise definition of the symbols in this formula is given following Eq. (9.62). The first term is the conventional spin–orbit interaction operator. The second and third terms are representing the spin–other-orbit and spin–spin interactions. The summations are over the valence-electron coordinates.

If we want to consider the spin-dependent effects in separation from the other relativistic effects in Eq. (5.9), then we can omit all but the spin-dependent terms from Eq. (5.9), and instead of Eq. (6.205), we can consider a Hamiltonian in which the operator H' is replaced by the operator H_s. If we assume that the two-electron terms in H_s are much smaller than the leading one-electron operator, then we can omit the second and third summations from H_s and keep only the first term. The total Hamiltonian will then be given by Eq. (6.205), with H' redefined as

$$H' = \sum_{i=1}^{n} f_i(r_i)\mathbf{s}_i \cdot \mathbf{L}_i. \qquad (6.208)$$

This operator contains the full spin–orbit interaction of the valence electrons in the field created by the bare nucleus, the core electrons, and the other valence electrons. The function f_i is defined by Eq. (9.65) and has the form

$$f_i(r) = \frac{\alpha^2}{2} \frac{1}{r} \frac{d}{dr} \left(-\frac{Z}{r} + U_C(r) + U_i(r) \right)$$

$$+ (\text{exchange corrections}). \qquad (6.209)$$

In this formula, α is the fine-structure constant, U_C is the potential of the atomic core, and U_i is the potential of the valence electrons with the ith valence orbital omitted. Both U_C and U_i are Hartree-type potentials without exchange. The terms indicated by "exchange corrections" are representing the effects of the core/valence and valence/valence exchange interactions.

Therefore, the Schroedinger equation of the problem will take the form

$$H\Psi = E\Psi, \qquad (6.210)$$

where

$$H = H_0 + H', \qquad (6.211)$$

and H' is given by Eq. (6.208).

Approximate solutions of Eq. (6.210) can be constructed as follows. If the spin–orbit interaction is weak, that is, if the resulting energy is much smaller than the energy of the electrostatic interaction, then H' can be viewed as a perturbation. Let E_0 be the eigenvalue of the unperturbed Hamiltonian, that is, let E_0 be the solution of the Schroedinger equation

$$H_0\Psi_0 = E_0\Psi_0, \tag{6.212}$$

In the first order of perturbation theory, we get the energy of the atom as

$$E = E_0 + E', \tag{6.213}$$

where

$$E' = \langle \Psi_0 | H' | \Psi_0 \rangle. \tag{6.214}$$

Equation (6.212) is the nonrelativistic Schroedinger equation. The eigenfunctions and eigenvalues of this equation are generally not known, although we have seen that E_0 can be estimated fairly accurately. However, we can approximate E_0 by the HF energy, E_F, and Ψ_0 by Ψ_{HF}, which is the HF wave function. Thus, we can put

$$E = E_F + E', \tag{6.215}$$

where now

$$E' = \langle \Psi_{HF} | H' | \Psi_{HF} \rangle. \tag{6.216}$$

As we have seen in Sec. 3.1 the symmetry properties of the nonrelativistic HF model are determined by the symmetry properties of the Hamiltonian given in Eq. (6.206). Because that Hamiltonian commutes with the operators L^2, S^2, L_z, and S_z, the HF total wave function was constructed in such a way that L, S, M_L, and M_S were good quantum numbers. In computing the matrix components of the spin–orbit coupling, that is, the expectation value of the operator in Eq. (6.208), we must take into account that the Hamiltonian in Eq. (6.208) does not commute with L^2, S^2, L_z, and S_z, but only with J^2 and J_z where $\mathbf{J} = \mathbf{L} + \mathbf{S}$ is the total angular momentum, and J_z is its z-component. Therefore, the LS-coupled HF functions in Eq. (6.216) do not have the same symmetry properties as the operator H'.

If we want to proceed consistently in the construction of the energy expression, Eq. (6.216), we must calculate the expectation value of H', given by Eq. (6.216), in such a way that the wave function that is used in the calculation will be an eigenfunction of J^2 and J_z. Then, clearly, E' will be a function of J, that is, we will have a J-dependent energy expression. Forming a linear combination of wave functions, we can construct eigenfunctions of J^2

and J_z in the form

$$\psi(J, M) = \sum_{M_L M_S} C(J, M; M_L, M_S)\varphi(M_L M_S), \tag{6.217}$$

where C is the Clebsch–Gordan coefficient. ψ is an eigenfunction of J^2 and J_z and φ is an eigenfunction of L_z, and S_z. In the construction of matrix components of H', this type of wave functions must be used. The expectation value computed this way will be J-dependent but not M-dependent: it is easy to show that the expectation value will be the same regardless of which M value is in the $\psi(J, M)$.

The range of the possible J values is given by the quantum mechanical rule

$$J = |L - S|, |L - S| + 1, \ldots, (L + S) - 1, (L + S). \tag{6.218}$$

Because the spin–orbit interaction term will be J-dependent, we see from this equation that the LS term will split into $(2S + 1)$ or $(2L + 1)$ levels, depending on whether $S < L$ or $L < S$. If $S < L$, then we can rewrite the sequence as

$$(L - S), (L - S) + 1, \ldots, (L - S) + (2S - 1), (L - S) + 2S, \tag{6.219}$$

which is clearly a set of $(2S + 1)$ terms. Thus, for $S < L$, the series will contain $(2S + 1)$ terms; likewise, for $L < S$, there will be $(2L + 1)$ terms.

It is easy to construct the expression in Eq. (6.216) for atoms with one valence electron. The spin–orbit coupling vanishes for an atomic core with complete groups. This statement is true if the core is represented by an HF determinant built from central-field spin-orbitals. We obtain, for an atom with one valence electron,

$$E' = D(l_v, j_v)\zeta(n_v l_v), \tag{6.220}$$

where $\zeta(n_v l_v)$ is generated by the radial part of the operator H', and $D(l_v j_v)$ is coming from the $(\mathbf{S} \cdot \mathbf{L})$ part. We obtain

$$\zeta(n_v l_v) = \frac{\alpha^2}{2} \int \frac{1}{r} \frac{dV}{dr} P_{n_v l_v}^2 \, dr, \tag{6.221}$$

where V is the potential in the formula, Eq. (6.209), and $P_{n_v l_v}$ is the radial part of the valence orbital. For the angular coefficient, we get

$$D(l_v j_v) = \tfrac{1}{2}[j_v(j_v + 1) - l_v(l_v + 1) - s_v(s_v + 1)]. \tag{6.222}$$

Because we are talking about one valence electron, we have $s_v = \frac{1}{2}$. For j_v, we get

$$j_v = l_v \pm \tfrac{1}{2}, \qquad \text{for } l_v \neq 0; \tag{6.223}$$

$$j_v = \tfrac{1}{2}, \qquad \text{for } l_v = 0. \tag{6.224}$$

Clearly, $D = 0$ for $l_v = 0$; there will be no spin–orbit splitting for s levels. Otherwise, all terms will be doublets. By putting Eq. (6.220) into Eq. (6.215), we get

$$E = E_{HF} + E' = E_{av} + D(l_v j_v)\zeta(n_v l_v) \tag{6.225}$$

$$E = E_{av} + \tfrac{1}{2}\big[j_v(j_v + 1) - l_v(l_v + 1) - s_v(s_v + 1)\big]$$

$$\times \frac{\alpha^2}{2} \int \frac{1}{r}\frac{dV}{dr} P^2_{n_v l_v}(r)\, dr, \tag{6.226}$$

where we have taken into account that for one valence electron outside of complete groups, $E_{HF} = E_{av}$. We note that for one valence electron, we have for the total quantum numbers

$$L = l_v, \qquad S = s_v, \qquad J = j_v. \tag{6.227}$$

For a configuration in which we have several valence electrons in several incomplete groups, the derivation of the matrix component of Eq. (6.216) is more complicated, but the Racah method yields the necessary expressions. A general formula is given by Cowan [Ref. 58, p. 335]. We have

$$E' = \sum_i D_i \zeta_i, \tag{6.228}$$

where the summation is over the incomplete groups of valence electrons. For each such group, we have the radial integral similar to Eq. (6.222), that is,

$$\zeta_i = \zeta_i(n_i l_i) = \frac{\alpha^2}{2} \int \frac{1}{r}\frac{dV}{dr} P^2_{n_i l_i}(r)\, dr, \tag{6.229}$$

where $P_{n_i l_i}$ is the HF orbital. D_i is a coefficient resulting from the angular and spin integrations. Cowan gives explicit expressions for these coefficients, for several different coupling schemes, in terms of the $6j$ symbols and Racah's $V^{(11)}$ coefficients. We have, for the general matrix component of the angular part of H', which is diagonal with respect to J, but may be off-diagonal with respect to α, L, and S:

$$D_i(\alpha SLJ; \alpha'S'L'J) = (-1)^{S'+L-J}\begin{Bmatrix} S & L & J \\ L' & S' & 1 \end{Bmatrix}$$

$$\times [l_i(l_i + 1)(2l_i + 1)]^{1/2}(\alpha SL\|V^{(11)}\|\alpha'S'L'). \tag{6.230}$$

This is the coefficient for the incomplete $(n_i l_i)$ group, and in Eq. (6.228), we have one such coefficient for each incomplete group. Although not indicated in the notation, the $V^{(11)}$ coefficients will depend on the configuration parameters $l_i^{(q_i)}$, that is, we will have different $V^{(11)}$ coefficients for p^2, d^2, d^3, and so on.

Finally, we ask the question: What will be the transition probabilities if spin–orbit coupling is taken into account? The Hamiltonian operator for this case will be

$$H = H_0 + H' + H'', \qquad (6.231)$$

where H_0 is the nonrelativistic Hamiltonian, Eq. (6.206), H' is the spin–orbit interaction, Eq. (6.208), and H'' is the operator representing the external radiation field, Eq. (6.137). The consistent thing to do would be to view $(H_0 + H')$ as an "unperturbed" Hamiltonian and H'', the radiation field, as a perturbation. Repeating the derivation presented in Appendix J, we would end up with the formula given by Eq. (6.138), except that now the wave functions that appear in Eq. (6.141) would be the exact eigenfunctions of $H_0 + H'$, rather than the eigenfunctions of H_0. As we have seen, however, the eigenfunctions of H_0 diagonolize L^2, S^2, L_z, and S_z and the eigenfunctions of $H_0 + H'$ diagonolize J^2 and J_z. A meaningful approximation can be constructed if we use radiation wave functions that diagonolize L^2, S^2, J^2, and J_z for the calculation of the matrix components of the dipole. The good quantum numbers for such a wave function are L, S, J, and M. These functions, of course, do not have the symmetry that the eigenfunctions of $H_0 + H'$ should have, but such an approximation is meaningful as long as the spin–orbit coupling is not too strong relative to the electrostatic interaction.

We have presented formulas for the selection rules in this case; they are given by Eqs. (6.184)–(6.186). We repeat these formulas here for the sake of completeness:

$$\Delta L = L' - L = 0, \pm 1; \qquad (6.232)$$

$$\Delta J = J' - J = 0, \pm 1; \qquad (6.233)$$

$$\Delta S = 0. \qquad (6.234)$$

The matrix components of the dipole radiation and the formulas for the oscillator strengths can be constructed on the basis of the formulas given by eqs. (6.187)–(6.189).

We emphasize that the previous selection rules are approximate and valid only for weak spin–orbit coupling. According to these selection rules, L and S are good quantum numbers; this is true only approximately, as we see from Eq. (6.230). That formula shows that the spin–orbit operator has off-diagonal matrix components between different (LS) and $(L'S')$ terms. From this, it follows that the selection rules given before break down when the spin–orbit interaction energy cannot be considered to be small relative to the electrostatic interaction.

6.7. ZEEMAN EFFECT

We talk about a Zeeman effect when atoms are placed in a magnetic field. Let us construct the Hamiltonian for such a case. Let H_0 be the nonrelativistic Hamiltonian given by Eq. (6.209) and H' be the spin–orbit coupling operator given by Eq. (6.210). Let H'' represent the external magnetic field. Then, our Hamiltonian becomes

$$H = H_0 + H' + H'', \tag{6.235}$$

and the Schroedinger equation takes the form

$$H\Psi = (H_0 + H' + H'')\Psi = E\Psi. \tag{6.236}$$

The complete relativistic Hamiltonian in the presence of an electromagnetic field is given by Eq. (4.73). In that equation, the interaction with the external magnetic field is given by the operator H_1. For this operator, we get, from Eqs. (4.15) and (4.30), by neglecting the A^2 term,

$$H_1 = H_1' + H_1'' = \sum_{j=1}^{N} \left(\frac{e}{mc} \mathbf{A}_j \cdot \mathbf{p}_j + \frac{e\hbar}{2mc} (\boldsymbol{\sigma}_j \cdot \mathscr{H}_j) \right). \tag{6.237}$$

Here $\mathbf{A}_j$ is the vector potential acting on the jth particle and $\mathscr{H}_j$ is the magnetic field. The magnetic field is connected to the vector potential by Eq. (4.2), that is, we have

$$\mathscr{H} = \operatorname{curl} \mathbf{A} = \nabla \times \mathbf{A}. \tag{6.238}$$

For a general homogeneous magnetic field, we have the vector potential

$$\mathbf{A} = \tfrac{1}{2}(\mathscr{H} \times \mathbf{r}), \tag{6.239}$$

where $\mathbf{r}$ is the distance from the nucleus. Using this expression, we get for Eq. (6.237)

$$H_1 = \sum_j \left(\frac{e}{2mc} \mathscr{H}_j \cdot \mathbf{L}_j + \frac{e\hbar}{2mc} (\boldsymbol{\sigma}_j \cdot \mathscr{H}_j) \right), \tag{6.240}$$

where $\mathbf{L}_j$ is the angular momentum of the jth particle. Using Eq. (4.45), we obtain, by making the change $H_1 \rightarrow H''$ in the notation,

$$H'' = \sum_j \varepsilon \mathscr{H}_j \cdot (\mathbf{L}_j + 2\mathbf{s}_j), \tag{6.241}$$

where

$$\varepsilon = \frac{e}{mc}.$$ (6.242)

Thus, the complete Hamiltonian, including the spin–orbit coupling and the external magnetic field, becomes

$$H = H_0 + H' + H'',$$ (6.243)

where

$$H_0 = \sum_j f_j + \sum_{j,k} \frac{e^2}{r_{jk}},$$ (6.244)

$$H' = \sum_j \frac{1}{2m^2c^2} \frac{1}{r_j} \frac{dV(j)}{dr_j} (\mathbf{L}_j \cdot \mathbf{s}_j),$$ (6.245)

$$H'' = \sum_j \varepsilon \mathcal{H}_j \cdot (\mathbf{L}_j + 2\mathbf{s}_j).$$ (6.246)

We construct approximation solutions of eq. (6.236) by viewing $H_0 + H'$ as the unperturbed Hamiltonian and H'' as a perturbation. We obtain this way

$$E = E_0 + E' + E'',$$ (6.247)

where $E_0 + E'$ is the unperturbed energy, including the spin–orbit interaction, and

$$E'' = \langle \Psi_0 | H'' | \Psi_0 \rangle.$$ (6.248)

Here Ψ_0 is the eigenfunction of $H_0 + H'$, and, as we see from the discussions of Sec. 6.6, the good quantum numbers for this function are J and M. Putting the z direction into $\mathcal{H}$, we get

$$H'' = \sum_j \varepsilon \mathcal{H} (L_{jz} + 2s_{jz}),$$ (6.249)

where $\mathcal{H}$ is the magnitude of the homogeneous field. By introducing the total angular momentum, $\mathbf{J} = \mathbf{L} + \mathbf{s}$, the operator H'' becomes

$$H'' = \varepsilon \mathcal{H} (J_z + s_z)$$ (6.250)

On the basis of the work of Güttinger and Pauli,[64] Slater has shown that the diagonal matrix component of the operator in Eq. (6.250) is given by the formula [Slater II, 197]

$$\langle JM|(L_z + 2s_z)|JM\rangle = gM\hbar, \tag{6.251}$$

where

$$g = 1 + \frac{J(J+1) - L(L+1) + S(S+1)}{2J(J+1)}. \tag{6.252}$$

Thus, the magnetic energy becomes

$$E'' = \langle\Psi_0|H''|\Psi_0\rangle = \varepsilon\mathcal{H}gM\hbar. \tag{6.253}$$

Using the formula for the Bohr magneton, Eq. (4.36), we get $\varepsilon = \mu/\hbar$ and so

$$E'' = \mu\mathcal{H}gM. \tag{6.254}$$

In this formula, μ is the Bohr magneton, $\mathcal{H}$ is the magnetic field, and g is the *Lande g factor* given by Eq. (6.252). M is the magnetic quantum number related to the operator of the z-component of the total angular momentum.

As we see from the previous equation, in a magnetic field, each J level will split into $2J + 1$ separate levels according to the $2J + 1$ different values of the magnetic quantum number M. The selection rules for the transition between such levels will be

$$\Delta M = M' - M = 0, \pm 1. \tag{6.255}$$

(There is no $M = 0 \rightarrow M' = 0$ transition for $\Delta J = 0$.)

A slightly more accurate formulation can be given by introducing the orbital and spin g factors. Let us go back to Eqs. (4.36) and (4.37). In these equations, we have the Bohr magneton, $\mu = e\hbar/2mc$, and we have the connection between the spin magnetic dipole moment $\boldsymbol{\mu}_s$ and $\boldsymbol{\sigma}$. Using Eq. (4.45), we get

$$\boldsymbol{\mu}_s = -\mu\boldsymbol{\sigma} = -\mu 2\mathbf{s}/\hbar. \tag{6.256}$$

This can be generalized into the formula

$$\boldsymbol{\mu}_s = -g_s\mu\mathbf{s}/\hbar, \tag{6.257}$$

where g_s is the spin g factor. Ordinarily, it is assumed that $g_s = 2$. Now, as we see from Eq. (6.240), the interaction of the spin with the magnetic field is given by a term of the type

$$\left(\frac{e\hbar}{2mc}\right)\boldsymbol{\sigma}\cdot\mathcal{H} = \mu\boldsymbol{\sigma}\cdot\mathcal{H} = \mu 2\mathbf{s}\cdot\mathcal{H}/\hbar, \tag{6.258}$$

and replacing the factor 2 by g_s, we get

$$\left(\frac{e\hbar}{2mc}\right)\boldsymbol{\sigma}\cdot\mathscr{H} = g_s\mu\mathbf{s}\cdot\mathscr{H}/\hbar. \tag{6.259}$$

Rewriting the first term of Eq. (6.240), we get

$$\left(\frac{e}{2mc}\right)\mathbf{L}\cdot\mathscr{H} = \frac{\mu}{\hbar}g_l\mathbf{L}\cdot\mathscr{H}, \tag{6.260}$$

where $g_l = 1$. Using these expressions, we get for H''

$$H'' = \sum_j \varepsilon\mathscr{H}_j\cdot(g_l\mathbf{L}_j + g_s\mathbf{s}_j), \tag{6.261}$$

where

$$\varepsilon = \frac{\mu}{\hbar} = \frac{e}{2mc}, \tag{6.262}$$

that is, we get the same factor as in Eq. (6.241). In fact, we recover that equation by putting $g_s = 2$. Introducing the total angular momentum, we get, instead of Eq. (6.250),

$$H'' = \varepsilon\mathscr{H}(J_z + (g_s - 1)s_z), \tag{6.263}$$

where we have put $g_l = 1$. If we use the usual definitions for the total angular momentum and spin operators, then we get

$$H'' = \varepsilon\mathscr{H}(g_l L_z + g_s s_z). \tag{6.264}$$

The diagonal matrix component of this operator becomes

$$\langle JM|(g_l L_z + g_s s_z)|JM\rangle = gM\hbar, \tag{6.265}$$

which has the same form as Eq. (6.251), but now we get for the Landé g factor:

$$g = g_l\frac{J(J+1) + L(L+1) - S(S+1)}{2J(J+1)}$$
$$+ g_s\frac{J(J+1) + S(S+1) - L(L+1)}{2J(J+1)}. \tag{6.266}$$

We recover the g factor in Eq. (6.252) by putting $g_l = 1$ and $g_s = 2$. The magnetic energy is again given by Eq. (6.253), but now the g factor is given by Eq. (6.266). If we put $g_l = 1$, but leave g_s in the formula, we obtain

$$g = 1 + (g_s - 1)\frac{J(J + 1) - L(L + 1) + S(S + 1)}{2J(J + 1)} \tag{6.267}$$

Because g depends on L, S, and J, we can write

$$g = g(L, S, J). \tag{6.268}$$

The situation with respect to g_s is as follows. Classical experiments, for example, the Stern–Gerlach experiment, yielded the value $g_s = 2$. However, if quantum electrodynamical effects are taken into account, the theory gives a g_s value slightly larger than 2. Cowan gives the value [Ref. 58, p. 442]

$$g_s = 2.0023192. \tag{6.269}$$

Although the previous formulas are sufficient for our discussion, we give, for the sake of completeness, the Racah formulation also. Let $|\alpha LSJM\rangle$ be an eigenstate with the quantum numbers L, S, J, and M. The symbol α combines all other parameters that might be necessary to identify the state. Then, using Racah's method, we get for the matrix component of the operator in Eq. (6.263)

$$\frac{1}{\hbar}\langle\alpha LSJM|(J_z + (g_s - 1)s_z)|\alpha'L'S'J'M'\rangle$$

$$= M\delta(b, b') - \delta(\alpha LSM; \alpha'L'S'M')(g_s - 1)$$

$$\times(-1)^{L+S+M}\sqrt{(2J + 1)(2J' + 1)}\sqrt{S(S + 1)(2S + 1)}$$

$$\times\begin{pmatrix} J & 1 & J' \\ -M & 0 & M \end{pmatrix}\begin{Bmatrix} L & S & J \\ 1 & J' & S \end{Bmatrix}, \tag{6.270}$$

where b means the full set of parameters $(\alpha LSJM)$; and $(\cdots)$ and $\{\cdots\}$ are the $3j$ and $6j$ symbols, respectively. Evaluation of these symbols yields

$$\frac{1}{\hbar}\langle\alpha LSJM|J_z + (g_s - 1)s_z|\alpha LSJM\rangle = Mg(LSJ), \tag{6.271}$$

where $g(LSJ)$ is given by Eq. (6.267). We obtained, of course, the same result as in Eq. (6.265).

In Sec. 6.1, we have mentioned that for some of the lanthanide atoms, it is necessary to construct wave functions that are the linear combinations of functions with the same J, but with different (LS) combinations. We have mentioned that this necessitates the construction of g factors for such combined states. Let the state in question be $|\alpha JM\rangle$. Then we put

$$|\alpha JM\rangle = \sum_{\gamma SL} |\gamma SLJM\rangle\langle\gamma SLJM|\alpha JM\rangle. \qquad (6.272)$$

This is a finite expansion in terms of the states $|\gamma SLJM\rangle$. These are wave functions for which $(SLJM)$ are good quantum numbers and γ indicates, like α, all other parameters that may be needed to identify the state $|\gamma SLJM\rangle$. The summation is for the necessary (γSL) combinations.

The g factor belonging to such a linear combination is constructed as follows. We calculate the matrix component of the operator in Eq. (6.263) with respect to the previous expansion. Using Eqs. (6.270) and (6.271), we get

$$\frac{1}{\hbar}\langle\alpha JM|J_z + (g_s - 1)S_z|\alpha JM\rangle$$

$$= \sum_{\gamma SL}\sum_{\gamma'S'L'} \langle\gamma SLJM|J_z + (g_s - 1)s_z|\gamma'S'L'JM\rangle$$

$$\times\langle\gamma SLJM|\alpha JM\rangle\langle\alpha JM|\gamma'S'L'JM\rangle$$

$$= \sum_{\gamma SL} \langle\gamma SLJM|J_z + (g_s - 1)s_z|\gamma SLJM\rangle$$

$$\times|\langle\gamma SLJM|\alpha JM\rangle|^2$$

$$= \sum_{\gamma SL} g(SLJ)M|\langle\gamma SLJM|\alpha JM\rangle|^2. \qquad (6.273)$$

In this formula, $g(SLJ)$ is the same as in Eq. (6.267). Thus, for a wave function that is characterized by (αJ), we obtain the result

$$\varepsilon\mathcal{H}\langle\alpha JM|J_z + (g_s - 1)s_z|\alpha JM\rangle = g(\alpha JM)M\mu\mathcal{H}, \qquad (6.274)$$

where

$$g(\alpha JM) = \sum_{\gamma SL} g(SLJ)|\langle\gamma SLJM|\alpha JM\rangle|^2, \qquad (6.275)$$

that is, for the matrix component with respect to a linear combination like the one given in Eq. (6.272), we obtain the same result as for a single wave function except the Lande g factor is now itself a linear combination of terms built from the g factors with different LS values and from coefficients that are figuring also in the expansion of Eq. (6.272).

Finally, we note that transitions in the presence of a magnetic field can easily be discussed in terms of Eq. (6.189). That formula gives the matrix components of the operator with respect to (JM) and $(J'M')$ states. The associated selection rules are given by Eqs. (6.184)–(6.186) and by Eq. (6.190).

6.8. SUMMARY AND DISCUSSION

Reviewing the contents of this chapter, we recall that in Sec. 6.1, we started by discussing the salient features of atomic energy levels. We pointed out that the energy levels are classified by

1. configuration and parity;
2. the characteristics of multiplet structure, that is, by the LS values;
3. the characteristics of the fine structure, that is, by the J value; and
4. the Lande g factor, which is derived from the study of the Zeeman effect.

Thus, as we have seen, most atomic energy levels are classified by (LSJ) and by the Lande g factor. The parity of each level is also given. The tables list the experimentally measured values of the energy levels.

Strictly speaking, from the data given in the Atomic Energy Level Tables, only the numerical values of the levels can be considered as experimental data. The classification of the levels involves the application of the Hartree–Fock model, or, more accurately, the application of a one-electron independent-particle model in the (LSJ) coupling. We review here the assumptions that lead to the classifications displayed in the tables.

Let us consider the exact Schroedinger equation of an atom without an external field. The Hamiltonian operator of an N-electron atom must be invariant for an inversion of particle coordinates; from this, it follows that the wave functions of the atom must be *even* or *odd* functions of the coordinates. As we have seen in Sec. 6.5, dipole transitions are possible only between states of different parity. Thus, denoting the even states by g and the odd states by u, we obtain that the transitions must obey the rule

$$g \leftrightarrow u. \qquad (6.276)$$

The exact Hamiltonian must also be invariant under a rotation of the coordinate system. From this, it follows that the eigenfunctions of the atom must also be eigenfunctions of the operator J_{op}^2, where J_{op} is the operator of the total angular momentum. Thus, we must have

$$J_{op}^2 \psi = J(J + 1)\hbar^2\psi, \qquad (6.277)$$

where ψ is the eigenfunction of the exact Hamiltonian operator, and J is the quantum number associated with the operator J_{op}^2.

Thus, as we see by comparing the basic theoretical information with the classification scheme presented before, only the parity and the J value of every level are determined by the exact Schroedinger equation. The concept of configuration and the (LS) numbers involve the introduction of models.

If we place the atom into a magnetic field, then the Hamiltonian will have rotational symmetry around the homogeneous magnetic field. The energy levels can be characterized by the magnetic quantum number M, where M goes from $-J$ to $+J$. The angular momentum in the direction of the magnetic field will be $M\hbar$. The magnetic energy will be

$$E_{mag} = M\mathscr{H}g\mu, \tag{6.278}$$

where M is the magnetic quantum number, $\mathscr{H}$ is the magnetic field, g is the Lande g factor, and μ is the Bohr magneton. Thus, we see that No. 4 of the classification scheme also follows from the exact Schroedinger equation.

We now discuss how the classification scheme is built up and what are the approximations involved. We start with the nonrelativistic Hamiltonian operator given by Eq. (6.209):

$$H_0 = \sum_j f_j + \frac{1}{2}\sum_{j,k}\frac{e^2}{r_{jk}}, \tag{6.279}$$

where f_j is the one-electron operator of Eq. (6.208). This operator was the starting point of Hartree,[1] who selected a configuration and with it the parity for each energy level. The energy depends only on the configuration in this approximation and transitions are possible if they satisfy Eq. (6.276).

The next step was taken by Slater,[12] who has shown how to build up LS coupling. Slater pointed out that the operator in Eq. (6.279) commutes with the operators L^2, S^2, L_z, and S_z. Slater used perturbation theory to construct energy expressions that are based on wave functions diagonalizing L^2, S^2, L_z, and S_z. In this approximation, the single energy level of the Hartree method breaks up into multiplets, that is, into levels with different LS combinations. (We note that in most texts, the description of this step is incorrect. It is stated generally that the multiplet structure is the result of the electron–electron interaction being taken into consideration. Not true: the electron–electron interaction is built into the Hartree method, which is clearly seen from the discussion in Sec. 2.2. The multiplet structure is the result of constructing wave functions that are eigenfunctions of L^2, S^2, L_z, and S_z. If these wave functions are used in the construction of the energy expression, the energy expression will become automatically (LS)-dependent because there will be different wave functions for different (LS) combinations.)

The LS coupling formulated by Slater did not explain fine structure. Let us consider now the Hamiltonian in which spin–orbit coupling is taken into account. We have

$$H = H_0 + H', \tag{6.280}$$

where H_0 is given by Eq. (6.279), and H' is the operator given in Eq. (6.210):

$$H' = + \sum_{j=1}^{N} \frac{e}{2m^2c^2} \frac{1}{r_j} \frac{dV(j)}{dr_j} (\mathbf{L}_j \cdot \mathbf{s}_j). \tag{6.281}$$

The Hamiltonian operator in Eq. (6.280) commutes with J^2 and J_z, but not with L^2, S^2, L_z, and S_z. A meaningful bridge between the nonrelativistic Hamiltonian, H_0, and the spin–orbit Hamiltonian, $H = H_0 + H'$, can be established by recalling that H_0, besides commuting with L^2, S^2, L_z, and S_z, also commutes with L^2, S^2, J^2, and J_z. The expectation value of the Hamiltonian H_0 leads to the same energies regardless of which of the two sets of angular-momentum operators are used.

Now as we have seen in the short survey of the exact properties at the beginning of this section, it is actually only J, which is a good quantum number and not the set (LSJ). However, as long as the spin–orbit interaction is small relative to the electrostatic interaction, we can calculate the spin–orbit energy with respect to wave functions that diagonalize the operators L^2, S^2, J^2, and J_z. This set does not commute with the operator $H_0 + H'$, but it "almost" commutes with it, that is, quantities like $L^2H - HL^2$ are small. In this approximation, the multiplet with an (LS) combination breaks up into $2S + 1$ or $2L + 1$ levels, depending whether $S < L$ or $L < S$. In this way, we obtain the fine structure. The energy levels will be $(2J + 1)$-fold degenerate.

Finally, if we place the atoms into a magnetic field, our Hamiltonian becomes

$$H = H_0 + H' + H'', \tag{6.282}$$

where H_0 is given by Eq. (6.279), H' by Eq. (6.281), and H'' is the operator that we have written down in Eq. (6.263):

$$H'' = \frac{\mu}{\hbar} \mathcal{H} (J_z + (g_s - 1)s_z), \tag{6.283}$$

where μ is the Bohr magneton, $\mathcal{H}$ is the magnetic field, and g_s is the spin g factor.

The diagonal matrix component of H'' leads to the energy

$$E'' = gM\mathcal{H}\mu, \tag{6.284}$$

where g is the Lande g factor given by Eq. (6.267). From this equation, we see that in the magnetic field, the single J level breaks up into $(2J + 1)$ levels according to the values of the magnetic quantum number M.

Now we see that the four steps in the classification of spectra that we presented at the beginning of the section are matched by the following four stages of the theoretical buildup:

1. *Hartree Model.* Energy depends only on configuration. Parity of levels established.
2. *Slater's Theory* of multiplet structure. LS coupling: energy depends on (LS).
3. *Spin–Orbit Interaction*: (LSJ) coupling. Energy depends on L, S, and J.
4. *Zeeman Effect.* Energy levels split into $(2J + 1)$ terms.

The discussion presented here is valid for the vast majority of spectra. Exceptions are the lanthanides and the noble gases. As we have seen, in the case of the lanthanides, the reliable data are the J value and the Lande g factor. Classification requires five different coupling schemes.

It is interesting to note that up to this point, the whole discussion concerned mainly the angular momentum and spin operators that commute with the Hamiltonian at various stages of approximations. Let us turn our attention now to the Hamiltonian operator, that is, let us turn to the Hartree–Fock model. Let us ask the question: How accurate is this model in the reproduction of the empirical values of the energy levels?

We do not present here any numbers because we want to postpone the discussion of calculations until after the presentation of the pseudopotential model. We can anticipate the discussion, however, by quoting here some of the conclusions. In Sec. 6.3, we developed the HF model for one valence electron in the frozen-core approximation. As we show later, this model can be exactly transformed into a pseudopotential formalism. Solutions of the pseudopotential equations that are exactly equivalent to the frozen-core HF solutions were constructed for the ground and excited states of a number of atoms (Sec. 7.2/D).

Comparing the calculations with the empirical data, we will conclude that in many cases, the results are only fair approximations to the empirical energy levels. Thus, we are compelled to observe that the HF model, in the case of the optical spectra, gives only a fair approximation to the exact Schroedinger equation. The reason for this is, of course, the lack of *correlation* effects in the HF model. In the next chapter, we begin the buildup of the effective Hamiltonian theory for valence electrons, which, among other things, will enable us to take into account correlation effects.

One final word about selection rules. In Sec. 6.5, we discussed transition probabilities and we established selection rules for the successive stages of approximations. Here we only want to emphasize again that each set of selection rules is valid only for the approximation in which the quantum numbers appearing in the selection rules are good quantum numbers. The selection rules become invalid when the quantum numbers in the rules cease to be good quantum numbers. For example, as long as the spin–orbit coupling is weak, we can formulate selection rules for L, S, and J; however, the L and S cease to be good quantum numbers and selection rules for them become invalid when the spin–orbit coupling cannot be considered small relative to the electrostatic interaction.

CHAPTER 7

EFFECTIVE HAMILTONIAN THEORY FOR ONE VALENCE ELECTRON

7.1. INTRODUCTION: GENERAL DEFINITION OF EFFECTIVE HAMILTONIANS

Introduction of the concept of an effective Hamiltonian; four reasons for the construction of effective Hamiltonians for the valence electrons; the basic task of effective Hamiltonian theory.

7.2. ONE-ELECTRON PSEUDOPOTENTIAL THEORY

A. The basic idea and the physical meaning of the pseudopotential; elimination of the core energies from the Schroedinger equation; comparison with the Hartree potential. B. Transformation of the frozen-core HF equations into pseudopotential equations, the effective pseudopotential Hamiltonian for a valence electron outside complete groups; the dependence of the effective Hamiltonian on the quantum numbers of the valence electron. C. The pseudopotential effective Hamiltonian for arbitrary core orbitals; the most general form of the pseudopotential operator; the elimination of indeterminacy by additional mathematical conditions. D. Presentation and discussion of calculations made with the HF-equivalent pseudopotential model; the conclusions drawn from the calculations.

7.3. CORE / VALENCE CORRELATION EFFECTS

A. Discussion of the importance of correlation effects in the construction of effective Hamiltonians. B. Two methods for the construction of an effective Hamiltonian with core / valence correlation; discussions of the properties of the effective Hamiltonian and its limitations. C. Transformation of the effective Hamiltonian into a pseudopotential operator. D. Analytic expressions for the polarization potential and for the pseudopotential. E. Determination of an effective potential from the energy levels; Prokofjew's and Bottcher's models.

7.1. INTRODUCTION: GENERAL DEFINITION
OF EFFECTIVE HAMILTONIANS

Let us consider an atom with N core electrons and n valence electrons. Let the Hamiltonian of this atom be H. We can formulate the Hamiltonian in a relativistic or nonrelativistic fashion; within both categories, we can choose from several different approximations. All these Hamiltonians will have one thing in common: they will be $(N + n)$-electron operators, that is, they will refer to the atom as a whole and there will be no distinction between valence and core electrons in the form of these operators.

In contrast, we now define an *effective Hamiltonian* for the valence electrons as follows. We define an operator that depends only on the coordinates of the n valence electrons. Let the space/spin coordinates of the valence electrons be $q_1, q_2, q_3, \ldots, q_n$. Let the effective Hamiltonian be H_{eff} and its eigenfunction and eigenvalue be Ψ and E, respectively. We define an effective Schroedinger equation by writing

$$H_{eff}(q_1 q_2 \cdots q_n)\Psi(q_1 q_2 \cdots q_n) = E\Psi(q_1 q_2 \cdots q_n). \quad (7.1)$$

As we have emphasized in the notation, both H_{eff} and Ψ are n-electron functions in contrast to the original Hamiltonian and its eigenfunctions, which refer to the whole atom, and, therefore, are $(N + n)$-electron functions.

Where are the core electrons in this picture? We say that H_{eff} depends on the wave function of the core *parametrically*. By this, we may mean a dependence on some numerical parameters of the core, but more often we mean a functional dependence on the wave function(s) of the core electrons. Thus:

> An effective Hamiltonian is an n-electron operator acting only on functions of the coordinates of the valence electrons. However, this does not mean that the effective Hamiltonian is independent of the core electrons; it is independent of the coordinates of the core electrons, but it does depend on the wave function(s) of the core electrons parametrically.

We distinguish two basic types of effective Hamiltonians. The first type will be called a *conventional* effective Hamiltonian. By this we mean a Hamiltonian whose eigenfunctions must be orthogonalized to the core wave functions. The second type will be called a *pseudopotential* effective Hamiltonian. The eigenfunctions of these operators need not be orthogonalized to the core functions; one can work with such a Hamiltonian "as if the core would not exist." From the definitions, we can see that the pseudopotential Hamiltonians are the "real thing"; these are the Hamiltonians that present the problem in such a way that the core electrons are *almost* eliminated from the picture. In fact, the core electrons influence the problem only through their "parameteric" presence in the effective pseudopotential Hamiltonian operator.

What is the motivation for the development of effective Hamiltonians? We present here four arguments that make it desirable that such Hamiltonians be developed:

1. In Sec. 6.1, one of our first observations was that the atomic spectra are generated by valence electrons. In fact, we observed that in each atom, the spectra are generated by a small number of electrons; we defined the valence electrons as those that generate the spectra. If we are able to develop an effective Hamiltonian for the valence electrons, our problem is reduced from an $(N + n)$-electron problem to an n-electron problem with $n \ll (N + n)$.

2. In the presentation of Hartree–Fock-equivalent pseudopotential calculations, we will show that *correlation* effects between the core and valence electrons are important. We have seen in Sec. 5.4/B that correlation effects are also large between the valence electrons. Thus, it is accurate to say that correlation effects, between the valence and core electrons and between the valence electrons themselves, are important. Thus, it is desirable to develop a "valence-only" theory in which the valence and core electrons are clearly separated and correlation effects between the two groups and between the valence electrons themselves can be taken into account.

3. In discussing the lanthanide atoms in Sec. 6.1, we observed that the simple (LSJ) coupling scheme often breaks down here. We have seen that the classification of lanthanide spectra required the use of five different coupling schemes. The development of effective Hamiltonians is extremely useful for these atoms. Treating the valence electrons separately from the core electrons is useful not only for the treatment of intravalence correlation effects, but also for the development of special coupling schemes for the valence electrons.

4. Last but not least, apart from any utilitarian considerations, the development of effective Hamiltonians is interesting per se; some of the mathematical methods of effective Hamiltonian theory are among the most interesting applications of quantum mechanics. In our opinion, this argument alone would be a sufficient reason for the construction of effective Hamiltonians.

The basic task of effective Hamiltonian theory can be formulated as follows. We want to construct H_{eff} in such a way that *for all practical purposes*, Eq. (7.1) would be *equivalent* to the corresponding full $(N + n)$-electron Schroedinger equation. This is, of course, the most ideal formulation of an effective Hamiltonian theory. In practice, we must be often satisfied with partial, and/or approximate equivalence between the complete Schroedinger equation and the effective Schroedinger equation.

In this chapter, we discuss the simple case when we have only one valence electron. The more interesting and challenging problem of several valence electrons will be attacked in the next chapter.

7.2. ONE-ELECTRON PSEUDOPOTENTIAL THEORY

A. The Basic Idea of Pseudopotentials

The idea of pseudopotentials was introduced into quantum mechanics by Hellmann.[67] Formulating the idea for one valence electron Hellmann has shown, using the Thomas–Fermi model, that the Pauli exclusion principle for the valence electron, that is, the requirement of orthogonality of the valence electron wave function with respect to the core orbitals, can be replaced by a nonclassical potential (Abstossungspotential), which is now called the pseudopotential (PP). Thus, Hellmann suggested for one valence electron the effective Hamiltonian

$$H_H = -\frac{1}{2}\Delta + V + V_P = -\frac{1}{2}\Delta - \frac{z}{r} + A\frac{e^{-\kappa r}}{r}, \qquad (7.2)$$

where z is the ionic charge of the core, that is, if the nucleus has Z positive elementary charges and the core contains N electrons, then $z = Z - N$. Constants A and κ are determined from the requirement that the effective Hamiltonian should reproduce the empirical optical spectrum of the valence electron as closely as possible. In Eq. (7.2), the analytic expression is meant to approximate the sum of $V + V_P$, where V represents the electrostatic, exchange, and correlation interactions between the core and the valence electron, and V_P is the pseudopotential. The effective Schroedinger equation is

$$H_H \psi_v = \varepsilon_v \psi_v, \qquad (7.3)$$

and this equation can be solved "as if the core would not exist," that is, the eigenfunctions of this equation need not satisfy any orthogonality requirements with respect to the core orbitals. The eigenfunctions of H_H are called pseudowave functions or pseudo-orbitals (POs).

With this idea, Hellmann succeeded to reduce an $(N + 1)$-electron problem to a one-electron problem. We note that the analytic expression in Eq. (7.2) is of exceptional simplicity and was an inspired guess of Hellmann's; this does not mean, however, that the theory was also that simple [see the author's book on the subject, Szasz, 1, 50, 79]. We note also that the Hamiltonian given by Eq. (7.2) is semi-empirical; this property, however, is not inherent in the pseudopotential concept.

The formula given by Eq. (7.2) does not mean much for a reader who learns about PP theory for the first time. The physical meaning of this idea

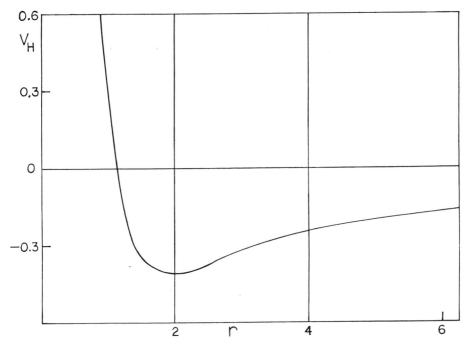

Fig. 7.1. The Hellmann potential for the valence electron of the Na atom.

becomes clear, however, if we look at Fig. 7.1. In that figure, we plotted the Hellmann potential, $V_H = (-z + Ae^{-\kappa r})/r$, for the valence electron of the neutral Na atom. Let us compare this diagram with Fig. 2.2. In Fig. 2.2, we have a typical Hartree potential: the electrostatic potential in the Hg atom. As we see, the Hartree potential is everywhere negative, that is, attractive; close to the nucleus, it approximates the potential of the nucleus and for large r, it is Coulombic. The structure of the Hellmann potential is totally different. For small r, the Hellmann potential has a high positive barrier; this is what keeps the valence electron out of the atomic core. The ascending part of the barrier is located approximately at the outer surface of the core. Outside of the core, the Hellmann potential has a deep negative well; the maximum of the valence-electron wave function will be within this well. Outside of the well, the Hellmann potential is Coulombic as the Hartree potential. We can say that the structure of the Hellmann potential shows the barrier/well/Coulombic sequence.

The physical meaning of the pseudopotential theory is fully explained by this diagram. As we said earlier, the idea is to replace the Pauli exclusion principle, that is, the orthogonality requirements, by a potential that is called the pseudopotential. This is a positive potential, which, when added to the Hartree–Fock potential, changes that into a potential of the type that we

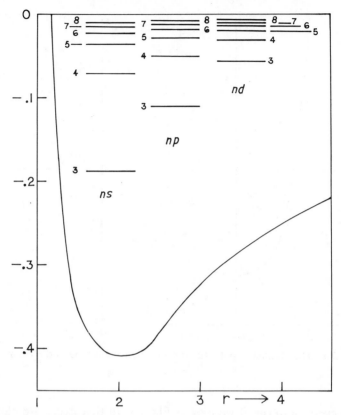

Fig. 7.2. The Hellmann potential for the valence electron of the sodium atom with the first few lines of the empirical optical spectrum.

have in Fig. 7.1. Instead of being negative (attractive) everywhere, this new potential has a strong positive barrier at the atomic core; this barrier is what keeps the valence electron out of the core, that is, this barrier is what takes the place of the Pauli exclusion principle.

In Fig. 2.2, we placed into the diagram, besides the potential, the energy levels of the core electrons in the Hg atom. In Fig. 7.2, we have the same Hellmann potential as in Fig. 7.1, but on a different scale, and into this diagram we placed, in a fashion analogous to Fig. 2.2, the first few energy levels of the s, p, and d series of the Na valence electron. These are the empirical levels; both the potential and the levels are in atomic units.

A comparison between Fig. 2.2 and Fig. 7.2 is very instructive. In the Hartree potential, the lowest level is the $1s$; that is, in this potential, we have the full spectrum and the lowest level is the lowest *core* level. In the Hellmann potential of Fig. 7.2, the lowest level is the lowest *valence* level;

there is no possibility for core energies here. We can see this clearly from the HF values for the orbital parameters[41]:

$$\varepsilon(2p) = -1.51814 \text{ a.u.}$$

$$\varepsilon(2s) = -2.79702 \text{ a.u.}$$

$$\varepsilon(1s) = -40.4785 \text{ a.u.}$$

Looking at the diagram, we see that the minimum of the Hellmann potential is at about -0.41 a.u. Such a potential could not possibly accommodate even the highest of core levels, which is at -1.5 a.u. Thus, the pseudopotential is shifting the lowest energy level from the lowest core level to the lowest valence level. This means that if the pseudopotential is added to the Hartree potential, the resulting spectrum does not have the energy levels of the core electrons; it is restricted to the energy levels of the valence electron. For the Schroedinger equation, Eq. (7.3), the core electrons "do not exist." Thus, the Hamiltonian operator suggested by Hellmann, Eq. (7.2), can be viewed as the prototype of effective Hamiltonians for atoms with one valence electron.

B. Hartree – Fock-Equivalent Pseudopotentials

We now develop the mathematical apparatus of the pseudopotential theory. The first step is to transform the frozen-core HF theory into a pseudopotential formalism. This transformation was carried out by Szepfalusy[68]; we present here his work. We start with Fock's formulation of the frozen-core approximation that was presented in Sec. 6.3.

Let us consider an atom with $N + 1$ electrons. The HF wave function of the core is, in the absence of the valence electron,

$$\psi = (N!)^{-1/2} \det[\varphi_1, \ldots, \varphi_N]. \tag{7.4}$$

As we have seen, the HF orbitals can be considered orthonormal:

$$\langle \varphi_i \varphi_k \rangle = \delta_{ik}. \tag{7.5}$$

The Hamiltonian for the N core electrons is given by

$$H = \sum_{j=1}^{N} f_j + \frac{1}{2} \sum_{j,k=1}^{N} \frac{1}{r_{jk}}. \tag{7.6}$$

The expectation value of this Hamiltonian with respect to the wave function in Eq. (7.4) is

$$E_C = \langle \psi | H | \psi \rangle. \tag{7.7}$$

Variation of this expression with respect to the one-electron orbitals under the subsidiary conditions given by Eq. (7.5) leads to the HF equation:

$$H_F \varphi_k = \varepsilon_k \varphi_k \qquad (k = 1, \ldots, N), \tag{7.8}$$

where

$$H_F = f + U, \tag{7.9a}$$

and

$$U = \sum_{j=1}^{N} U_j. \tag{7.9b}$$

U_j is the HF potential generated by the orbital φ_j. This is linear operator with the kernel

$$\langle q | U_j | q' \rangle = \int \frac{|\varphi_j(q'')|^2 \, dq''}{|\mathbf{r} - \mathbf{r}''|} \delta(q - q') - \frac{\varphi_j(q)\varphi_j^*(q')}{|\mathbf{r} - \mathbf{r}'|}, \tag{7.9c}$$

from which we get, by integration, the usual formula, Eq. (6.37c). Next, we add the valence electron to the core and write the wave function of the atom in the form

$$\psi = [(N+1)!]^{-1/2} \det[\varphi_1, \ldots, \varphi_N, \varphi_v], \tag{7.10}$$

where φ_v is the valence orbital. Without loss of generality, it can be assumed that φ_v is orthogonal to the core orbitals:

$$\langle \varphi_i | \varphi_v \rangle = 0 \qquad (i = 1, 2, \ldots, N), \tag{7.11}$$

and it is normalized

$$\langle \varphi_v | \varphi_v \rangle = 1. \tag{7.12}$$

The Hamiltonian operator for the $(N + 1)$-electron system is

$$H = \sum_{j=1}^{N+1} f_j + \frac{1}{2} \sum_{j,k=1}^{N+1} \frac{1}{r_{jk}}, \tag{7.13}$$

and for the total energy, we get with this operator:

$$E_T = \langle \psi | H \psi \rangle, \tag{7.14}$$

where ψ is given by Eq. (7.10).

Variation of E_T with respect to the valence orbital φ_v leads to the equation

$$H_F \varphi_v = \varepsilon_v \varphi_v, \tag{7.15}$$

where we have kept the core orbitals fixed in the variation according to the frozen-core approximation. The Lagrangian multipliers resulting from Eq. (7.11) can be eliminated by a unitary transformation. We arrived at Eq. (6.51) of Sec. 6.3.

The Hamiltonian in Eq. (7.15) is a "conventional" effective Hamiltonian because the solutions of this equation must be orthogonal to the core orbitals. We now transform this equation into a pseudopotential equation. According to App. A, we can think of φ_v as Schmidt orthogonalized to the core orbitals, and we have seen that this type of orthogonalization does not change the total determinantal wave function. Therefore, let us put

$$\varphi_v = A_0 \left(\psi_v - \sum_{i=1}^{N} \alpha_i \varphi_i \right). \tag{7.16}$$

Here ψ_v is the pseudo-orbital that does not need to satisfy any orthogonality conditions with respect to the core orbitals. The α_i constants are the coefficients of Schmidt orthogonalization, that is, they have the form

$$\alpha_i = \langle \varphi_i | \varphi_v \rangle \qquad (i = 1, 2, \ldots, N), \tag{7.17}$$

and A_0 is a normalization constant:

$$A_0 = \left(1 - \sum_{i=1}^{N} |\alpha_i|^2 \right)^{-1/2}. \tag{7.18}$$

Next, we substitute φ_v, as given by Eq. (7.16), into Eq. (7.15). We get

$$H_F \varphi_v = H_F \psi_v - H_F \sum_{i=1}^{N} \alpha_i \varphi_i$$

$$= \varepsilon_v \varphi_v = \varepsilon_v \psi_v - \varepsilon_v \sum_{i=1}^{N} \alpha_i \varphi_i. \tag{7.19}$$

Let us introduce the linear operator Ω with the definition

$$\langle q | \Omega | q' \rangle = \sum_{i=1}^{N} \langle q | \varphi_i \rangle \langle \varphi_i | q' \rangle. \tag{7.20}$$

Using this operator, we can rewrite Eq. (7.19) as follows:

$$H_F \psi_v - H_F \Omega \psi_v = \varepsilon_v \psi_v - \varepsilon_v \Omega \psi_v, \tag{7.21}$$

and so we get

$$H_F \psi_v + (\varepsilon_v - H_F) \Omega \psi_v = \varepsilon_v \psi_v. \tag{7.22}$$

Let the linear operator V_P be defined as

$$V_P = (\varepsilon_v - H_F) \Omega. \tag{7.23}$$

By using this operator, the equation for the pseudo-orbital becomes

$$(H_F + V_P) \psi_v = \varepsilon_v \psi_v. \tag{7.24}$$

The operator $H_F + V_P$ in this equation is the effective pseudopotential Hamiltonian, and V_P, in Eq. (7.23), is the pseudopotential (operator).

Taking into account that the core orbitals satisfy the HF equation, Eq. (7.8), we obtain

$$\begin{aligned}
H_F \Omega &= H_F \sum_{i=1}^{N} |\varphi_i\rangle\langle\varphi_i| \\
&= \sum_{i=1}^{N} \varepsilon_i |\varphi_i\rangle\langle\varphi_i|,
\end{aligned} \tag{7.25}$$

and using this relationship, we can write Eq. (7.23) in the form

$$V_P = \sum_{i=1}^{N} (\varepsilon_v - \varepsilon_i) |\varphi_i\rangle\langle\varphi_i|. \tag{7.26}$$

Let us write down the total potential in Eq. (7.24). Adding the Hartree–Fock potential given by Eq. (7.9c) to the pseudopotential given by Eq. (7.26), we get

$$\begin{aligned}
\langle q|U + V_P|q'\rangle &= \sum_{j=1}^{N} \langle q|U_j|q'\rangle + \langle q|V_P|q'\rangle \\
&= \sum_{j=1}^{N} \int \frac{|\langle q''|\varphi_j\rangle|^2 \, dq''}{|\mathbf{r} - \mathbf{r}''|} \delta(q - q') \\
&\quad - \sum_{j=1}^{N} \frac{\langle q|\varphi_j\rangle\langle\varphi_j|q'\rangle}{|r - r'|} \\
&\quad + \sum_{j=1}^{N} (\varepsilon_v - \varepsilon_j)\langle q|\varphi_j\rangle\langle\varphi_j|q'\rangle.
\end{aligned} \tag{7.27}$$

Thus, the total potential in the effective pseudopotential Hamiltonian consists of the electrostatic and exchange potentials of the core plus the pseudopotential, which is the third term in Eq. (7.27). Like the exchange potential, the pseudopotential is a linear operator, that is, a nonlocal potential. The total potential will be called the *modified* potential, meaning the HF potential modified by the addition of the pseudopotential V_P. We put

$$V_M = -\frac{Z}{r} + U + V_P. \tag{7.28}$$

As we have emphasized before, the pseudopotential given by Eq. (7.23) is a linear operator. It was shown by Phillips and Kleinman[69] that a local pseudopotential can be defined by localizing the pseudopotential operator in the same way as Slater localized the exchange operator (Sec. 2.4). We put

$$V_P = \frac{(\varepsilon_v - H_F)\Omega\psi_v}{\psi_v}$$

$$= \sum_{j=1}^{N} \frac{(\varepsilon_v - \varepsilon_i)|\varphi_i\rangle\langle\varphi_i|\psi_v\rangle}{\psi_v}, \tag{7.29}$$

where we have used Eqs. (7.23) and (7.26), respectively. These potentials, which are now ordinary local potential functions, are called Phillips–Kleinman (PK) type pseudopotentials.

Next, we want to see the form of the modified potential for the case when the core consists of complete groups. Our formula is Eq. (7.27). Let us put the core orbitals into central-field form:

$$\varphi_j(q) = \frac{\hat{P}_{n_j l_j}(r)}{r} Y_{l_j m_{l_j}}(\vartheta, \varphi)\eta_{m_{s_j}}(\sigma). \tag{7.30}$$

Here we introduce the notation according to which the radial part of the core orbitals will be denoted by $\hat{P}_{nl}$ and the valence-electron pseudo-orbitals will be P_{nl}. Because we are dealing with a core of N electrons from which $N/2$ will have positive and $N/2$ negative spins, we can utilize the derivations of Sec. 2.3. In that section, we have seen that the equation for the spatial part of the HF orbital of an electron with either positive or negative spin is given by Eq. (2.89). The Hamiltonian operator in Eq. (2.89) is the same as our H_F in Eq. (7.24), that is, it is the same after the spin integration is carried out in Eq. (7.24). In order to transfer the results of Sec. 2.3 to our present derivation, let us put

$$\psi_v = \frac{P_{n_v l_v}(r)}{r} Y_{l_v m_{l_v}}(\vartheta, \varphi)\eta_{m_{s_v}}(\sigma). \tag{7.31}$$

In order to get the spatial part of $H_F\psi_v$, which we need for Eq. (7.24), we simply take Eq. (2.97a) with the substitution $(n_i l_i) \rightarrow (n_v l_v)$. Thus, we get

$$
\begin{cases}
-\frac{1}{2r}\frac{d^2}{dr^2}P_{n_v l_v} + \frac{l_v(l_v+1)}{2r^2}\frac{P_{n_v l_v}}{r} - \frac{Z}{r}\frac{P_{n_v l_v}}{r} \\
\\
+ \sum_{n_j l_j} 2(2l_j+1)\frac{1}{r}Y_0(n_j l_j, n_j l_j|r)\frac{P_{n_v l_v}}{r} \\
\\
- \sum_{n_j l_j k} \sqrt{\frac{2l_j+1}{2l_v+1}}\, c^k(l_v 0, l_j 0)\frac{1}{r}Y_k(n_j l_j n_v l_v|r)\frac{\hat{P}_{n_j l_j}}{r}
\end{cases} Y_{lm_{l_v}}. \quad (7.32)
$$

Next, we calculate $V_P\psi_v$ with V_P given by Eq. (7.26). For φ_j, we use Eq. (7.30), and for ψ_v, we use Eq. (7.31). Then we get

$$
V_P\psi_v = \sum_{j=1}^{N}(\varepsilon_v - \varepsilon_j)\langle q|\varphi_j\rangle\langle \varphi_j|\psi_v\rangle
$$

$$
= \sum_{n_j l_j m_{l_j} m_{s_j}}(\varepsilon_{n_v l_v} - \varepsilon_{n_j l_j})\frac{\hat{P}_{n_j l_j}(r)}{r}Y_{l_j m_{l_j}}(\vartheta\varphi)\eta_{m_{s_j}}(\sigma)
$$

$$
\times \int \hat{P}_{n_j l_j}(r)P_{n_v l_v}(r)\,dr \int Y^*_{l_j m_{l_j}}(\vartheta\varphi)Y_{l_v m_{l_v}}(\vartheta\varphi)\,d\omega
$$

$$
\times \int \eta^*_{m_{s_j}}(\sigma)\eta_{m_{s_v}}(\sigma)\,d\sigma. \quad (7.33)
$$

Carrying out the integration with respect to the spin and the angles, we get

$$
V_P\psi_v = \sum_{n_j}(\varepsilon_{n_v l_v} - \varepsilon_{n_j l_v})\frac{\hat{P}_{n_j l_v}}{r}Y_{l_v m_{l_v}}(\vartheta\varphi)
$$

$$
\times \eta_{m_{s_v}}(\sigma)\int \hat{P}_{n_j l_v}P_{n_v l_v}\,dr. \quad (7.34)
$$

In order to get Eq. (7.24), we combine Eqs. (7.32) and (7.34). From Eq. (7.34), we can omit the spin function because the HF expression, Eq.

(7.32), is spinless. We get this way

$$
\left\{ -\frac{1}{2r}\frac{d^2}{dr^2}P_{n_v l_v} + \frac{l_v(l_v+1)}{2r^2}\frac{P_{n_v l_v}}{r} - \frac{Z}{r}\frac{P_{n_v l_v}}{r} \right.
$$

$$
+ \sum_{n_j l_j} 2(2l_j + 1)\frac{1}{r}Y_0(n_j l_j, n_j l_j | r)\frac{P_{n_v l_v}}{r}
$$

$$
- \sum_{n_j l_j k} \sqrt{\frac{2l_j+1}{2l_v+1}}\, c^k(l_v 0, l_j 0)\frac{1}{r}Y_k(n_j l_j n_v l_v | r)\frac{\hat{P}_{n_j l_j}}{r}
$$

$$
+ \sum_{n_j} (\varepsilon_{n_v l_v} - \varepsilon_{n_j l_v})\frac{\hat{P}_{n_j l_v}}{r}
$$

$$
\left. \times \int \hat{P}_{n_j l_v} P_{n_v l_v}\, dr \right\} Y_{l_v m_{l_v}} = \varepsilon_{n_v l_v}\frac{P_{n_v l_v}}{r}Y_{l_v m_{l_v}}. \qquad (7.35)
$$

Multiplying by $r/Y_{l_v m_{l_v}}$, we obtain the equation for the radial part of the valence electron pseudo-orbital. Let U be the nonlocal potential in Eq. (2.98) and let V_P now be defined as the operator

$$
V_P(r)P_{n_v l_v}(r) = \sum_{n_j} (\varepsilon_{n_v l_v} - \varepsilon_{n_j l_v})\hat{P}_{n_j l_v}\int \hat{P}_{n_j l_v} P_{n_v l_v}\, dr. \qquad (7.36)
$$

Using these notations, we obtain

$$
\left\{ -\frac{1}{2}\frac{d^2}{dr^2} + \frac{l_v(l_v+1)}{2r^2} - \frac{Z}{r} + U(r) + V_P(r) \right\} P_{n_v l_v}(r)
$$

$$
= \varepsilon_{n_v l_v} P_{n_v l_v}(r). \qquad (7.37)
$$

We can write this equation easily into a form in which the operator character of U and V_P is emphasized. From Eq. (2.98), we obtain for the kernel of the operator U:

$$
\langle r|U|r'\rangle = \sum_{n_j l_j} 2(2l_j + 1)\frac{1}{r}Y_0(n_j l_j n_j l_j | r)\,\delta(r - r')
$$

$$
- \sum_{n_j l_j k} \sqrt{\frac{2l_j+1}{2l_v+1}}\, c^k(l_v 0, l_j 0)\hat{P}_{n_j l_j}(r)\hat{P}_{n_j l_j}(r')L_k(r, r'), \qquad (7.38)
$$

where $L_k(r, r')$ is defined by Eq. (2.72). For the kernel of the pseudopotential operator, we get from Eq. (7.36)

$$\langle r|V_P|r'\rangle = \sum_{n_j} (\varepsilon_{n_v l_v} - \varepsilon_{n_j l_v}) \hat{P}_{n_j l_v}(r) \hat{P}_{n_j l_v}(r'). \qquad (7.39)$$

In Eq. (7.37), both the HF potential and the pseudopotential are nonlocal. We can write these operators into local form using Slater's expression for the exchange potential and the PK form for the pseudopotential. Then we get

$$\left\{ -\frac{1}{2}\frac{d^2}{dr^2} + \frac{l_v(l_v + 1)}{2r^2} - \frac{Z}{r} + \tilde{U}(r) + \tilde{V}_P(r) \right\} P_{n_v l_v}$$

$$= \varepsilon_{n_v l_v} P_{n_v l_v}(r). \qquad (7.40)$$

In this equation, both $\tilde{U}$ and $\tilde{V}_P$ are ordinary local potentials, and we have

$$\tilde{U}(r) = \frac{\int \langle r|U|r'\rangle P_{n_v l_v}(r')\,dr'}{P_{n_v l_v}(r)}, \qquad (7.41)$$

and

$$\tilde{V}_P(r) = \frac{\int \langle r|V_P|r'\rangle P_{n_v l_v}(r')\,dr'}{P_{n_v l_v}(r)}. \qquad (7.42)$$

It is clear that Eq. (7.42) is identical with the PK pseudopotential given by Eq. (7.29).

Next, we discuss the structure of these potentials. In Eq. (7.38), the summations over $n_j l_j$ is over all complete groups in the core. For k, we have the limits $0 \le k \le l_j + l_v$. In the pseudopotential, Eq. (7.39), the summation over n_j means the summation over those core orbitals that have the same azimuthal quantum number as the valence electron, that is, have the quantum number l_v.

It is interesting to note that the pseudopotential operator does not change the angular and spin part of the wave function on which it operates. We see this clearly from Eq. (7.34). The angular and spin part of ψ_v is $Y_{l_v m_{l_v}} \eta_{m_{s_v}}$; we have the same on the right side of Eq. (7.34). Thus, the pseudopotential operator has the property of operating only on the radial part of the wave function to which it is applied. As we have seen in Sec. 2.3, the exchange operator has the same property. This is, of course, because the core consists of complete electron groups.

It is an important point to establish the dependence, if any, of the complete potential on the valence-electron data. For this purpose, Eq. (7.27) is the best suited. As we see, the electrostatic and exchange potentials, the first two terms in Eq. (7.27), depend only on the core orbitals. The last term, the pseudopotential, depends on the core orbitals, on the orbital parameters

ε_j of the core electrons, and on the orbital parameter of the valence electron, ε_v. Thus, the pseudopotential will depend on $(n_v l_v)$ because ε_v depends on n_v and l_v.

Now from Eq. (7.34), we see that there will be a different summation in the potential for different l_v's. That is, for $l_v = 0$, the summation will be over the s electrons of the core; for $l_v = 1$, over the p electrons, and so on. For valence states with the same l_v but different n_v, the potential will be different, but the dependence on n_v will not be strong. We can see this as follows. In the expression

$$\varepsilon_{n_v l_v} - \varepsilon_{n_j l_v}, \tag{7.43}$$

the summation will be over n_j. For fixed l_v, the summation will be the same, that is, the $\varepsilon_{n_v l_v}$ numbers will be the same. For two states that have the same l_v but different n_v's, the core orbitals in the summation will be the same; the potential will be different in the two cases because of the different $\varepsilon_{n_v l_v}$'s. But $\varepsilon_{n_v l_v}$ is the orbital parameter of the valence electron and that number is always much smaller in absolute value than the orbital parameters of the core electrons. Therefore, the summation will be dominated by the core electrons and because the summation over the core parameters is the same for two valence states with different n_v's, the dependence of the pseudopotential on n_v will be a weak dependence. We can sum up:

The pseudopotential depends strongly on the azimuthal quantum number l_v of the valence electron. There will be a different pseudopotential for each l_v. The pseudopotential will also depend on n_v, but only very slightly. That is, the pseudopotentials for two valence states with the same l_v but different n_v will be very nearly identical.

Turning now to the HF potential, Eq. (7.38), we see clearly that it depends only on l_v. This is true also for the azimuthal part of the Laplacian, $l_v(l_v + 1)/2r^2$. Therefore, we can make a statement that is quite fundamental for the theory of atomic spectra. Let us consider Eq. (7.37). In that equation, the operators U and V_P are defined by Eqs. (7.38) and (7.39), respectively. Summing up our discussion, we can say that

The effective Hamiltonian operator in Eq. (7.37) depends, apart from a very weak dependence on n_v, only on the azimuthal quantum number l_v of the valence electron. Thus, in a very good approximation, the radial functions $P_{n_v l_v}$ for a fixed l_v will be the eigenfunctions of the same operator.

In Sec. 6.1, we described how the energy levels of each series have a Rydberglike structure. A series has a common l_v and the Rydberglike structure is the direct consequence of these states being the eigenstates of the same operator. Thus, as we see, the pseudopotential model in the frozen-core approximation reproduces one of the most important features of the atomic spectra.

At the last part of this section, we now show that the spectrum of the effective pseudopotential Hamiltonian in Eq. (7.24) is identical with the valence spectrum of the HF Hamiltonian, H_F. As we have seen from Eqs. (7.8) and (7.15), the HF Hamiltonian reproduces the whole spectrum, core and valence. Equation (7.8) gives the core energies and wave functions:

$$H_F \varphi_i = \varepsilon_i \varphi_i \qquad (i = 1, 2, \ldots, N), \tag{7.44}$$

and Eq. (7.15) generates the valence spectrum and valence orbitals:

$$H_F \varphi_v = \varepsilon_v \varphi_v \qquad (v = N + 1, \ldots, \infty). \tag{7.45}$$

It is the essence of the frozen-core approximation that H_F is the same operator in these two equations. Now consider Eq. (7.24):

$$(H_F + V_P)\psi = \varepsilon \psi, \tag{7.46}$$

with V_P given by Eq. (7.26). Let us expand ψ in terms of the complete set of solutions generated by H_F:

$$\psi = \sum_{j=1}^{\infty} c_j \varphi_j, \tag{7.47}$$

and let us put this expression into Eq. (7.46). We obtain

$$(H_F + V_P)\psi = (H_F + V_P) \sum_j c_j \varphi_j$$

$$= \sum_{j=1}^{\infty} c_j H_F \varphi_j + \sum_{j=1}^{\infty} \sum_{k=1}^{N} c_j (\varepsilon_v - \varepsilon_k) |\varphi_k\rangle \langle \varphi_k | \varphi_j \rangle$$

$$= \varepsilon \sum_{j=1}^{\infty} c_j \varphi_j. \tag{7.48}$$

Multiply from the left by φ_i^* and integrate. First, let φ_i be a core state. Then we get

$$c_i \varepsilon_i + \sum_{j=1}^{\infty} \sum_{k=1}^{N} c_j (\varepsilon_v - \varepsilon_k) \langle \varphi_i | \varphi_k \rangle \langle \varphi_k | \varphi_j \rangle$$

$$= c_i \varepsilon_i + c_i (\varepsilon_v - \varepsilon_i) = \varepsilon c_i, \tag{7.49}$$

from which we get

$$\varepsilon = \varepsilon_v. \tag{7.50}$$

Next, let φ_i be a valence state. Then, in the second term of Eq. (7.49), the term $\langle \varphi_i | \varphi_k \rangle$ will be zero because φ_k is a core state. Therefore, we obtain

$$c_i \varepsilon_i \equiv c_v \varepsilon_v = \varepsilon c_v, \qquad (7.51)$$

where we have changed the notation from i to v. We get again that

$$\varepsilon = \varepsilon_v. \qquad (7.52)$$

Thus, as we see from this argument, the eigenvalue of the pseudopotential equation, Eq. (7.46), is always a valence-level energy parameter. In contrast to the HF Hamiltonian, H_F, which describes the whole spectrum, core and valence, the pseudopotential Hamiltonian $H_F + V_P$ generates only the valence spectrum. The core spectrum does not exist for this operator.[†] This is what we have described qualitatively at the end of Sec. 7.2/A.

C. The Pseudopotential Method of Weeks and Rice

In the preceding section, we have formulated the pseudopotential formalism for atoms with one valence electron outside complete groups. Mathematically, this meant the pseudopotential transformation of an HF equation from which the Lagrangian multipliers had been eliminated. There are situations, however, in HF theory in which the Lagrangians cannot be eliminated. Such a situation occurs, for example, when we consider an atom with several valence electrons in an incomplete group. As we have seen in Sec. 3.2, in this case, the Lagrangians cannot be eliminated from the HF equations of the valence electrons. Furthermore, in this case, the core orbitals are not the eigenfunctions of the valence-electron Hamiltonian, therefore, we cannot use Eq. (7.25), which leads to the final simple form of the pseudopotential given by Eq. (7.26).

To cover this problem as well as other cases, Weeks and Rice[70] developed a pseudopotential formalism that contains only a minimum amount of initial assumptions and for this reason has a wide applicability. The formalism can be applied to one-electron problems as well as many-electron situations. Here we consider only the one-electron formulation. Let H be a one-electron Hamiltonian of which we assume that it is Hermitian but otherwise arbitrary. We assume that this Hamiltonian represents a valence electron. Let the atomic core be represented by the orbitals $\varphi_1, \ldots, \varphi_N$. We assume that these orbitals are orthogonal and normalized but otherwise arbitrary. We do not assume any connection between H and the core orbitals.

Let the eigenfunction and eigenvalue of the valence electron be φ_v and ε_v, respectively. We assume that this is a conventional problem, that is, the

[†]Under certain circumstances, pseudopotential Hamiltonians might have core solutions, but these can easily be eliminated from the calculations [Szasz, 32].

eigenfunction φ_v of the valence electron must be orthogonal to the core orbitals. In other words, φ_v is subject to the condition

$$\langle \varphi_i | \varphi_v \rangle = 0 \quad \text{(for all } v \text{ and for } i = 1, 2, \ldots, N), \quad (7.53)$$

and it is normalized.

We assume also that the energy of the valence electron is given by

$$\varepsilon_v = \langle \varphi_v | H | \varphi_v \rangle. \quad (7.54)$$

Variation of this expression with respect to φ_v gives the Schroedinger equation for the valence electron. The variation must be carried out under the subsidiary conditions given by Eq. (7.53) and under the normalization requirement. We obtain, by carrying out the variation,

$$H\varphi_v = \varepsilon_v \varphi_v + \sum_{i=1}^{N} \lambda_{iv} \varphi_i, \quad (7.55)$$

where the λ_{iv} are Lagrangian multipliers. Using the orthonormality of the core orbitals, we get

$$\lambda_{iv} = \langle \varphi_i | H | \varphi_v \rangle. \quad (7.56)$$

Let us recall that in Eq. (7.20), we introduced the orthogonality projection operator Ω with the definition

$$\Omega = \sum_{i=1}^{N} |\varphi_i\rangle\langle\varphi_i|. \quad (7.57)$$

Let

$$\Pi = 1 - \Omega. \quad (7.58)$$

It is easy to show that the following relationships are satisfied:

$$\Omega\varphi_k = \varphi_k \quad (k = 1, \ldots, N); \quad (7.59a)$$
$$\Omega\varphi_v = 0 \quad (v = N + 1, \ldots); \quad (7.59b)$$
$$\Pi\varphi_k = 0 \quad (k = 1, \ldots, N); \quad (7.59c)$$
$$\Pi\varphi_v = \varphi_v \quad (v = N + 1, \ldots). \quad (7.59d)$$

In these equations, φ_k is one of the core orbitals and φ_v is any orbital orthogonal to the core orbitals.

Using the orthogonality projection operators, we can rewrite Eq. (7.55) as follows:

$$H\varphi_v = \varepsilon_v \varphi_v + \Omega H\varphi_v, \quad (7.60)$$

or

$$(1 - \Omega)H\varphi_v = \Pi H\varphi_v = \varepsilon_v\varphi_v, \qquad (7.61)$$

from which we see that the valence orbitals are eigenfunctions of the non-Hermitian operator

$$H' = \Pi H, \qquad (7.62)$$

and we can write

$$H'\varphi_v = \varepsilon_v\varphi_v. \qquad (7.63)$$

H' is non-Hermitian because, in general, the operators Π and H, which are Hermitian individually, will not commute. This is the consequence of not having assumed any connection between the orbitals from which Π is built and the Hamiltonian operator H.

Let us verify now that the solutions of Eq. (7.63) will be orthogonal to the core orbitals. Multiply Eq. (7.63) from the left by φ_k $(k = 1,\ldots,N)$, and integrate. We get

$$\langle\varphi_k|H'|\varphi_v\rangle = \varepsilon_v\langle\varphi_k|\varphi_v\rangle. \qquad (7.64)$$

Using the Hermitian property of Π, we get

$$
\begin{aligned}
\langle\varphi_k|H'|\varphi_v\rangle &= \langle\Pi\varphi_k|H|\varphi_v\rangle \\
&= \langle(1 - \Omega)\varphi_k|H|\varphi_v\rangle = \langle(\varphi_k - \varphi_k)|H|\varphi_v\rangle \\
&= 0, \qquad (7.65)
\end{aligned}
$$

where we have used Eq. (7.59a). Using Eq. (7.65) in Eq. (7.64), we obtain

$$\langle\varphi_k|\varphi_v\rangle = 0, \qquad (7.66)$$

that is, the eigenfunctions of Eq. (7.63) will be orthogonal to the core orbitals.

Next, we show that the eigenfunctions of Eq. (7.63) belonging to different eigenvalues will be orthogonal despite of H' not being Hermitian. Let us consider Eq. (7.63) and let us consider

$$H'\varphi_{v'} = \varepsilon_{v'}\varphi_{v'} \qquad (v' \neq v). \qquad (7.67)$$

Multiply Eq. (7.63) from the left by φ_v^*, Eq. (7.67) by $\varphi_{v'}^*$, and integrate. Subtracting Eq. (7.63) from Eq. (7.67), we obtain

$$
\begin{aligned}
\langle\varphi_v|H'|\varphi_{v'}\rangle &- \langle\varphi_{v'}|H'|\varphi_v\rangle \\
&= (\varepsilon_{v'} - \varepsilon_v)\langle\varphi_{v'}|\varphi_v\rangle. \qquad (7.68)
\end{aligned}
$$

Using again the Hermitian property of Π, we obtain

$$\langle \varphi_v | \Pi H | \varphi_{v'} \rangle - \langle \varphi_{v'} | \Pi H | \varphi_v \rangle$$

$$= \langle \Pi \varphi_v | H | \varphi_{v'} \rangle - \langle \Pi \varphi_{v'} | H | \varphi_v \rangle$$

$$= \langle \varphi_v | H | \varphi_{v'} \rangle - \langle \varphi_{v'} | H | \varphi_v \rangle = 0. \qquad (7.69)$$

In the last line, we used the Hermitian character of H. We also used Eq. (7.59d), which is certainly permissible since the eigenfunctions φ_v are orthogonal to the core orbitals. Using the last equation, we get from Eq. (7.68)

$$\langle \varphi_{v'} | \varphi_v \rangle = 0 \qquad (v' \neq v). \qquad (7.70)$$

Now we transform Eq. (7.63) into a pseudopotential equation. To achieve this transformation, we put

$$\varphi_v = (1 - \Omega) \psi_v, \qquad (7.71)$$

where ψ_v is the pseudo-orbital. This is, of course, the same transformation as in Eq. (7.16), which we used in Eq. (7.15). (The valence orbital is not normalized here.) Substituting this expression into Eq. (7.63), we obtain

$$H' \varphi_v = H'(1 - \Omega) \psi_v = \varepsilon_v (1 - \Omega) \psi_v, \qquad (7.72)$$

or

$$H'(1 - \Omega) \psi_v = (1 - \Omega) H (1 - \Omega) \psi_v$$

$$= \varepsilon_v (1 - \Omega) \psi_v. \qquad (7.73)$$

From this, we get the pseudopotential equation

$$(H + V_P) \psi_v = \varepsilon_v \psi_v, \qquad (7.74)$$

where V_P is the Weeks–Rice pseudopotential[70]:

$$V_P = -\Omega H - H\Omega + \Omega H \Omega + \varepsilon_v \Omega. \qquad (7.75)$$

This can be written in the alternative form

$$V_P = -\Omega(H - \varepsilon_v) - \Pi H \Omega. \qquad (7.76)$$

We now proceed to analyze the properties of Eq. (7.74). First, we show that both the core orbitals, φ_k $(k = 1, \ldots, N)$, and the valence orbital, φ_v, are solutions of Eq. (7.74) with the valence eigenvalue ε_v which occurs in

Eq. (7.75). First, consider φ_k. We get

$$
\begin{aligned}
V_P\varphi_k &= -\Omega(H - \varepsilon_v)\varphi_k - \Pi H(\Omega\varphi_k) \\
&= -\Omega(H - \varepsilon_v)\varphi_k - \Pi H\varphi_k \\
&= -\Omega(H - \varepsilon_v)\varphi_k - (1 - \Omega)H\varphi_k \\
&= \varepsilon_v(\Omega\varphi_k) - H\varphi_k = (\varepsilon_v - H)\varphi_k.
\end{aligned} \tag{7.77}
$$

Here we have used twice the relationship $\Omega\varphi_k = \varphi_k$. Next consider

$$
(H + V_P)\varphi_k = (H + \varepsilon_v - H)\varphi_k = \varepsilon_v\varphi_k. \tag{7.78}
$$

Now put φ_v into the pseudopotential equation. First, we get

$$
\begin{aligned}
V_P\varphi_v &= -\Omega(H - \varepsilon_v)\varphi_v - \Pi H(\Omega\varphi_v) \\
&= (\varepsilon_v - H)\varphi_v + \varepsilon_v(\Omega\varphi_v) \\
&= (\varepsilon_v - H)\varphi_v,
\end{aligned} \tag{7.79}
$$

where we have used Eqs. (7.60) and (7.59b). Using this equation, we get

$$
(H + V_P)\varphi_v = (H + \varepsilon_v - H)\varphi_v = \varepsilon_v\varphi_v. \tag{7.80}
$$

Therefore, because both φ_k and φ_v are eigenfunctions of the pseudopotential Hamiltonian, we can form the linear combination

$$
\psi_v = \varphi_v + \sum_{k=1}^{N} \alpha_k\varphi_k, \tag{7.81}
$$

where α_k is an arbitrary constant. From the preceding argument, we see that ψ_v in Eq. (7.81) will be a solution of Eq. (7.74) with the eigenvalue ε_v. The constants are arbitrary, therefore, we get an infinite number of POs, each satisfying Eq. (7.74) with the same eigenvalue ε_v. We note that Eq. (7.81) is the same as Eq. (7.71) because we can rearrange Eq. (7.81) and get

$$
\varphi_v = \psi_v - \sum_{k=1}^{N} \alpha_k\varphi_k = (1 - \Omega)\psi_v. \tag{7.82}
$$

We note also that while there are infinitely many pseudo-orbitals that satisfy the PP equation, each will reproduce the same uniquely defined valence orbital φ_v through the relationship given by Eq. (7.82). It is worthwhile to look at this argument for a moment. We have seen that there are infinitely many α_k's and, therefore, infinitely many ψ_v's that can be used in Eq. (7.81).

Let us select any one of these and let us put the ψ_v into Eq. (7.82). Then we get

$$(1 - \Omega)\psi_v = (1 - \Omega)\left(\varphi_v + \sum_{k=1}^{N} \alpha_k \varphi_k\right)$$

$$= \varphi_v + \sum_{k=1}^{N} \alpha_k \varphi_k - \sum_{k=1}^{N} \alpha_k (\Omega \varphi_k)$$

$$= \varphi_v + \sum_{k=1}^{N} \alpha_k \varphi_k - \sum_{k=1}^{N} \alpha_k \varphi_k = \varphi_v, \qquad (7.83)$$

which means that no matter which α_k's are used in the ψ_v, the result is always φ_v because the α_k's are subtracted out by the projection operator.

This property of the pseudopotentials, which we call "pseudopotential indeterminacy," was discovered by Cohen and Heine.[71] They pointed out that the indeterminacy can be eliminated and the PPs made uniquely defined by imposing some additional mathematical requirement on the pseudopotential Hamiltonian.

In order to bring out the indeterminacy of the pseudopotentials, we show now[53] that V_P can be written in the form

$$V_P = \sum_{i=1}^{N} |\varphi_i\rangle\langle F_i| - \Pi H \Omega, \qquad (7.84)$$

where the F_i functions are arbitrary. We recover Eq. (7.76) by putting

$$\langle F_i| = -\langle\varphi_i|(H - \varepsilon_v). \qquad (7.85)$$

We now prove that V_P is the correct pseudopotential regardless of the choice of the F_i. The proof goes as follows. Let ψ_v be a solution of

$$(H + V_P)\psi_v = \varepsilon_v \psi_v. \qquad (7.86)$$

Then, we state that $\varphi_v = \Pi\psi_v$ is the solution of Eq. (7.63) regardless of the F_i's that were used.

Multiply Eq. (7.86) from the left by Π:

$$(\Pi H + \Pi V_P)\psi_v = [\Pi H + \Pi(-\Omega H - H\Omega + \Omega H\Omega + \varepsilon_v\Omega)]\psi_v$$

$$= \varepsilon_v \Pi\psi_v. \qquad (7.87)$$

From the definitions, it follows that

$$\Pi\Omega = (1 - \Omega)\Omega = \Omega - \Omega^2 = \Omega - \Omega = 0. \qquad (7.88)$$

Thus, we get from Eq. (7.87)

$$(\Pi H + \Pi V_P)\psi_v = (\Pi H - \Pi H\Omega)\psi_v$$
$$= \Pi H(1 - \Omega)\psi_v = \Pi H\varphi_v = \varepsilon_v \Pi\psi_v. \qquad (7.89)$$

For the right side, we get $\varepsilon_v \varphi_v$, and so the equation becomes

$$\Pi H\varphi_v = H'\varphi_v = \varepsilon_v \varphi_v, \qquad (7.90)$$

which is Eq. (7.63), Q.E.D.

One way of determining the functions F_i is to impose the additional mathematical requirement that the PO should minimize the expectation value of the kinetic energy:

$$E^K = \langle \psi_v | t | \psi_v \rangle, \qquad (7.91)$$

where t is the kinetic energy operator, and ψ_v is normalized. This requirement leads to the following form for the F_i:

$$\langle F_i | = \langle \varphi_i | (\varepsilon_v - U - E^K). \qquad (7.92)$$

The pseudopotential chosen this way produces a very smooth pseudo-orbital. Here the operator U is the total potential in H, including the nuclear potential, that is,

$$H = t + U. \qquad (7.93)$$

For the PP, we get with Eq. (7.92)

$$V_P = \Omega(\varepsilon_v - U - E^K) - \Pi H\Omega. \qquad (7.94)$$

For the derivation of this expression, as well as for other choices for the F_i, we refer to the author's book [Szasz, 27, 39].

It is easy to prove that if the F_i's are properly chosen, the pseudo-orbitals, which are the solutions of Eq. (7.74), are uniquely determined; in other words, the proper choice of the F_i's removes the pseudopotential indeterminacy. Here we summarize the proof of this statement; the details can be found in the author's book [Szasz, 40]. The pseudo-orbital has the form

$$\psi_v = \varphi_v + \sum_{i=1}^{N} \alpha_i \varphi_i. \qquad (7.95)$$

In order to satisfy the PP equation, Eq. (7.74), the α_i must be the solutions of the following system of inhomogeneous linear equations:

$$\sum_{j=1}^{N} M_{ij}\alpha_j = N_i \qquad (i = 1, \dots, N). \tag{7.96}$$

Here the matrix components are defined as follows:

$$M_{ij} = \langle F_i | \varphi_j \rangle - \varepsilon_v \delta_{ij}, \tag{7.97}$$

and

$$N_i = -\langle F_i | \varphi_v \rangle. \tag{7.98}$$

The system given by Eq. (7.96) will have unique solutions if the system determinant is not zero:

$$\det M_{ij} \neq 0, \tag{7.99}$$

and if the inhomogeneity terms are not zero:

$$N_i \neq 0 \qquad (i = 1, \dots, N). \tag{7.100}$$

Clearly, these conditions impose restrictions on the choice of the F_i's. For example, one of the "forbidden" choices is

$$F_i = C_i \langle \varphi_i |, \tag{7.101}$$

where C_i is a constant. This choice leads to vanishing inhomogeneity terms according to Eq. (7.98).

Finally, we note that we recover from the Weeks–Rice theory, the Hartree–Fock-equivalent PP formalism by imposing on the Hamiltonian operator the condition that the core orbitals be the eigenfunctions of this operator. We put

$$H\varphi_i = \varepsilon_i \varphi_i. \tag{7.102}$$

It is easy to see that if this condition is imposed, the Hamiltonian and Ω will commute [Szasz, 271]:

$$H\Omega - \Omega H = 0, \tag{7.103}$$

and, of course,

$$H\Pi - \Pi H = 0. \tag{7.104}$$

Then, we get for the second term of the pseudopotential in Eq. (7.76),

$$\Pi H\Omega = H\Pi\Omega = H(1 - \Omega)\Omega = 0. \tag{7.105}$$

Thus, we get for the PP from Eq. (7.76)

$$V_P = -\Omega(H - \varepsilon_v) = -(H - \varepsilon_v)\Omega$$
$$= (\varepsilon_v - H)\Omega, \tag{7.106}$$

which is identical to the expression in Eq. (7.23). Taking into account Eq. (7.102), we arrive back at the expression given by Eq. (7.26).

D. Calculations with HF-Equivalent Pseudopotentials

We now review some pseudopotential calculations made with the formalism developed in Sec. 7.2/B. Because the formalism developed in that section is equivalent to the frozen-core HF model that we have discussed in Sec. 6.3, these calculations represent the test of both methods.

Pseudopotential calculations were made by Szasz and McGinn[72] for the ground states of the one-valence-electron atoms Li, Na, K, Rb, Be$^+$, Mg$^+$, Ca$^+$, Al^{2+}, Cu, and Zn$^+$. Calculations were made for the low-lying excited states of the same atoms by McGinn.[73] The calculations were made by obtaining exact solutions of Eq. (7.40) by numerical integration. In this equation, the potentials are defined by Eqs. (7.41) and (7.42). The core orbitals and orbital parameters were kept fixed in the calculations; they were taken from previous analytic HF calculations.[42] In the course of the calculations, the pseudo-orbitals and the orbital parameters of the valence electron were obtained. With these, the local pseudopotential, Eq. (7.42), and the localized HF potential, Eq. (7.41), were constructed. For reference purposes, we write down here some formulas. Equation (7.40) has the form

$$\left\{ -\frac{1}{2}\frac{d^2}{dr^2} + V_{\text{eff}}(r) \right\} P_{n_v l_v} = \varepsilon_{n_v l_v} P_{n_v l_v}(r), \tag{7.107}$$

where V_{eff} is the effective potential that is defined as

$$V_{\text{eff}} = \frac{l_v(l_v + 1)}{2r^2} - \frac{Z}{r} + \tilde{U}(r) + \tilde{V}_P(r). \tag{7.108}$$

In this formula, $\tilde{U}(r)$ is the localized HF potential given by Eqs. (7.38) and (7.41), and $\tilde{V}_P$ is the localized pseudopotential given by Eqs. (7.39) and (7.42). We referred to the "modified potential," which was defined as

$$V_M = -\frac{Z}{r} + \tilde{U}(r) + \tilde{V}_P(r), \tag{7.109}$$

that is, this is the localized HF potential modified by the addition of the localized pseudopotential. The term $l_v(l_v + 1)/2r^2$ is the azimuthal part of the kinetic energy.

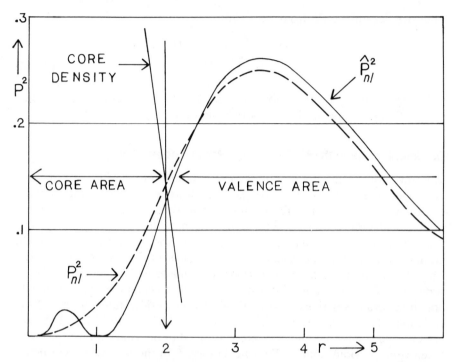

Fig. 7.3. The radial density of the valence electron of the Na atom. P_{nl}^2 is the pseudo-density and $\hat{P}_{nl}^2$ is the Hartree–Fock density.

First, the calculations yielded the pseudo-orbitals. The radial density, derived from a typical PO is shown in Fig. 7.3. In this figure, we have P_{nl}^2 for the $3s$ state of the Na atom. The solid line is the HF density[41] and the dashed line is the pseudo-density. Also, we have the line indicating the tail end of the core radial density in the diagram. The reason why this is a steep line instead of a slowly declining exponential is that is is drawn on the same scale as the valence densities; on such a scale, the core density, which is much larger than the valence density, appears as a rapidly ascending curve. The arrow indicates the point where the pseudo-density and the core density cross each other.

The diagram shows the character of the PO. In contrast to the HF radial density, which has one zero point at about $r = 1$ a.u., the pseudo-density is a nodeless Slater-type function. Outside of the core, the HF density and the pseudo-density run side by side except the pseudo-density is a little lower; this is coming from the normalization. Inside of the core, the pseudo-density is the average of the HF density. It is generally true that in the ground state of the valence electron the pseudo-orbitals are nodeless Slater-type functions.

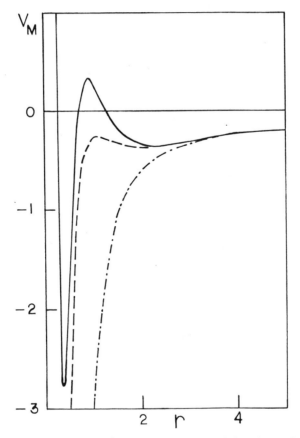

Fig. 7.4. The modified potentials (Hartree–Fock potential and pseudopotential) for the valence electron of the Na atom. Solid line: the s potential. Dashed line: the p potential. Dot–dash line: the Hartree–Fock potential.

Next, in Figs. 7.4 and 7.5, we have the modified potentials. These functions are defined by Eq. (7.109). In Fig. 7.4, we have the modified potentials for the $3s$ and $3p$ states of the Na atom plus the HF potential $(-Z/r + \bar{U})$ for the $3s$ state. In Fig. 7.5, we have the s-, p-, and d-modified potentials plus the HF potential in the ground state for the Cu atom. As we see, the s potential of the Na and the s and p potentials of the Cu show the same general structure that we have observed in the Hellmann potential, Fig. 7.1 (except for the detail that the Na s potential has a second smaller barrier outside of the core). By this we mean that the basic structure of these potentials is the barrier/well/Coulombic sequence.

Some of the modified potentials do not show this sequence. The p potential of the Na and the d potential of the Cu do not have the positive

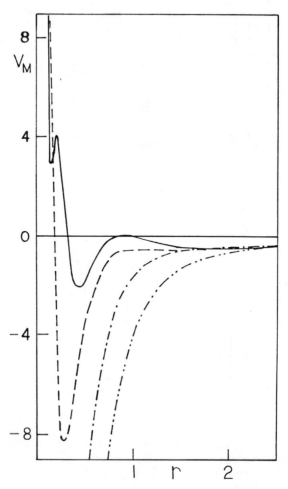

Fig. 7.5. The modified potentials (Hartree–Fock potential and pseudopotential) for the (single) valence electron of the Cu atom. Solid line: the s potential. Dashed line: the p potential. Dot–dash line: the d potential. Two dots–dash line: the Hartree–Fock potential.

barrier at the atomic core. The reason for this is that, as can be shown, the pseudopotential represents the radial part of the valence electron kinetic energy [Szasz, 53]. The azimuthal part is given by the term $l_v(l_v + 1)/2r^2$. If we add this function to the modified potential, that is, if we consider the effective potential, Eq. (7.108), that potential will always have a high potential barrier at the atomic core. Earlier, we said that the pseudopotential represents the Pauli exclusion principle for a nonorthogonal orbital. We must make our language a little more precise: the pseudopotential plus the term

TABLE 7.1. The Optical Energy Levels of the Atoms Li, Na, and K

Atom	n	S states Calc.	Obs.	Δ	P states Calc.	Obs.	Δ	D states Calc.	Obs.	Δ
Li	2	0.19657	0.19816	0.81	0.12863	0.13025	1.24			
	3	0.07384	0.07419	0.47	0.05677	0.05724	0.82	0.05556	0.05561	0.09
	4	0.03849	0.03862	0.37	0.03178	0.03198	0.63	0.03125	0.03128	0.10
	5	0.02358	0.02364	0.25	0.02028	0.02038	0.49	0.02000	0.02001	0.05
	6	0.01591	0.01595	0.25	0.01405	0.01411	0.43	0.01389	0.01390	0.07
	7	0.01146	0.01148	0.17	0.01031	0.01034	0.29	0.01020	0.01021	0.10
Na	3	0.18108	0.18886	4.12	0.10940	0.11156	1.94	0.05567	0.05594	0.48
	4	0.07002	0.07158	2.18	0.05031	0.05094	1.24	0.03132	0.03144	0.38
	5	0.03701	0.03759	1.54	0.02893	0.02920	0.92	0.02004	0.02011	0.35
	6	0.02286	0.02313	1.16	0.01878	0.01892	0.74	0.01391	0.01395	0.29
	7	0.01551	0.01566	0.96	0.01317	0.01325	0.60	0.01022	0.01025	0.29
	8	0.01121	0.01131	0.88	0.00975	0.00980	0.51	0.00782	0.00782	0.26
K	3							0.05812	0.06139	5.33
	4	0.14669	0.15952	8.04	0.09547	0.10022	4.74	0.03278	0.03468	5.22
	5	0.06090	0.06371	4.41	0.04554	0.04693	2.96	0.02096	0.02198	4.64
	6	0.03336	0.03444	3.14	0.02676	0.02737	2.23	0.01449	0.01510	4.04
	7	0.02105	0.02158	2.46	0.01762	0.01794	1.78	0.01060	0.01099	3.55
	8	0.01448	0.01478	2.03	0.01248	0.01267	1.50	0.00808	0.00834	3.12
	9	0.01057	0.01076	1.77	0.00931	0.00943	1.27			

$l_v(l_v + 1)/2r^2$ together represent the Pauli principle. The pseudopotential is the radial part of the kinetic energy and the $l_v(l_v + 1)/2r^2$ is the azimuthal part. If this term would be added to the p potential of the Na and to the d potential of the Cu, the resulting effective potentials would show the barrier/well/Coulombic sequence.

Review of the computed potentials revealed that the modified potentials are strongly l_v-dependent, but, for a fixed l_v, they show little dependence on n_v. This is in agreement with our conclusions reached earlier by the analysis of the mathematical structure of the potentials.

In Tables 7.1–7.4, we have the computed and empirical energy levels of the atoms considered. The tables contain the ground states and the first few excited states. The energies are in atomic units and Δ is the deviation of the computed values from the experimental[†] in percents. Here we recall that we are talking about HF-equivalent pseudopotential calculations, that is, these calculations represent a test of the HF method in the frozen-core approximation.

[†]The tables contain the simple average for the empirical values of the doublets.

TABLE 7.2. The Optical Energy Levels of the Rb, Cu, and Be$^+$ Atoms

Atom	n	S states Calc.	Obs.	Δ	P states Calc.	Obs.	Δ	D states Calc.	Obs.	Δ
Rb	4							0.06007	0.06532	8.04
	5	0.13794	0.15351	10.14	0.09026	0.09565	5.64	0.03398	0.03640	6.65
	6	0.05832	0.06177	5.59	0.04369	0.04528	3.51	0.02157	0.02279	5.35
	7	0.03228	0.03362	3.99	0.02590	0.02660	2.63	0.01485	0.01554	4.44
	8	0.02050	0.02116	3.12	0.01716	0.01753	2.11	0.01083	0.01125	3.73
	9	0.01417	0.01454	2.54	0.01220	0.01242	1.77	0.00824	0.00852	3.29
	10	0.01038	0.01061	2.17	0.00912	0.00927	1.62			
Cu	4	0.23389	0.28394	17.63	0.12226	0.14424	15.24	0.05512	0.05640	2.27
	5	0.08007	0.08739	8.38	0.05419	0.05893	8.04	0.03100	0.03157	1.81
	6	0.04066	0.04314	5.75	0.03059	0.03377	9.42	0.01986	0.02015	1.44
	7	0.02458	0.02572	4.43	0.01965	0.02111	6.92	0.01380	0.01397	1.22
	8	0.01646	0.01708	3.63	0.01368	0.01375	0.51	0.01015	0.01026	1.07
	9	0.01179	0.01216	3.04	0.01007	0.01020	1.27	0.00777	0.00785	1.02
Be$^+$	2	0.66484	0.66928	0.66	0.51941	0.52378	0.83			
	3	0.26647	0.26725	0.29	0.22841	0.22958	0.51	0.22229	0.22249	0.09
	4	0.14288	0.14315	0.19	0.12766	0.12814	0.37	0.12504	0.12512	0.06
	5	0.08894	0.08906	0.13	0.08137	0.08162	0.31	0.08002	0.08006	0.05
	6	0.06065	0.06071	0.10	0.05635	0.05648	0.23	0.05557	0.05559	0.04
	7	0.04399	0.04400	0.02	0.04132	0.04140	0.19	0.04083	0.04084	0.02

TABLE 7.3. The Optical Energy Levels of the Mg$^+$ and Ca$^+$ Atoms

Atom	n	S states Calc.	Obs.	Δ	P states Calc.	Obs.	Δ	D states Calc.	Obs.	Δ
Mg$^+$	3	0.54851	0.55255	0.73	0.38338	0.38981	1.65	0.22482	0.22680	0.87
	4	0.23334	0.23448	0.49	0.18322	0.18513	1.03	0.12648	0.12738	0.71
	5	0.12936	0.12977	0.32	0.10763	0.10847	0.77	0.08085	0.08132	0.58
	6	0.08215	0.08233	0.22	0.07082	0.07127	0.63	0.05607	0.05635	0.50
	7	0.05678	0.05688	0.18	0.05014				0.04115	0.04132	0.41
	8	0.04158	0.04164	0.14	0.03736					
Ca$^+$	3							0.33249	0.37393	11.08
	4	0.41386	0.43627	5.14	0.30994	0.32098	3.44	0.16908	0.17724	4.60
	5	0.19234	0.19857	3.14	0.15673	0.16027	2.21	0.10149	0.10490	3.25
	6	0.11159	0.11423	2.31	0.09519	0.09680	1.66	0.06759	0.06936	2.55
	7	0.07291	0.07426	1.82	0.06402			0.04823	0.04926	2.09
	8	0.05137	0.05215	1.50	0.04602			0.03613	0.03678	1.77
	9	0.03814			0.03468					

TABLE 7.4. The Optical Energy Levels of the Zn$^+$ and Al^{++} Atoms

Atom	n	S states Calc.	S states Obs.	Δ	P states Calc.	P states Obs.	Δ	D states Calc.	D states Obs.	Δ
Zn$^+$	4	0.60984	0.66018	7.63	0.40600	0.43729	7.16	0.21372	0.21851	2.19
	5	0.24628	0.25722	4.25	0.19015	0.19776	3.85	0.12063	0.12261	1.61
	6	0.13425	0.13848	3.05	0.11062	0.11376	2.76	0.07759	0.07862	1.31
	7	0.08454	0.08662	2.40	0.07239	0.07500	3.48	0.05411	0.05471	1.10
	8	0.05812	0.05930	1.99	0.05106	0.05051	-1.09	0.03989	0.04026	0.92
	9	0.04241	0.04314	1.69	0.03795			0.03062	0.03087	0.81
Al^{++}	3	1.04240	1.04549	0.30	0.79002	0.80035	1.29	0.51201	0.51714	0.99
	4	0.46907	0.47064	0.33	0.38760	0.39086	0.83	0.28785	0.29010	0.78
	5	0.26733	0.26800	0.25	0.23090	0.23240	0.65	0.18371	0.18489	0.64
	6	0.17267	0.17303	0.21	0.15329	0.15411	0.53	0.12725	0.12794	0.54
	7	0.12071	0.12097	0.21	0.10918	0.10978	0.55	0.09329	0.09374	0.48
	8	0.08912			0.08171					

First, we observe that by comparing our results with the ionization potential tables in Sec. 5.4/A, we are able to gauge the accuracy of the frozen-core approximation. In Table 5.5, we have the first ionization potentials for neutral atoms. In the frozen-core approximation, the orbital parameter is equal to the binding energy. Thus, we get an idea on the accuracy of frozen core by comparing the deviations given in Table 5.5 with the deviations given in Tables 7.1–7.4. The former are the "exact" HF values; the latter are the frozen-core values. We get the following numbers: Li ($2s$), $\Delta = 0$ (0.81); Na ($3s$), $\Delta = 3.7$ (4.12); K ($4s$), $\Delta = 7.85$ (8.04); Rb ($5s$), $\Delta = 10.9$ (10.14); Cu ($4s$), $\Delta = 21.8$ (17.6). Here the numbers mean percents and the values in parentheses are the frozen-core values.

Of course, this is a very small sample. Nevertheless, treating these numbers with due caution, we may say that the frozen core appears to be an accurate approximation except for the very small core of the Li and for the "loose" $(3d)^{10}$ group of the Cu. That the frozen core does not work well for very small cores is plausible. It may also be somewhat inaccurate in other cases where the outermost group of the core has a "loose" structure.

Looking at the calculated energy levels, we see that generally the results are reasonable, although there are considerable deviations from the empirical values. In each series, the deviation is the largest in the state with the lowest n_v. As we go to higher n_v values, the deviations diminish but only slowly; they are still not negligibly small at the higher excited states.

In Fig. 7.6, we plotted the Δ values in percents versus the principal quantum numbers for the s, p, and d series of the Rb atom. As we go to higher n_v values, the valence electron moves further out from the core; thus, by plotting n_v on the abscissa, we are demonstrating how the deviations

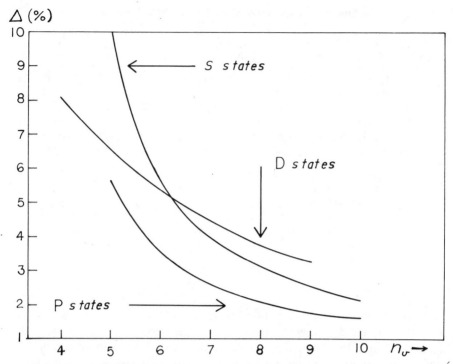

Fig. 7.6. The deviation of the computed energies from the experimental versus the principal quantum number for the valence electron of the Rb atom.

decrease as the distance of the valence electron from the core increases. The diagram shows that at first, the deviation decreases rapidly at low n_v values, that is, it decreases rapidly at the vicinity of the core; however, the decrease is very slow for larger n_v values. If we attribute the deviation to the core/valence correlation effects, then we can say that this effect is large, in each series, in the state with the lowest n_v, decreases rapidly for the next few n_v values and it is a slowly declining function for higher n_v values.

Finally, the integrals necessary for the calculation of transition probabilities and oscillator strengths were computed for those transitions that are permitted by the selection rules. The quantities computed are defined by Eqs. (6.152) and (6.153). The valence electron orbitals in these formulas are the orthogonalized HF orbitals; in the pseudopotential calculations, these integrals were computed by orthogonalizing the pseudo-orbitals to the core functions. The interested reader can look up these values in the original publications. Here we quote only the results for the Kuhn–Thomas sum rule, Eq. (6.155). Adding up the transitions $3s \rightarrow np$ for $2, \ldots, 8$, the results were Na $= 0.9435$, Mg$^+ = 0.8992$, and Al$^{++} = 0.8439$. These results are quite good considering that only seven transitions were added.

We sum up. Hartree–Fock-equivalent pseudopotential calculations yielded the following results:

1. In the ground state of one-valence-electron atoms, the pseudo-orbitals are nodeless Slater-type functions. The valence-electron radial density, $P^2_{n_v l_v}$, runs close to the HF density outside the core and averages the HF density inside the core.

2. The structure of the modified potentials, defined by Eq. (7.109), generally shows the barrier/well/Coulombic sequence. If the modified potential does not show this structure, the effective potential, Eq. (7.108), will be a function showing the barrier/well/Coulombic sequence.

3. The modified potentials depend strongly on l_v, but only very slightly on n_v. Thus, in a very good approximation, the modified potentials are the same for states with common l_v but different n_v.

4. The deviation of the calculated energy values from the empirical is fairly large in each series for the state with the lowest n_v and declines slowly as the valence electron moves away from the atomic core into higher excited states.

7.3. CORE / VALENCE CORRELATION EFFECTS

A. Introduction

We have seen in the preceding section that the HF-equivalent pseudopotential calculations yielded rather poor results for the energy levels of atoms with one valence electron. Therefore, it is necessary, at this point, to expand the framework of the discussions and to step outside of the boundaries of the HF model. We want to do this by introducing correlation effects. As we have discussed in Sec. 2.1, the basic idea of the HF model is that the electron feels the presence of other electrons only through an average potential. In reality, the eigenfunction of the exact Hamiltonian will be such that the position of each electron will depend on the momentary positions of all other electrons. Consequently, the real potential felt by the electron will not be the average potential of the HF model, but it will depend on the position of all other electrons.

In this section, we investigate the electron correlation between a single valence electron and the atomic core in the field of which the valence electron is moving. There are two correlation effects to be considered. First, the wave function of the core will depend on the position of the valence electron. This dependence will make the core-density distribution and the core potential different from the corresponding HF functions; this difference will be called the *polarization potential*. Second, we want to take into account core–core correlations, that is, we want to take into account the fact that the correlation between the core electrons can make the potential of the core, in which the valence electron is moving, different from the HF potential. We show that this effect can be built into the electrostatic, exchange, and polarization potentials of the core.

At the beginning of this chapter, we listed four reasons for the development of effective Hamiltonians. In connection with the electron correlation, we want to introduce a fifth reason. In many *molecular* and *solid-state* studies, the systems under consideration—molecules and solids—retain, to a certain extent, atomic structures. Primarily, we talk about systems in which the *cores* of the constituent atoms retain the structure that they have in the free atoms and only the wave functions of the valence electrons acquire molecular and crystalline characteristics. In such problems, it is necessary to have a very accurate description of the core-valence interaction. If the atomic core is the same in a molecule or solid as it is in a free atom, then the effective potential of the core electrons will also be the same in the molecule or solid as it is in the free atom. In order to study the structure of such a molecule or solid, an effective Hamiltonian for the valence electrons of the free atom must be first set up. Thus, as we see, the first step in the study of a molecular or solid-state problem might be the construction of an accurate effective Hamiltonian for the corresponding free atoms. The motivation for the construction of effective atomic Hamiltonians comes, in such cases, from molecular and solid-state physics, that is, from outside of atomic physics proper. This is the fifth reason for the construction of effective atomic Hamiltonians.

Here we want to emphasize the need for very accurate effective potentials and effective Hamiltonians. At the present stage of the development of quantum physics, the molecular and solid-state scientists require effective Hamiltonians of the highest accuracy. It is not a matter of indifference whether, for example, correlation effects are represented in an effective potential with high accuracy or only qualitatively. It is known that accurate molecular and solid-state calculations can be done only with highly accurate atomic effective Hamiltonians. We should keep in mind this point in the forthcoming discussions.

B. Effective Hamiltonian with Core Polarization

There are two theoretical studies for the construction of effective Hamiltonians, which include core polarization. The first is the method of Callaway;[74] the second is the derivation of Bottcher and Dolgarno.[75] We present both in some detail here followed by a discussion in which we compare the two methods.

We start with Callaway's method. In the HF approximation, we have represented an $(N + 1)$-electron atom by the wave function given by

Eq. (7.10):

$$\psi = [(N + 1)!]^{-1/2} \det[\varphi_1 \cdots \varphi_N \varphi_v], \qquad (7.110)$$

where $\varphi_1 \cdots \varphi_N$ were the core orbitals, and φ_v was the valence wave function. We introduce now correlation by replacing the one-electron core orbitals with the antisymmetric N-electron core function Φ_C, which is not necessarily an antisymmetrized product of one-electron orbitals. We put

$$\Phi_T = \tilde{A}\{\Phi_C(1,\ldots,N)\psi_v(N+1)\}, \qquad (7.111)$$

where $\tilde{A}$ is a partial antisymmetrizer operator, and ψ_v is the valence wave function. We assume that both Φ_C and ψ_v are spatial functions, that is, we assume that spin properties need not be considered in this derivation. Let $\mathbf{r}_C$ mean the coordinates of all core electrons and let $\mathbf{r}_v$ be the position of the valence electron. We introduce core/valence correlation by putting

$$\Phi_C = \Phi_C(\mathbf{r}_C,\mathbf{r}_v), \qquad (7.112)$$

that is, we assume that Φ_C will depend on the valence electron position $\mathbf{r}_v$. We assume that the density distribution of the core will depend on the position of the valence electron.

The Hamiltonian of the system is given as

$$H = H_C(\mathbf{r}_C) + H_v(\mathbf{r}_v) + H_i(\mathbf{r}_C,\mathbf{r}_v), \qquad (7.113)$$

where H_C is the Hamiltonian of the core,

$$H_C(\mathbf{r}_C) = \sum_{i=1}^{N} (t_i + g_i) + \frac{1}{2} \sum_{i,j=1}^{N} \frac{1}{r_{ij}}, \qquad (7.114)$$

H_v is the one-electron part of the valence Hamiltonian,

$$H_v(r_v) = t_v + g_v, \qquad (7.115)$$

and H_i is the valence/core interaction,

$$H_i(\mathbf{r}_C,\mathbf{r}_v) = \sum_{j=1}^{N} \frac{1}{r_{jv}}. \qquad (7.116)$$

t is the kinetic energy and g is the potential of the nucleus.

We now introduce the approximate wave function for the whole atom:

$$\Phi_T = \Phi_C(\mathbf{r}_C,\mathbf{r}_v)\psi_v(\mathbf{r}_v) \qquad (7.117)$$

In this function, we omit the antisymmetrization between core and valence. We assume that ψ_v is normalized and Φ_C is normalized for all values of $\mathbf{r}_v$:

$$\int \Phi_C^*(\mathbf{r}_C, \mathbf{r}_v) \Phi_C(\mathbf{r}_C, \mathbf{r}_v) \, dv_C = 1, \qquad (7.118)$$

where dv_C is integration over all core-electron coordinates.

The Schroedinger equation for Φ_T is as follows:

$$H\Phi_T = (H_C + H_v + H_i)(\Phi_C \psi_v) = E_T(\Phi_C \psi_v), \qquad (7.119)$$

where E_T is the total energy. Multiplying the equation from the left by Φ_C^* and integrating over the core, we get

$$\left[\int \Phi_C^*(H_C + H_i)\Phi_C \, dv_C \right] \psi_v + \int \Phi_C^* H_v(\Phi_C \psi_v) \, dv_C = E_T \psi_v, \quad (7.120)$$

where we have used Eq. (7.118).

Let us assume that Φ_C satisfies the equation

$$(H_C + H_i)\Phi_C = E_C(\mathbf{r}_v)\Phi_C. \qquad (7.121)$$

Here E_C is the energy of the core, which will depend on $\mathbf{r}_v$. Substituting Eq. (7.121) into (7.120), we get

$$\int \Phi_C^* H_v(\Phi_C \psi_v) \, dv_C = (E_T - E_C(\mathbf{r}_v))\psi_v. \qquad (7.122)$$

Taking into account the form of H_v and the normalization of Φ_C, we get

$$\int \Phi_C^* H_v(\Phi_C \psi_v) \, dv_C = (t_v + g_v)\psi_v + \left(\int \Phi_C^* t_v \Phi_C \, dv_C \right)\psi_v, \quad (7.123)$$

and substituting this into Eq. (7.122), we obtain

$$\left(t_v + g_v + \int \Phi_C^* t_v \Phi_C \, dv_C + E_C(\mathbf{r}_v) - E_T \right)\psi_v = 0, \qquad (7.124)$$

which is the Schroedinger equation for ψ_v. As we see, the equation contains an effective potential. The determination of this potential boils down to the determination of $E_C(\mathbf{r}_v)$, that is, to solving Eq. (7.121).

For the solution of Eq. (7.121), we utilize the HF perturbation theory developed in Sec. 2.5. First, we put Φ_C into the form

$$\Phi_C(\mathbf{r}_C, \mathbf{r}_v) = \det[u_1(\mathbf{r}_1, \mathbf{r}_v)u_2(\mathbf{r}_2, \mathbf{r}_v) \cdots u_N(\mathbf{r}_N, \mathbf{r}_v)]. \qquad (7.125)$$

We form the expectation value of $H_C + H_i$ with this wave function and vary the one-electron orbitals u_j. This procedure yields the HF equations in the presence of a perturbation that is the operator H_i. We get

$$
\left(t_1 + g_1 + V(\mathbf{r}_1, \mathbf{r}_v) - A(\mathbf{r}, \mathbf{r}_v) + \frac{1}{r_{1v}} \right) u_j(\mathbf{r}_1, \mathbf{r}_v) = \varepsilon_j(r_v) u_j(\mathbf{r}_1, \mathbf{r}_v)
$$

$$(j = 1, \ldots, N). \quad (7.126)$$

Here V and A are the electrostatic exchange potentials, respectively:

$$
V(\mathbf{r}_1, \mathbf{r}_v) = \sum_{j=1}^{N} \int \frac{|u_j(\mathbf{r}_2, \mathbf{r}_v)|^2}{r_{12}} \, dv_2, \quad (7.127)
$$

and

$$
A(\mathbf{r}_1, \mathbf{r}_v) f(\mathbf{r}_1, \mathbf{r}_v) = \sum_{\substack{j=1 \\ \text{(parallel spins)}}}^{N/2} \int \frac{u_j(\mathbf{r}_1, \mathbf{r}_v) u_j^*(\mathbf{r}_2, \mathbf{r}_v) f(\mathbf{r}_2, \mathbf{r}_v) \, dv_2}{r_{12}} \quad (7.128)
$$

The HF equations in Eq. (7.126) are the same as in Eq. (2.141) with $V(1) = 1/r_{1v}$. Treating this interaction as a perturbation, we obtain the equations developed in Sec. 2.5. We first use Eqs. (2.146)–(2.147):

$$
u_j = u_j^{(0)} + \lambda u_j^{(1)} + \lambda^2 u_j^{(2)} + \cdots, \quad (7.129)
$$

$$
\varepsilon_j = \varepsilon_j^{(0)} + \lambda \varepsilon_j^{(1)} + \lambda^2 \varepsilon_j^{(2)} + \cdots. \quad (7.130)
$$

For the zeroth-order solutions, we get the unperturbed HF equations, Eq. (2.152),

$$
H_F u_j^{(0)} = \varepsilon_j^{(0)} u_j^{(0)} \quad (j = 1, \ldots, N), \quad (7.131)
$$

and for the first-order correlation, we have Eq. (2.154),

$$
\left(H_F - \varepsilon_j^{(0)} \right) u_j^{(1)} = \left(\varepsilon_j^{(1)} - X - V \right) u_j^{(0)}. \quad (7.132)
$$

Callaway has shown that the terms designated by X can be omitted in a good approximation. Then we obtain

$$
\left(H_F - \varepsilon_j^{(0)} \right) u_j^{(1)} = \left(\varepsilon_j^{(1)} - V \right) u_j^{(0)}. \quad (7.133)
$$

For $\varepsilon_j^{(1)}$, we obtain, from Eq. (2.156), by again omitting the terms designated by X,

$$
\varepsilon_j^{(1)} = \langle u_j^{(0)} | V | u_j^{(0)} \rangle, \quad (7.134)
$$

and for $\varepsilon_j^{(2)}$, we get

$$\varepsilon_j^{(2)} = \langle u_j^{(0)}|V|u_j^{(1)}\rangle. \tag{7.135}$$

We obtain for the total energy of the atom, up to and including the second-order corrections to the orbital parameters, from Eq. (2.165), the expression

$$E = E_0 + \sum_{j=1}^{N} \langle u_j^{(0)}|V|u_j^{(0)}\rangle + \sum_{j=1}^{N} \langle u_j^{(0)}|u_j^{(1)}\rangle, \tag{7.136}$$

where E_0 is the unperturbed HF energy of the core. This is our $E_C(\mathbf{r}_v)$, that is, this is the perturbed energy of the core and the $\mathbf{r}_v$-dependence becomes evident if we restore the original notation $V = 1/r_{1v}$. Then we get

$$E_C(\mathbf{r}_v) = E_0 + \sum_{j=1}^{N} \langle u_j^{(0)}|\frac{1}{r_{1v}}|u_j^{(0)}\rangle + \sum_{j=1}^{N} \langle u_j^{(0)}|\frac{1}{r_{1v}}|u_j^{(1)}\rangle. \tag{7.137}$$

Let us define the polarization potential as follows:

$$V_C(\mathbf{r}_v) = \sum_{j=1}^{N} \langle u_j^{(0)}|\frac{1}{r_{1v}}|u_j^{(1)}\rangle + \int \Phi_C^* t_v \Phi_C \, dv_C. \tag{7.138}$$

The second term in Eq. (7.137) is clearly the HF electrostatic potential. We did not get an exchange operator here because we omitted the antisymmetrization in Eq. (7.117). Restoring the exchange potential into Eq. (7.137) and using Eq. (7.138), we obtain for the Schroedinger equation of the valence electron, Eq. (7.124),

$$H_{\text{eff}}\psi_v = \varepsilon_v\psi_v, \tag{7.139}$$

where

$$H_{\text{eff}} = t_v + g_v + U + V_C, \tag{7.140}$$

and

$$\varepsilon_v = E_T - E_0. \tag{7.141}$$

U is the complete HF potential of the core, electrostatic and exchange, and V_C is the polarization potential given by Eq. (7.138).

Comparing our results with Eq. (7.8), we can write

$$H_{\text{eff}} = H_F + V_C, \tag{7.142}$$

and our effective Schroedinger equation becomes

$$(H_F + V_C)\psi_v = \varepsilon_v \psi_v. \tag{7.143}$$

We note here that the correlation effects between the core electrons are not taken into account. This can be seen because the equation contains the uncorrelated core orbitals built into the HF potential U.

Callaway computed V_C as follows. V_C is defined by Eq. (7.138). First, it can be shown that if Φ_C is given by Eq. (7.125), the second term of Eq. (7.138) vanishes. In order to calculate the first term, we need the functions $u_j^{(1)}(\mathbf{r}_1, \mathbf{r}_v)$. These are the solutions of Eq. (7.133). Writing that equation out in detail, we get

$$\left(H_F(\mathbf{r}_1) - \varepsilon_j^{(0)}\right)u_j^{(1)}(\mathbf{r}_1, \mathbf{r}_v) = \left(\varepsilon_j^{(1)} - \frac{1}{r_{1v}}\right)u_j^{(0)}(\mathbf{r}_1) \qquad (j = 1, \ldots, N).$$
$$\tag{7.144}$$

Let us expand $1/r_{1v}$:

$$\frac{1}{r_{1v}} = \sum_{k=0}^{\infty} L_k(r_1, r_v) P_k(\cos \theta), \tag{7.145}$$

where

$$L_k(r_1, r_v) = \frac{r_<^k}{r_>^{k+1}}, \tag{7.146}$$

and P_k is the Legendre polynomial with θ being the angle between $\mathbf{r}_1$ and $\mathbf{r}_v$.

The system of the HF-type equations in Eq. (7.144) can be solved and the $u_j^{(1)}$ determined. The solution is complicated by Eq. (7.145) because separate solutions must be computed for $r_1 < r_v$ and for $r_1 > r_v$; the two solutions must be matched at $r_1 = r_v$.

In order to simplify the problem, Callaway assumed that only the $r_1 < r_v$ region need be considered because the valence electron is mostly outside the core. If we introduce this approximation, then

$$\frac{1}{r_{1v}} = \sum_{k=0}^{\infty} \frac{r_1^k}{r_v^{k+1}} P_k(\cos \theta). \tag{7.147}$$

Also, from Eq. (7.134), we see that $\varepsilon_j^{(1)}$ is the electrostatic potential generated by the core orbital φ_i. This is the HF potential seen by the valence electron, that is, $\varepsilon_j^{(1)}$ is a function of $\mathbf{r}_v$. If we restrict the solution of Eq. (7.144) to the $r_1 < r_v$ region, we can omit the potential $\varepsilon_j^{(1)}(\mathbf{r}_v)$. Thus, in this approximation, we get

$$\left(H_F(\mathbf{r}_1) - \varepsilon_j^{(0)}\right)u_j^{(1)}(\mathbf{r}_1, \mathbf{r}_v) = -\sum_{k=0}^{\infty} \frac{r_1^k}{r_v^{k+1}} P_k(\cos\theta)u_j^{(0)}(\mathbf{r}_1), \quad (7.148)$$

and the equation is valid for $r_1 < r_v$. Having obtained the $u_j^{(1)}$ functions from this equation, we get from Eq. (7.138)

$$\begin{aligned}
V_C(\mathbf{r}_v) &= \sum_{j=1}^{N} \langle u_j^{(0)}| \frac{1}{r_{1v}} |u_j^{(1)}\rangle \\
&= \sum_{j=1}^{N} \sum_{k=0}^{\infty} \frac{1}{r_v^{k+1}} \int_{(r_1 < r_v)} u_j^{(0)*}(\mathbf{r}_1) r_1^k P_k(\cos\theta) u_j^{(1)}(\mathbf{r}_1, \mathbf{r}_v) \, d\mathbf{r}_1 \\
&= \sum_{k=0}^{\infty} \frac{1}{r_v^{k+1}} F_k(r_v),
\end{aligned} \qquad (7.149)$$

where

$$F_k(r_v) = \sum_{j=1}^{N} \int_{(r_1 < r_v)} u_j^{(0)*}(\mathbf{r}_1) r_1^k P_k(\cos\theta) u_j^{(1)}(\mathbf{r}_1, \mathbf{r}_v) \, d\mathbf{r}_1. \quad (7.150)$$

Another form can be derived by going back to Eq. (7.144), from which we obtain, by omitting $\varepsilon_j^{(1)}$ but not yet making the approximation $r_1 < r_v$,

$$\left(H_F - \varepsilon_j^{(0)}\right)u_j^{(1)} = -\frac{1}{r_{1v}}u_j^{(0)}, \qquad (7.151)$$

and so

$$u_j^{(1)} = -\frac{1 - P_j^{(0)}}{H_F - \varepsilon_j^{(0)}}\left[\frac{1}{r_{1v}}u_j^{(0)}\right], \qquad (7.152)$$

where $P_j^{(0)} = |u_j^{(0)}\rangle\langle u_j^{(0)}|$. By putting this into Eq. (7.138), we get

$$V_C(\mathbf{r}_v) = -\sum_{j=1}^{N} \langle u_j^{(0)}| \frac{1}{r_{1v}}\left(\frac{1 - P_j^{(0)}}{H_F - \varepsilon_j^{(0)}}\right)\frac{1}{r_{1v}}|u_j^{(0)}\rangle. \qquad (7.153)$$

Using the closure relationship for the unperturbed HF orbitals

$$\sum_{j=1}^{\infty} |u_j^{(0)}\rangle\langle u_j^{(0)}| = 1, \qquad (7.154)$$

we get for Eq. (7.153)

$$V_C(\mathbf{r}_v) = - \sum_{\substack{i=1 \\ (j \neq i)}}^{N} \sum_{j=1}^{\infty} \langle u_i^{(0)} | \frac{1}{r_{1v}} \left(\frac{|u_j^{(0)}\rangle\langle u_j^{(0)}|}{\varepsilon_j^{(0)} - \varepsilon_i^{(0)}} \right) \frac{1}{r_{1v}} |u_i^{(0)}\rangle, \qquad (7.155)$$

which is the standard expression for the second-order perturbation energy. Now again making the approximation $r_1 < r_v$ and omitting the $k = 0$ term from the expansion of Eq. (7.147), we get from Eq. (7.153)

$$V_C(r_v) = - \sum_{k=1}^{\infty} \frac{1}{r_v^{2k+2}} F_k(r_v), \qquad (7.156)$$

where

$$F_k(r_v) = \sum_{i=1}^{N} \int_{(r_1 < r_v)} u_i^{(0)*}(\mathbf{r}_1) r_1^k P_k(\cos \theta_1) \Lambda_i(\mathbf{r}_1)$$
$$\times \left[r_1^k P_k(\cos \theta_1) u_i^{(0)}(\mathbf{r}_1) \right] d\mathbf{r}_1. \qquad (7.157)$$

Here $\Lambda_i(r_1)$ is an integral operator operating on functions of $\mathbf{r}_1$ and defined as

$$\Lambda_i = \frac{1 - P_i^{(0)}}{H_F - \varepsilon_i^{(0)}} = \sum_{\substack{j=1 \\ (j \neq i)}}^{\infty} \frac{|u_j^{(0)}\rangle\langle u_j^{(0)}|}{\varepsilon_j^{(0)} - \varepsilon_i^{(0)}}. \qquad (7.158)$$

We now turn to the derivation of Bottcher and Dolgarno.[75] The starting point is again the complete Hamiltonian given by Eq. (7.113). Let us consider the first term of that Hamiltonian, H_C. Let the eigenfunctions of H_C be ϕ_λ, that is, let ϕ_λ satisfy the Schroedinger equation

$$H_C \phi_\lambda = E_\lambda \phi_\lambda. \qquad (7.159)$$

Let v be the effective potential for the valence electron and let us assume that the valence-electron orbitals are ψ_γ. Thus, for the valence electron we have the equation

$$H_{\text{eff}} \psi_\gamma = \varepsilon_\gamma \psi_\gamma, \qquad (7.160)$$

where

$$H_{\text{eff}} = H_v + v = t + g + v. \qquad (7.160a)$$

The total Hamiltonian can be written as

$$H = H_C + H_v + H_i = H_C + H_{\text{eff}} + \Delta V, \qquad (7.161)$$

where

$$\Delta V = H_i - v. \tag{7.162}$$

We consider $H_C + H_{\text{eff}}$ as the unperturbed Hamiltonian and ΔV as the perturbation. The complete set of unperturbed solutions is

$$\Phi(\lambda, \gamma) = \phi_\lambda \psi_\gamma, \tag{7.163}$$

and we have

$$(H_C + H_{\text{eff}})\Phi(\lambda, \gamma) = (E_\lambda + \varepsilon_\gamma)\Phi(\lambda, \gamma). \tag{7.164}$$

Let $\Psi(\lambda, \gamma)$ be the solution of the complete problem, that is, let

$$(H_C + H_{\text{eff}} + \Delta V)\Psi(\lambda, \gamma) = E_{\lambda\gamma}\Psi(\lambda, \gamma). \tag{7.165}$$

We want to evaluate $\Psi(\lambda, \gamma)$ in the first order of perturbation theory. Let λ_0 characterize the spherically symmetric ground state of the core. The wave function in the first order of perturbation theory is given by

$$\Psi(\lambda_0, \gamma_0) = \Phi(\lambda_0, \gamma_0) + \sum_{\substack{(\lambda\gamma) \\ \neq (\lambda_0\gamma_0)}} \frac{\langle \lambda\gamma | \Delta V | \lambda_0 \gamma_0 \rangle}{(E_{\lambda_0} + \varepsilon_{\gamma_0}) - (E_\lambda + \varepsilon_\gamma)} \Phi(\lambda, \gamma). \tag{7.166}$$

Let us introduce the approximation

$$E_{\lambda_0} - E_\lambda \gg \varepsilon_{\gamma_0} - \varepsilon_\gamma. \tag{7.167}$$

Omitting the $(\varepsilon_{\gamma_0} - \varepsilon_\gamma)$ from the denominator of the expression in Eq. (7.166), we obtain

$$\Psi(\lambda_0, \gamma_0) = \Phi(\lambda_0, \gamma_0) + \sum_{\substack{(\lambda\gamma) \\ \neq (\lambda_0\gamma_0)}} \frac{\langle \lambda\gamma | \Delta V | \lambda_0 \gamma_0 \rangle}{E_{\lambda_0} - E_\lambda} \Phi(\lambda, \gamma). \tag{7.168}$$

This can also be written in the form

$$|\Psi(\lambda_0, \gamma_0)\rangle = |\Phi(\lambda_0, \gamma_0)\rangle$$
$$+ \sum_{\substack{(\lambda\gamma) \\ \neq (\lambda_0\gamma_0)}} \frac{|\Phi(\lambda, \gamma)\rangle\langle\Phi(\lambda, \gamma)|}{E_{\lambda_0} - E_\lambda} \Delta V |\Phi(\lambda_0, \gamma_0)\rangle. \tag{7.169}$$

Let us write out the projection operator in detail. According to Eq. (7.163), we have

$$|\Phi(\lambda,\gamma)\rangle\langle\Phi(\lambda,\gamma)| = (|\phi_\lambda\rangle\langle\phi_\lambda|)(|\psi_\gamma\rangle\langle\psi_\gamma|). \qquad (7.170)$$

Using this equation, we can write

$$|\Psi(\lambda_0,\gamma_0)\rangle = |\Phi(\lambda_0,\gamma_0)\rangle + \sum_{\lambda \neq \lambda_0} \frac{|\phi_\lambda\rangle\langle\phi_\lambda|}{E_{\lambda_0} - E_\lambda}$$

$$\times \left[\sum_\gamma |\psi_\gamma\rangle\langle\psi_\gamma| - |\psi_{\gamma_0}\rangle\langle\psi_{\gamma_0}| \right] \Delta V |\Phi(\lambda_0,\gamma_0)\rangle. \quad (7.171)$$

The first term in square brackets is equal to 1 because of the closure relationship. The second term can be omitted approximately because the integral

$$\langle\psi_{\gamma_0}|\Delta V|\psi_{\gamma_0}\rangle = \langle\psi_{\gamma_0}|H_i - v|\psi_{\gamma_0}\rangle \qquad (7.171a)$$

can be considered small; we have introduced the effective potential v to replace the actual interaction H_i. Thus,

$$\langle\psi_{\gamma_0}|H_i|\psi_{\gamma_0}\rangle \approx \langle\psi_{\gamma_0}|v|\psi_{\gamma_0}\rangle, \qquad (7.171b)$$

and using this, we obtain

$$|\Psi(\lambda_0,\gamma_0)\rangle = |\Phi(\lambda_0,\gamma_0)\rangle + G_0 \Delta V |\Phi(\lambda_0,\gamma_0)\rangle, \qquad (7.171c)$$

where

$$G_0 = \sum_{\lambda \neq \lambda_0} \frac{|\phi_\lambda\rangle\langle\phi_\lambda|}{E_{\lambda_0} - E_\lambda}. \qquad (7.172)$$

The eigenfunctions in this operator are the solutions of the core equation, Eq. (7.159), and the zero index reminds us that ϕ_{λ_0} is the spherically symmetric ground state.

Let us now go back to Eq. (7.164). That equation is the *model equation*, meaning that it is this equation that defines the model potential v for the valence electron. Let us now impose the requirement that the eigenvalues of the model equation be identical with the eigenvalues of the exact equation, Eq. (7.165). In the formula:

$$E_{\lambda_0} + \varepsilon_{\gamma_0} = E_{\lambda_0\gamma_0}. \qquad (7.173)$$

This is actually a condition imposed on ε_{γ_0} because E_{λ_0} and $E_{\lambda_0\gamma_0}$ are determined by the Hamiltonians H_C and $(H_C + H_v + H_i)$, respectively. Substituting the condition into Eq. (7.165), we get

$$(H_C + H_{\text{eff}} + \Delta V)\Psi(\lambda_0, \gamma_0) = (E_{\lambda_0} + \varepsilon_{\gamma_0})\Psi(\lambda_0, \gamma_0). \quad (7.174)$$

Now substitute Eq. (7.171c) into this equation, that is, assume that the identity with the exact spectrum is enforced even in the first order of perturbation theory. We obtain

$$\left[H_C + H_{\text{eff}} + \Delta V - (E_{\lambda_0} + \varepsilon_{\gamma_0}) \right](1 + G_0 \Delta V)\Phi(\lambda_0, \gamma_0) = 0. \quad (7.175)$$

By multiplying from the left by the conjugate complex of the function in Eq. (7.171c) and integrating, we obtain

$$\langle \Delta V \rangle + 2\langle \Delta V G_0 \, \Delta V \rangle + \langle \Delta V G_0 \big[H_C - E_{\lambda_0} + H_{\text{eff}} - \varepsilon_{\gamma_0} \big] G_0 \, \Delta V \rangle$$

$$+ \langle \Delta V G_0 \, \Delta V G_0 \, \Delta V \rangle = 0, \quad (7.176)$$

where the bracket notation means expectation value with respect to $\Phi(\lambda_0, \gamma_0)$, that is,

$$\langle \cdots \rangle = \langle \Phi(\lambda_0, \gamma_0)| \cdots |\Phi(\lambda_0, \gamma_0)\rangle. \quad (7.177)$$

Let us recall that λ_0 is the spherically symmetric ground state of the core and γ_0 is an arbitrary valence state. We have the following relationships:

$$(H_C - E_{\lambda_0})G_0 = (H_C - E_{\lambda_0}) \sum_{\lambda \neq \lambda_0} \frac{|\phi_\lambda\rangle\langle\phi_\lambda|}{E_{\lambda_0} - E_\lambda}$$

$$= \sum_{\lambda \neq \lambda_0} \frac{E_\lambda - E_{\lambda_0}}{E_{\lambda_0} - E_\lambda}|\phi_\lambda\rangle\langle\phi_\lambda|$$

$$= -\sum_{\lambda \neq \lambda_0} |\phi_\lambda\rangle\langle\phi_\lambda| = -1 + |\phi_{\lambda_0}\rangle\langle\phi_{\lambda_0}|, \quad (7.178)$$

where we have invoked the closure relationship. In the expression of Eq. (7.176), we have $G_0(H_C - E_{\lambda_0})G_0$. Using Eq. (7.172), we get

$$G_0(H_C - E_{\lambda_0})G_0 = \sum_{\lambda \neq \lambda_0} \frac{|\phi_\lambda\rangle\langle\phi_\lambda|}{E_{\lambda_0} - E_\lambda}(-1 + |\phi_{\lambda_0}\rangle\langle\phi_{\lambda_0}|) = -G_0. \quad (7.178a)$$

Also, we have

$$G_0 v = v G_0, \quad (7.179)$$

which follows because G_0 operates only on the core-electron coordinates and v depends only on the valence-electron position. Finally,

$$G_0 \Phi(\lambda_0, \gamma_0) = \sum_{\lambda \neq \lambda_0} \frac{|\phi_\lambda\rangle\langle\phi_\lambda|}{E_{\lambda_0} - E_\lambda} |\phi_{\lambda_0}\rangle |\psi_{\lambda_0}\rangle = 0. \qquad (7.180)$$

Using the last three relationships and taking into account the form of ΔV as given by Eq. (7.162), we obtain, after a lengthy but straightforward derivation,

$$\langle v \rangle = \langle H_i \rangle + \langle H_i G_0 H_i \rangle + \langle H_i G_0 [H_{\text{eff}} - \varepsilon_{\gamma_0}] G_0 H_i \rangle + \langle H_i G_0 H_i G_0 H_i \rangle. \qquad (7.181)$$

This equation is valid for any ψ_{γ_0}. Therefore, we can write

$$v \approx \langle \phi_{\lambda_0} | H_i + H_i G_0 H_i + H_i G_0 G_0 [H_{\text{eff}} H_i - H_i H_{\text{eff}}]$$
$$+ H_i G_0 H_i G_0 H_i | \phi_{\lambda_0} \rangle. \qquad (7.182)$$

The term with the commutator is obtained as follows. From Eq. (7.179), it follows that

$$G_0 H_{\text{eff}} = H_{\text{eff}} G_0, \qquad (7.183)$$

because these operators operate on different coordinates. Thus, we get for the third term of Eq. (7.181)

$$\langle H_i G_0 | H_{\text{eff}} - \varepsilon_{\gamma_0} | G_0 H_i \rangle = \langle H_i G_0 G_0 | H_{\text{eff}} - \varepsilon_{\gamma_0} | H_i \rangle. \qquad (7.184)$$

It is easy to see that

$$\langle H_i G_0 G_0 H_i [H_{\text{eff}} - \varepsilon_{\gamma_0}] \rangle = \langle H_i G_0 G_0 H_i [H_{\text{eff}} - \varepsilon_{\gamma_0}] | \phi_{\lambda_0} \rangle | \psi_{\gamma_0} \rangle = 0. \qquad (7.185)$$

Adding the negative of this term, which is zero, to the expression in Eq. (7.184), we get

$$\langle H_i G_0 G_0 [H_{\text{eff}} - \varepsilon_{\gamma_0}] H_i \rangle - \langle H_i G_0 G_0 H_i [H_{\text{eff}} - \varepsilon_{\gamma_0}] \rangle$$
$$= \langle H_i G_0 G_0 [H_{\text{eff}} H_i - H_i H_{\text{eff}}] \rangle, \qquad (7.186)$$

from which we get the third term in Eq. (7.182).

Now we evaluate the expression for the model potential, Eq. (7.182). Clearly, the first term,

$$V(r_v) = \langle \phi_{\lambda_0} | H_i | \phi_{\lambda_0} \rangle = \sum_{j=1}^{N} \langle \phi_{\lambda_0} | \frac{1}{r_{jv}} | \phi_{\lambda_0} \rangle, \qquad (7.187)$$

is the electrostatic potential of the core. In order to evaluate the terms with G_0, we assume again that only the region $r_{core} < r_v$ needs to be considered, that is, we use the expansion of Eq. (7.147). We get

$$H_i = \sum_{j=1}^{N} \frac{1}{r_{jv}} = \sum_{j=1}^{N} \sum_{k=1}^{\infty} \frac{r_j^k}{r_v^{k+1}} P_k(\cos \theta_{jv}),$$ (7.188)

where we have omitted the term with $k = 0$, as before, and θ_{jv} is the angle between $\mathbf{r}_j$ and $\mathbf{r}_v$. We obtain, by inserting the expression in Eq. (7.188) into the second term of Eq. (7.182),

$$\langle \phi_{\lambda_0} | H_i G_0 H_i | \phi_{\lambda_0} \rangle = - \sum_{k=1}^{\infty} \frac{\alpha^{(k)}/2}{r_v^{2k+2}},$$ (7.189)

where

$$\alpha^{(k)} = 2 \sum_{j=1}^{N} \sum_{j'=1}^{N} \langle \phi_{\lambda_0} | r_j^k P_k(\cos \theta_j) G_0 r_{j'}^k P_k(\cos \theta_{j'}) | \phi_{\lambda_0} \rangle.$$ (7.189a)

Here $\alpha^{(k)}$ is the 2^k-pole polarizability of the atomic core, that is, the leading term with $k = 1$ is the dipole polarizability, the next with $k = 2$ is the quadrupole polarizability, and so on.[†] For the third term of Eq. (7.182), Bottcher and Dolgarno obtained

$$\langle \phi_{\lambda_0} | H_i G_0 G_0 [H_{eff} H_i - H_i H_{eff}] | \phi_{\lambda_0} \rangle = \frac{1}{2} \sum_{k=1}^{\infty} (k+1)(k+2) \frac{\beta^{(k)}}{r_v^{2k+4}},$$ (7.190)

where $\beta^{(k)}$ is called dynamic polarizability and is given as

$$\beta^{(k)} = \sum_{j, j'=1}^{N} \langle \phi_{\lambda_0} | r_j^k P_k(\cos \theta_j) G_0 G_0 r_{j'}^k P_k(\cos \theta_{j'}) | \phi_{\lambda_0} \rangle.$$ (7.191)

The derivation of Bottcher and Dolgarno was given for an arbitrary number of valence electrons; we have simplified it here for one valence electron. The result for n_v valence electrons is as follows. The polarization potential, that

[†] In Eq. (7.189a) the operator G_0 is -1 times the expression in Eq. (7.172), that is, it is the same as the operator in Eq. (7.158).

is, the v without the electrostatic interaction is given by

$$
V_P = \sum_{v=1}^{n_v} V_C(r_v) - \sum_{v<v'} \sum_{k=1}^{\infty} \left\{ \alpha^{(k)} - \frac{(k+1)(k+2)}{2} \beta^{(k)} \left(\frac{1}{r_v^2} + \frac{1}{r_{v'}^2} \right) \right\}
$$
$$
\times \frac{P_k(\cos \theta_{vv'})}{(r_v r_{v'})^{k+1}}. \tag{7.192}
$$

In this formula, $V_C(r_v)$ is the one-electron polarization potential and the double summation represents the 2-electron polarization potentials. The summation over v is for all valence electrons and the double summation $(v < v')$ is the summation over all valence-electron pairs. $\theta_{vv'}$ is the angle between $\mathbf{r}_v$ and $\mathbf{r}_{v'}$. For the $V_C(r_v)$, we collect the terms in Eqs. (7.189) and (7.190):

$$
V_C(r_v) = -\sum_{k=1}^{\infty} \frac{1}{2} \alpha^{(k)} \frac{1}{r_v^{2k+2}} + \sum_{k=1}^{\infty} \frac{(k+1)(k+2)}{2} \frac{\beta^{(k)}}{r_v^{2k+4}}. \tag{7.193}
$$

and $\alpha^{(k)}$ and $\beta^{(k)}$ are given by Eqs. (7.189a) and (7.191), respectively.

Equation (7.192) represents the formula for the complete polarization potential for an arbitrary number of valence electrons. It is valid in such a way that this expression represents the perturbation energy up to and including second-order terms.

Now we want to write down the complete effective Hamiltonian for n_v valence electrons. We get

$$
H_{\text{eff}} = T + G + \tilde{U} + V_P + Q, \tag{7.194}
$$

where $T + G$ are the kinectic energies and nuclear potentials, respectively:

$$
T = \sum_{v=1}^{n_v} t_v, \tag{7.195}
$$

$$
G = \sum_{v=1}^{n_v} g_v, \tag{7.196}
$$

and the $\tilde{U}$ is the complete core/valence electrostatic and exchange interaction. We use the tilde in the notation to indicate that these operators may include the core–core correlation, that is, they are not necessarily the ordinary HF-type electrostatic and exchange potentials like the expression in Eq. (7.9b). As we see from Eq. (7.187), the wave functions in that potential are the solutions of Eq. (7.159), that is, these are wave functions including all core–core correlations. We did not get any core/valence exchange operator in this derivation because we used the approximation of Eq. (7.163); we restored this interaction in $\tilde{U}$ because, obviously, we would have obtained this term had we antisymmetrized the products $\phi_\lambda \psi_\gamma$. The V_P is the complete polarization potential given by Eq. (7.192), and Q is the electrostatic potential of the valence electrons,

$$
Q = \frac{1}{2} \sum_{v=1}^{n_v} \sum_{\substack{v'=1 \\ (v' \neq v)}}^{n_v} \frac{1}{r_{vv'}}. \tag{7.197}
$$

For one valence electron, we get

$$
H_{\text{eff}} = t_v + g_v + v, \tag{7.198}
$$

where

$$
v = \tilde{U} + V_C(r_v). \tag{7.199}
$$

Here $\tilde{U}$ is the one-electron electrostatic and exchange potential of the core and $V_C(r_v)$ is the one-electron polarization potential given by Eq. (7.193). Both $\tilde{U}$ and V_C may contain core–core correlation effects.

We now review the theory of core/valence correlation as was presented in this section. Reviewing the derivation of Callaway and the derivation of Bottcher and Dolgarno, we realize that the physical assumptions are identical. In both derivations, the core/valence exchange was omitted. This step probably has little effect on the polarization potential and the relevant operator can easily be restored into the effective Hamiltonian as we have restored it. A more serious restriction is that in both derivations, we imposed the condition

$$
r_i < r_v \quad (i = 1, \ldots, N), \tag{7.200}
$$

where r_i is the distance from the nucleus of any core electron, and r_v is the radial coordinate of the valence electron. (The approximation in Eq. (7.167) is equivalent to this condition.) *Thus, both derivations are valid only outside the atomic core.* This is conspicuous in Eq. (7.193). If V_C would extend into the core, we would get negative infinity at $r = 0$. Thus, in practical applications, V_C must be cut off at the surface of the atomic core.

We note that Callaway's derivation is more practical because a computational procedure is given for the calculation of V_C. The procedure consists of the calculation of $u_j^{(1)}$ from the HF-type equations given in Eq. (7.148). On the other hand, the derivation of Bottcher and Dolgarno is much stronger in the sense that the core wave functions are the eigenfunctions of the exact Hamiltonian, whereas in Callaway's derivation, the core is represented by the unperturbed HF wave functions. Thus, the effective Hamiltonian given by Bottcher and Dolgarno contains, in principle, all core–core correlation effects.

We sum up. In a very good approximation, it is possible to define for a single valence electron the effective Schroedinger equation

$$H_{\text{eff}}\psi_\gamma = \varepsilon_\gamma\psi_\gamma, \tag{7.201}$$

where the effective Hamiltonian is given by Eqs. (7.198) and (7.199). The Hamiltonian contains, in principle, all core–core correlation effects. The core/valence correlation is given up to the second order of the perturbation theory. The polarization potential is valid only outside the atomic core. This is a conventional Hamiltonian, that is, the eigenfunctions must be orthogonalized to the core orbitals or they must have the necessary number of nodes. For n_v valence electrons, we get again Eq. (7.201), but now ψ_γ is an n_v-electron wave function and H_{eff} is given by Eq. (7.194).

In connection with Eq. (7.192), we note that for the actual construction of that potential, the 2^k-pole polarizibilities, $\alpha^{(k)}$, and the dynamic polarizabilities, $\beta^{(k)}$, are needed. The dipole polarizabilities, that is, the $\alpha^{(k)}$ constants with $k = 1$, which form the leading term in the potential, have been calculated using a variation-perturbation method by Fraga, Karkowski, and Saxena. In such a method, the calculation requires only the HF wave functions for the ground state. The calculated values for all neutral atoms and for many positive ions are tabulated in Fraga's book [FKS, 319].

C. Pseudopotentials with Core Polarization

In the introduction to this chapter, we stated that there are two kinds of effective Hamiltonians: the first kind, which we called conventional, required that the valence-electron wave functions be orthogonalized to the core orbitals; the second kind, which we called pseudopotential Hamiltonians, do not require any kind of orthogonalization. We have seen in Sec. 7.2 how the conventional one-electron effective Hamiltonians can be transformed into pseudopotential Hamiltonians.

In the preceding section, we formulated two conventional effective Hamiltonians for one valence electron. The equations had the form

$$H_{\text{eff}}\hat{\psi}_v = \varepsilon_v\hat{\psi}_v, \tag{7.202}$$

where we have changed the notation in such a way as to indicate with the caret that these eigenfunctions must be orthogonal to the core orbitals. For Callaway's effective Hamiltonian, we had the defining equation, Eq. (7.142),

$$H_{\text{eff}} = H_F + V_C = t + g + U + V_C, \tag{7.203}$$

where H_F was the HF Hamiltonian, and V_C was the polarization potential. For the effective Hamiltonian of Bottcher and Dolgarno, we derived the expression given by Eq. (7.198):

$$H_{\text{eff}} = t + g + \tilde{U} + \tilde{V}_C, \tag{7.204}$$

where $\tilde{U}$ is the electrostatic and exchange potential of the core, and $\tilde{V}_C$ is the polarization potential. We changed the notation of Eq. (7.199) in such a way as to indicate with the tilde that both $\tilde{U}$ and $\tilde{V}_C$ contain, in principle, all core–core correlation effects.

The significance of these Hamiltonians is that their eigenvalue spectrum should approximate closely the empirical spectrum because both contain the core polarization and, in addition, the expression in Eq. (7.204) also contains the core–core correlation effects. However, these are conventional effective Hamiltonians, so their eigenfunctions must be orthogonal to the core wave functions.

We now show that both equations can be subjected to a pseudopotential transformation. Let us denote the core solutions of Eq. (7.202) by $\hat{\psi}_1, \ldots, \hat{\psi}_N$ and let the valence solutions be $\hat{\psi}_v$. These core solutions certainly exist if V_C is cut off inside the core; in fact, for the Hamiltonian of Eq. (7.203), the core solutions will be very close to the HF orbitals.

Let Ω be the usual projection operator formed from the core orbitals:

$$\Omega = \sum_{i=1}^{N} |\hat{\psi}_i\rangle\langle\hat{\psi}_i|. \tag{7.205}$$

In Sec. 7.2/C, we have seen that the most general one-electron pseudopotential transformation is the Weeks–Rice transformation. We obtain, by using that transformation, the pseudopotential equation

$$(H_{\text{eff}} + V_P)\psi_v = \varepsilon_v \psi_v, \tag{7.206}$$

where V_P is given by Eq. (7.84):

$$V_P = \sum_{i=1}^{N} |\hat{\psi}_i\rangle\langle F_i| - (1 - \Omega)H_{\text{eff}}\Omega, \tag{7.207}$$

and ψ_v is now pseudo-orbital that does not need to satisfy any orthogonality conditions. We have seen that the eigenvalues of Eq. (7.206) reproduce the spectrum of Eq. (7.202) *exactly*, for any set of F_i functions.

The pseudopotential in Eq. (7.207) can be simplified by recalling that Ω is built from the eigenfunctions of H_{eff}, which means that H_{eff} and Ω commute. Thus, the second term of the pseudopotential vanishes. Also, we can make the pseudo-orbitals unique by imposing a mathematical condition on the F_i functions. One of the possibilities is to demand that the kinetic energy be minimized by the pseudo-orbital. This requirement leads to Eq. (7.92) for the F_i functions. We obtain different results for the two effective Hamiltonians under discussion. For Callaway's effective Hamiltonian, Eq. (7.203), we get

$$\langle F_i| = \langle \hat{\psi}_i|\left(\varepsilon_v - U - V_C - E^K\right), \tag{7.208}$$

and so the pseudopotential becomes

$$V_P = \Omega\left(\varepsilon_v - U - V_C - E^K\right). \tag{7.209}$$

For the Bottcher–Dolgarno Hamiltonian, Eq. (7.204), we get

$$\langle F_i| = \langle \hat{\psi}_i|\left(\varepsilon_v - \tilde{U} - \tilde{V}_C - E^K\right), \tag{7.210}$$

and for the pseudopotential, we have

$$V_P = \Omega\left(\varepsilon_v - \tilde{U} - \tilde{V}_C - E^K\right). \tag{7.211}$$

Both U and $\tilde{U}$ contain the nuclear potential g.

In summing up, we can say that the pseudopotential equation, Eq. (7.206), has the properties so that

1. it reproduces the spectra of the conventional Hamiltonians exactly;
2. the eigenvalues will closely approximate the empirical energy levels because the Hamiltonians contain core polarization and the Bottcher–Dolgarno Hamiltonian can contain also core–core correlation;
3. the pseudo-orbitals will be uniquely defined because of the additional mathematical condition imposed on the pseudopotential; and
4. the pseudo-orbitals do not need to satisfy any orthogonality conditions; the equation can be solved "as if the core did not exist."

D. Model Pseudopotentials

Using the results of the last section, we are able to put the effective pseudopotential Hamiltonians in such a mathematical form that is very convenient in atomic, molecular, and solid-state applications. We have seen that Eq. (7.206) is a pseudopotential equation that can be expected to describe the empirical spectrum with very high accuracy. In order to make it also *simple* mathematically, we now put the equation into *model* form. By *modelization*, we mean the replacement of the exact potentials by simple analytic expressions. These analytic expressions can be simple, but they should represent the exact potentials with accuracy. We emphasize that the concept of modelization is *not* synonymous with making the potentials semiempirical. There are model potentials that are semiempirical, but there are others that rest on theoretical consideration and do not need the introduction of empirical data. Some of the model potentials are adjusted to the HF potentials; these can be called semitheoretical.

The effective Hamiltonian that we want to modelize is given by Eq. (7.204). It contains the polarization potential $\tilde{V}_C$. Let us first take a look at the analytic expressions that can be used for polarization potentials.

The first analytic expression for polarization potentials was given by Bethe.[76] For the Li atom, the core consists of two $(1s)$ electrons. If the wave functions of these electrons are given in analytic form, then solutions of Eq. (7.151) can also be obtained in analytic form, and for the polarization potential, we obtain, using Bethe's procedure,[74]

$$V_C(x) = -\frac{9}{x^4}\left[1 - \frac{1}{3}e^{-2x}\left(1 + 2x + 6x^2 + \frac{20}{3}x^3 + \frac{4}{3}x^4\right)\right.$$
$$\left. -\frac{2}{3}e^{-4x}(1+x)^4\right]. \quad (7.212)$$

Here $x = Zr_v$, where r_v is the valence-electron radial coordinate. We note that in Eq. (7.212), the $r_v < r_i$ region is not omitted; therefore, this potential shows the structure of V_C inside the core also.

As we see, V_C exhibits the following asymptotic behavior:

$$\lim_{x \to 0} V_C = 0, \quad (7.213)$$

$$\lim_{x \to \infty} V_C = -\frac{9}{x^4}. \quad (7.214)$$

The polarization potential has a negative maximum within the core and goes to zero at the origin. This is plausible, because, as the valence electron penetrates the core, there is "less and less to polarize." For large x, the potential goes like $-1/x^4$.

Next, Callaway obtained[74] polarization potentials by using Sternheimer's HF calculations.[77] In these calculations, Sternheimer solved Eq. (7.148), which contains the $r_i < r_v$ approximation. Thus, polarization potentials obtained by Callaway are, strictly speaking, valid only outside the core. These potentials were put into analytic form by Szasz and McGinn and were incorporated into pseudopotential theory [Szasz, 265]. It is interesting to compare the core/valence correlation computed by Callaway with the empirical data. Here are Callaway's results[74] for the correlation energies in the ground states of some alkali atoms with the empirical in parentheses (atomic units): E (Li) = -0.0037 (-0.0016); E (Na) = -0.0065 (-0.0078); E (K) = -0.0176 (-0.0128). The results are meaningful but not very accurate.

Model expressions for $\tilde{V}_C$ were suggested by Bardsley[78] and by Dolgarno, Bottcher, and Victor.[79] Bardsley's expression is

$$\tilde{V}_C = -\frac{\alpha_d}{2(r^2 + d^2)^2} - \frac{\alpha_q}{2(r^2 + d^2)^3}. \qquad (7.215)$$

Here α_d and α_q are the dipole and quadrupole polarizabilities, respectively. As we see, this expression represents the two leading terms of the first sum of Eq. (7.193). In reality, Eq. (7.193) is valid only outside the core. In the $\tilde{V}_C$ above d is a cutoff radius that prevents $\tilde{V}_C$ from going to negative infinity at $r = 0$. We have

$$\lim_{r \to 0} \tilde{V}_C = -\frac{\alpha_d}{2d^4} - \frac{\alpha_q}{2d^6} = \text{constant}. \qquad (7.216)$$

By introducing d, Bardsley extended the domain of integration to the core area with $\tilde{V}_C$ being approximately valid there. For large r, the expression for $\tilde{V}_C$ exhibits the correct asymptotic behavior.

The formula suggested by Dolgarno, Bottcher, and Victor is

$$\tilde{V}_C = -\frac{\alpha_d}{2r^4}(1 - e^{-6\kappa r}) - \frac{\alpha_q}{2r^6}(1 - e^{-8\kappa r}), \qquad (7.217)$$

where α_d and α_q again are the polarizabilities. The parameter κ again is a cutoff parameter preventing $\tilde{V}_C$ to become infinite at the origin. We have

$$\lim_{r \to 0} \tilde{V}_C = 0, \qquad (7.218)$$

thus the expression in Eq. (7.217) has the correct asymptotic behavior for both $r \to 0$ and $r \to \infty$.

We now turn to the effective Hamiltonian in Eq. (7.204). There are many model potentials for this Hamiltonian. There is an extensive discussion of these in the author's book [Szasz, 79] along with a survey of the analytic formulas. Here we give only two typical expressions, both suggested by Bardsley. In the first, the potential of Eq. (7.204) is written in the following form:

$$g + \tilde{U} + \tilde{V}_C = \sum_{l=0}^{\infty} \left\{ -\frac{Z - N}{r} + A_l e^{-\kappa_l r} \right\} \Omega_l + \tilde{V}_C, \qquad (7.219)$$

where $\tilde{V}_C$ is identical with the expression in Eq. (7.215). This is a *semilocal* potential in which the strong l dependence of the pseudopotential, which we discussed before, is taken into account. Ω_l is the angular projection operator:

$$\Omega_l = \sum_{m=-l}^{+l} |Y_{lm}\rangle\langle Y_{lm}|. \tag{7.220}$$

As we see, Ω_l selects the l-component of the function on which it operates and we have a different modified potential for each l in accordance with the theory. A_l and κ_l are adjustable parameters determined by the requirement that the Schroedinger equation with the potential of Eq. (7.219) in it should reproduce exactly the lowest energy levels of the valence electron. It turns out that after the parameters are determined this way, the higher excited levels are reproduced within a 0.1% accuracy. This is certainly an improvement over HF-equivalent pseudopotentials.

For molecular calculations, it is often advantageous to have Gaussian functions in the modified potential. For this, Bardsley suggested

$$g + \tilde{U} + \tilde{V}_C = \sum_{l=0}^{\infty}\left\{-\frac{Z-N}{r} + A_l r^2 e^{-\kappa_l r^2}\right\}\Omega_l + \tilde{V}_C, \tag{7.221}$$

which differs from the potential in Eq. (7.213) only in that it is written in a form that is mathematically more convenient in molecular calculations. The method for the determination of the parameters is the same for the two potentials and they reproduce the valence-electron spectrum with the same accuracy.

E. Inverse Potential Methods

In the preceding section, we have given a very short discussion of what we have called model pseudopotentials. As we see from the text and even more from the much more detailed discussion in the author's book [Szasz, 79], the model potential method is an ad hoc procedure for the construction of pseudopotentials. The usual method is to choose an analytic form on the basis of the exact theory and/or on the basis of physical plausibility and then to demand that this analytic potential, when inserted into the Schroedinger equation, should reproduce the empirical (or the HF) spectrum. The matching of the spectrum is accomplished by changing the parameters of the analytic potential until the calculated eigenvalues of the Schroedinger equation approximate closely the empirical values.

There is nothing inherently wrong with this procedure, although its ad hoc character means that we can never be sure beforehand that a given analytic expression will provide the necessary accuracy in the reproduction of the energy levels. It is desirable to have other methods that would be different in that they would provide convergent procedures for the calculation of energy levels. By a convergent procedure, we mean a method with which the energy levels could be calculated with any desirable accuracy. Such procedures come under the heading of inverse potential methods. Instead of determining the energy levels for a given potential, in an inverse potential method, we determine the potential for a set of given empirical (or HF) levels.

In this section, we present two such methods. Both methods possess some interesting features. First, we discuss the model potential that was constructed by Prokofjew[80] on the basis of some earlier work of Sugiura and Thomas.[81] The second procedure we discuss is the model potential method of Bottcher.[82] Both potentials are conventional, that is, the eigenfunctions of the corresponding effective Hamiltonians will be orthogonal to each other. (The potentials are not pseudopotentials.)

Let us consider first Prokofjew's method. The interesting feature of this method is that it rests on the Bohr theory. In the general formulation of that theory, a system with f degrees of freedom must satisfy the quantum conditions

$$\oint p_i \, dq_i = n_i h \qquad (i = 1, \dots, f), \tag{7.222}$$

where q_i is any coordinate representing a degree of freedom of the system, and p_i is the corresponding momentum. The number n_i is a quantum number corresponding to the degree of freedom. The previous equations define an infinite number of quantized motions. The energies of these motions are the energy levels of the system. Let us consider an electron in an arbitrary atom. The radial momentum, p_r, is quantized according to the equation

$$\oint p_r \, dr = n - \left(l + \tfrac{1}{2} \right), \tag{7.223}$$

where (nl) are the principal and azimuthal quantum numbers, and we substituted $l + \tfrac{1}{2}$ for the k of the Bohr theory. The quantization of the azimuthal component of the momentum leads to the relationship

$$L = \left(l + \tfrac{1}{2} \right) \hbar, \tag{7.224}$$

where L is the angular momentum. The energy levels of the electron are determined by Eq. (7.223).

In order to get an expression for p_r, we use the classical formula for the energy

$$E = \frac{1}{2m} \left(p_r^2 + \frac{L^2}{r^2} \right) + V(r), \tag{7.225}$$

from which we get

$$p_r = \left\{ 2m(E - V) - \frac{L^2}{r^2} \right\}^{1/2}.$$ (7.226)

Using Eq. (7.224), we obtain

$$p_r = \left\{ 2m(E - V) - \frac{\left(l + \frac{1}{2}\right)^2 \hbar^2}{r^2} \right\}^{1/2}.$$ (7.227)

Next, we identify E with the energy level of the electron, ε_{nl}. Also, let

$$Q(r) = -r^2 V(r).$$ (7.228)

Then we get

$$p_r = \left\{ 2m\left(-|\varepsilon_{nl}| + \frac{Q}{r^2} \right) - \frac{\left(l + \frac{1}{2}\right)^2 \hbar^2}{r^2} \right\}^{1/2}$$

$$= \frac{\sqrt{2m}}{r} \left\{ Q(r) - |\varepsilon_{nl}|r^2 - \frac{1}{2m}\left(l + \frac{1}{2}\right)^2 \hbar^2 \right\}^{1/2}.$$ (7.229)

Changing to atomic units, we get

$$p_r = \frac{\sqrt{2}}{r} \left\{ Q(r) - |\varepsilon_{nl}|r^2 - \frac{1}{2}\left(l + \frac{1}{2}\right)^2 \right\}^{1/2}.$$ (7.230)

Let us introduce the function

$$X_{nl}(r) = |\varepsilon_{nl}|r^2 + \frac{1}{2}\left(l + \frac{1}{2}\right)^2.$$ (7.231)

Using this symbol, we get for p_r

$$p_r = \frac{\sqrt{2}}{r} \{ Q(r) - X_{nl}(r) \}^{1/2}.$$ (7.232)

Inserting this into Eq. (7.223), we obtain

$$\oint p_r \, dr = \sqrt{2} \int_{r_{min}}^{r_{max}} [Q(r) - X_{nl}(r)]^{1/2} \frac{dr}{r}$$
$$= n - (l + \tfrac{1}{2}). \qquad (7.233)$$

This is the equation for the determination of the energies ε_{nl} if potential Q is known. Prokofjew reversed the process: he has shown that the infinite set of conditions for $n = 1, 2, \ldots$ and $l = 0, 1, \ldots, n - 1$, given by Eq. (7.233), can be used for the determination of the potential if we use the empirical energy levels (core and valence) as input.

The calculation of the potential was carried out for the Na atom, and it takes place as follows. For large r, we have $V = -1/r$, so we put

$$Q(r) = -r^2 \left(-\frac{1}{r} \right) = r. \qquad (7.234)$$

For the vicinity of the nucleus, we have $V = -11/r$, because for Na, $Z = 11$. Thus, we put

$$Q(r) = -r^2 \left(-\frac{11}{r} \right) = 11r. \qquad (7.235)$$

These two functions must be joined smoothly for intermediate r values. From the definitions of Q and X_{nl}, we see that both are positive for all values of r. The integral in Eq. (7.233) is meaningful only if $Q - X_{nl}$ is also positive. For any ε_{nl}, this will be accomplished only for r values between r_{min} and r_{max}, which are the limits of integration. In classical theory, these limits are, of course, the turning points. Here we can use these limits for the determination of the potential in the range

$$r_{min} \le r \le r_{max}. \qquad (7.236)$$

Thus, the r scale is divided into sections, and in each section, the potential Q is assumed to have the form

$$Q = \alpha r^2 + \beta r + \gamma, \qquad (7.237)$$

which corresponds to

$$V = -\frac{Q}{r^2} = -\left(\alpha + \frac{\beta}{r} + \frac{\gamma}{r^2} \right). \qquad (7.238)$$

The parameters α, β, and γ are determined by the requirements that Q and its derivative should be continuous at the points joining two sections, and for each section, the integral in Eq. (7.233) should have the correct value, that is, the value $n - (l + \frac{1}{2})$. The method is, of course, not restricted to the simple choice given by Eq. (7.237). Any function can be used for Q as long as it is continuous at the joining points and satisfies the integral condition, Eq. (7.233).

For the Na atom, Prokofjew obtained the potential described as follows:

Range	$Q(r)$
$0 \quad < r < 0.01$	$= 11r,$
$0.01 < r < 0.15$	$= -26.4r^2 + 11.53r - 0.0264,$
$0.15 < r < 1.00$	$= -2.84r^2 + 4.46r + 0.5275,$
$1.00 < r < 1.55$	$= -1.508r^2 - 4.236r + 4.876,$
$1.55 < r < 3.30$	$= 0.1196r^2 + 0.2072r + 1.319,$
$3.30 < r < 6.74$	$= 0.0005r^2 + 0.9933r + 0.0222,$
$6.74 < r < \infty$	$= r.$

As the next step, Prokofjew inserted this potential into the Schroedinger equation, that is, he considered

$$(t + V(r))\psi_\gamma = \varepsilon_\gamma \psi_\gamma, \qquad (7.239)$$

where $V = -Q/r^2$, with Q given as before. The equation was not solved exactly, but approximate wave functions were constructed using the empirical values for the energy levels, and with these wave functions, some oscillator strengths were evaluated. Prokofjew has shown that the agreement between calculated and empirical data is quite good.

Let us now review the physical meaning of this procedure. As the method is formulated, Eq. (7.239) is not equivalent to any HF equation. In the construction of the potential, the empirical binding energies are used: consequently, the potential includes all correlation effects between the selected electron and the other electrons in the atom. The effective Hamiltonian that comes close to Eq. (7.239) is the Hamiltonian in Eq. (7.201). Let us write down that equation here. We had

$$H_{\text{eff}}\psi_\gamma = \varepsilon_\gamma \psi_\gamma, \qquad (7.240)$$

where

$$H_{\text{eff}} = t + g + \tilde{U} + \tilde{V}_C. \qquad (7.241)$$

Here $\tilde{U}$ and $\tilde{V}_C$ are the complete HF and polarization potentials, including core–core correlation effects. Denoting the Prokofjew potential by V_{PR}, we can write Eq. (7.239) into the form of Eq. (7.240) with

$$H_{eff} = t + V_{PR}. \qquad (7.242)$$

Thus, the connection between the theory and the model potential of Prokofjew is given by the substitution

$$g + \tilde{U} + \tilde{V}_C \to V_{PR}. \qquad (7.243)$$

If the Prokofjew potential can be constructed in such a way that it reproduces the energy levels, core and valence, with high accuracy, then it is superior to the combination $g + \tilde{U} + \tilde{V}_C$, because, as we have seen, this combination reproduces only the valence levels because $\tilde{V}_C$ is valid only outside the core. We can say with accuracy that the quantum theory of electronic structure, as it was presented up to this point, does not provide a theoretical background for the Prokofjew potential. A theoretical background can be established, however, by using density-functional formalism.

We proceed now to Bottcher's method. Here we start with Eq. (7.241), and we put for the potential

$$g + \tilde{U} + \tilde{V}_C \to U_0 + \sum_j c_j U_j \qquad (7.244)$$

In this expression, U_0 is the HF potential plus the polarization potential. The second term represents a correction that, in principle, should compensate for all approximations that were made in the derivation of $g + \tilde{U} + \tilde{V}_C$. For example, the correction term should represent the higher-order terms that were omitted from $\tilde{V}_C$. By increasing j, that is, by increasing the number of terms in the correction, one should be able to reproduce all energy levels of the electron with arbitrary accuracy. Thus, this method provides, in principle, a convergent inverse potential procedure for the construction of an accurate effective potential.

We quote the results of the Bottcher for the valence electron of the Na atom as a juxtaposition to Prokofjew's expression. He obtained

$$V = V_{HF} - \frac{0.4724}{r^4}\left[1 - e^{-r^6}\right] - \frac{3.18006}{r^6}\left[1 - e^{-r^8}\right]$$

$$+ 0.162339 e^{-r} - 0.034902 \, r e^{-r}. \qquad (7.245)$$

In this potential, the second and third terms are the polarization potentials with the correct asymptotic behavior for $r \to 0$. For V_{HF}, Bottcher used an expression without an exchange potential; therefore, the exchange effects are incorporated into the last two terms. The potential is meant to reproduce only the valence levels. The eigenvalues obtained were (empirical values in parenthesis)

$\varepsilon\,(3s) = 0.18899\ (0.18886)$.
$\varepsilon\,(4s) = 0.07169\ (0.07158)$.
$\varepsilon\,(5s) = 0.03764\ (0.03758)$.

$\varepsilon\,(3p) = 0.11123\ (0.11155)$.
$\varepsilon\,(4p) = 0.05082\ (0.05094)$.
$\varepsilon\,(5p) = 0.02914\ (0.02919)$.

$\varepsilon\,(3d) = 0.05621\ (0.05594)$.
$\varepsilon\,(4d) = 0.03160\ (0.03144)$.
$\varepsilon\,(5d) = 0.02020\ (0.02011)$.

As we see from the numbers, the calculated values approximate the empirical quite well. Thus, the expression in Eq. (7.245) represents an accurate model potential for the valence levels of the Na atom.

CHAPTER 8

EFFECTIVE HAMILTONIANS
FOR SEVERAL VALENCE ELECTRONS

---

tions for the ground states of negative alkali ions using effective pseudopotential Hamiltonian; discussion of the results; the importance of correlation effects.

8.1. INTRODUCTION

We now proceed to the development of effective Hamiltonians for atoms with an arbitrary number of valence electrons. As we have outlined in Sects. 7.1 and 7.3/A, our goal is to construct effective Hamiltonians that reproduce accurately the properties of valence electrons. The main problem here is the question of correlation effects. In order to develop accurate methods, the correlation effects must be included in the effective Hamiltonians. There are three groups of effects if we divide the atom into core and valence electrons. Let us write down the sequence of HF orbitals for illustration:

$$\underbrace{\varphi_1, \varphi_2, \ldots, \varphi_N}_{Core\ Orbitals} \qquad \underbrace{\varphi_{v_1}, \varphi_{v_2}, \ldots, \varphi_{v_n}}_{Valence\ Orbitals}$$

In this scheme, $\varphi_1, \ldots, \varphi_N$ are the core orbitals and $\varphi_{v_1}, \ldots, \varphi_{v_n}$ are the valence wave functions. The three groups of correlation effects are

1. core–core correlation;
2. core–valence correlation;
3. valence–valence correlation.

Experience shows that in atoms with several valence electrons, the third effect is the most important, the second is quite important, and the first is negligible. Thus, in constructing an effective Hamiltonian, it is absolutely necessary to take into account number 3; number 2 should be included if possible and number 1 can be omitted.

We must now revise the concept of the "frozen core." In Sec. 6.3, we defined the frozen-core approximation as one in which we first calculate the HF orbitals of the positive ion in the absence of a valence electron. Then, we add the valence electron to the core and assume that the addition of the valence electron does not have an effect on the core, that is, we assume that the core is "frozen." In the forthcoming discussions, this approximation is called the "positive-ion approximation." Another approximation, which we call the "neutral-atom approximation," is defined as follows. In the neutral-atom approximation, an HF calculation is performed first for the neutral atom, that is, we calculate a full set of HF orbitals, core plus valence. Then, we remove the valence orbitals and replace them with a correlated wave function or we place the atomic core into a molecule or a solid. In either case, we will assume that the core orbitals remain the same that they were in the neutral atom; again, the core is "frozen," but now it is the distribution of the neutral atom and not the distribution of the positive ion that is "frozen."

In summing up, we say the following:

1. In the positive-ion approximation, the core is frozen into what it was in the positive ion.
2. In the neutral-atom approximation, the core is frozen into what it was in the free neutral atom.

8.2. THE METHOD OF FOCK, VESELOV, AND PETRASHEN

Let us consider an atom with N core and two valence electrons. The Hamiltonian is

$$H = \sum_{i=1}^{N+2} (t_i + g_i) + \frac{1}{2} \sum_{i,j=1}^{N+2} \frac{1}{r_{ij}}, \tag{8.1}$$

where we use our usual notation. In the HF model, the single-determinantal wave function is

$$\Psi_F = [(N+2)!]^{-1/2} \det[\varphi_1, \ldots, \varphi_N \varphi_{N+1} \varphi_{N+2}]. \tag{8.2}$$

Here the core orbitals, $\varphi_1, \ldots, \varphi_N$, are orthonormal,

$$\langle \varphi_i | \varphi_k \rangle = \delta_{ik}, \tag{8.3}$$

and the valence orbitals are orthogonal to the core orbitals:

$$\langle \varphi_i | \varphi_{N+1} \rangle = \langle \varphi_i | \varphi_{N+2} \rangle = 0 \quad (i = 1, \ldots, N), \tag{8.4}$$

as well as to each other.

The basic idea of the theory of Fock, Veselov, and Petrashen (FVP) is as follows. In order to introduce electron correlation between the two valence electrons, let us replace the orbitals φ_{N+1} and φ_{N+2}, in the wave function above, by the arbitrary antisymmetric two-electron function, $\Phi(q_1, q_2)$. Thus, let us replace Eq. (8.2) by

$$\Psi = [(N+2)!]^{-1/2} \tilde{A}\{\det[\varphi_1(1) \cdots \varphi_N(N)]\Phi(N+1, N+2)\}, \tag{8.5}$$

where $\tilde{A}$ is a partial antisymmetrizer operator.

As the next step, FVP formed the expectation value of the Hamiltonian of Eq. (8.1) with Ψ and varied the resulting energy expression with respect to Φ and with respect to $\varphi_1 \cdots \varphi_N$. Later, the present author modified the procedure[84] by keeping the core orbitals fixed and varying only with respect to Φ. With this step, the method of FVP becomes an effective Hamiltonian theory. This is the theory, the FVP method modified by the author, that we present here [Szasz, 120].

The wave function, Eq. (8.5), is a combination of one- and two-electron functions. This fact necessitates the introduction of a new orthogonality concept. Generally, we call two wave functions orthogonal if

$$\int \psi^*(1,\ldots,n)\varphi(1,\ldots,n)\,dq_1\,dq_2\,\cdots\,dq_n = 0. \tag{8.6}$$

In this relationship, both ψ and φ are n-electron functions and we integrate over all coordinates. Equation (8.3) is a special case of Eq. (8.6). Let us call this *common* (or ordinary) orthogonality. In contrast, we say that the two-electron function, $\Phi(1,2)$, is *strong-orthogonal* to the orbital $\varphi_k(1)$ if we have the following relationship:

$$\int \varphi_k^*(q_1)\Phi(q_1,q_2)\,dq_1 = 0. \tag{8.7}$$

This orthogonality is "strong" because it is a functional relationship; the equation holds for any value of q_2. We subject the two-electron function in Eq. (8.5) to the condition that it be strong-orthogonal to all core orbitals:

$$\int \varphi_k^*(1)\Phi(1,2)\,dq_1 = 0 \qquad (k = 1,\ldots,N). \tag{8.8}$$

The strong-orthogonality condition appears to be mathematically complicated, but, in fact, it can be satisfied easily. Let us define the one-electron operator Ω_1 by

$$\Omega_1 = \sum_{j=1}^{N} \langle 1|\varphi_j\rangle\langle\varphi_j|, \tag{8.9}$$

and let

$$\Pi_1 = 1 - \Omega_1. \tag{8.10}$$

Let the two-electron operators Ω and Π be defined as

$$\Pi = \Pi_1\Pi_2 = 1 - \Omega = (1 - \Omega_1)(1 - \Omega_2). \tag{8.11}$$

In this chapter, operators without subscript will always mean n-electron operators ($n = 2, 3, 4, \ldots$).

If $\Phi_0(1,2)$ is an arbitrary two-electron function, then we can easily generate the strong-orthogonal two-electron function $\Phi(1,2)$, putting

$$\Phi(1,2) = \Pi\Phi_0(1,2). \tag{8.12}$$

Indeed, it is easy to show that

$$\int \varphi_k^*(1)\Phi(1,2)\, dq_1 = \int \varphi_k^*(1)\big[\Pi\Phi_0(1,2)\big]\, dq_1 = 0, \tag{8.13}$$

regardless of the form of $\Phi_0(1,2)$ [Szasz, 125].

The most important property of the strong-orthogonalization is the following. Let us form our new trial function, Eq. (8.5), with an arbitrary two-electron function Φ_0. We state that the total wave function Ψ does not change if we replace Φ_0 by the strong-orthogonal Φ defined by Eq. (8.12), that is, we state that

$$\Psi = \big[(N+2)!\big]^{-1/2} \tilde{A}\{\det[\varphi_1(1) \cdots \varphi_N(N)]\Phi_0(N+1, N+2)\}$$
$$= \big[(N+2)!\big]^{-1/2} \tilde{A}\{\det[\varphi_1(1) \cdots \varphi_N(N)]\Phi(N+1, N+2)\}. \tag{8.14}$$

The simple proof of this theorem, which is fundamental for effective Hamiltonian theory, can be found in the author's book [Szasz, 279]. We have seen that the orthogonality of the core orbitals, Eq. (8.3), does not restrict the generality of a determinantal function. Now we see that the strong-orthogonality does not restrict the generality of the total wave function Ψ. Thus, the conditions in Eqs. (8.3) and (8.8) do not in any way restrict the generality of the total wave function given by Eq. (8.5).

Now we construct the expectation value of the Hamiltonian given by Eq. (8.1) with respect to the wave function given by Eq. (8.5). We obtain [Szasz, 281]

$$E_T = \frac{\langle \Psi|H|\Psi\rangle}{\langle \Psi|\Psi\rangle} = \langle \Phi|H_{12}|\Phi\rangle + E_C, \tag{8.15}$$

where

$$H_{12} \equiv H_1 + H_2 + \frac{1}{r_{12}}, \tag{8.16}$$

and

$$H_i = t_i + g_i + U_C(i), \tag{8.17}$$

with

$$U_C(i) = \sum_{j=1}^{N} U_j(i). \tag{8.18}$$

Here U_j is the HF potential, electrostatic and exchange, generated by the core orbital φ_j. The potential U_C is the total potential of the core. E_C is the HF energy of the core formed with the core orbitals $\varphi_1 \cdots \varphi_N$. Denoting the energy of the two valence electrons by E_v, we get

$$E_T = E_v + E_C, \qquad (8.19)$$

where

$$E_v = \langle \Phi | H_{12} | \Phi \rangle. \qquad (8.20)$$

Thus, we have a clear separation of core and valence electrons just as we have it in all forms of the HF model. This is accomplished by the strong-orthogonality, Eq. (8.8).

Now, following the author's prescription,[84] we vary the total energy of the atom with respect to the two-electron function Φ. Subsidiary conditions are the normalization of Φ and the strong-orthogonality, Eq. (8.8). The core orbitals are kept fixed and unspecified in the variation, subjected only to the condition given by Eq. (8.3). In this way, we obtain the effective Hamiltonian for the best two-electron function for a set of fixed unspecified core orbitals:

$$\left(\hat{H}_1 + \hat{H}_2 + \Pi \frac{1}{r_{12}} \right) \Phi = E_v \Phi, \qquad (8.21)$$

where

$$\hat{H}_i = \Pi H_i, \qquad (8.22)$$

and H_i is given by Eq. (8.17) [Szasz, 283]. Thus, our effective Schroedinger equation for the two valence electrons is

$$H_{\text{eff}} \Phi = E_v \Phi, \qquad (8.23)$$

where

$$H_{\text{eff}} = \hat{H}_1 + \hat{H}_2 + \Pi \frac{1}{r_{12}}. \qquad (8.24)$$

Next we show two important properties of the effective Hamiltonian. First, we prove that the solutions of Eq. (8.23) are strong-orthogonal to all core orbitals. This is necessary to keep the valence electrons out of the core. Multiply Eq. (8.21) from the left by $\varphi_k^*(1)$ and integrate over q_1. Then we get

$$\langle \varphi_k | \Pi H_{12} | \Phi \rangle = E_v \langle \varphi_k | \Phi \rangle, \qquad (8.25)$$

where $\langle \cdot \cdot \cdot \rangle$ means integration with respect to q_1. Further,

$$\langle \varphi_k | \Pi H_{12} | \Phi \rangle = \langle \varphi_k | (1 - \Omega_1)(1 - \Omega_2) H_{12} | \Phi \rangle$$
$$= \langle (1 - \Omega_1) \varphi_k(1) | (1 - \Omega_2) H_{12} | \Phi \rangle = 0, \quad (8.26)$$

where we have used the Hermitian character of $1 - \Omega_1$ and we used Eq. (7.59c). The last equation proves that

$$\langle \varphi_k | \Phi \rangle = 0 \quad (k = 1, \ldots, N). \quad (8.27)$$

Next, we prove that two solutions of Eq. (8.23), belonging to different eigenvalues, are orthogonal. Let us write down Eq. (8.23) first for the valence-electron state α:

$$H_{\text{eff}} \Phi_\alpha = E_\alpha \Phi_\alpha, \quad (8.28)$$

and then the conjugate complex equation for the state β:

$$H_{\text{eff}}^* \Phi_\beta^* = E_\beta \Phi_\beta^*. \quad (8.29)$$

Multiply Eq. (8.28) by Φ_β^*, Eq. (8.29) by Φ_α, and integrate. Subtract Eq. (8.29) from Eq. (8.28). Then we get

$$\langle \Phi_\beta | H_{\text{eff}} | \Phi_\alpha \rangle - \langle \Phi_\alpha | H_{\text{eff}} | \Phi_\beta \rangle^* = (E_\alpha - E_\beta) \langle \Phi_\beta | \Phi_\alpha \rangle. \quad (8.30)$$

Now consider

$$\langle \Phi_\beta | H_{\text{eff}} | \Phi_\alpha \rangle = \langle \Phi_\beta | \Pi H_{12} | \Phi_\alpha \rangle$$
$$= \langle \Pi \Phi_\beta | H_{12} | \Phi_\alpha \rangle$$
$$= \langle \Phi_\beta | H_{12} | \Phi_\alpha \rangle, \quad (8.31)$$

where we have used the rule that if a function Φ_β is strong-orthogonal to the core orbitals then [Szasz, 277]

$$\Pi \Phi_\beta = \Phi_\beta. \quad (8.32)$$

Likewise, we get

$$\langle \Phi_\alpha | H_{\text{eff}} | \Phi_\beta \rangle^* = \langle \Phi_\alpha | \Pi H_{12} \Phi_\beta \rangle^*$$
$$= \langle \Pi \Phi_\alpha | H_{12} | \Phi_\beta \rangle^*$$
$$= \langle \Phi_\alpha | H_{12} | \Phi_\beta \rangle^*$$
$$= \langle \Phi_\beta | H_{12} | \Phi_\alpha \rangle, \quad (8.33)$$

where we have used Eq. (8.32) and the Hermitian character of H_{12}. Inserting Eqs. (8.31) and (8.33) into Eq. (8.30), we obtain

$$\langle \Phi_\beta | \Phi_\alpha \rangle = 0, \qquad \text{if } E_\alpha \neq E_\beta, \tag{8.34}$$

which is what we wanted to prove. In the derivation, we assumed that the eigenvalues are real despite that H_{eff} is not Hermitian. This can be proved easily. Let the conjugate complex of Eq. (8.23) be

$$H_{\text{eff}}^* \Phi^* = E_v^* \Phi^*. \tag{8.35}$$

Multiply Eq. (8.23) by Φ^*, Eq. (8.35) by Φ, and integrate. Subtract Eq. (8.23). Then we obtain

$$\langle \Phi | H_{\text{eff}} | \Phi \rangle - \langle \Phi | H_{\text{eff}} | \Phi \rangle^* = (E_v - E_v^*) \langle \Phi | \Phi \rangle$$
$$= E_v - E_v^*. \tag{8.36}$$

Using the same arguments that we used in the derivation of Eq. (8.34), that is, the Hermitian character of Π and H_{12} and the relationship given by Eq. (8.32), we obtain from Eq. (8.36)

$$E_v = E_v^*. \tag{8.37}$$

We note that the orthogonality in Eq. (8.34) is the *common* orthogonality, that is, beginning with Eq. (8.28) all integrations were for all valence-electron coordinates.

Next, we discuss the parametric dependence of the effective Hamiltonian on the core orbitals. Specifically, we want to show that the effective Hamiltonian takes quite different forms for different sets of core orbitals. Let us assume first that we apply the positive-ion approximation. That means that we start with the total wave function given by Eq. (7.4) and assume that the core orbitals satisfy Eq. (7.8), that is,

$$H_F \varphi_k = \varepsilon_k \varphi_k \qquad (k = 1, \ldots, N). \tag{8.38}$$

Here the HF Hamiltonian is given by

$$H_F = t + g + U, \tag{8.39}$$

where U is the electrostatic and exchange potential of all electrons. We have

$$U = \sum_{j=1}^{N} U_j, \tag{8.40}$$

where

$$\langle 1|U_j|2 \rangle = \int \frac{|\varphi_j(q_3)|^2 \, dq_3}{|\mathbf{r}_1 - \mathbf{r}_3|} \delta(\mathbf{r}_1 - \mathbf{r}_2) - \frac{\varphi_j(1)\varphi_j(2)}{|\mathbf{r}_1 - \mathbf{r}_2|}, \qquad (8.41)$$

is the kernel of the operator generated by the orbital φ_j. Comparing Eq. (8.39) with Eq. (8.17), we see that the HF Hamiltonian of Eq. (8.38) is formally identical with the operator H_i in the effective Hamiltonian, Eq. (8.16). From this, it follows that Π_i and H_i will commute because we have seen that the Hamiltonian commutes with the Ω_i operator, which is formed from its eigenfunctions. Thus, we have

$$\Pi_i H_i = H_i \Pi_i, \qquad (8.42)$$

and so, using Eq. (8.32), we obtain

$$H_{\text{eff}} \Phi = E\Phi, \qquad (8.43)$$

with

$$H_{\text{eff}} = H_F(1) + H_F(2) + \Pi \frac{1}{r_{12}}. \qquad (8.44)$$

This is a much simpler operator than the one in Eq. (8.24) because the projection operators disappeared from the first two terms.

Next, let us consider the neutral-atom approximation. In that case, we start with the Hamiltonian in Eq. (8.1) and our total wave function is given by Eq. (8.2). The HF equations will be

$$H_F \varphi_i = \varepsilon_i \varphi_i \qquad (i = 1, \ldots, N, N+1, N+2). \qquad (8.45)$$

The HF Hamiltonian again has the form given by Eq. (8.39), but now

$$U = U_C + U_v, \qquad (8.46)$$

where U_C is given by Eq. (8.40) and

$$U_v = U_{N+1} + U_{N+2}. \qquad (8.47)$$

The HF Hamiltonian again commutes with Π_i, but it is not the HF Hamiltonian that we have in the effective Hamiltonian. We have

$$\left(\hat{H}_1 + \hat{H}_2 + \Pi \frac{1}{r_{12}} - E \right) \Phi$$

$$= \Pi \left\{ H_1 + H_1 + \frac{1}{r_{12}} - E \right\} \Phi$$

$$= \Pi \left\{ H_F(1) + H_F(2) + \frac{1}{r_{12}} - U_v(1) - U_v(2) - E \right\} \Phi = 0. \quad (8.48)$$

From this, we obtain, by using Eq. (8.32),

$$H_{\text{eff}} \Phi = E\Phi, \quad (8.49)$$

where

$$H_{\text{eff}} = H_F(1) + H_F(2) + \Pi S_{12} \quad (8.50)$$

and

$$S_{12} \equiv \frac{1}{r_{12}} - U_v(1) - U_v(2), \quad (8.51)$$

or

$$S_{12} = \frac{1}{r_{12}} - U_{N+1}(1) - U_{N+2}(1) - U_{N+1}(2) - U_{N+2}(2). \quad (8.52)$$

Comparing Eq. (8.50) with Eq. (8.44), we see that the two effective Hamiltonians are quite different because the different choice of the core orbitals resulted in a different interaction term; $1/r_{12}$ of Eq. (8.44) is replaced by S_{12} in Eq. (8.50). The meaning of H_F is also quite different in the two Hamiltonians. Thus, we see that the choice of core orbitals influences the form of the effective Hamiltonian quite considerably.

At this point, we want to clarify the structure of the effective Hamiltonians. The effective Hamiltonians always have two groups of terms. The first group represents the interaction potentials between the core and valence electrons. (In addition, this group has the usual kinetic-energy operators and nuclear potentials.) In Eqs. (8.44) and (8.50), this group is given by the first two terms. The second group represents the interaction between the valence electrons themselves. In our equations, these are the terms with $1/r_{12}$ and S_{12}, respectively. Both groups may contain projection operators. The interaction between the core and valence electrons is always represented by one-electron operators, whereas the intravalence interaction contains two-electron operators as well.

The generalization of the FVP method to n valence electrons is straightforward [Szasz, 178]. The starting point is the Hamiltonian

$$H = \sum_{i=1}^{N+n} (t_i + g_i) + \frac{1}{2} \sum_{i,j=1}^{N+n} \frac{1}{r_{ij}}. \tag{8.53}$$

The total wave function of the atom is

$$\Psi = [(N+n)!]^{-1/2} \tilde{A}\{\Phi(1,2,\ldots,n)$$
$$\times \det[\varphi_1(n+1)\varphi_2(n+2),\ldots,\varphi_N(n+N)]\}. \tag{8.54}$$

Here Φ is an arbitrary n-electron wave function, $\varphi_1,\ldots,\varphi_N$ are the core orbitals, and $\tilde{A}$ is an antisymmetrizer operator. We again have the projection operators given by Eqs. (8.9) and (8.10) and we put

$$\Pi = \Pi_1\Pi_2 \cdots \Pi_n, \tag{8.55}$$

and

$$\Omega = 1 - \Pi. \tag{8.56}$$

These are now n-electron operators. If $\Phi_0(1,2,\ldots,n)$ is an arbitrary n-electron wave function, then we can strong-orthogonalize Φ_0 to the core orbitals by putting

$$\Phi = \Pi\Phi_0. \tag{8.57}$$

Without restricting the generality of the total wave function given by Eq. (8.54), we can assume that Φ satisfies the strong-orthogonality condition:

$$\int \varphi_k^*(1)\Phi(1,2,\ldots,n)\, dq_1 = 0 \qquad (k = 1,2,\ldots,N). \tag{8.58}$$

The effective n-electron Schroedinger equation becomes

$$H_{\text{eff}}\Phi = E\Phi, \tag{8.59}$$

where

$$H_{\text{eff}} = \Pi H^{(n)}. \tag{8.60}$$

Here Π is given by Eq. (8.55) and

$$H^{(n)} = \sum_{i=1}^{n} H_i + \frac{1}{2} \sum_{i,j=1}^{n} \frac{1}{r_{ij}}, \tag{8.61}$$

where H_i is given by Eq. (8.17). All properties of the two-electron effective Hamiltonian can be transferred, *mutatis mutandis*, to the *n*-electron effective Hamiltonian.

8.3. THE METHOD OF BOTTCHER AND DOLGARNO

In Sec. 7.3/B, we have seen that an effective Hamiltonian had been formulated for *n* valence electrons by Bottcher and Dolgarno (BD). For the many-valence-electron case, the formula for the effective Hamiltonian was given by Eq. (7.194). It is easy to synchronize the notation of that equation with the notation of the preceding section. It is clear that we can write

$$T + G + \tilde{U} = \sum_{i=1}^{n} H_i, \tag{8.62}$$

where H_i is given by Eq. (8.17). The second term of Eq. (8.61) is what is denoted by Q in Eq. (7.194). V_P is given by Eq. (7.192), which we can write in the form

$$V_P = \sum_{i=1}^{n} V_C(r_i) + \sum_{i<j=1}^{n} V_C(\mathbf{r}_i, \mathbf{r}_j). \tag{8.63}$$

Let

$$\hat{Q} \equiv \frac{1}{2} \sum_{i<j=1}^{n} \left(\frac{1}{r_{ij}} + V_C(i,j) \right). \tag{8.64}$$

By using these notations, the effective Hamiltonian of BD, Eq. (7.194), becomes

$$H = \sum_{i=1}^{n} \left(H_i + V_C(i) \right) + \hat{Q}. \tag{8.65}$$

As we see, this operator differs from Eq. (8.61) only in the presence of polarization terms $V_C(i)$ and $V_C(i,j)$. We obtain the effective Schroedinger equation in the form

$$(H - E)\Phi = 0, \tag{8.66}$$

which is now an *n*-electron equation for the valence electrons with a set of arbitrary fixed-core orbitals (these are in the operators H_i, Eq. (8.62)).

When we discussed the derivation of BD, we pointed out that the total wave function was not antisymmetrized, that is, it did not satisfy the Pauli exclusion principle. We mentioned that the omission of the antisymmetry requirement probably does not appreciably affect the form of the polarization potential. However, the overall form of the effective Hamiltonian is profoundly influenced by the lack of antisymmetrization. Because the Pauli exclusion principle is not introduced between the core and valence electrons, there is no orthogonality requirement imposed on the eigenfunctions of Eq. (8.66). In fact, by solving that equation with the variation method, there is nothing that prevents the valence electrons from "sinking into the core," that is, there is nothing that prevents the valence-electron energy from converging toward core energies.

This shortcoming can be remedied easily. We revary Eq. (8.66), demanding that the eigenfunctions be strong-orthogonal to the core orbitals. Let $\varphi_1, \ldots, \varphi_N$ be a set of unspecified core orbitals subjected only to the orthonormality condition. Let us form the projection operators given by Eqs. (8.55) and (8.56) with these orbitals. What will be the effective Hamiltonian after we revary Eq. (8.66)?

We could go through the variation procedure, taking into account the strong orthogonality, by just repeating the derivation that was used to derive the effective Hamiltonian of FVP, Eq. (8.60). This derivation can be found in the author's book [Szasz, 283]. However, it is easy to see that there is no need for any derivation. We obtain the same result by simply multiplying the Hamiltonian in Eq. (8.66) from the left by the projection operator Π given by Eq. (8.55). Thus, we obtain the correct effective Hamiltonian from Eq. (8.66) by writing

$$H_{eff} = \Pi H$$

$$= \Pi \left\{ \sum_{i=1}^{n} \left(H_i + V_C(i) \right) + \hat{Q} \right\}, \qquad (8.67)$$

where $\hat{Q}$ is given by Eq. (8.64). The effective Schroedinger equation becomes

$$H_{eff} \Phi = E \Phi. \qquad (8.68)$$

It is easy to show that the eigenfunctions of this equation will be strong-orthogonal to the core orbitals. Thus, they also will satisfy the relationship

$$\Pi \Phi = \Phi. \qquad (8.69)$$

It can be shown just as easily that the eigenvalues of Eq. (8.68) will be real and the eigenfunctions belonging to different eigenvalues will be orthogonal. Thus, this equation will have all the characteristics of the FVP equation, Eq. (8.59).

We have now at our disposal two effective Hamiltonians for the determination of the valence-electron wave functions. These are the operator of Fock, Veselov, and Petrashen and the operator derived by Bottcher and Dolgarno. The formulas are in Eqs. (8.60) and (8.67), respectively. These two

Hamiltonians are what we called conventional, that is, they are operators whose eigenfunctions are strong-orthogonal to the core orbitals.

In the next section, we show that effective Hamiltonian theory can be simplified decisively by the introduction of pseudopotentials. We show that both Hamiltonians, the effective operator of FVP as well as the operator of BD, can be transformed exactly into pseudopotential Hamiltonians and that this transformation simplifies the formalism considerably.

8.4. THE PSEUDOPOTENTIAL TRANSFORMATION OF SZASZ AND BROWN

The effective Hamiltonians derived in the preceding two sections can be used in any atomic, molecular, or solid-state calculation. However, such calculations are complicated by the fact that the eigenfunctions of these Hamiltonians must be orthogonalized to the core orbitals, that is, they must be put in the form of Eq. (8.57). Such orthogonalization might slow down the calculations considerably, especially in molecular studies.

We now show that the orthogonality requirement can be transformed into pseudopotentials and that this transformation is exact, that is, it does not involve any further approximations. We emphasize this because many-electron pseudopotential theory is *not* a straightforward generalization of the one-electron theory. In Chapter 7, we have seen that exact pseudopotential transformations are possible in the one-electron effective Hamiltonian theory. From this, it does not automatically follow, however, that exact pseudopotential transformations are also possible in the many-electron effective Hamiltonian theory. We present here the exact transformation based on the work of Szasz and Brown[85] [Szasz, 138, 183].

Before carrying out the pseudopotential transformation, we must make a choice for the core orbitals. As we will see presently, the transformation depends, to a considerable degree, on the choice of core orbitals. In order to keep the discussion general, let us consider an atom with n valence electrons. Let us assume that these are in an arbitrary configuration and are subjected to an arbitrary angular momentum and spin coupling. Let φ_i $(i = 1, \ldots, N)$ be the core orbitals and φ_v $(v = 1, \ldots, n)$ be the valence orbitals. The HF equation for one of the valence orbitals, φ_v, can be written in the form

$$H_v \varphi_v = \varepsilon_v \varphi_v + \sum_{i=1}^{N} \lambda_{iv} \varphi_i, \qquad (8.70)$$

where λ_{iv} is a Lagrangian multiplier, and H_v is given by

$$H_v = t + g + U_C + U_v, \qquad (8.71)$$

where U_C and U_v are HF potential operators representing the core and the valence electrons, respectively. The core–valence separation in the HF potential is always possible regardless of the configuration and regardless of the coupling. Introducing our one-electron projection operators, Ω and Π, we obtain

$$\hat{H}_v \varphi_v = \varepsilon_v \varphi_v, \qquad (8.72)$$

where

$$\hat{H}_v = \Pi H_v. \qquad (8.73)$$

We have seen in Sec. 7.2/C that the exact one-electron pseudopotential transformation of an equation like Eq. (8.72) is given by

$$(H_v + V_P)\psi_v = \varepsilon_v \psi_v, \qquad (8.74)$$

where V_P is the Weeks–Rice one-electron pseudopotential given by Eq. (7.76) or by Eq. (7.84). Writing down the latter equation, we have

$$V_P = \sum_{i=1}^{N} |\varphi_i\rangle\langle F_i| - \Pi H_v \Omega. \qquad (8.75)$$

Clearly, Eq. (8.72) is the many-electron generalization of the one-electron equation, Eq. (7.63). Indeed, if we have only one valence electron, then $U_v = 0$ and we can identify the H_v of Eq. (8.71) with the H in Eq. (7.54). Or, we can keep U_v and think of Eq. (7.63) not as an equation for the single valence electron, but as the equation for one of the several valence electrons. Thus, most of the formalism of Sec. 7.2/C can be directly used here.

We must decide at this point whether we use the positive-ion or the neutral-atom approximation. The effective Hamiltonians, Eqs. (8.61) and (8.67), are valid for both cases. These Hamiltonians both contain the core orbitals parametrically; the positive-ion and neutral-atom approximations will be distinguished by the different choices for the core orbitals. Let us consider the neutral-atom approximation. In that case, the core orbitals must be solutions of the HF equations for the neutral atom. These HF equations are given by Eq. (8.72); the corresponding PP equations are given by Eq. (8.74). We introduce the neutral-atom approximation not by taking the individual HF orbitals from a neutral-atom calculation, but by taking the operator H_i, which occurs both in Eqs. (8.61) and (8.67), from Eq. (8.74), that is, from the expression in Eq. (8.71). We have

$$(t + g + U_C + U_v + V_P)\psi_v = \varepsilon_v \psi_v. \qquad (8.76)$$

Introducing the modified potential, V_M, by Eq. (7.28), we can write

$$V_M = g + U_C + V_P, \qquad (8.77)$$

and so we get

$$(t + V_M + U_v)\psi_v = \varepsilon_v \psi_v. \qquad (8.78)$$

After this equation is solved, we can obtain from this the modified potential in the form of a local (or semilocal) potential. We get

$$V_M = \frac{(\varepsilon_v - t - U_v)\psi_v}{\psi_v}. \qquad (8.79)$$

We are using the neutral-atom approximation if we use this expression in a many-electron effective Hamiltonian. It is easy to see that the formal transition to the positive-ion approximation is accomplished by putting

$$U_v = 0. \qquad (8.80)$$

We emphasize that we are talking here about the core potential. In the effective Hamiltonians, the core–valence interaction potential is represented by the operator H_i, which is, according to Eq. (8.17),

$$H_i = t_i + g_i + U_C(i). \qquad (8.81)$$

Thus, in the effective Schroedinger equations, Eqs. (8.59) and (8.68), the core–valence interactions are one-electron operators despite these equations being valid for the correlated many-electron wave functions of the valence electrons. We can use the modified potential V_M in the many-electron effective Hamiltonians, that is, we can use a one-electron interaction operator in the many-electron Hamiltonians, because *the core potential is always a one-electron operator regardless whether the valence-electron wave function is correlated or not.*

We return now to the pseudopotential transformation of the n-electron effective Hamiltonians. The transformation can be carried out on the FVP equation, Eq. (8.59), or on the BD equation, Eq. (8.68). We will work with the FVP equation; the transformation of the BD equation would lead to the same results, except we would get the polarization potentials as additional terms. These terms can be added to our results later. Let Ψ be an n-electron pseudo-wave function connected to the Φ of Eq. (8.59) by the relationship

$$\Phi = \Pi\Psi, \qquad (8.82)$$

where Π is the n-electron projection operator given by Eq. (8.55). We also have

$$\Omega = 1 - \Pi. \tag{8.83}$$

We state that the pseudopotential equation, exactly equivalent to Eq. (8.59), is given by

$$H_{\text{eff}}\Psi = E\Psi, \tag{8.84}$$

where

$$H_{\text{eff}} = \sum_{i=1}^{n} \{H_v(i) + V_P(i)\} + S\Pi. \tag{8.85}$$

The operator H_v is given by Eq. (8.71). The connection between H_v and the H_i in Eq. (8.61) is given by

$$H_v(i) = H_i + U_v(i), \tag{8.86}$$

which we also see from Eq. (8.81). V_P is the pseudopotential given by Eq. (8.75). Operator S is given by

$$S = Q - \sum_{i=1}^{n} U_v(i), \tag{8.87}$$

and Q is shorthand for the electrostatic interaction of the valence electrons, that is, for the last term in Eq. (8.61).

We now prove that Eq. (8.84) is equivalent to Eq. (8.59). Multiply the equation from the left by Π. Then we get

$$\Pi H_{\text{eff}}\Psi = \Pi \left(\sum_{i=1}^{n} [H_v(i) + V_P(i)] \right) + \Pi S \Pi \Psi$$
$$= E\Pi\Psi. \tag{8.88}$$

Let us consider the effect of operator Π on the expression in braces. We get

$$\Pi \left\{ \sum_{i=1}^{n} [H_v(i) + V_P(i)] \right\}$$
$$= \sum_{i=1}^{n} \left\{ \Pi H_v(i) + \Pi \left[\sum_{j=1}^{N} \langle i|j\rangle\langle F_i| - \Pi_i H_v(i)\Omega_i \right] \right\}. \tag{8.89}$$

From Eq. (8.55), it follows that Π annihilates any core orbitals and that it has the property [Szasz, 277]

$$\Pi\Pi_i = \Pi, \tag{8.90}$$

because Π_i is idempotent. Thus, we obtain

$$\Pi \left\{ \sum_{i=1}^{n} [H_v(i) + V_P(i)] \right\} = \sum_{i=1}^{n} \Pi H_v(i)(1 - \Omega_i)$$

$$= \sum_{i=1}^{n} \Pi H_v(i)\Pi_i. \qquad (8.91)$$

Taking into account that $H_v(i)$ commutes with any Π_k except Π_i and that operators Π_k are idempotent, we get

$$\Pi \left\{ \sum_{i=1}^{n} [H_v(i) + V_P(i)] \right\} = \sum_{i=1}^{n} \Pi H_v(i)\Pi. \qquad (8.92)$$

Inserting this expression into Eq. (8.88), we get

$$\Pi H_{\text{eff}} \Psi = \sum_{i=1}^{n} \Pi H_v(i)\Pi\Psi + \Pi S \Pi \Psi$$

$$= E\Pi\Psi, \qquad (8.93)$$

or

$$\Pi \left\{ \sum_{i=1}^{n} H_v(i) + S \right\}(\Pi\Psi) = E(\Pi\Psi). \qquad (8.94)$$

Taking into account Eqs. (8.86) and (8.87) and using the relationship given by Eq. (8.82), we get

$$\Pi \left\{ \sum_{i=1}^{n} H_i + Q \right\}\Phi = E\Phi, \qquad (8.95)$$

which is identical with Eq. (8.59), Q.E.D.

It is an important point that in the pseudopotential Hamiltonian, we have the operator S instead of the Q that we have in Eq. (8.61). This is the result of the neutral atom approximation. After Eq. (8.86), we have written "the V_P is given by Eq. (8.75)." In Eq. (8.75), we have H_v and not H_i; this means that the presence of U_v in the one-electron pseudopotential is the reason why we must have the operator S and not Q in the Hamiltonian, Eq. (8.85). From this, it follows that in the positive-ion approximation, in which we have Eq. (8.80), we can put

$$S \equiv Q, \qquad (8.96)$$

$$H_v \equiv H_i, \qquad (8.97)$$

and for the PP effective Hamiltonian, we get

$$H_{\text{eff}} = \sum_{i=1}^{n} \{H_i + V_P(i)\} + Q\Pi, \qquad (8.98)$$

where now

$$V_P(i) = \sum_{j=1}^{n} \langle i|\varphi_j\rangle\langle F_j| - \Pi_i H_i \Omega_i. \qquad (8.99)$$

It is evident that if we want to introduce core/valence correlation into the effective Hamiltonians, we must add the one-electron polarization potentials, $V_C(i)$, to the one-electron operators and the two-electron polarization potentials, $V_C(i, j)$ to the two-electron operator Q. This change would be equivalent to the pseudopotential transformation of the BD effective Hamiltonian given by Eq. (8.67).

The main difference between the pseudopotential Hamiltonian, Eq. (8.85), and the conventional Hamiltonian, Eq. (8.60), is that in the conventional Hamiltonian, we have Π in front of the operator, whereas in the pseudopotential Hamiltonian, we do not have Π in the front, but we have the one-electron pseudopotentials, V_P, in the Hamiltonian. This is plausible: we have seen that in order to strong-orthogonalize the eigenfunctions of a many-electron Hamiltonian, we have to multiply the Hamiltonian from the left by Π. We also have seen, in the discussions of one-electron pseudopotential theory, that the orthogonalization can be replaced exactly by the introduction of pseudopotentials. Thus, if we remove operator Π from the front, we must have pseudopotentials in the Hamiltonian. An additional finesse of the transformation is that the removal of a *many-electron* projection operator is compensated for by the introduction of *one-electron* pseudopotentials. (The moving of Π from the front of Q to the back of S is a mathematically important detail, but it does not have such a plausible interpretation.)

The operator in Eq. (8.85) is the centerpiece of effective Hamiltonian theory. In view of its importance for atomic, molecular, and solid-state physics, we summarize here the properties of this effective Hamiltonian. First, we write down our results in a detailed form. The effective Schroedinger equation for n valence electrons is given by

$$H_{\text{eff}}\Psi(1, 2, \ldots, n) = E\Psi(1, 2, \ldots, n), \qquad (8.100)$$

where H_{eff} is the Hamiltonian

$$H_{\text{eff}} = \sum_{i=1}^{n} \{H_v(i) + V_P(i)\} + S\Pi. \qquad (8.101)$$

Here H_v is the one-electron Hamiltonian:

$$H_v = t + g + U_C + U_v, \tag{8.102}$$

and V_P is the one-electron pseudopotential:

$$V_P = \sum_{j=1}^{N} |\varphi_j\rangle\langle F_j| - \Pi H_v \Omega. \tag{8.103}$$

The operator S is given by

$$S = Q - \sum_{i=1}^{n} U_v(i), \tag{8.104}$$

where

$$Q = \frac{1}{2} \sum_{j,k=1}^{n} \frac{1}{r_{jk}}. \tag{8.105}$$

The properties of H_{eff} are as follows:

1. The Hamiltonian is *exactly* equivalent to the conventional effective Hamiltonian given by Eq. (8.60). This means that the eigenvalues of the two operators are equal and the eigenfunctions of the pseudopotential Hamiltonian reproduce the eigenfunctions of the conventional Hamiltonian through the relationship

$$\Phi = \Pi \Psi. \tag{8.106}$$

2. The valence–valence correlation effects are included in Eq. (8.100) exactly, that is, the exact solutions of that equation will contain all intravalence correlation effects. The core–valence correlation effects can be taken into account approximately if we introduce the polarization potentials, that is, if we carry out the pseudopotential transformation on the BD equation, Eq. (8.68), instead of on the FVP equation, Eq. (8.59).
3. The equation is self-contained, that is, it can be solved as if the core would not exist.
4. The effective Hamiltonian is non-Hermitian. Throughout the discussions of this book, we encountered Hamiltonian operators that are not Hermitian. This is a frequent occurrence in atomic structure theory and we summarize the properties of these operators in App. K. The most important property is that these operators form a special class of non-Hermitian operators whose eigenvalues are always real.

5. Whatever coupling scheme is used for the whole atom, the same coupling scheme will apply to the valence-electron function Ψ. The effective Hamiltonian, Eq. (8.101), represents a core with complete groups; therefore, it will not change the angular-momentum properties of Ψ. (We have seen that in the case of a core with complete groups, the one-electron electrostatic, exchange, and pseudopotentials do not change the angular and spin part of a one-electron orbital $\psi(q)$.)

6. The effective pseudopotential Hamiltonian can be transferred, almost without change, into a molecule or solid [Szasz, 200], that is, a molecular effective Hamiltonian can be built up from atomic effective Hamiltonians almost without any changes in the latter.

We now discuss some simplifications that can be carried out in H_{eff}. Let us consider the last term in the operator given by Eq. (8.101), the term that contains the projection operator Π. This term is very inconvenient for calculations because every application of the one-electron projection operator Π_i increases the number of dimensions in the integrations (makes a two-electron integral out of a one-electron integral). Thus, the presence of the n-electron projection operator is a serious obstacle to calculations.

Let us write down the effective Hamiltonian again and write out the detailed form of $S\Pi$. We get

$$
\begin{aligned}
H_{\text{eff}} &= \sum_{i=1}^{n} \{ H_v(i) + V_P(i) \} + S(1 - \Omega) \\
&= \sum_{i=1}^{n} \{ H_v(i) + V_P(i) \} + Q - \sum_{i=1}^{n} U_v(i) - S\Omega. \quad (8.107)
\end{aligned}
$$

Taking into account Eq. (8.86), we get

$$
H_{\text{eff}} = \sum_{i=1}^{n} \{ H_i + V_P(i) \} + Q - S\Omega. \quad (8.108)
$$

Thus, the mathematically offending term is $-S\Omega$, where S is given by Eq. (8.104). In App. L, we show[†] that in good approximation, we can write

$$
Q - S\Omega \approx \hat{\eta} Q, \quad (8.109)
$$

where $\hat{\eta}$ is a constant of the form

$$
\hat{\eta} = 1 + \eta, \quad (8.110)
$$

[†]The derivations of App. L are presented here for the first time. The results embodied in Eqs. (8.110), (8.111), (8.113), and (8.115) were presented in the author's book [Szasz, 187] along with some references to earlier work on the problem. We also presented a numerical discussion of η and $\hat{\eta}$, showing the magnitude of these constants in certain special cases.

and η is given by

$$\eta = 1 - [\langle\Pi\rangle_1\langle\Pi\rangle_2 \cdots \langle\Pi\rangle_n]. \tag{8.111}$$

Here $\langle\Pi\rangle_k$ is the expectation value of the one-electron projection operator Π with respect to ψ_k, the pseudo-orbital solution of Eq. (8.78), that is,

$$\langle\Pi\rangle_k = \langle\psi_k|\Pi|\psi_k\rangle = 1 - \langle\psi_k|\Omega|\psi_k\rangle. \tag{8.112}$$

Thus, in the neutral-atom approximation, we get

$$H_{\text{eff}} = \sum_{i=1}^{n} \{H_i + V_P(i)\} + \hat{\eta}Q. \tag{8.113}$$

In the positive-ion approximation, we have $S \equiv Q$, and, as we show in App. L, we get

$$Q - Q\Omega \approx (1 - \eta)Q, \tag{8.114}$$

and so the effective Hamiltonian becomes

$$H_{\text{eff}} = \sum_{i=1}^{n} \{H_i + V_P(i)\} + (1 - \eta)Q. \tag{8.115}$$

The derivation of these approximations are given in App. L. Here we want to discuss the physical meaning of these approximations. Let us consider, first, Eq. (8.114), which says that the expectation value of $Q\Omega$ that must be subtracted from the expectation value of Q in order to satisfy the Pauli exclusion principle can be expressed as $\eta\langle Q\rangle$, where η is a constant given by Eq. (8.111). The physical meaning of η is simple. We have

$$\begin{aligned}
\eta &= 1 - \langle\Pi\rangle_1\langle\Pi\rangle_2 \cdots \langle\Pi\rangle_n \\
&= 1 - \{(1 - \langle\Omega\rangle_1)(1 - \langle\Omega\rangle_2) \cdots (1 - \langle\Omega\rangle_n)\}. \tag{8.116}
\end{aligned}$$

The expectation values of Ω with respect to the pseudo-orbitals, $\psi_1, \psi_2, \ldots, \psi_n$, are the quantities denoted here by $\langle\Omega\rangle_k$. These quantities represent the core penetration by the pseudo-orbitals and they are generally small numbers. If the pseudo-orbitals would be replaced by the HF orbitals, $\varphi_1 \cdots \varphi_n$, the expectation values would be zero. Thus, we have

$$\lim_{\langle\Omega\rangle_k \to 0} \eta = 0. \tag{8.117}$$

In general, we see from the formula for η

$$0 \le \eta \le 1. \tag{8.118}$$

Thus, the meaning of this approximation is that instead of subtracting $\langle Q\Omega \rangle$ from $\langle Q \rangle$, we can subtract $\eta \langle Q \rangle$, where η is given by Eq. (8.111) and subjected to Eq. (8.118). This is what we would expect physically because the subtraction of the "core part," $\langle Q\Omega \rangle$, would certainly diminish $\langle Q \rangle$ and this is exactly what is accomplished by multiplying $\langle Q \rangle$ with the constant $1 - \eta$. It should be noted, however, that this approximation provides a meaningful reduction in the value of $\langle Q \rangle$, but, of course, the constant $1 - \eta$ does not provide the "push" out of the core that is provided by the n-electron projection operator Π. That "push" would move the valence electrons farther away from the core, thereby changing their total binding energy. If we replace Π by $1 - \eta$, this "push" will be missing from the effective Hamiltonian.

The physical meaning of $\hat{\eta}$ in Eq. (8.109) is not so straightforward because $\hat{\eta}$ gives the ratio

$$\hat{\eta} = \frac{\langle Q \rangle - \langle S\Omega \rangle}{\langle Q \rangle}, \qquad (8.119)$$

which is a comparison between different operators. It is shown in App. L how this comes about. In any case, as we see from the definition of $\hat{\eta}$ and from Eq. (8.118), we have

$$1 \leqq \hat{\eta} \leqq 2. \qquad (8.119a)$$

In an actual calculation, η and $\hat{\eta}$ can be either computed from Eq. (8.111) or can be treated as adjustable parameters. In most calculations, it is customary to put

$$\eta = 0, \qquad \hat{\eta} = 1. \qquad (8.119b)$$

As we see from Eqs. (8.113) and (8.115), after the elimination of the projection operators, the only quantity needed for the calculations is the operator $H_i + V_P(i)$. This operator can be generated by solving the one-electron pseudopotential equations for the valence electrons, Eq. (8.78). Having solved those equations, we obtain the local (or semi-local) modified potential from Eq. (8.79). Thus, we get, in the neutral-atom approximation,

$$H_{\text{eff}} = \sum_{i=1}^{n} \{t_i + V_M(i)\} + \hat{\eta}Q, \qquad (8.120)$$

where we used the relationship

$$H_i + V_P(i) = t_i + g_i + U_C(i) + V_P(i)$$

$$= t_i + V_M, \qquad (8.121)$$

and V_M must satisfy the pseudopotential equation

$$(t + V_M + U_v)\psi_v = \varepsilon_v \psi_v. \tag{8.122}$$

In the positive-ion approximation, we get

$$H_{\text{eff}} = \sum_{i=1}^{n} \{t_i + V_M(i)\} + (1 - \eta)Q, \tag{8.123}$$

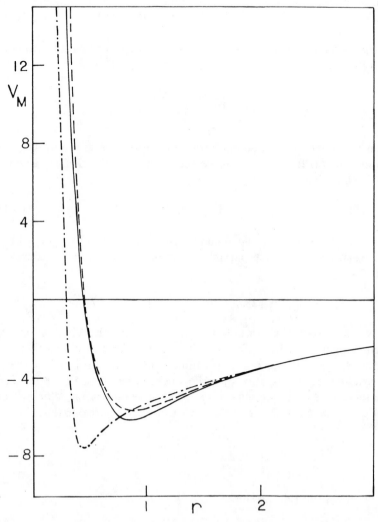

Fig. 8.1. The modified potentials (Hartree–Fock potential and pseudopotential) for a valence electron of the Br atom. Solid line: the s potential. Dashed line: the p potential. Dot–dash line: the d potential.

where V_M must satisfy the pseudopotential equation

$$(t + V_M)\psi_v = \varepsilon_v\psi_v. \tag{8.124}$$

In both cases, V_M is given by Eq. (8.79), where $U_v \equiv 0$ for the positive-ion approximation. Ab initio pseudopotentials were constructed first by Szasz and McGinn.[72] The resultant modified potentials, which are in the positive-ion approximation, were discussed in Sec. 7.2. For atoms with several valence electrons, Kahn, Baybutt, and Truhlar constructed very smooth ab initio modified potentials; these are in the neutral-atom approximation [Szasz, 153]. The modified potentials for the valence electrons of the Br atom are illustrated in Fig. 8.1.

Another method for constructing V_M is to put this potential into analytic form with adjustable parameters and determine the parameters from the requirement that Eq. (8.122) or Eq. (8.124) should reproduce the HF orbital parameters. The choice of the analytic form is guided by the form of the HF-equivalent modified potentials, that is, by the requirement that the potential should have the barrier/well/Coulombic sequence [Szasz, 79, 163].

Finally, last but not least, we register a caveat about effective Hamiltonian operators. Generally, the eigenfunctions of these operators cannot be used for the calculation of expectation values of operators other than the Hamiltonian, that is, the energy operator. The reason for this is that in the derivations of the effective Hamiltonians, we have repeatedly transferred relationships between expectation values to the operators themselves. Thus, strictly speaking, the derivations are valid for the expectation values (of the original Hamiltonian), but not for the effective operators themselves. Another reason is that the eigenfunctions of Eq. (8.100) are pseudo-wave functions. In order to calculate the expectation values of N-electron operators, the n-electron pseudo-wave functions must be strong-orthogonalized to the core orbitals and placed into the total wave function given by Eq. (8.54). It is that wave function that must be used for the calculation of expectation values and not the pseudo-wave function.

8.5. REPRESENTATIVE CALCULATIONS

When we are trying to demonstrate the use of effective Hamiltonians in atomic structure theory, we are faced with the difficulty that a very high percentage of the work concerning effective Hamiltonians was done for molecular applications. There are many molecular calculations with effective Hamiltonians that we could discuss [Szasz, 195], but those are outside of the scope of this book. From the relatively few atomic calculations, we have selected for presentation McGinn's work,[87] which is a Hartree–Fock-equivalent pseudopotential calculation for two valence electrons, and the work of Bardsley et al.,[88] which is a pseudopotential calculation with the valence–

valence correlation taken into account. From these two calculations, we can see the useful simplicity of the effective Hamiltonian methods and we get an idea about their accuracy.

We start with the presentation of McGinn's work. Let us consider an atom with N core electrons in complete groups plus two valence electrons. If the core orbitals are $\varphi_1, \varphi_2, \ldots, \varphi_N$ and valence electrons are represented by the function $\Phi(1, 2)$, then the expectation value of the Hamiltonian is given by Eq. (8.15):

$$E_T = E_C + \langle \Phi(1, 2)|H_{12}|\Phi(1, 2)\rangle, \qquad (8.125)$$

where E_T is the total energy, E_C is the core energy, and

$$H_{12} = H_1 + H_2 + \frac{1}{r_{12}}, \qquad (8.126)$$

with

$$H_i = H_F(i) = t_i + g_i + U(i), \qquad (8.127)$$

where

$$U(i) = \sum_{k=1}^{N} U_k(i). \qquad (8.128)$$

In this equation, U_k is the HF potential generated by the orbital φ_k. The total wave function for the atom underlying Eq. (8.125) is given by Eq. (8.5). The one-electron orbitals and the valence-electron function are arbitrary, apart from the conditions given by Eqs. (8.3) and (8.8).

We start by putting, for the two-electron function of the valence electron,

$$\Phi(1, 2) = \frac{1}{\sqrt{2}} \begin{vmatrix} \varphi_{N+1}(1) & \varphi_{N+1}(2) \\ \varphi_{N+2}(1) & \varphi_{N+2}(2) \end{vmatrix}, \qquad (8.129)$$

where the spin-orbitals φ_{N+1} and φ_{N+2} are normalized and orthogonal to each other as well as to the core orbitals. By putting Eq. (8.129) into Eq. (8.125), we obtain

$$\begin{aligned} E_F = E_T - E_C = {} & \langle \varphi_{N+1}|H_F|\varphi_{N+1}\rangle + \langle \varphi_{N+2}|H_F|\varphi_{N+2}\rangle \\ & + \langle \varphi_{N+1}(1)\varphi_{N+2}(2)|\frac{1}{r_{12}}|\varphi_{N+1}(1)\varphi_{N+2}(2)\rangle \\ & + \langle \varphi_{N+1}(1)\varphi_{N+2}(2)|\frac{1}{r_{12}}|\varphi_{N+1}(2)\varphi_{N+2}(1)\rangle. \end{aligned} \qquad (8.130)$$

Let us assume that the valence orbitals have the form

$$\left.\begin{aligned} \varphi_{N+1} &= \psi_{ns}\alpha, \\ \varphi_{N+2} &= \psi_{ns}\beta, \end{aligned}\right\} \tag{8.130a}$$

where α and β are the up and down spin functions. Substituting these expressions into Eq. (8.130), we can carry out the spin integrations and obtain

$$
\begin{aligned}
E_F = &\ 2\langle \psi_{ns}|H_F|\psi_{ns}\rangle \\
&+ \langle \psi_{ns}(1)\psi_{ns}(2)| \frac{1}{r_{12}} |\psi_{ns}(1)\psi_{ns}(2)\rangle,
\end{aligned}
\tag{8.131}
$$

where

$$H_F = t + g + V - A. \tag{8.132}$$

In this equation, V and A are the spinless electrostatic and exchange potential operators, respectively, defined as

$$V(1) = 2\sum_i \int \frac{|\psi_i(2)|^2 \, dv_2}{r_{12}} \tag{8.133}$$

and

$$A(1)f(1) = \sum_i \int \frac{\psi_i(1)\psi_i^*(2)f(2)\,dv_2}{r_{12}}, \tag{8.134}$$

where ψ_i is the spatial part of the core orbitals, and the summation is over all core orbitals with different spatial parts; that is, because the spatial parts of the orbitals with up and down spin are identical, the summation is over orbitals with parallel spins.

We obtain the HF equation for ψ_{ns} by varying E_F with respect to ψ_{ns} under the subsidiary conditions of normalization and orthogonality to the spatial parts of the core orbitals.

The result is

$$H_F\psi_{ns} + V_0\psi_{ns} = \varepsilon_{ns}\psi_{ns} + \sum_i \lambda_i \psi_i, \tag{8.135}$$

where

$$V_0(1) = \int \frac{|\psi_{ns}(2)|^2 \, dv_2}{r_{12}}, \tag{8.136}$$

and λ_i is a Lagrangian multiplier. It is easy to see that

$$\lambda_i = \langle \psi_i | H_F + V_0 | \psi_{ns} \rangle. \tag{8.137}$$

Multiplying Eq. (8.135) from the left by ψ_{ns}^* and integrating, we obtain

$$\langle \psi_{ns} | H_F | \psi_{ns} \rangle + \langle \psi_{ns} | V_0 | \psi_{ns} \rangle = \varepsilon_{ns}. \tag{8.138}$$

From Eq. (8.131), we get

$$E_F = 2\langle \psi_{ns} | H_F | \psi_{ns} \rangle + \langle \psi_{ns} | V_0 | \psi_{ns} \rangle. \tag{8.139}$$

From Eq. (8.138), we obtain

$$2\langle \psi_{ns} | H_F | \psi_{ns} \rangle = 2\varepsilon_{ns} - 2\langle \psi_{ns} | V_0 | \psi_{ns} \rangle,$$

and substituting this into Eq. (8.139), we get, for the energy of the two valence electrons,

$$E_F = 2\varepsilon_{ns} - \langle \psi_{ns} | V_0 | \psi_{ns} \rangle. \tag{8.140}$$

Here we note that Eq. (8.135) is a special case of Eq. (8.70) with

$$U_v = V_0. \tag{8.141}$$

We introduce now the pseudo-orbital ψ_{ns}^0 with the relationship

$$\psi_{ns} = N_0 \left(\psi_{ns}^0 - \sum_i \alpha_i \psi_i \right), \tag{8.142}$$

where N_0 is a normalization constant and

$$\alpha_i = \langle \psi_i | \psi_{ns}^0 \rangle. \tag{8.143}$$

By putting Eq. (8.142) into Eq. (8.135), we obtain the equation for ψ_{ns}^0:

$$\left(H_F + V_0 + \tilde{V}_P \right) \psi_{ns}^0 = \varepsilon_{ns} \psi_{ns}^0, \tag{8.144}$$

where $\tilde{V}_P$ is a pseudopotential given by the formula

$$\tilde{V}_P = \sum_i \frac{[\alpha_i(\varepsilon_{ns} - \varepsilon_i - V_0) + C_i]\psi_i}{\psi_{ns}^0}, \tag{8.145}$$

with

$$C_i = -\langle \psi_i | V_0 | \psi_{ns}^0 \rangle + \sum_j \alpha_j \langle \psi_i | V_0 | \psi_j \rangle. \tag{8.146}$$

In obtaining Eq. (8.145), we have assumed the validity of the *positive-ion* approximation, that is, we have assumed that the core orbitals are the same as in the positive ion, which is obtained by the removal of the two valence electrons. In equation

$$H_F \psi_i = \varepsilon_i \psi_i, \tag{8.147}$$

and if this assumption is made, then Eq. (8.137) simplifies to

$$\lambda_i = \langle \psi_i | V_0 | \psi_{ns} \rangle. \tag{8.148}$$

Inspecting the pseudopotential given by Eq. (8.145), we observe that the first two terms form the Phillips–Kleinmann potential. The term with V_0, which is new here, represents the presence of a second valence electron. The coefficients C_i, which occur in the pseudopotential and which are given by Eq. (8.146), can be shown to be negligibly small. These coefficients are the result of the presence of Lagrangian multipliers in Eq. (8.135); the smallness of these shows that the multipliers are negligible.

McGinn solved Eq. (8.144), in a self-consistent fashion, for the ^1S, $(ns)^2$ ground states of the atoms Li^-, Na^-, Rb^-, Cu^-, Ag^-, Be, Mg, Ca, Zn, and Al^+. The equation must be solved by iteration because the potentials depend on both ψ_{ns}^0 and ε_{ns}. The core orbitals remained fixed in the calculations. The starting estimates for the pseudo-orbitals were obtained by taking the self-consistent results of the pseudopotential calculations for the atoms with one less valence electron. For example, the starting estimate for the pseudo-orbital of Li^- was the $(2s)$ pseudo-orbital of Li, the starting estimate for Ca was the $(4s)$ PO of Ca^+, and so on. Here McGinn used the results of previous pseudopotential calculations, which we discussed in Section 7.2/D.

The results are summarized in Table 8.1. The ionization energy is computed from the formula

$$I = -[E_F - E(N + 1)], \tag{8.149}$$

where $E(N + 1)$ is the binding energy of the valence electron in the $(N + 1)$-electron atom. The numbers for $E(N + 1)$ are taken from the one-valence-electron pseudopotential calculations. The difference between the calculated and observed ionization energies is due to the core–valence and valence–valence correlation effects, which were not taken into account here. The most striking example for the importance of the valence–valence correlation effects is the absence of binding (negative ionization energy) for the ions. No electron affinities were obtained by McGinn in these pseudopotential calculations, which are equivalent to the HF method. Thus, the binding of a second valence electron in these atoms is a correlation effect.

Next, McGinn developed the formalism for the $(nsml)$ ^1L and ^3L states of atoms with two valence electrons outside complete groups. The method is

TABLE 8.1. Energy Values of the Ground States of Atoms with Two Valence Electrons

Atom	n	ε_{ns}	E_F	$E(N+1)$	Ionization Energy Calc.	Ionization Energy Obs.
Be	2	−0.3085	−0.9597	−0.6648	0.2949	0.3426
Mg	3	−0.2568	−0.7956	−0.5485	0.2471	0.2810
Ca	4	−0.1940	−0.6008	−0.4139	0.1870	0.2247
Zn	4	−0.2890	−0.8882	−0.6098	0.2784	0.3454
Al$^+$	3	−0.6605	−1.6961	−1.0424	0.6537	0.6919
Li$^-$	2	−0.0145	−0.1921	−0.1966	−0.0044	0.0228
Na$^-$	3	−0.0132	−0.1774	−0.1811	−0.0037	0.0202
K$^-$	4	−0.0102	−0.1438	−0.1467	−0.0029	0.0184
Rb$^-$	5	−0.0095	−0.1354	−0.1379	−0.0025	0.0180
Cu$^-$	4	−0.0198	−0.2331	−0.2339	−0.0008	
Ag$^-$	5	−0.0181	−0.2141	−0.2140	0.0001	

exactly the same as in the preceding section. We use the expression in Eq. (8.125) as our starting point. The next step is to define the two-electron function for the valence electrons. We put

$$\Phi(1,2) = \tfrac{1}{2}\left[\psi_{ns}(1)\psi_{ml}(2) \pm \psi_{ns}(2)\psi_{ml}(1)\right]$$

$$\times\left[\alpha(1)\beta(2) \mp \alpha(2)\beta(1)\right], \qquad (8.150)$$

where the upper sign is for the singlet, and the lower sign for the triplet states. The ψ_{ns} and ψ_{ml} are normalized and orthogonal to the core states. Except for the ^1S states, they are also orthogonal to each other.

Next, the wave function is put into Eq. (8.125) and an energy expression, in terms of ψ_{ns} and ψ_{ml}, is obtained. The HF equations for these orbitals can be obtained by application of the variation principle.

Here McGinn introduced the extended frozen-core approximation (Sec. 6.4). He assumed that the ψ_{ns} is frozen and is identical with the valence orbital of the $(N + 1)$-electron atom. As an additional simplification, McGinn also assumed that the Lagrangian multipliers can be omitted. This was assumed on the basis of the work done for the $(ns)^2$ configuration.

Varying the energy expressions, under the assumption of these approximations, with respect to ψ_{ml}, one obtains the HF equations

$$(H_F + V_0 \mp A_0)\psi_{ml} = \varepsilon_{ml}\psi_{ml} + K\psi_{ns}, \qquad (8.151)$$

where H_F is given by Eq. (8.132), and V_0 by Eq. (8.136). The upper sign is for the triplet and the lower for the singlet states and

$$A_0(1)f(1) = \int \frac{\psi_{ns}(1)\psi_{ns}^*(2)f(2)\,dv_2}{r_{12}}. \tag{8.152}$$

Also

$$K = \langle \psi_{ns}|V_0|\psi_{ml}\rangle, \qquad \text{for } {}^1S; \\ = 0, \qquad\qquad \text{otherwise.} \Big\} \tag{8.153}$$

Next, we transform the HF equation into a pseudopotential equation by putting

$$\psi_{ml} = N_0\left(\psi_{ml}^0 - \sum_i \alpha_i \psi_i - \alpha_{ns}\psi_{ns}\right), \tag{8.154}$$

where the symbols mean the same as in Eq. (8.142), and

$$\alpha_{ns} = \langle \psi_{ns}|\psi_{ml}^0\rangle, \qquad \text{for } {}^3S; \\ = 0, \qquad\qquad \text{otherwise.} \tag{8.155}$$

By putting Eq. (8.154) into Eq. (8.151), we obtain the following equations for the Rydberg orbital.

For ${}^{1,3}L$ states, $L \neq S$,

$$\left(H_F + V_0 \mp A_0 + \tilde{V}_P\right)\psi_{ml}^0 = \varepsilon_{ml}\psi_{ml}^0. \tag{8.156}$$

For ${}^3S$ states

$$\left(H_F + V_0 - A_0 + \tilde{V}_P + \hat{V}_P\right)\psi_{ms}^0 = \varepsilon_{ms}\psi_{ms}^0. \tag{8.157}$$

For ${}^1S$ states,

$$\left(H_F + V_0 + A_0 + \tilde{V}_P + C\right)\psi_{ms}^0 = \varepsilon_{ms}\psi_{ms}^0. \tag{8.158}$$

In these equations, we have

$$\tilde{V}_P = \sum_i \alpha_i(\varepsilon_{ml} - \varepsilon_i - V_0 \pm A_0)\frac{\psi_i}{\psi_{ml}^0}, \tag{8.159}$$

$$\hat{V}_P = \alpha_{ns}(\varepsilon_{ms} - \varepsilon_{ns})\frac{\psi_{ns}}{\psi_{ms}^0}, \tag{8.160}$$

and

$$C = \frac{\left(-\langle \psi_{ns}|V_0|\psi_{ms}^0\rangle \psi_{ns} + \sum_i \alpha_i \langle \psi_{ns}|V_0|\psi_i\rangle \psi_{ns}\right)}{\psi_{ms}^0}. \quad (8.161)$$

In Eqs. (8.156) and Eq. (8.159), the upper sign is for the triplet, and the lower for the singlet states.

Equations (8.156), (8.157), and (8.158) are the one-electron pseudopotential equations for the Rydberg orbitals. Thus, by fixing one of the valence electrons in the ground state, the original two-valence-electron problem is transformed into a one-electron problem.

As we have mentioned before, the valence orbital was taken to be the same as in the $(N + 1)$-electron problem. Thus, the physical meaning of the pseudopotential equations is that the orbital parameters, which appear in the equations, represent the energies of the Rydberg orbital in the potential field of the fixed $(N + 1)$-electron atom. Therefore, in the framework of the approximations that were used by McGinn, the orbital parameters approximate the ionization energies for the corresponding Rydberg states.

The pseudopotential equations were solved by McGinn with a self-consistent iteration procedure for the $(nsml)^{1,3}S$ and $^{1,3}P$ states of Be for $m = 2 - 4$; for the $^{1,3}S$ and $^{1,3}P$ states of Ca for $m = 4 - 6$, and for the $^{1,3}D$ states of Ca with $m = 3 - 5$. In addition, some transition integrals were also computed using the HF orbitals generated from the pseudo-orbitals. In Table 8.2, we have the results for the s and p states of Be and the results for Ca. The results are compared with the empirical values of the ionization energy.

TABLE 8.2. The Ionization Energies of Various States of the Be and Ca Atoms

Beryllium			Calcium				
				Singlet		Triplet	
State	Calc.	Obs.	State	Calc.	Obs.	Calc.	Obs.
$2s2s'$ ^1S	0.29956	0.34262	$4s4s'$	0.19071	0.22465		
$2s2p$ ^3P	0.22731	0.24246	$4s5s$	0.06479	0.07285	0.07503	0.08095
$2s2p$ ^1P	0.11716	0.14867	$4s6s$	0.03482	0.03426	0.03826	0.04024
$2s3s$ ^3S	0.10009	0.10530	$4s4p$	0.08586	0.11688	0.14953	0.15511
$2s3s$ ^1S	0.08488	0.09348	$4s5p$	0.04378	0.03475	0.05500	0.05805
$2s3p$ ^3P	0.07144	0.07473	$4s6p$	0.02622	0.02447	0.03066	0.03090
$2s3p$ ^1P	0.05577	0.06851	$4s3d$	0.06412	0.12510	0.09284	0.13190
$2s4s$ ^3S	0.04721	0.04869	$4s4d$	0.03944	0.05471	0.04476	0.05263
$2s4s$ ^1S	0.04236	0.04532	$4s5d$	0.02479	0.02910	0.02658	0.02989
$2s4p$ ^3P	0.03725	0.04212					
$2s4p$ ^1P	0.03175	0.03629					

The agreement between the calculated and observed values is reasonably good; we must recall that the difference is caused by the correlation effects. Also, the additional assumption of the extended frozen-core approximation may contribute to the difference between computed and measured values. In order to check this last point, McGinn compared his results with ab initio HF calculations, that is, with HF calculations in which the "extended frozen core" was not used. The agreement was good in the lower Rydberg states and very good in the higher.

McGinn also applied the theory to the doubly excited states of atoms with two valence electrons. Investigated were the ($npms$) states of Be, Mg, Al$^+$, Si^{2+}, Ca, and Zn. Specifically, the states investigated were as follows:

Be:
  ($2pns$) for $n = 3 - 7$;
  ($3pns$) for $n = 4 - 7$;
  ($3snp$) for $n = 3 - 7$;
  ($4snp$) for $n = 4 - 7$;
Mg, Al$^+$, Si^{2+}:
  ($3pns$) for $n = 4 - 7$;
Ca:
  ($4pns$) for $n = 5 - 8$;
  ($5pns$) for $n = 6 - 8$;
  ($5snp$) for $n = 5 - 8$;
  ($6snp$) for $n = 6 - 7$;
Zn:
  ($4pns$) for $n = 5 - 8$.

In all these states, McGinn used the extended frozen-core approximation, that is, he fixed one of the orbitals to be identical with the valence orbital of the ($N + 1$)-electron system. Thus, for example, in calculating the Be ($2pns$) series, the ($2p$) orbital was frozen to be identical with the ($2p$) orbital of the Be$^+$ atom in the ($1s$)^2($2p$) configuration. In other words, the frozen orbitals define the ($N + 1$)-electron system that is obtained when the second valence electron is removed. The problem is again reduced to a one-electron problem this way.

The pseudopotential equations, which are necessary for the calculations, are derived in the same way as in the preceding calculations. The two-electron function for the valence electrons is given by

$$\Phi(1,2) = \tfrac{1}{2}\left[\psi_f(1)\psi_v(2) \pm \psi_f(2)\psi_v(1)\right]$$
$$\times \left[\alpha(1)\beta(2) \mp \alpha(2)\beta(1)\right], \qquad (8.162)$$

where the upper sign is for the singlet, and the lower for the triplet states. ψ_f is the orbital that is fixed and ψ_v is the orbital of the running electron.

Next, the HF equations are derived by keeping the core and ψ_f fixed. The Lagrangian multipliers are omitted. We obtain by this procedure

$$(H_F + V_0 \mp A_0)\psi_v = \varepsilon_v \psi_v, \tag{8.163}$$

where the upper sign is for the triplet, and the lower for the singlet state. H_F is defined by Eq. (8.132) and

$$V_0(1) = \int \frac{|\psi_f(2)|^2 \, dv_2}{r_{12}}, \tag{8.164}$$

$$A_0(1)f(1) = \int \frac{\psi_f(1)\psi_f^*(2)f(2) \, dv_2}{r_{12}}. \tag{8.165}$$

We introduce the pseudo-orbital by putting

$$\psi_v = N_0\left(\psi_v^0 - \sum_i \alpha_i \psi_i\right), \tag{8.166}$$

and substituting this into Eq. (8.163), we obtain

$$\left(H_F + V_0 \mp A_0 + \tilde{V}_P\right)\psi_v^0 = \varepsilon_v \psi_v^0, \tag{8.167}$$

where

$$\tilde{V}_P = \sum_i \frac{\alpha_i[\varepsilon_v - \varepsilon_i - V_0 \pm A_0]\psi_i}{\psi_v^0}. \tag{8.168}$$

In both equations, the upper sign is for the triplet and the lower sign is for the singlet state.

The calculations consisted of the exact numerical solutions of Eq. (8.167) with a self-consistent iteration method. The results for the Be and Ca atoms are in Table 8.3; those for Mg, Al$^+$ and Si^{2+} are in Table 8.4. In the tables, we find the absolute values of the orbital energies of the running electron, that is, the orbital parameter in Eq. (8.167). Comparison with empirical values is possible if we recall that ε_v is the energy of the system relative to the energy of the corresponding $(N + 1)$-electron atom. Thus, the comparison must be made between the calculated and observed difference between the indicated $^{1,3}P$ state of the $(N + 2)$-electron atom and the corresponding doublet state of the $(N + 1)$-electron atom. Such a comparison is presented in Table 8.5.

Reviewing the results of McGinn, we can say that for all three groups of atomic states, the computed values provide a reasonable approximation to the empirical data. The sizable gap between theory and experiment is caused, in every case, by the lack of electron correlation in the Hartree–Fock-equiv-

TABLE 8.3. Energy Levels of the Doubly Excited States of the Be and Ca Atoms

Configuration	^3P	^1P	Configuration	^3P	^1P
		Beryllium			
STATES WITH ^2P Be$^+$ LIMIT					
$2p3s$	0.09532	0.08458	$3p4s$	0.05415	0.04876
$2p4s$	0.04584	0.04261	$3p5s$	0.03033	0.02854
$2p5s$	0.02696	0.02555	$3p6s$	0.01946	0.01862
$2p6s$	0.01774	0.01700	$3p7s$	0.01356	0.01309
$2p7s$	0.01256	0.01212			
STATES WITH ^2S Be$^+$ LIMIT					
$3s3p$	0.11118	0.06637	$4s4p$	0.06347	0.04080
$3s4p$	0.04513	0.03699	$4s5p$	0.03051	0.02537
$3s5p$	0.02631	0.02315	$4s6p$	0.01930	0.01712
$3s6p$	0.01732	0.01575	$4s7p$	0.01340	0.01226
$3s7p$	0.01229	0.01138			
		Calcium			
STATES WITH ^2P Ca$^+$ LIMIT					
$4p5s$	0.07374	0.06841	$5p6s$	0.04407	0.04116
$4p6s$	0.03803	0.03632	$5p7s$	0.02587	0.02477
$4p7s$	0.02328	0.02251	$5p8s$	0.01704	0.01655
$4p8s$	0.01573	0.01531			
STATES WITH ^2S Ca$^+$ LIMIT					
$5s5p$	0.07391	0.05014	$6s6p$	0.04437	0.03260
$5s6p$	0.03500	0.02927	$6s7p$	0.02438	0.02081
$5s7p$	0.02156	0.01907			
$5s8p$	0.01470	0.01338			

alent pseudopotential calculations. The work of McGinn demonstrates the usefulness and simplicity of the extended frozen-core approximation.

We now proceed to the discussion of a calculation in which the n-electron effective pseudopotential Hamiltonian given by Eq. (8.123) was used. The Hamiltonian reads

$$H_{\text{eff}} = \sum_{i=1}^{n} \{t_i + V_m(i)\} + Q, \qquad (8.169)$$

TABLE 8.4. Energy Levels of the Doubly Excited States of the Mg, Al$^+$, and Si^{2+} Atoms

Configuration	Mg		Al$^+$		Si^{2+}	
	^3P	^1P	^3P	^1P	^3P	^1P
$3p4s$	0.0881	0.0813	0.2667	0.2545	0.5185	0.5020
$3p5s$	0.0434	0.0413	0.1419	0.1380	0.2869	0.2816
$3p6s$	0.0259	0.0250	0.0883	0.0865	0.1825	0.1801
$3p7s$	0.0173	0.0168	0.0602	0.0593	0.1263	0.1250

where we have put $\eta = 0$ in Eq. (8.123). Bardsley et al.[88] made calculations for the negative ions Li$^-$, Na$^-$, K$^-$, Rb$^-$, and Cs$^-$. These ions are treated as two-electron problems that are described by the equation

$$H_{\text{eff}}\Psi(1,2) = E\Psi(1,2). \tag{8.170}$$

The Hamiltonian is given by Eq. (8.169) with $n = 2$. The core–valence effective interaction potential is given by V_M, for which the expression given by Eq. (7.219) was chosen. As we see, the potential contains both the core–core correlation and the core–valence correlation effects (the last through the polarization potential $\tilde{V}_C$).

The most important feature of these calculations is that here the valance–valence correlation effects are fully taken into account. This is done by applying the configuration interaction (CI) method in the calculation of an approximate solution of Eq. (8.170). Using this method, Bardsley et al. obtained the following results for the electron affinities of the negative ions (empirical values in parentheses): $E(\text{Li}^-) = 0.62$ (0.62); $E(\text{Na}^-) = 0.55$ (0.55); $E(\text{K}^-) = 0.53$ (0.50); $E(\text{Rb}^-) = 0.51$ (0.49); and $E(\text{Cs}^-) = 0.49$ (0.47). These are energies in eV units.

These good results are clearly due to the proper treatment of the valence–valence correlation. As we see from Table 8.1, the Hartree–Fock-

TABLE 8.5. Doubly Excited States of the Be, Al$^+$, and Si^{2+} Atoms

Atom	Configuration	Energy	
		Calc.	Obs.
Be	$2p3s \; ^3P$	0.09532	0.09827
Al$^+$	$3p4s \; ^3P$	0.2667	0.2725
	$3p4s \; ^1P$	0.2545	0.2629
Si^{2+}	$3p4s \; ^3P$	0.5185	0.5243
	$3p4s \; ^1P$	0.5020	0.4820

equivalent pseudopotential calculations did not give binding for these ions. Thus, we conclude that the decisive characteristic that makes the effective Hamiltonian method highly accurate is that the core–core and core–valence correlation effects are built into the Hamiltonian and the valence–valence correlation effects can be fully taken into account by constructing approximate solutions for Eq. (8.170) in which these effects are included. Thus, if the analytic expressions for the core–valence interaction potentials, that is, the analytic expressions for V_M, are chosen carefully and the solution of the wave equation, Eq. (8.170), is undertaken with high accuracy, then the effective pseudopotential Hamiltonian method can be expected to yield highly accurate results.

CHAPTER 9

RELATIVISTIC EFFECTIVE HAMILTONIAN THEORY

9.1. INTRODUCTION

Discussion of the necessity for the introduction of relativistic effects into effective Hamiltonian theory; explanation of the nomenclature concerning one-, two-, and four-component models.

9.2. QUASIRELATIVISTIC HF MODEL

Formulation of an HF model in which the major relativistic effects are put into the form of a potential and the *LS* coupling format is maintained.

9.3. THE RELATIVISTIC EFFECTIVE HAMILTONIAN OF KAHN, HAY, AND COWAN

Formulation of a one-component effective Hamiltonian for the valence electrons in which relativistic effects are built into the core–valence interaction potential.

9.4. SPIN–ORBIT COUPLING IN EFFECTIVE HAMILTONIAN THEORY

The complete spin-dependent part of the Pauli Hamiltonian is analyzed and the part that must be written into the effective valence-electron Hamiltonian separated.

9.5. THE RELATIVISTIC EFFECTIVE HAMILTONIAN OF LEE, ERMLER, AND PITZER

Pseudopotential transformation with four-component wave functions; replacement of the four-component pseudopotential equation by a two-component equation; the effective two-component valence-electron Hamiltonian and its properties; results of atomic test calculations.

9.6. THE FOUR-COMPONENT EFFECTIVE HAMILTONIAN OF ISHIKAWA AND MALLI

Construction of an effective Hamiltonian with four-component pseudopotentials; difference between the pseudopotentials for the large and small components; the form of the effective potentials in practical calculations.

9.1. INTRODUCTION

When we first introduced relativistic effects into atomic theory in Chapter 4, we talked about conceptual and practical reasons for the introduction of these effects. Having built up a nonrelativistic effective Hamiltonian theory in the preceding sections, we must now ask the question: Are relativistic effects important for effective Hamiltonian theory?

At first sight, the answer to this question is in the negative. We pointed out in Chapter 6 that the major part of the atomic spectra can be explained in terms of the *LSJ* coupling, that is, by a theory that is nonrelativistic with the spin–orbit coupling taken into account by perturbation method. Thus, for an understanding of the optical spectra, which is the main task of effective Hamiltonian theory, we do not need relativity theory apart from the formula for the spin–orbit interaction.

The arguments about conceptual completeness, which we presented in Chapter 4, are, of course, valid here also. Thus, we could justify the relativistic treatment by citing the argument about the need for presenting a conceptually complete atomic theory. It turns out, however, that here an even more relevant argument presents itself. In order to overcome computational difficulties, scientists working on relativistic effective potential theory introduced some highly interesting ideas whose usefulness is not restricted to effective Hamiltonian theory. A case in point is the quasirelativistic model presented in Sec. 9.2. Indeed, one can safely say that the usefulness of the idea of incorporating relativistic effects into a one-component HF theory will not be restricted to effective Hamiltonian methods; in fact, this idea may prove to be even more important for the development of the all-electron HF model. Thus, we can say that although the introduction of relativistic effects is not vitally important for the understanding of the major features of atomic spectra, the introduction of these effects stimulated scientists to the development of new ideas whose main usefulness may very well lie outside the area of effective Hamiltonian theory.

At this point, it will be useful to explain the nomenclature used in this chapter. A one-component theory is based on one-electron wave functions given by Eq. (2.69). These are the usual central-field one-electron functions. A two-component theory is based on one-electron functions given by Eq. (H.3) from App. H. These are the Pauli-type functions and there is only one radial function for each electron. A four-component theory is based on one-electron wave functions of the type of Eq. (H.27). These are the four-component Dirac functions and there are *two* radial functions for each electron; we talk about large and small components denoted by P_A and Q_A, respectively. Thus, in this chapter, we discuss one-component theories with one radial function, two-component theories with one radial function, and four-component theories with two radial functions.

9.2. QUASIRELATIVISTIC HF MODEL

A one-component quasirelativistic HF model had been developed by Cowan and Griffin.[89] The goal of this work is to incorporate into the nonrelativistic HF formalism the major relativistic effects in such a way that the basic format of the nonrelativistic approach remains intact.

In developing the basic equations of the Cowan–Griffin (CG) model, we start with the Pauli equation, Eq. (H.1):

$$
\left\{ -\frac{\hbar^2}{2m}\Delta - eV - \frac{(-i\hbar\nabla)^4}{8m^3c^2} + \frac{e\hbar^2}{4m^2c^2}\frac{1}{r}\frac{dV}{dr}(\mathbf{r}\cdot\nabla) \right.
$$

$$
\left. -\frac{e}{2m^2c^2}\frac{1}{r}\frac{dV}{dr}(\mathbf{s}\cdot\mathbf{L}) \right\} u = Wu. \quad (9.1)
$$

In this equation, V is the potential in which the electron is moving. The first two terms are the nonrelativistic energies. The third term is the energy resulting from the relativistic increase of the electron mass, the fourth term is the relativistic interaction of the moving electron with the external electric field, and the last term is the spin–orbit interaction. The fourth term can be called the Darwin term because it was Darwin who has shown[17] that the two-component approximation of the Dirac equation contains this energy.

Cowan and Griffin introduced the following approximations in Eq. (9.1). They argued that the relativistic effects are the largest for s electrons for which the last term, the spin–orbit coupling, is zero. Thus, for s electrons, the relativistic effects are represented by the velocity term and the Darwin term. For terms other than s terms, that is, for $l \neq 0$, the spin–orbit coupling produces the j-dependence of the Hamiltonian. Cowan and Griffin omitted the spin–orbit term for $l \neq 0$, so as to obtain a j-independent simplified equation. This approximation was bolstered by the argument that the $(2j + 1)$-weighted average of the spin–orbit term is zero. Therefore, one can expect that the (nl)-dependent radial function, obtained from the simplified equation, will be a good approximation to the $(2j + 1)$-weighted average of the relativistic orbitals with $j = l - \frac{1}{2}$ and $j = l + \frac{1}{2}$.

Now let us take a look at the structure of the Darwin term. Introducing $P = rR$, we get

$$\frac{1}{r}\frac{dV}{dr}(\mathbf{r} \cdot \nabla)R = \frac{1}{r}\frac{dV}{dr}r\frac{d}{dr}\left(\frac{P}{r}\right)$$

$$= \frac{dV}{dr}\left[\frac{1}{r}\frac{dP}{dr} - \frac{P}{r^2}\right]$$

$$= \frac{1}{r}\frac{dV}{dr}\left[\frac{dP}{dr} - \frac{P}{r}\right]. \qquad (9.2)$$

When we transform Eq. (9.1) to standard form, the $1/r$ cancels; this can be seen clearly from Eq. (C.10) in App. C. From that equation, we get

$$-\frac{1}{2}\Delta u \rightarrow -\frac{1}{2r}\frac{d^2}{dr^2}P_{nl}, \qquad (9.3)$$

where we have omitted the angular part of Eq. (C.10). Thus, the radial part of the Darwin term becomes

$$\frac{dV}{dr}\left[\frac{dP}{dr} - \frac{P}{r}\right]. \qquad (9.4)$$

A further modification must be introduced because of a singularity problem. In the vicinity of the nucleus, the potential approaches Z/r and so we get

$$\frac{dV}{dr} = -\frac{Z}{r^2}. \qquad (9.5)$$

In Eq. (9.4), we have P/r. For s functions, that is, for $l = 0$, the radial function P goes like $r^l \approx$ constant, from which it follows that in the vicinity of the nucleus, we get

$$\frac{P}{r} \approx \frac{\text{const.}}{r}. \qquad (9.6)$$

Thus, the expression in Eq. (9.4) produces a singularity proportional to $1/r^3$. This is not integrable, because the volume element is proportional to r^2 only. It was shown by Bethe (Ref. 18, p. 307) that this problem is coming from the approximation by which we omitted the small components from the Dirac equation. In the derivation of the Pauli equation, we used the approximation given by Eq. (G.38):

$$E + eV + mc^2 \approx 2mc^2. \qquad (9.7)$$

If we *do not* make this approximation, then we get for the expression on the left side of Eq. (9.7)

$$E + eV + mc^2 = -eV + \frac{p^2}{2m} + mc^2 + eV + mc^2$$

$$= \frac{p^2}{2m} + 2mc^2, \tag{9.8}$$

where we have used Eq. (G.29). If we replace the relationship given by Eq. (9.7) with the relationship given by Eq. (9.8), the result will be that we have to make the substitution

$$2mc^2 \rightarrow 2mc^2 + \frac{p^2}{2m}. \tag{9.9}$$

Introducing this substitution into the Darwin term of Eq. (9.1), we get

$$\frac{e\hbar^2}{4m^2c^2} = \frac{e\hbar^2}{2m(2mc^2)} \rightarrow \frac{e\hbar^2}{2m\left(2mc^2 + \dfrac{p^2}{2m}\right)}. \tag{9.10}$$

The nonrelativistic energy, W, is given by

$$W = \frac{p^2}{2m} - eV. \tag{9.11}$$

From this, we obtain

$$2mc^2 + \frac{p^2}{2m} = 2mc^2 + W + eV$$

$$= 2mc^2\left[1 + \frac{1}{2mc^2}(W + eV)\right]. \tag{9.12}$$

Thus, the substitution given by Eq. (9.9) brings the Darwin term to the form

$$\frac{e\hbar^2}{4m^2c^2}\left[1 + \frac{1}{2mc^2}(W + eV)\right]^{-1} \frac{dV}{dr}\left[\frac{dP}{dr} - \frac{P}{r}\right]. \tag{9.13}$$

In the vicinity of the nucleus, we get $V \approx Z/r$. Thus, we obtain

$$\left[1 + \frac{1}{2mc^2}(W + eV)\right]^{-1} \sim \left[\text{const.} + \text{const.}\ \frac{Z}{r}\right]^{-1}$$

$$= \left[\frac{(\text{const.})r + (\text{const.})Z}{r}\right]^{-1}$$

$$= \frac{r}{(\text{const.})r + (\text{const.})Z}.$$

In the vicinity of $r = 0$, this expression will go to zero like r and thus it will compensate for the $1/r^3$ singularity, changing it to $1/r^2$, which is integrable.

Let us now introduce the fine-structure constant α, which is defined as

$$\alpha = \frac{e^2}{\hbar c} = \frac{1}{137.037}. \tag{9.14}$$

In atomic units, we have $e = \hbar = m = 1$ and

$$\alpha = \frac{1}{c}. \tag{9.15}$$

For the constants in Eq. (9.13), we get, in atomic units,

$$\frac{1}{2mc^2} = \frac{1}{2}\alpha^2, \tag{9.16}$$

$$\frac{e\hbar^2}{4m^2c^2} = \frac{1}{4}\alpha^2. \tag{9.17}$$

Finally, we introduce the potential into the relativistic velocity expression. Using Eq. (9.11), we get

$$p^2 = 2m(W + eV), \tag{9.18}$$

and

$$\frac{p^4}{8m^3c^2} = \frac{4m^2(W + eV)^2}{8m^3c^2} = \frac{\alpha^2}{2}(W + eV)^2. \tag{9.19}$$

We are now able to write the equation of Cowan and Griffin. Let the equation be written for the atomic orbital P_i, where i stands for $(n_i l_i)$. In Eq.

(9.1), V is the potential. Let V^i be the potential energy in atomic units for the orbital P_i, that is, let us change the notation according to

$$-eV \rightarrow V^i \quad (\text{a.u.})$$

Then, using this relationship and Eqs. (9.13), (9.16), (9.17), and (9.19), we obtain from Eq. (9.1)

$$
\left\{ -\frac{1}{2}\frac{d^2}{dr^2} + \frac{l_i(l_i+1)}{2r^2} + V^i(r) - \frac{\alpha^2}{2}(\varepsilon_i - V^i(r))^2 \right.
$$
$$
- \delta(l_i 0)\frac{\alpha^2}{4}\left[1 + \frac{\alpha^2}{2}(\varepsilon_i - V^i(r)) \right]^{-1}
$$
$$
\left. \times \frac{dV^i}{dr}\left[\left(\frac{dP_i}{dr}\bigg/ P_i \right) - \frac{1}{r} \right] \right\} P_i = \varepsilon_i P_i, \tag{9.20}
$$

where we have changed the nonrelativistic energy, W, of Eq. (9.1) into ε_i. All quantities are in atomic units and V^i is the potential energy.

We now repeat the approximations that lead to Eq. (9.20). The starting point is the Pauli equation, Eq. (9.1). From that equation, the last term, the spin–orbit interaction is omitted for all l_i values, that is, for all orbitals. This omission is an approximation for $l \neq 0$; for $l = 0$, the term is zero exactly. The fourth term of the Pauli equation, the Darwin term, is omitted for $l \neq 0$. For $l = 0$, the term is modified in such a way that no singularity occurs in the vicinity of the nucleus. The modification is not an approximation; on the contrary, the modification consists of falling back on a more exact expression than what was used in the derivation of the Pauli equation. Finally, the relativistic velocity term is kept unchanged for all l_i values.

Cowan and Griffin carried out self-consistent field calculations for a number of atoms using Eq. (9.20). For $V^i(r)$, they used the HF electrostatic potential augmented by an exchange term proportional to $\rho^{1/3}$, that is, they borrowed the potential from the X_α model. The constant of the exchange term was chosen to give good results for the atoms considered. The results were as follows. For the total energy of a number of atoms across the periodic system, CG obtained values quite close to the results of the HFD model. The calculated energies represent a considerable improvement over the nonrelativistic HF values. The results for the orbital parameters can be directly compared to the HFD values only for the s terms where there is no spin–orbit splitting. Calculations for the neutral U atom gave results very close to the HFD values in the case of s terms. Finally, the calculated values of spin–orbit parameters agreed well with the values obtained from the HFD model and in most cases represented a marked improvement over the nonrelativistic HF calculations.

9.3. THE RELATIVISTIC EFFECTIVE HAMILTONIAN OF KAHN, HAY, AND COWAN

A one-component relativistic effective Hamiltonian method was formulated by Kahn, Hay, and Cowan (KHC). This method rests on the quasirelativistic model presented in the preceding section. As we have seen, in that model, the major relativistic effects are incorporated into the effective potential that appears in the Schroedinger equation of the orbital P_{nl}. The orbitals themselves were of the central-field type, that is, the basic structure of the nonrelativistic HF model remained intact except for the change in the effective potential. The model is a one-component theory, that is, the potentials and the orbitals were j-independent. The spin–orbit interaction was omitted.

We have seen in Chapters 7 and 8 how an effective Hamiltonian method is formulated in the nonrelativistic case. We have seen that the first step is to separate the core and valence wave functions. The effective Hamiltonian for the valence electrons is obtained by varying the total energy of the atom with respect to the valence-electron wave function under the subsidiary conditions of strong-orthogonality and normalization. The core orbitals remain fixed in this procedure. The equation that emerges from the variation defines the effective Hamiltonian for a set of unspecified and fixed core orbitals. A salient feature of the procedure is that the interaction between the core and valence electrons is obtained as a one-electron operator depending functionally on the core orbitals.

In the method of KHC, the effective Hamiltonian retains its nonrelativistic form, that is, the valence-electron wave functions will be j-independent. However, the core–valence interaction potential will be relativistic in the sense that the orbitals that form the building blocks of the interaction potentials will be the solutions of the equations of Cowan and Griffin instead of being the solutions of the nonrelativistic HF equations. Thus, we can talk here about a relativistic core–valence interaction potential in a nonrelativistic effective Hamiltonian. As far as the mathematical details are concerned, the method is a straightforward application of the Kahn–Baybutt–Truhlar method[86] that we mentioned in Chapter 8 and discussed in detail in our book on pseudopotentials [Szasz, 153].

The core–valence interaction potential is set up as follows. In the Cowan and Griffin model, the wave functions of the valence electrons are solutions of Eq. (9.20). We can write that equation into the form

$$\left\{ -\frac{1}{2}\frac{d^2}{dr^2} + \frac{l(l+1)}{2r^2} + \hat{V}_{nl}(r) \right\} \hat{P}_{nl}(r) = \varepsilon_{nl}\hat{P}_{nl} + \sum_{n'} \lambda_{nl,n'l}\hat{P}_{n'l}. \quad (9.21)$$

In this equation, $\hat{V}_{nl}(r)$ is the full effective potential that has the form

$$\hat{V}_{nl}(r) = -\frac{Z}{r} + \hat{V}_{nl}^{HF} + \hat{V}_{nl}^{R}, \tag{9.22}$$

where $\hat{V}_{nl}^{HF}$ is the HF electrostatic and exchange potential, and $\hat{V}_{nl}^{R}$ represents the relativistic terms, that is, the mass-velocity and Darwin energies. The caret on $\hat{P}_{nl}$ indicates that the CG eigenfunctions are orthogonal to each other, and on $\hat{V}_{nl}$, it indicates that the potential will depend functionally on the orthogonal $\hat{P}_{nl}$ orbitals. $\lambda_{nl, n'l}$ is a Lagrangian multiplier ensuring the orthogonality between the orbitals. Equation (9.21) can be written as

$$\hat{H}_{nl}\hat{P}_{nl} = \varepsilon_{nl}\hat{P}_{nl} + \sum_{n'}\lambda_{nl, n'l}\hat{P}_{n'l}, \tag{9.23}$$

where the meaning of $\hat{H}_{nl}$ is given by the formula

$$\hat{H}_{nl} = -\frac{1}{2}\frac{d^2}{dr^2} + \frac{l(l+1)}{2r^2} + \hat{V}_{nl}$$

$$= -\frac{1}{2}\frac{d^2}{dr^2} + \frac{l(l+1)}{2r^2} - \frac{Z}{r} + \hat{V}_{nl}^{HF} + \hat{V}_{nl}^{R}. \tag{9.24}$$

We now introduce the projection operators Ω_{nl} and Π_{nl}. Let

$$\Omega_{nl} = \sum_{n'}|\hat{P}_{n'l}\rangle\langle\hat{P}_{n'l}|, \tag{9.25}$$

where the summation is over those orbitals that have the same azimuthal quantum number as the selected $\hat{P}_{nl}$. Also, let

$$\Pi_{nl} = 1 - \Omega_{nl}. \tag{9.26}$$

Using these operators, we can put Eq. (9.23) in the form

$$\hat{H}'_{nl}\hat{P}_{nl} = \varepsilon_{nl}\hat{P}_{nl}, \tag{9.27}$$

where

$$\hat{H}'_{nl} = \Pi_{nl}\hat{H}_{nl}. \tag{9.28}$$

Now we transform Eq. (9.27) into a pseudopotential equation. First, we put

$$\hat{P}_{nl} = (1 - \Omega_{nl})P_{nl}, \tag{9.29}$$

where P_{nl} is the pseudo-orbital. Using this relationship, we can write the equation for P_{nl} in the form

$$\left(\hat{H}_{nl} + \hat{V}_{nl}^P\right)P_{nl} = \varepsilon_{nl}P_{nl}, \tag{9.30}$$

where $\hat{H}_{nl}$ is the Hamiltonian of Eq. (9.24), and $\hat{V}_{nl}^P$ is the pseudopotential

$$\hat{V}_{nl}^P = -\Omega_{nl}\hat{H}_{nl} - \hat{H}_{nl}\Omega_{nl} + \Omega_{nl}\hat{H}_{nl}\Omega_{nl} + \varepsilon_{nl}\Omega_{nl}. \tag{9.31}$$

This is the same Weeks–Rice pseudopotential as in Eq. (7.75). As we have shown in Sec. 7.2/C, following Eq. (7.75), it is easy to prove that Eq. (9.30) is equivalent to the original Eqs. (9.27) and (9.23). Thus, the pseudopotential transformation is exact.

The pseudopotential equation for the pseudo-orbital P_{nl}, Eq. (9.30), can be simplified as follows. Using Eq. (9.24), we can write

$$\left\{-\frac{1}{2}\frac{d^2}{dr^2} + \frac{l(l+1)}{2r^2} - \frac{Z}{r} + \hat{U}_{nl}^{\text{core}} + \hat{W}_{nl}^{\text{val}}\right\}P_{nl} = \varepsilon_{nl}P_{nl}. \tag{9.32}$$

In this formula, $\hat{W}_{nl}^{\text{val}}$ contains the electrostatic and exchange potentials generated by the valence orbitals. Here the additional approximation is made that in this potential, the orthogonal valence orbitals are replaced by the nonorthogonal pseudo-orbitals. Denoting the potential by W_{nl}^{val}, which depends functionally on the pseudo-orbitals in the same way as $\hat{W}_{nl}^{\text{val}}$ depends functionally on the orthogonal orbitals, we put

$$\hat{W}_{nl}^{\text{val}} \rightarrow W_{nl}^{\text{val}}. \tag{9.33}$$

In Eq. (9.32), the potential $\hat{U}_{nl}^{\text{core}}$ contains the core contributions to the electrostatic and exchange potentials, the relativistic "potential," $\hat{V}_{nl}^R$, and the pseudopotential, Eq. (9.31).

Instead of directly evaluating the extremely complicated expressions in $\hat{U}_{nl}^{\text{core}}$, one can proceed as follows. Let $U_{nl}^{\text{core}}(r)$ be a local potential for (nl), which is defined in such a way that this potential is equivalent to $\hat{U}_{nl}^{\text{core}}$ at every point in space; that is, we demand that the eigenfunctions and eigenvalues of the equation

$$\left\{-\frac{1}{2}\frac{d^2}{dr^2} + \frac{l(l+1)}{2r^2} - \frac{Z}{r} + U_{nl}^{\text{core}}(r) + W_{nl}^{\text{val}}\right\}P_{nl} = \varepsilon_{nl}P_{nl} \tag{9.34}$$

be identical with the eigenfunctions and eigenvalues of Eq. (9.32) (after the approximation of Eq. (9.33) is carried out in Eq. (9.32)). Now because U_{nl}^{core} is a local potential, we can solve Eq. (9.34) for U_{nl}^{core}:

$$U_{nl}^{\text{core}} = \varepsilon_{nl} - \frac{l(l+1)}{2r^2} + \frac{Z}{r} + \frac{1}{2P_{nl}} \frac{d^2P_{nl}}{dr^2} - W_{nl}^{\text{val}} P_{nl}/P_{nl}. \quad (9.35)$$

(Note that W_{nl}^{val} is an operator.) Potential U_{nl}^{core} can be constructed from this formula without difficulty if the pseudo-orbitals are known. According to Eq. (9.29), the pseudo-orbitals are linear combinations of the orthogonal orbitals, that is, we can write the pseudo-orbital for the (nl) state in the form

$$P_{nl} = \sum_{n'} b_{nl,n'l} \hat{P}_{n'l}, \quad (9.36)$$

where the summation is for all orthogonal orbitals that have the same azimuthal quantum number as the selected valence state (the summation includes the orthogonal orbital with $n' = n$). In fact, P_{nl} is a solution of the pseudopotential equation with an arbitrary set of constants. For the procedure that produces a particularly smooth pseudo-orbital and a very smooth U_{nl}^{core}, we refer to the author's book [Szasz, 159].

The potential function is different for each (nl). As we have seen in Sec. 7.2, the n dependence can be ignored and only the l dependence need be considered. Let the largest azimuthal quantum number in the core be l_m and let $L = l_m + 1$. Then, the effective potential that is valid for any l, including those that do not occur in the core, can be written in the form

$$U^{\text{core}} = U_L^{\text{core}} + \sum_{l=0}^{L-1} [U_l^{\text{core}} - U_L^{\text{core}}] \Omega_l. \quad (9.37)$$

In this formula, U_l^{core} is defined for $l \leq l_m$ by Eq. (9.35). For those l values that do not occur in the core, that is, for $l > l_m$, the pseudopotential vanishes and so, instead of Eq. (9.35), we obtain

$$U_l^{\text{core}} = \varepsilon_{nl} - \frac{l(l+1)}{2r^2} + \frac{Z}{r} + \frac{1}{2\hat{P}_{nl}} \frac{d^2\hat{P}_{nl}}{dr^2} - \frac{\hat{W}_{nl}^{\text{val}} \hat{P}_{nl}}{\hat{P}_{nl}}. \quad (9.38)$$

In Eq. (9.37), Ω_l is a projection operator defined as

$$\Omega_l = \sum_{m=-l}^{+l} |Y_{lm}\rangle\langle Y_{lm}|, \quad (9.39)$$

where Y_{lm} is the normalized spherical harmonics. As we see, U^{core} is a semilocal potential, that is, it is an operator whose components are, for each l, local potentials. For the construction of Eq. (9.37), we refer to our book on pseudopotentials [Szasz, 158].

We are now in the position to formulate the effective Hamiltonian for n valence electrons. Denoting the kinetic energy and the nuclear potential by t and g, respectively, we can write

$$H_{\text{eff}} = \sum_{j=1}^{n} \left[t_j + g_j + U^{\text{core}}(j) \right] + \frac{1}{2} \sum_{j,k=1}^{n} \frac{1}{r_{jk}}, \qquad (9.40)$$

where U^{core} is given by Eq. (9.37). The Schroedinger equation for the n valence electrons can be written as

$$H_{\text{eff}}\Psi(1,2,\ldots,n) = E\Psi(1,2,\ldots,n), \qquad (9.41)$$

where we have emphasized, by the notation, that this is an n-electron equation, that is, an equation for the valence electrons only.

We summarize the basic properties of the effective Hamiltonian. In this model, we have a nonrelativistic Hamiltonian, that is, a Hamiltonian that does not contain the spin–orbit interaction and is j-independent. However, the major part of the relativistic effects is built into the core–valence interaction operator, which is given by Eq. (9.37). This operator is constructed in such a way that when it is placed into the Schroedinger equation, that equation becomes equivalent to the equation of the Cowan–Griffin quasirelativistic model. Thus, the effective potential of Eq. (9.37) can be called the Cowan–Griffin-equivalent core potential. We have seen also that the effective core potential contains the pseudopotential that replaces the orthogonality requirement between core and valence electrons. This means that Eq. (9.41) can be solved "as if the core did not exist." The presence of the pseudopotentials prevents the valence electrons from falling into the core. It is also clear from the derivation that this model represents, in our terminology, the "neutral-atom approximation." In this model, this phrase means that the electrostatic and exchange potentials as well as the "relativistic potential" are frozen as they were in the Cowan–Griffin model for the neutral atom.

As the first test of the model, Kahn, Hay, and Cowan carried out calculations for the valence electrons of the U atom. Computed were the total energy of the valence electrons, the orbital parameters and expectation values of powers of r. In all cases, the computed values agreed well with the results of all-electron calculations.

In principle, the Hamiltonian of Eq. (9.40) can be transferred to any molecular calculation that deals with molecules in which the inner electron orbitals preserve their atomic character. In practice, the effective core-valence potentials must be put into such form that provides a manageable basis for an accurate calculation of molecular wave functions. The most convenient method is to put the core potential into the form of an expansion in terms of simple analytic functions. The task of putting the potentials into a directly usable form in molecular calculations was carried out by Hay and Wadt in a series of papers.[90] For a large number of atoms, Hay and Wadt constructed effective core potentials that include relativistic effects. The potentials were represented by analytic fits consisting of powers of r multiplied by Gaussian ($e^{-\alpha r^2}$)-type functions. These are the analytic forms most convenient in molecular calculations.

9.4. SPIN – ORBIT COUPLING IN EFFECTIVE HAMILTONIAN THEORY

In analyzing the general structure of the optical spectra in Chapter 6, we have observed that the J dependence, that is, the fine structure of the spectra, is an important feature that must be built into atomic structure theory. The fine structure arises because of spin–orbit coupling. Thus, it is clear that a usable effective Hamiltonian theory must take the spin–orbit interaction into account. The effective Hamiltonian theory presented in the preceding sections did not contain the spin–orbit effect. Our goal in this section is to show how this effect can be built into the theory.

The first theoretical investigation of the spin–orbit interaction was carried out by Thomas.[91] We discussed his formula in connection with the two-component Dirac equation at the end of App. G. Thomas's formula was derived for an H-like atom with one electron. For a many-electron atom, the spin–orbit interaction was investigated by Blume and Watson.[92] Their study was done for an atom with N electrons in complete groups and n valence electrons in one incomplete group. The model used was the HF approximation with LS coupling. More recently, Kahn presented[93] a formula for the spin–orbit interaction in the framework of an effective Hamiltonian theory. Kahn considered an atom with N core electrons and n valence electrons. He derived an expression that must be placed into the effective Hamiltonian for the valence electrons. Our discussion, presented in this section, is constructed on the basis of the work of Blume and Watson as well as of the work of Kahn.

Our starting point is Eq. (5.9), which is the relativistic part of the Breit–Pauli Hamiltonian operator for an atom with N electrons. Let us write down here those terms of the Hamiltonian that depend on the spin operator. The terms are the third, the sixth and the seventh. Writing down those operators for two electrons, we substitute $\mathbf{s} = 2\boldsymbol{\sigma}$ and introduce atomic units. In atomic units, the fine-structure constant, α, becomes $1/c$. Thus, the spin-dependent terms take the form

$$H_s = -\frac{\alpha^2}{2}[\mathscr{E}_1 \cdot (s_1 \times p_1)] - \frac{\alpha^2}{2}[\mathscr{E}_2 \cdot (s_2 \times p_2)]$$

$$+\frac{\alpha^2}{r_{12}^3}[(r_{12} \times p_2) \cdot s_1 + (r_{21} \times p_1) \cdot s_2]$$

$$+\alpha^2\left[-\frac{8\pi}{3}(s_1 \cdot s_2)\,\delta(r_{12}) \right.$$

$$\left. +\frac{1}{r_{12}^3}\left((s_1 \cdot s_2) - \frac{3(s_1 \cdot r_{12})(s_2 \cdot r_{12})}{r_{12}^2} \right) \right]. \qquad (9.42)$$

In deriving Eq. (5.9), we assumed that the external electric and magnetic fields vanish. This terminology must be clarified here. The electric field, $\mathscr{E}$ in Eq. (9.42), is defined by Eqs. (4.1) and (4.10). From these equations, we see that they take fully into account external electric and magnetic fields as well as the electric field of the nucleus. However, in writing down the spin-dependent operators, we must remember that every electron, besides moving in the field of the nucleus, is also exposed to the electric field of the other electrons. This field will be considered "internal," that is, when we say that the external electric field vanishes, we mean an electric field over and above the field of the nucleus and the field of the other electrons.

We take into account the vanishing of the external magnetic field by putting $A = 0$ in Eq. (4.1). However, in Eq. (4.10), we do not put $\varphi = 0$ despite the vanishing of the external field. We put the field of the other electrons into φ, thereby taking into account the effect of this field on the spin interactions.

Thus, we put for $\mathscr{E}_1$

$$\mathscr{E}_1 = \frac{Z}{r_1^2}\frac{r_1}{r_1} - \frac{1}{r_{12}^2}\frac{r_{12}}{r_{12}}, \qquad (9.43)$$

where r_1/r_1 and r_{12}/r_{12} are the unit vectors pointing into the direction of r_1 and r_{12}, respectively. The second term is the Coulombic force exerted by electron "two" on electron "one." The vector $-r_{12} = r_2 - r_1$ points from electron "one" to electron "two" as it should in the case of a Coulombic force of negative charge. Similarly, we get

$$\mathscr{E}_2 = \frac{Z}{r_2^2}\frac{r_2}{r_2} - \frac{1}{r_{12}^2}\frac{r_{21}}{r_{12}}. \qquad (9.44)$$

By putting these expression into Eq. (9.42), we get

$$
H_s = -\frac{\alpha^2}{2}\left[\left(\frac{Z}{r_1^2}\hat{r}_1 - \frac{r_{12}}{r_{12}^3}\right)\cdot(s_1 \times p_1)\right]
$$

$$
-\frac{\alpha^2}{2}\left[\left(\frac{Z}{r_2^2}\hat{r}_2 - \frac{r_{21}}{r_{12}^3}\right)\cdot(s_2 \times p_2)\right]
$$

$$
+\frac{\alpha^2}{r_{12}^3}\left[(r_{12} \times p_2)\cdot s_1 + (r_{21} \times p_1)\cdot s_2\right]
$$

$$
+\alpha^2\left[-\frac{8\pi}{3}(s_1\cdot s_2)\,\delta(r_{12})\right.
$$

$$
\left.+\frac{1}{r_{12}^3}\left((s_1\cdot s_2) - \frac{3(s_1\cdot r_{12})(s_2\cdot r_{12})}{r_{12}^2}\right)\right]. \quad (9.45)
$$

Using the vector identity of Eq. (G.59) from App. G, we get

$$
-r_{12}\cdot(s_1 \times p_1) = s_1\cdot(r_{12} \times p_1), \quad (9.46)
$$

and using the relationship and a similar one for electron "two," we obtain

$$
-\frac{\alpha^2}{2}\left[-\frac{r_{12}}{r_{12}^3}\cdot(s_1 \times p_1) - \frac{r_{21}}{r_{12}^3}\cdot(s_2 \times p_2)\right]
$$

$$
+\frac{\alpha^2}{r_{12}^3}\left[(r_{12} \times p_2)\cdot s_1 + (r_{21} \times p_1)\cdot s_2\right]
$$

$$
= -\frac{\alpha^2}{2r_{12}^3}\left[(r_{12} \times p_1)\cdot s_1 + (r_{21} \times p_2)\cdot s_2\right]
$$

$$
+\frac{\alpha^2}{r_{12}^3}\left[(r_{12} \times p_2)\cdot s_1 + (r_{21} \times p_1)\cdot s_2\right]
$$

$$
= -\frac{\alpha^2}{2r_{12}^3}\left[(r_{12} \times p_1)\cdot(s_1 + 2s_2) + (r_{21} \times p_2)\cdot(s_2 + 2s_1)\right]. \quad (9.47)
$$

In the last line, we used the relationship $r_{ik} = -r_{ki}$. By putting this expression into Eq. (9.45), we get

$$
H_s = \frac{\alpha^2}{2} \left[\frac{Z}{r_1^3} (\mathbf{r}_1 \times \mathbf{p}_1) \cdot \mathbf{s}_1 + \frac{Z}{r_2^3} (\mathbf{r}_2 \times \mathbf{p}_2) \cdot \mathbf{s}_2 \right.
$$

$$
\left. - \frac{1}{r_{12}^3} (\mathbf{r}_{12} \times \mathbf{p}_1) \cdot (\mathbf{s}_1 + 2\mathbf{s}_2) - \frac{1}{r_{12}^3} (\mathbf{r}_{21} \times \mathbf{p}_2) \cdot (\mathbf{s}_2 + 2\mathbf{s}_1) \right]
$$

$$
+ \alpha^2 \left[- \frac{8\pi}{3} (\mathbf{s}_1 \cdot \mathbf{s}_2) \delta(\mathbf{r}_{12}) \right.
$$

$$
\left. + \frac{1}{r_{12}^3} \left((\mathbf{s}_1 \cdot \mathbf{s}_2) - \frac{3(\mathbf{s}_1 \cdot \mathbf{r}_{12})(\mathbf{s}_2 \cdot \mathbf{r}_{12})}{r_{12}^2} \right) \right], \tag{9.48}
$$

where we have used again the vector identity, Eq. (9.46).

Equation (9.48) gives the complete spin-dependent part of the Breit–Pauli Hamiltonian operator for two electrons in the field of the nucleus. Each term has a clear physical meaning. The first two terms are the spin–orbit interactions derived by Thomas for an electron in an H-like atom. As we discuss at the end of App. G, the Thomas expression, Eq. (G.65), is

$$
H = \frac{\alpha^2}{2} \frac{1}{r} \frac{dV}{dr} (\mathbf{s} \cdot \mathbf{L}). \tag{9.49}
$$

If we put $V = -Z/r$ for the potential energy, then

$$
\frac{dV}{dr} = + \frac{Z}{r^2}, \tag{9.49a}
$$

and the angular momentum is given by

$$
\mathbf{L} = \mathbf{r} \times \mathbf{p}. \tag{9.49b}
$$

Using these expressions in Eq. (9.49), we obtain the first two terms of Eq. (9.48). Thus, it is clear that the first two terms of Eq. (9.48) are the spin–orbit interactions of the two electrons in the field of the bare nucleus.

Next let us consider the third term of Eq. (9.48). We write this term in two parts as follows:

$$
V_{so}^1 = - \frac{1}{r_{12}^3} (\mathbf{r}_{12} \times \mathbf{p}_1) \cdot \mathbf{s}_1, \tag{9.50}
$$

and

$$V_{so}^2 = -\frac{1}{r_{12}^3}(\mathbf{r}_{12} \times \mathbf{p}_1) \cdot 2\mathbf{s}_2. \tag{9.51}$$

These two terms have quite different physical meanings. The first term, V_{so}^1, has the same structure as the first term of Eq. (9.48). In fact, whereas the first term of Eq. (9.48) is the spin–orbit interaction in the field of the bare nucleus, V_{so}^1 is the spin–orbit interaction of the first electron in the field of the second. On the other hand, V_{so}^2 is the interaction of the spin–magnetic moment of the second electron with the magnetic field created by the current of the first electron. Thus, V_{so}^1 is a spin–orbit interaction and V_{so}^2 is a spin–other-orbit interaction. Quite naturally, the fourth term of Eq. (9.48) is the same as the third with the roles of the electrons reversed. Finally, the fifth term is clearly the spin–spin interaction.

We write now Eq. (9.48) for a many-electron atom. We obtain

$$H_s = \frac{\alpha^2}{2} \sum_i \frac{Z}{r_i^3}(\mathbf{r}_i \times \mathbf{p}_i) \cdot \mathbf{s}_i - \frac{\alpha^2}{2} \sum_{\substack{i,j \\ (i \neq j)}} \frac{1}{r_{ij}^3}(\mathbf{r}_{ij} \times \mathbf{p}_i) \cdot (\mathbf{s}_i + 2\mathbf{s}_j)$$

$$+ \alpha^2 \sum_{\substack{ij \\ (i \neq j)}} \left\{ -\frac{8\pi}{3}(\mathbf{s}_i \cdot \mathbf{s}_j)\, \delta(\mathbf{r}_{ij}) \right.$$

$$\left. + \frac{1}{r_{ij}^3}\left[(\mathbf{s}_i \cdot \mathbf{s}_j) - \frac{3(\mathbf{s}_i \cdot \mathbf{r}_{ij})(\mathbf{s}_j \cdot \mathbf{r}_{ij})}{r_{ij}^2} \right] \right\}. \tag{9.52}$$

In this expression, the second term is written in an asymmetric form as a result of which the summations over i and j run independently for all electrons. Summing up this way, we get the correct symmetric expression of Eq. (9.48).

Next, we consider an atom with N core electrons in complete groups and n valence electrons in an arbitrary configuration. We work with a one-component approximation, which may be either the conventional HF model or the Cowan–Griffin model. We assume LS coupling. The core electrons will be represented by a determinant built from central-field-type one-electron orbitals. The valence electrons will be represented by an n-electron function Φ that may be correlated.

Blume and Watson established that the summations over electrons in complete groups vanish in the first and third terms of Eq. (9.52). Therefore, the equation can be rewritten as follows:

$$H_s = \frac{\alpha^2}{2} \sum_{\substack{i \\ \text{(valence)}}} \frac{Z}{r_i^3}(\mathbf{r}_i \times \mathbf{p}_i) \cdot \mathbf{s}_i - \frac{\alpha^2}{2} \sum_{\substack{i,j \\ (i \neq j)}} \frac{1}{r_{ij}^3}(\mathbf{r}_{ij} \times \mathbf{p}_i) \cdot (\mathbf{s}_i + 2\mathbf{s}_j)$$

$$+ \alpha^2 \sum_{\substack{i,j \\ \text{(valence pairs, } i \neq j)}} \left\{ -\frac{8\pi}{3}(\mathbf{s}_i \cdot \mathbf{s}_j)\,\delta(\mathbf{r}_{ij}) \right.$$

$$\left. + \frac{1}{r_{ij}^3}\left[(\mathbf{s}_i \cdot \mathbf{s}_j) - \frac{3(\mathbf{s}_i \cdot \mathbf{r}_{ij})(\mathbf{s}_j \cdot \mathbf{r}_{ij})}{r_{ij}^2} \right] \right\}. \quad (9.53)$$

As we have indicated, the second term of the previous expression consists of two kinds of interactions. The term containing $\mathbf{s}_i$ is the spin–orbit interaction of electron i in the field created by electron j. The term containing $\mathbf{s}_j$ is the interaction of the spin–magnetic moment of electron j with the magnetic field of electron i. Generally, one can say that the spin–orbit interaction is a larger effect than the spin–other-orbit interaction although we cannot state that the latter is negligibly small relative to the former.

Let us consider now the spin–orbit interaction of one of the valence electrons in the field of the core electrons, that is, let us consider the expression

$$V_{so} = -\frac{\alpha^2}{2} \sum_{i=N+1}^{N+n} \sum_{j=1}^{N} \frac{1}{r_{ij}^3}(\mathbf{r}_{ij} \times \mathbf{p}_i) \cdot \mathbf{s}_i, \quad (9.54)$$

where the summation over j runs from 1 to N, that is, over the indices of the core electrons and the summation over i runs from $N + 1$ to $N + n$, that is, over the range of the valence electrons. We show in App. M that this term can be combined with the first summation of Eq. (9.53) and the two can be written, in a good approximation, in the form

$$\frac{\alpha^2}{2} \sum_{\substack{i \\ \text{(valence)}}} \left[\nabla_i\left(-\frac{Z}{r_i} + V_C(i) \right) \times \mathbf{p}_i \right] \cdot \mathbf{s}_i, \quad (9.55)$$

where V_C is the electrostatic potential of the core electrons:

$$V_C(1) = \sum_k \int \frac{\varphi_k^*(2)\varphi_k(2)}{|\mathbf{r}_1 - \mathbf{r}_2|}\, dq_2. \quad (9.56)$$

Thus, replacing the core–valence part of the second sum in Eq. (9.53) by the expression given by Eq. (9.55), we obtain for the spin-dependent part of the effective valence-electron Hamiltonian the formula

$$
H_s = \frac{\alpha^2}{2} \sum_{\substack{i=1 \\ (\text{valence})}}^{n} \left[\nabla_i \left(-\frac{Z}{r_i} + V_C(i) \right) \times \mathbf{p}_i \right] \cdot \mathbf{s}_i
$$

$$
- \frac{\alpha^2}{2} \sum_{\substack{i,j=1 \\ (\text{valence}, i \neq j)}}^{n} \frac{1}{r_{ij}^3} (\mathbf{r}_{ij} \times \mathbf{p}_i) \cdot (\mathbf{s}_i + 2\mathbf{s}_j)
$$

$$
+ \alpha^2 \sum_{\substack{i,j \\ (\text{valence pairs}, j \neq i)}} \left\{ -\frac{8\pi}{3} (\mathbf{s}_i \cdot \mathbf{s}_j) \delta(\mathbf{r}_{ij}) \right.
$$

$$
\left. + \frac{1}{r_{ij}^3} \left[(\mathbf{s}_i \cdot \mathbf{s}_j) - \frac{3(\mathbf{s}_i \cdot \mathbf{r}_{ij})(\mathbf{s}_j \cdot \mathbf{r}_{ij})}{r_{ij}^2} \right] \right\}. \quad (9.57)
$$

All summations in this Hamiltonian are for the valence electrons only. In the second summation, i and j are running independently for all valence electrons. In the third term, the summation is for all distinct valence-electron pairs.

Equation (9.57) defines the spin-dependent part of the effective valence-electron Hamiltonian. Each term has a clear physical meaning. The first term is the spin–orbit interaction of the valence electrons in the field of the nucleus plus the core electrons. The second term represents the spin–orbit interaction of the valence electrons in the field of the other valence electrons and the spin–other-orbit interaction of the valence electrons. The last term is the spin–spin interaction of the valence electrons.

The first term of operator H_s can be brought to the conventional form of the spin–orbit coupling operators. We have derived a formula for the electrostatic potential of the core in App. D. According to Eq. (D.21), the electrostatic potential of a core with complete groups depends only on coordinate r. Thus, we can write

$$
V_C(\mathbf{r}) = V_C(r). \quad (9.58)
$$

This being the case, we can write ($V_C \equiv U_C$)

$$
\nabla \left(-\frac{Z}{r} + U_C(\mathbf{r}) \right) = \frac{d}{dr} \left(-\frac{Z}{r} + U_C(r) \right) \frac{\mathbf{r}}{r}
$$

$$
= \frac{1}{r} \frac{d}{dr} \left(-\frac{Z}{r} + U_C(r) \right) \mathbf{r}, \quad (9.59)
$$

and using this and the definition of the angular momentum, we get the first term of Eq. (9.57) in the form

$$\frac{\alpha^2}{2} \sum_{i=1}^{n} \left[\frac{1}{r_i} \frac{d}{dr_i} \left(-\frac{Z}{r_i} + U_C(r_i) \right) \mathbf{s}_i \cdot \mathbf{L}_i \right]. \qquad (9.60)$$

This formula is remarkably similar to Thomas's original expression, which was derived for a one-electron atom. The difference between Thomas's formula and the formula here is that instead of the bare-nucleus potential that occurred in Thomas's formula, here we have the Hartree potential:

$$V_H(i) = -\frac{Z}{r_i} + U_C(r_i). \qquad (9.61)$$

Thus, we see that in calculating the spin–orbit coupling of a valence electron in a many-electron atom, we must replace the bare-nucleus potential by the Hartree potential. The effect of the core electrons is that they reduce the size of the spin–orbit coupling constant. This is plausible because, in the original Hartree model, which we described in Sec. 2.1, the electron was assumed to move in the Hartree potential. If we construct a semiclassical derivation for the valence electron of an atom with N core electrons, a semiclassical derivation similar to Thomas's derivation, then, it is plausible that we have to replace the bare-nucleus potential with the potential of Eq. (9.61). In an H-like atom, the electron moves in the bare-nucleus potential; therefore, that potential will appear in the spin–orbit coupling operator. In a many-electron atom, the electron is supposed to move in the Hartree potential; therefore, it is the Hartree potential that will appear in the spin–orbit coupling operator.

We have pointed out in App. M that in the derivation of Eq. (9.60), the core–valence exchange was neglected and if this effect were taken into account, there would be an additional term, besides U_C, in the operator. Formulas for the additional term were derived for the case of a single incomplete group by Blume and Watson. Kahn has derived a general formula for the additional term. The main characteristic of this term is that the core–valence exchange tends to reduce further the bare-nucleus potential providing additional shielding similar to the shielding provided by U_C.

Summarizing the arguments leading to Eq. (9.60), we can say that in constructing a spin–orbit interaction operator for a valence electron of an atom with many core electrons and several valence electrons, it is a good approximation to assume that the potential appearing in the operator will be the electrostatic potential of the bare nucleus reduced by the potential of the core electrons. It is also evident from the discussions that the core–valence exchange will have an effect on the potential. This effect will be a reduction of the coupling constant similar to but smaller than the reduction produced by core potential U_C.

Turning to the second term of Eq. (9.57), we note that Blume and Watson have shown that part of this term can also be written in the form that is proportional to $s_i \cdot L_i$, where the index i is a valence electron index. In fact, we can see easily that this statement is correct. In this term, in which the summations are over the valence electrons, we have the spin–orbit interaction of a valence electron in the field of the other valence electrons. Constructing a derivation for the valence electrons similar to the derivation that was constructed for the core and that we presented in App. M, we can show easily that the spin–orbit coupling in the field of the valence electrons must have the same mathematical form as the spin–orbit coupling in the field of the core electrons. The only difference between these two effects is that in the latter, we would have the electrostatic potential of the core electrons, whereas in the former, we have the electrostatic potential of the valence electrons.

We now summarize our results. The spin-dependent part of an effective valence-electron Hamiltonian is given by the formula

$$H_s = \sum_{i=1}^{n} f_i(r_i) s_i \cdot L_i + \sum_{\substack{i,j=1 \\ (i \neq j)}}^{n} g_{so}(ij) + \sum_{\substack{i,j=1 \\ (i \neq j)}}^{n} h_{so}(ij). \qquad (9.62)$$

The operator consists of two parts. The first part is a sum of one-electron operators of the conventional spin–orbit type. The function $f_i(r)$ has the form

$$f_i(r) = \frac{\alpha^2}{2} \frac{1}{r} \frac{d}{dr} \left(-\frac{Z}{r} + U_C(r) \right)$$

$$+ \text{(valence electron contribution)}$$

$$+ \text{(core–valence exchange contribution)}. \qquad (9.63)$$

The second part of Eq. (9.62) consists of two-particle interactions. The terms denoted by $g_{so}(ij)$ represent the spin–other-orbit interactions. The terms denoted by $h_{so}(ij)$ represent the spin–spin interactions. The second summation of Eq. (9.62), that is, the summation over the g_{so} terms, is the second term of Eq. (9.57) minus the terms in $f_i(r)$ that are designated as "valence-electron contributions." In other words, the second summation of Eq. (9.62) represents that part of the second summation of Eq. (9.57) which cannot be written in the form proportional to $(s_i \cdot L_i)$. Finally the last term in Eq. (9.62) is identical with the last term of Eq. (9.57).

The term denoted by "valence electron contribution" in Eq. (9.63) can easily be written down if we adopt the approach which we called the "neutral atom approximation." In the effective Hamiltonian theory the valence electrons are represented by an arbitrary, correlated, n-electron function. We have denoted this function by Φ and it is displayed in Eq. (M.2). Let us assume however, that the electrostatic potential of the valence electrons can be determined from the formula

$$U_i(1) = \sum_{\substack{j=1 \\ (j \neq i)}}^{n} \int \frac{\varphi_j^*(2)\varphi_j(2)}{r_{12}} \, dq_2. \tag{9.64}$$

This is the potential acting on the valence electron orbital i. Using this in Eq. (9.63), we obtain

$$f_i(r) = \frac{\alpha^2}{2} \frac{1}{r} \frac{d}{dr} \left(-\frac{Z}{r} + U_C(r) + U_i(r) \right)$$
$$+ (\text{exchange corrections}). \tag{9.65}$$

Thus, as we see from this formula, what we get into the one-electron part of the spin-interaction operator is the complete Hartree potential augmented by corrections representing the effects of the core–valence and valence–valence exchange interactions. This formula is derived in the neutral-atom approximation, that is, in the derivation of Eq. (9.65), the valence electrons are represented by a Hartree–Fock-type determinantal wave function.

9.5. THE RELATIVISTIC EFFECTIVE HAMILTONIAN OF LEE, ERMLER, AND PITZER

At approximately the same time as the work of Kahn, Hay, and Cowan was published, another relativistic effective Hamiltonian model was developed by Lee, Ermler, and Pitzer (LEP). The purpose of this work was to incorporate relativistic effects into the effective Hamiltonian theory without imposing the constraint that the model should retain the basic structure of the LS coupling.

It is relatively straightforward, *in principle*, to build up a relativistic effective Hamiltonian theory. We have seen that the HFD model is a straightforward generalization of the nonrelativistic HF model. Logically, we would expect that a relativistic effective Hamiltonian theory would be a straightforward generalization of the Fock–Veselov–Petrashen model, which we discussed in Sec. 8.2. Whether such a relativistic model would be the simplest possible is another question that we will discuss in what follows.

Lee, Ermler, and Pitzer started with the Hamiltonian, given by Eq. (5.5), which is

$$H_B = \sum_{k=1}^{N+n} H_D(k) + \frac{1}{2} \sum_{k,j=1}^{N+n} \frac{1}{r_{kj}}, \tag{9.66}$$

where we have changed Eq. (5.5) in such a way that the Hamiltonian is now written down for an atom with N core and n valence electrons. H_D is the Dirac Hamiltonian operator given by Eq. (4.28), which, in atomic units, is

$$H_D(k) = -c\boldsymbol{\alpha}(k) \cdot \mathbf{p}_k - \beta c^2 - V(k), \qquad (9.67)$$

where $\boldsymbol{\alpha}$ and β are the Dirac matrices, $V(k)$ is the potential in which the kth electron is moving, and $\mathbf{p}_k$ is the momentum. The operator H_B is the relativistic counterpart of the nonrelativistic operator given by Eq. (8.53).

For the wave function of the atom, LEP chose the expression

$$\Psi_T = \bar{A}\left\{\frac{1}{\sqrt{N!}}\,\frac{1}{\sqrt{n!}}\,\det\left[\hat{\psi}_1(1)\hat{\psi}_2(2)\,\cdots\,\hat{\psi}_n(n)\right]\right.$$
$$\left. \times \det\left[\varphi_1(n+1)\varphi_2(n+2)\,\cdots\,\varphi_N(n+N)\right]\right\}, \qquad (9.68)$$

which is the generalization of Eq. (8.54). The core electrons are represented by $\varphi_1, \ldots, \varphi_N$ and the valence electrons by $\hat{\psi}_1, \ldots, \hat{\psi}_n$. In contrast to the nonrelativistic case, these are four-component wave functions of the form given by Eq. (4.102):

$$\varphi_{n\kappa m_j} = \begin{pmatrix} i(Q_{n\kappa}/r)u_{-\kappa m_j} \\ (P_{n\kappa}/r)u_{\kappa m_j} \end{pmatrix}. \qquad (9.69)$$

In this formula, $P_{n\kappa}$ and $Q_{n\kappa}$ are the radial functions for the large and small components, respectively, $u_{\kappa m_j}$ is the function given by Eq. (4.101), and κ is the quantum number replacing l and j.

The one-electron wave functions in Eq. (9.68) are orthonormal. In analogy to Eq. (8.15), we obtain for the expectation value of the Hamiltonian

$$E_T = \frac{\langle \Psi_T|H_B|\Psi_T\rangle}{\langle \Psi_T|\Psi_T\rangle} = \langle \Phi|H_v|\Phi\rangle + E_C, \qquad (9.70)$$

where E_C is the HFD energy of the core, Φ is the valence-electron wave function given by

$$\Phi = \frac{1}{\sqrt{n!}}\,\det\left[\hat{\psi}_1 \,\cdots\, \hat{\psi}_n\right], \qquad (9.71)$$

and H_v is the operator

$$H_v = \sum_{k=1}^{n} H(k) + \frac{1}{2}\sum_{k,j=1}^{n}\frac{1}{r_{kj}}, \qquad (9.72)$$

where

$$H(k) = H_D(k) + U_C(k). \qquad (9.73)$$

In this formula, H_D is the Dirac one-electron operator, Eq. (9.67), and U_C is the core potential. This last is an operator of the form

$$U_C = \sum_{A=1}^{N} U_A$$

$$U_A(1)f(1) = \int \frac{|\varphi_A(2)|^2 \, dq_2}{r_{12}} f(1) - \varphi_A(1) \int \frac{\varphi_A^*(2)f(2) \, dq_2}{r_{12}}. \qquad (9.74)$$

As we see, U_C is closely analogous to Eq. (8.18) except now the one-electron orbitals are four-component functions.

Let us consider now a single electron outside complete groups. In this case, $n = 1$ and we get

$$E_T = E_v + E_C, \qquad (9.75)$$

where

$$E_v = \langle \hat{\psi} | H | \hat{\psi} \rangle \qquad (9.76)$$

is the energy of the valence electron. Varying E_v with respect to $\hat{\psi}$ with fixed core orbitals, we get

$$(H_D + U_C)\hat{\psi} = \varepsilon \hat{\psi} + \sum_{A=1}^{N} \lambda_A \varphi_A, \qquad (9.77)$$

where the λ_A are Lagrangian multipliers taking care of the orthogonality between the valence orbital and the core orbitals. Introducing the projection operators Ω and Π with definitions analogous to Eqs. (8.9) and (8.10), we get

$$\Omega = \sum_{A=1}^{N} |\varphi_A\rangle\langle\varphi_A|, \qquad (9.78)$$

and

$$\Pi = 1 - \Omega. \qquad (9.79)$$

Using these operators, we obtain for Eq. (9.77)

$$\hat{H}\hat{\psi} = \varepsilon \hat{\psi}, \qquad (9.80)$$

with

$$\hat{H} \equiv \Pi(H_D + U_C). \tag{9.81}$$

We want to transform now Eq. (9.80) into a pseudopotential equation. In Sec. 7.2/C, we have shown how a general one-electron valence-electron equation can be transformed into a pseudopotential equation with the Weeks–Rice procedure. We see now that our equations here are closely analogous to the nonrelativistic case with the four-component wave functions and operators taking the place of the one-component quantities of the nonrelativistic theory. Consequently, the Weeks–Rice pseudopotential transformation can be transferred, *mutatis mutandis*, to the relativistic formulas.

It is easy to see that the pseudopotential equation taking the place of Eq. (9.80) will be

$$(H + V_P)\psi = \varepsilon\psi. \tag{9.82}$$

In this formula, ψ is a pseudo-orbital that does not need to be orthogonalized to the core orbitals. Naturally, ψ is a four-component function given by

$$\hat{\psi} = (1 - \Omega)\psi. \tag{9.83}$$

The operator H is given by Eq. (9.73) and V_P is a four-component pseudopotential given by Eq. (7.75):

$$V_P = -\Omega H - H\Omega + \Omega H\Omega + \varepsilon\Omega, \tag{9.84}$$

where Ω is now the four-component projection operator given by Eq. (9.78).

It was shown by LEP that Eq. (9.82) leads to two equations for the large and small radial components of the pseudo-orbitals. At this point, it became evident that to replace the orthogonality requirement with respect to the core functions, two different pseudopotentials are needed for the small and large components of the pseudo-orbitals.

In order to avoid this kind of complication, Lee, Ermler, and Pitzer observed that the significance of the small components in the HFD calculations declines as one proceeds from inner core orbitals to the outer core orbitals and to the valence orbitals. In other words, the properties of the valence electrons depend on the small component of the HFD wave function to a much lesser degree than the properties of the inner core electrons depend on the small components of their wave functions. It is plausible to expect that the valence-electron properties can be accurately described by a two-component wave function with just one radial component. Because the purpose of the effective Hamiltonian theory is to construct a Schroedinger equation for the valence electrons, Lee, Ermler, and Pitzer suggested replacing Eq. (9.82) by a two-component equation with only one radial orbital.

The reduction of the model to a two-component theory is carried out in a fashion closely analogous to the reduction of the H-atom Dirac equation. We see in App. G how the four-component Dirac equation can be reduced to the two-component Pauli equation. Essentially, what we do there is to replace the Dirac one-electron Hamiltonian, given by Eq. (9.67), by the Pauli Hamiltonian, given by Eq. (G.56). In the absence of an external electromagnetic field, the Pauli Hamiltonian is

$$H_P = -\tfrac{1}{2}\Delta - eV + H_m + H_d + H_s, \qquad (9.85)$$

where V is the potential in which the electron is moving, H_m is the relativistic velocity term, H_d is the Darwin term, and H_s is the spin–orbit coupling. These expressions are given by Eq. (G.56). H_P is a two-component Hamiltonian whose eigenfunctions have the form of Eq. (H.3).

We now replace, in Eq. (9.73), the Dirac 4-component operator H_D by the Pauli two-component operator H_P. Thus, we get

$$H(k) = H_P(k) + U_C(k), \qquad (9.86)$$

and Eq. (9.72), the valence Hamiltonian, becomes

$$H_v = \sum_{k=1}^{n} (H_P(k) + U_C(k)) + \frac{1}{2} \sum_{k,j=1}^{n} \frac{1}{r_{kj}}. \qquad (9.87)$$

The core potential, U_C, is meant to be a two-component operator here. It is obtained from Eqs. (9.73) and (9.74) by omitting the small components from φ_A.

We can now repeat the derivation beginning with Eq. (9.75). The only change that we must carry out is to replace the four-component quantities with the two-component functions. Instead of Eq. (9.77), we get now

$$(H_P + U_C)\hat{\psi} = \varepsilon\hat{\psi} + \sum_{A=1}^{N} \lambda_A\varphi_A, \qquad (9.88)$$

which is now a two-component equation. Replacing the four-component functions by two-component quantities in Eq. (9.78), we get the projection operators into two-component form. Finally, we end up with Eq. (9.82):

$$(H + V_P)\psi = \varepsilon\psi, \qquad (9.89)$$

but now H is given by Eq. (9.86). V_P is given by Eq. (9.84), but all operators now are two-component quantities.

It is easy to get the pseudopotentials from this equation. Using Eqs. (9.85) and (9.86), we obtain

$$\{-\tfrac{1}{2}\Delta - eV + H_m + H_d + H_s + U_C + V_P\}\psi = \varepsilon\psi. \qquad (9.90)$$

Separating out the angular and spin parts, we obtain for the radial part

$$\left\{ -\frac{1}{2}\frac{d^2}{dr^2} + \frac{l(l+1)}{2r^2} - \frac{Z}{r} + H_m(r) + H_d(r) \right.$$

$$\left. + H_s(r) + U_C(r) + V_P(r) \right\} P_{nlj}(r) = \varepsilon_{nlj} P_{nlj}(r), \quad (9.91)$$

where $P_{nlj}(r)$ is the radial part of the Pauli eigenfunction, Eq. (H.3).

Now let $V_P(nlj|r)$ be a local potential and let us replace the pseudopotential operator V_P in Eq. (9.91) by the local potential. We define $V_P(nlj|r)$ by demanding that for any n, l, and j, the Schroedinger equation with the local potential should reproduce the eigenfunctions and eigenvalues of Eq. (9.91) *exactly*. Thus, the pseudo-orbital will be the eigenfunction of the equation

$$\left\{ -\frac{1}{2}\frac{d^2}{dr^2} + \frac{l(l+1)}{2r^2} - \frac{Z}{r} + H_m(r) + H_d(r) + H_s(r) \right.$$

$$\left. + U_C(r) + V_P(nlj|r) \right\} P_{nlj} = \varepsilon_{nlj} P_{nlj}. \quad (9.92)$$

Because V_P here is a local potential, we can solve the equation for V_P and get

$$V_P(nlj|r) = \varepsilon_{nlj} - \frac{l(l+1)}{2r^2} + \frac{Z}{r}$$

$$- \frac{1}{P_{nlj}}\left\{ -\frac{1}{2}\frac{d^2}{dr^2} + H_m + H_d + H_s + U_C \right\} P_{nlj}. \quad (9.93)$$

We obtain the pseudopotential function from this equation if we have the pseudo-orbitals P_{nlj}. The pseudo-orbitals are the solutions of Eq. (9.89) and it is known from pseudopotential theory [Szasz, 35] that the pseudo-orbitals are of the form

$$\psi = \sum_{A=1}^{N} \alpha_A \varphi_A + \hat{\psi}. \qquad (9.94)$$

In this formula, φ_A is the core orbital and $\hat{\psi}$ is the valence-electron wave function, which is the solution of Eq. (9.88). The caret ($\hat{\ }$) indicates that this is an HF-type orbital, orthogonal to the core functions, whereas ψ is a pseudo-orbital that does not need to satisfy any orthogonality conditions. We obtain for the radial part

$$P_{nlj} = \hat{P}_{nlj} + \sum_{\substack{n' \\ (\text{core})}} \alpha_{n'lj}\hat{P}_{n'lj},\tag{9.95}$$

where the summation is over those core states that have the same l and j as the valence orbital. The $\hat{P}_{n'lj}$ are the radial parts of the core orbitals φ_A. The methods for the determination of the constants α_A are described in the author's book [Szasz, 40, 157].

The question now arises: How to choose the core orbitals? A consistent procedure would be to construct Pauli-type solutions for the core states. LEP have postulated that the radial parts of the large components of the Hartree–Fock–Dirac wave functions can be used for the core orbitals as well as for $\hat{P}_{nlj}$. This is, in fact, a plausible choice because we have obtained the Hamiltonian H_P by approximating the Dirac Hamiltonian with a two-component operator. Thus, it is plausible that the eigenfunctions of H_P will be well approximated by the large components of the HFD solutions. The radial parts of the large components are available from HFD calculations.

Two further modifications can be made at this point. If we have several valence electrons, the potential of the "other" valence orbitals must be taken into account. Also, in determining the core–valence interaction, the terms H_m, H_d, and H_s can be neglected. Thus, we obtain

$$V_P(nlj|r) = \varepsilon_{nlj} - \frac{l(l+1)}{2r^2} + \frac{Z}{r}$$
$$- \frac{1}{P_{nlj}}\left\{-\frac{1}{2}\frac{d^2}{dr^2} + U_C + U_v\right\}P_{nlj},\tag{9.96}$$

where U_v is the potential (including exchange) of the other valence electrons.

Next, we want to put V_P of Eq. (9.90) into a general form. The physical meaning of the pseudopotential is that it replaces the orthogonality requirement with respect to the core orbitals. In this model, our wave functions, core and valence, have the form of Eq. (H.3). As we see from Eq. (H.21), we must have orthogonality with respect to all four quantum numbers, n, l, j, and m_j. The angular and spin parts of the orbital will ensure orthogonality with respect to l, j, and m_j. Thus, the pseudopotential replaces the orthogonality requirement with respect to the principal quantum number n.

As we see from Eq. (9.96), we must construct different local pseudopotentials for each (nlj) combination. Let $\Omega(lj)$ be a projection operator that selects the potential with the correct (lj) values. We have

$$\Omega(lj) = \sum_{m_j=-j}^{+j} |ljm_j\rangle\langle ljm_j|, \tag{9.97}$$

where $|ljm_j\rangle$ is the angular–spin part of Eq. (H.3). The pseudopotential V_P in Eq. (9.90) will have the form

$$V_P = \sum_l \sum_j V_P(nlj|r)\Omega(lj), \tag{9.98}$$

where $V_P(nlj|r)$ is given by Eqs. (9.95) and (9.96). The summation over j for any given l is for $l + \frac{1}{2}$ and $l - \frac{1}{2}$. The summation over l is, in principle, from 0 to ∞, but in fact can be essentially restricted to the range $0 \le l \le l_m + 1$, where l_m is the largest l occurring in the core [Szasz, 159].

Having determined the core–valence interaction potential with the method of Lee, Ermler, and Pitzer, we are now able to write down the effective Hamiltonian for the valence electrons. This will be the operator given by Eq. (9.87) augmented by the pseudopotential. Thus, we get

$$H_v = \sum_{k=1}^{n} \{H_P(k) + U_C(k) + V_P(k)\} + \frac{1}{2}\sum_{k,j=1}^{n}\frac{1}{r_{kj}}. \tag{9.99}$$

In this formula, H_P is the Pauli operator given by Eq. (9.85), U_C is the core potential given by Eqs. (9.73) and (9.74), and V_P is the pseudopotential given by Eq. (9.98), where the radial part is given by Eq. (9.96). The core orbitals occurring here are the large components of the HFD orbitals. The Schroedinger equation for the n valence electrons is given by

$$H_v\Psi(1,2,\ldots,n) = E\Psi(1,2,\ldots,n). \tag{9.100}$$

The solutions of this equation do not need to satisfy any orthogonality conditions; the equation can be treated as if the core did not exist.

In order to test the method, LEP suggested the following procedure. Calculate the valence-electron properties from the effective Hamiltonian above by writing the valence-electron wave function in the form of a determinant

$$\Psi = \frac{1}{\sqrt{n!}}\det[\psi_1(1)\psi_2(2)\cdots\psi_n(n)], \tag{9.101}$$

where $\psi_1,\ldots,\psi_n$ are pseudo-orbitals, that is, they are functions that do not have to satisfy any orthogonality conditions with respect to the core functions. The orbitals are orthonormal among themselves:

$$\langle\psi_A|\psi_B\rangle = \delta(A,B), \tag{9.102}$$

where A and B stand for the four quantum numbers, n, l, j, and m_j. Function ψ_A has the two-component form given by Eq. (H.3):

$$\psi_{nljm_j}(r, \vartheta, \varphi, \sigma)$$

$$= \frac{P_{nlj}(r)}{r} \times \sum_{m_s} C\left(l\tfrac{1}{2}j; (m_j - m_s), m_s\right) Y_l(m_j - m_s)(\vartheta, \varphi) \eta_{m_s}(\sigma).$$

$$(9.103)$$

The properties of this function are discussed in App. H.

Next, LEP determined the form of the equations for P_{nlj}. The equations for these functions can be easily derived from the HFD equations in the nonrelativistic limit. We are considering n valence electrons in an arbitrary configuration forming several complete or incomplete groups. For this case, the HFD equations, for the average energy of the configuration, were given in Eqs. (4.207) and (4.208). We write down those equations here as our starting points:

$$\frac{dP_A}{dr} + \frac{\kappa_A}{r}P_A - \left[2c + \frac{1}{c}\left(\frac{Y^E(A|r)}{r} - \varepsilon_A\right)\right]Q_A + W_Q(A|r)$$

$$= - \sum_{(B \neq A)} \frac{\varepsilon_{AB}}{c} q(B)\, \delta(\kappa_A\kappa_B)Q_B, \qquad (9.104)$$

and

$$\frac{dQ_A}{dr} - \frac{\kappa_A}{r}Q_A + \frac{1}{c}\left(\frac{Y^E(A|r)}{r} - \varepsilon_A\right)P_A - W_P(A|r)$$

$$= \sum_{B \neq A} \frac{\varepsilon_{AB}}{c} q(B)\delta(\kappa_A\kappa_B)Q_B. \qquad (9.105)$$

In these equations, we have the potential function $(1/r)Y^E(A|r)$, which was defined by Eq. (4.209):

$$\frac{1}{r}Y^E(A|r) = \frac{Z}{r} - \sum_{\substack{A' \\ \text{(valence)}}} q(A')\frac{1}{r}Y_0^E(A'A'|r)$$

$$+ \frac{1}{r}Y_0(AA|r) + \sum_{k>0}[q(A) - 1]\frac{2j_A + 1}{4j_A}\Gamma_{j_Aj_A}^k \frac{1}{r}Y_k^E(AA|r).$$

$$(9.106)$$

In these equations, A stands for $(n_A l_A j_A)$, which are the quantum numbers of the selected group. The summation over B is over all other valence groups. Throughout the derivation, we should keep in mind that we can treat the valence electrons as if the core does not exist, that is, P_A and Q_A do not have to be orthogonal to the core orbitals. The functions W_Q and W_P are given by Eqs. (4.164) and (4.165). We have

$$W_Q(A|r) = -\frac{1}{rc} \sum_{B \neq A} \sum_k \frac{q(B)}{2} \Gamma^k_{j_A j_B} Y^E_k(AB|r) Q_B(r), \quad (9.107)$$

and

$$W_P(A|r) = -\frac{1}{rc} \sum_{B \neq A} \sum_k \frac{q(B)}{2} \Gamma^k_{j_A j_B} Y^E_k(AB|r) P_B(r). \quad (9.108)$$

We now introduce the $c \to \infty$ limit. Then, obviously, we get in Eq. (9.104)

$$\frac{1}{c}\left(\frac{Y^E(A|r)}{r} - \varepsilon_A \right) \ll 2c. \quad (9.109)$$

Also, with $c \to \infty$, we can omit the right side of Eq. (9.104) and we get

$$\frac{dP_A}{dr} + \frac{\kappa_A}{r} P_A - 2cQ_A + W_Q(A|r) = 0. \quad (9.110)$$

Rearranging, we get

$$2cQ_A = \frac{dP_A}{dr} + \frac{\kappa_A}{r} P_A + W_Q(A|r). \quad (9.111)$$

Differentiating, we get

$$2c\frac{dQ_A}{dr} = \frac{d^2P_A}{dr^2} - \frac{\kappa_A}{r^2} P_A + \frac{\kappa_A}{r}\frac{dP_A}{dr} + \frac{dW_Q(A|r)}{dr}. \quad (9.112)$$

From Eq. (9.105), we obtain

$$2c\frac{dQ_A}{dr} - \frac{2c\kappa_A}{r} Q_A + 2\left(\frac{Y^E(A|r)}{r} - \varepsilon_A \right) P_A - 2cW_P(A|r)$$
$$= \sum_{B \neq A} 2\varepsilon_{AB} q(B)\, \delta(\kappa_A \kappa_B) P_B. \quad (9.113)$$

Now put Eqs. (9.111) and (9.112) into Eq. (9.113). The result is

$$\frac{d^2P_A}{dr^2} - \frac{\kappa_A}{r^2} P_A + \frac{\kappa_A}{r}\frac{dP_A}{dr} + \frac{dW_Q(A|r)}{dr}$$
$$- \frac{\kappa_A}{r}\left(\frac{dP_A}{dr} + \frac{\kappa_A}{r} P_A + W_Q(A|r) \right)$$
$$+ 2\left(\frac{Y^E(A|r)}{r} - \varepsilon_A \right) P_A - 2cW_P(A|r)$$
$$= \sum_{B \neq A} 2\varepsilon_{AB} q(B)\, \delta(\kappa_A \kappa_B) P_B. \quad (9.114)$$

In this formula, we have

$$-\frac{\kappa_A + \kappa_A^2}{r^2}P_A = -\frac{\kappa_A(\kappa_A + 1)}{r^2}P_A. \qquad (9.115)$$

From Eq. (9.107), we see that

$$\frac{dW_Q(A|r)}{dr} \to 0 \quad \text{as} \quad c \to \infty, \qquad (9.116)$$

and also

$$\frac{\kappa_A}{r}W_Q(A|r) \to 0 \quad \text{as} \quad c \to \infty. \qquad (9.117)$$

Taking these relationships into account, we obtain from Eq. (9.114)

$$\frac{d^2P_A}{dr^2} - \frac{\kappa_A(\kappa_A + 1)}{r^2}P_A + 2\left(\frac{Y^E(A|r)}{r} - \varepsilon_A\right)P_A$$

$$+ \frac{1}{r}\sum_{B \neq A}\sum_k q(B)\Gamma_{j_Aj_B}^k Y_k^E(AB|r)P_B$$

$$= \sum_{B \neq A} 2\varepsilon_{AB}q(B)\,\delta(\kappa_A\kappa_B)P_B. \qquad (9.118)$$

Multiplying this equation by $-\frac{1}{2}$, in order to bring it to standard form, we obtain

$$-\frac{1}{2}\frac{d^2P_A}{dr^2} + \frac{\kappa_A(\kappa_A + 1)}{2r^2} - \frac{1}{r}Y^E(A|r)P_A$$

$$- \sum_{B \neq A}\sum_k \frac{q(B)}{2}\Gamma_{j_Aj_B}^k \frac{1}{r}Y_k^E(AB|r)P_B$$

$$= -\varepsilon_A P_A - \sum_{B \neq A}\varepsilon_{AB}q(B)\,\delta(\kappa_A\kappa_B)P_B. \qquad (9.119)$$

In order to see the structure of this equation clearly, we take into account the fact that, as we see from Eqs. (4.94) and (4.95),

$$\kappa_A(\kappa_A + 1) = l_A(l_A + 1), \qquad (9.120)$$

regardless whether j_A is generated by l_A or $\bar{l}_A$. Also, we have seen that the energy parameters in Eqs. (4.207) and (4.208) were the negatives of the usual parameters. Thus, let us make the notational changes

$$\varepsilon_A \to -\varepsilon_A, \qquad (9.121)$$

and

$$\lambda_{AB} = -\varepsilon_{AB}q(B). \tag{9.122}$$

Finally, we must introduce the core–valence interaction into the equation because as it now stands, it contains only the nuclear potential in Y^E. According to Eq. (9.99), the core–valence interaction is given by $U_C + V_P$. Thus, we obtain for Eq. (9.119)

$$\left\{ -\frac{1}{2}\frac{d^2}{dr^2} + \frac{l(l_A + 1)}{2r^2} - \frac{Z}{r} + U_C + U_v + V_P \right\} P_A$$

$$= \varepsilon_A P_A + \sum_{B \neq A} \lambda_{AB}\, \delta(\kappa_A \kappa_B) P_B. \tag{9.123}$$

In this formula, U_v means the potential of the valence electrons. From Eqs. (9.106) and (9.119), we obtain

$$U_v(A|r) = \sum_{\substack{A' \\ (\text{valence})}} q(A')\frac{1}{r}Y_0^E(A'A'|r)$$

$$-\frac{1}{r}Y_0(AA|r) - \sum_{k>0}[q(A) - 1]\frac{2j_A + 1}{4j_A}\Gamma_{j_A j_A}^k \frac{1}{r}Y_k^E(AA|r)$$

$$-\sum_{B \neq A}\sum_k \frac{q(B)}{2}\Gamma_{j_A j_B}^k \frac{1}{r}Y_k^E(AB|r)\frac{P_B}{P_A}. \tag{9.124}$$

This is the function that occurs also in the pseudopotential, Eq. (9.96). Denoting the core–valence interaction by V_C^{EP}, we get

$$\left\{ -\frac{1}{2}\frac{d^2}{dr^2} + \frac{l_A(l_A + 1)}{2r^2} + V_C^{EP} + U_v \right\} P_A$$

$$= \varepsilon_A P_A + \sum_{B \neq A} \lambda_{AB}\, \delta(\kappa_A \kappa_B) P_B, \tag{9.125}$$

where

$$V_C^{EP} = -\frac{Z}{r} + U_C + V_P. \tag{9.126}$$

We have a HF-type equation for the selected valence-electron pseudo-orbital P_A. This equation can be treated as if the core does not exist; the core influences the calculations only through the fixed core–valence interaction potential V_C^{EP}, which also contains the pseudopotential, V_P, replacing the orthogonality requirement with respect to the core orbitals. The Lagrangian multipliers λ_{AB} ensure orthogonality with respect to other valence orbitals, that is, valence orbitals that have the same l_A and j_A as the selected valence orbital.

Lee, Ermler, and Pitzer developed this method primarily for molecular calculations, because, as we mentioned before, the atomic effective Hamiltonian can be transferred, in a very good approximation, unchanged, into a certain class of molecular Hamiltonians [Szasz, 200]. Nevertheless, they have made test calculations for atoms that we want to discuss here. Calculations were made for the Xe atom viewing the $(5s)^2(5p)^6$ subshell as the valence shell. LEP calculated the total energy of the valence electrons and obtained

$$E_T = -421.8 \; eV \; (-415.4 \; eV),$$

where, in parentheses, we have the relativistic HF value calculated by Fraga, Karkowski, and Saxena [FKS, 259]. We obtained this value by adding the first eight ionization potentials of FKS. For the orbital parameters, LEP obtained

$\varepsilon(5s) = 0.997 \; (1.010; \; 0.8602);$
$\varepsilon(5\bar{p}) = 0.481 \; (0.493; \; 0.4941);$
$\varepsilon(5p) = 0.430 \; (0.440; \; 0.4459).$

In this set, in every line, we have first the result of LEP. The first number within parentheses is the value obtained by Deslaux with the all-electron HFD model.[33] The second number is the experimental value of Lotz.[49] These energies are the negatives of the orbital energies in atomic units.

Next, LEP made calculations for the Au atom. Calculations were made by viewing different number of electrons as valence electrons. We quote the results of calculations in which 33 electrons were treated as valence electrons, which means the following groups:

$$(4f)^{14}(5s)^2(5p)^6(5d)^{10}(6s).$$

The results for the orbital energies were, with the HFD and experimental values in parentheses,

$\varepsilon(4\bar{f}) = 3.9555 \; (3.8675; \; 3.3453);$
$\varepsilon(4f) = 3.8123 \; (3.7202; \; 3.1982);$
$\varepsilon(5s) = 4.6494 \; (4.6873; \; 4.1908);$
$\varepsilon(5\bar{p}) = 3.2376 \; (3.1893; \; 2.7938);$
$\varepsilon(4p) = 2.5760 \; (2.5588; \; 2.2424);$
$\varepsilon(5\bar{d}) = 0.4972 \; (0.4934; \; 0.4595);$
$\varepsilon(5d) = 0.4254 \; (0.4286; \; 0.4080);$
$\varepsilon(6s) = 0.3438 \; (0.2919; \; 0.3393).$

These numbers are again the negatives of the orbital energies in atomic units.

We summarize the method of Lee, Ermler and Pitzer. An effective Hamiltonian operator has been formulated for the valence electrons. The relativistic effects were incorporated into the core/valence interaction potential by using the large components of the HFD functions. The valence Hamiltonian is a nonrelativistic operator although it gives j-dependent orbital parameters because of the (jj) coupling which is used in the derivation of the HF equations for the valence electron pseudo-orbitals. Test calculations for the Xe and Au atoms have shown that the calculated values of orbital parameters were reasonably close to the HFD values.

9.6. THE FOUR-COMPONENT EFFECTIVE HAMILTONIAN OF ISHIKAWA AND MALLI

A four-component relativistic effective Hamiltonian theory, which is the analog of the Hartree–Fock–Dirac method, has been developed by Ishikawa and Malli (IM).[95] In order to build up the theory, we can adopt, without any change, that part of the preceding section that extends from the beginning of the section to Eq. (9.74). Up to that point, the two models are identical.[†] Following Eq. (9.74), however, LEP approximated the four-component formalism with a two-component model. In the presentation of IM, no such approximation is made; the whole model is developed in 4-component form.

Thus, we adopt here the formulas between Eqs. (9.66) and (9.74). The effective Hamiltonian of Eq. (9.72) is given by

$$H_v = \sum_{k=1}^{n} H(k) + \frac{1}{2} \sum_{k,j=1}^{n} \frac{1}{r_{kj}}. \tag{9.127}$$

This is a valence-electron Hamiltonian that, when averaged with respect to a determinantal wave function built from four-component one-electron functions, gives the energy of the valence electrons.

Ishikawa and Malli transformed H_v into a pseudopotential Hamiltonian by postulating that the latter should have the form

$$H_P = \sum_{k=1}^{n} \left[c\boldsymbol{\alpha}(k) \cdot \mathbf{p}_k + \beta c^2 + V_C^{EP}(k) \right] + \frac{1}{2} \sum_{k,j=1}^{n} \frac{1}{r_{kj}}, \tag{9.128}$$

where the first two terms are from the Dirac Hamiltonian, Eq. (9.67). The

[†] The HFD formalism of IM differs from our previous discussions in some minor details. For example, the Dirac vector matrices are -1 times the matrices we used before.

third term of the Dirac Hamiltonian, the potential, is incorporated into V_C^{EP}. This last expression represents the full core–valence interaction, including the pseudopotential that replaces the orthogonality requirements.

In order to understand the structure of V_C^{EP}, we must go back to basic pseudopotential theory. We have quoted a formula for the full core–valence interaction potential, in Eq. (9.37), for the case of a one-component theory. For a detailed discussion, see also the author's book [Szasz, 159]. Equation (9.37) reads

$$U^{\text{core}} = U_L^{\text{core}} + \sum_{l=0}^{L-1} [U_l^{\text{core}} - U_L^{\text{core}}]\Omega_l \qquad (9.129)$$

where $L = l_m + 1$, with l_m being the largest azimuthal quantum number occurring in the core. Ishikawa and Malli generalized this expression for the four-component theory as follows. Let l_m again be the largest azimuthal quantum number in the core and let $L = l_m + 1$. The relativistic quantum number κ was defined by Eq. (4.95). Let K be a relativistic quantum number defined as

$$\begin{aligned} K &= -\left(J + \tfrac{1}{2}\right) && \text{if } J = L + \tfrac{1}{2}; \\ K &= J + \tfrac{1}{2} && \text{if } J = L - \tfrac{1}{2}. \end{aligned} \qquad (9.130)$$

Thus, K is the quantum number determining the two J values associated with L, which is the largest azimuthal quantum number of the core plus 1.

Ishikawa and Malli defined two potential functions for the large and small components. These are

$$\begin{aligned} U_{\kappa-K}^P &= U_\kappa^P - U_K^P, \\ U_{-(\kappa-K)}^Q &= U_{-\kappa}^Q - U_{-K}^Q. \end{aligned} \qquad (9.131)$$

Here U^P is for the large component and U^Q is for the small. Then, the core–valence interaction potential for the large component becomes

$$U_K^P + \sum_{\kappa,m_j} U_{\kappa-K}^P |\kappa m_j\rangle\langle\kappa m_j|, \qquad (9.132)$$

and for the small component, we get

$$U_{-K}^Q + \sum_{\kappa,m_j} U_{-(\kappa-K)}^Q |-\kappa m_j\rangle\langle -\kappa m_j|. \qquad (9.133)$$

In these formulas, the projection operator is the same as in Eq. (9.97), that is,

$$\sum_{m_j} |\kappa m_j\rangle\langle\kappa m_j| = \sum_{m_j} |ljm_j\rangle\langle ljm_j| = \Omega(lj). \qquad (9.134)$$

In order to compare Eq. (9.132) with Eq. (9.129), let us consider an example. Let us consider the Cu atom with the ground state

$$(1s+)^2(2s+)^2(2p-)^2(2p+)^4(3s+)^2$$
$$\times(3p-)^2(3p+)^4(3d-)^4(3d+)^6(4s+).$$

The largest azimuthal quantum number occurring in the core is $l_m = 2$. Thus, $L = 3$ and for K, we get the values

$$\begin{align} K &= -4 \quad & \text{if } J = L + \tfrac{1}{2} = \tfrac{7}{2}; \\ K &= 3 \quad & \text{if } J = L - \tfrac{1}{2} = \tfrac{5}{2}. \end{align} \qquad (9.135)$$

We get, with Eq. (9.131), for the effective potential of the large component,

$$U_K^P + \sum_{\kappa, m_j} \left[U_\kappa^P - U_K^P \right] |\kappa m_j\rangle\langle\kappa m_j|. \qquad (9.136)$$

The summation over κ is for those values that occur in the core. From Table 4.1, we see that the κ values occurring in the core will be $-1, +1, -2, +2,$ and -3. For each κ value, the m_j will run from $-j$ to $+j$, that is,

$$-\left(|\kappa| - \tfrac{1}{2}\right) \le m_j \le |\kappa| - \tfrac{1}{2}. \qquad (9.137)$$

Now let us assume that the valence electron is in the $(4p+)$ state, that is, $n_v = 4$, $l_v = 1$, $j_v = 3/2$, and $\kappa_v = -2$. The projection operator will select the potential

$$U_K^P + \left[U_{-2}^P - U_K^P \right] = U_{-2}^P, \qquad (9.138)$$

that is, it will select the potential for $l = 1$ and $j = 3/2$, as it should. Thus, the expression in Eq. (9.136) is the straightforward generalization of Eq. (9.129).

A similar argument applies for the small component. We obtain the core–valence interaction in the form

$$U_{-K}^Q + \sum_{\kappa m_j} \left[U_{-\kappa}^Q - U_{-K}^Q \right] |-\kappa m_j\rangle\langle-\kappa m_j|. \qquad (9.139)$$

The small component of the $(4p +)$ state has the quantum number $-\kappa_v = 2$. The projection operator will select the potential

$$U^Q_{-K} + U^Q_{+2} - U^Q_{-K} = U^Q_{+2}, \tag{9.140}$$

which is the correct term.

Next, we put V^{EP}_C in matrix form:

$$V^{EP}_C = \begin{pmatrix} U^P_K(r) & & & \\ & U^P_K(r) & & \\ & & U^Q_{-K}(r) & \\ & & & U^Q_{-K}(r) \end{pmatrix}$$

$$+ \begin{pmatrix} \sum_{\kappa m_j} U^P_{\kappa-K}(r)|\kappa m_j\rangle\langle\kappa m_j| & 0 \\ 0 & \sum_{\kappa m_j} U^Q_{-(\kappa-K)}(r)|-\kappa m_j\rangle\langle -\kappa m_j| \end{pmatrix} \tag{9.141}$$

where, in order to clarify the notation, we recall that the projection operator $|\kappa m_j\rangle\langle\kappa m_j|$ is a matrix; this can be seen clearly from Eq. (H.19).

We now place V^{EP}_C from Eq. (9.141) into the Hamiltonian of Eq. (9.128). Then, we form the expectation value of H_P with respect to a determinantal function built from the four-component one-electron functions

$$\psi_{n\kappa m_j} = \begin{pmatrix} (P_{n\kappa}/r)u_{\kappa m_j} \\ (iQ_{n\kappa}/r)u_{-\kappa m_j} \end{pmatrix}. \tag{9.142}$$

This is the usual Dirac-type function with $P_{n\kappa}$ and $Q_{n\kappa}$ being the large and small components, respectively. The energy minimum principle leads to the following two equations for $P_{n\kappa}$ and $Q_{n\kappa}$:

$$\frac{dP_A}{dr} + \frac{\kappa_A}{r}P_A - \left[\frac{2}{\alpha} + \alpha\left(V_A + U^Q_{-K} + U^Q_{-(\kappa-K)} - \varepsilon_A\right)\right]Q_A$$

$$+ X^Q_A = \sum_{B\neq A} \lambda_{AB}Q_B, \tag{9.143}$$

and

$$\frac{dQ_A}{dr} - \frac{\kappa_A}{r}Q_A + \alpha\left(V_A + U^P_K + U^P_{\kappa-K} - \varepsilon_A\right)P_A + X^P_A$$

$$= \sum_{B\neq A} \lambda_{AB}P_B. \tag{9.144}$$

Apart from differences in the notation, these two equations are essentially the same as Eqs. (9.104) and (9.105). The constants λ_{AB} are Lagrangian multipliers taking care of the orthogonality between the valence electrons themselves. The main difference between these equations and Eqs. (9.104) and (9.105) is the presence of the core–valence interaction terms in Eqs. (9.143) and (9.144). These terms, coming from V_C^{EP} contain the nuclear potential, the HF potential of the core including exchange, and the pseudopotential replacing the orthogonality requirement between valence and core orbitals. The functions V_A and X_A^Q represent, respectively, the electrostatic and exchange interactions of P_A with the rest of the valence orbitals; functions V_A and X_A^P represent the same for Q_A. The conspicuous feature of these equations is that the core–valence interactions are different functions in the two equations. It was observed by Lee, Ermler, and Pitzer that to replace the orthogonality requirement with respect to the core functions, different pseudopotentials are needed for the large and small components.

Ishikawa and Malli used this model to create core–valence interaction potentials for molecular calculations. They have put U_K^P and $U_{\kappa-K}^P$ and the corresponding functions for the small component into analytic forms containing adjustable parameters. Then, Eqs. (9.143) and (9.144) were solved, with the analytic core–valence interactions in the equations, for the valence electrons of several atoms. The parameters of the potential were adjusted in such a way that the equations reproduced, in a very good approximation, the HFD orbital parameters. At the same time, the pseudo-orbitals obtained were what are called, in pseudopotential theory, coreless HF orbitals [Szasz, 168]. These are particularly suitable for the construction of smooth pseudopotentials. In order to simplify the calculations, IM assumed that the two potential functions, Eqs. (9.132) and (9.133), can be represented by the same analytic expression. This is an approximation whose results were that whereas the lowest P_A was nodeless, the lowest Q_A was not. In view of the theory of pseudopotentials, this is hardly surprising. The pseudopotential is strongly l-dependent. In the Dirac theory, the large and small components belong to the same j but not to the same l. Thus, a pseudopotential that removes the nodes of P_A cannot be expected to do the same for Q_A.

Apart from molecular calculations, what we can say about the significance of this model for atomic theory is that with this model, it was shown that although it is more complicated than the one-component and two-component models, the construction of a four-component effective Hamiltonian is possible in the framework of the relativistic pseudopotential theory.

APPENDIX A

SINGLE-DETERMINANTAL WAVE FUNCTIONS AND THE DIAGONAL MATRIX COMPONENTS OF THE HAMILTONIAN WITH RESPECT TO THEM

In this appendix, we study some properties of single-determinantal wave functions built from one-electron orbitals. These properties are the orthogonality of the one-electron orbitals and the normalization of the determinantal function. We also derive formulas for the diagonal matrix components of the Hamiltonian with respect to determinantal functions.

Let us write down here Fock's single-determinantal wave function:

$$
\Psi_T = \begin{vmatrix} \varphi_1(1) & \varphi_1(2) & \cdots & \varphi_1(N) \\ \varphi_2(1) & \varphi_2(2) & \cdots & \varphi_2(N) \\ \vdots & \vdots & & \vdots \\ \varphi_N(1) & \varphi_N(2) & \cdots & \varphi_N(N) \end{vmatrix}, \tag{A.1}
$$

which is, as yet, not normalized. We assume that the orbitals are normalized but not orthogonal. We introduce Schmidt orthogonalization and put

$$
\bar{\varphi}_2 = \varphi_2 - c_{21}\varphi_1, \tag{A.2}
$$

where $\bar{\varphi}_2$ will be orthogonal to φ_1 if

$$
c_{21} = \langle \varphi_1 | \varphi_2 \rangle, \tag{A.3}
$$

and we have taken into account that φ_1 is normalized. Next, we normalize $\bar{\varphi}_2$ by putting

$$
\bar{\varphi}_2 = A_2(\varphi_2 - c_{21}\varphi_1), \tag{A.4}
$$

where $\bar{\varphi}_2$ will be normalized if

$$A_2 = \left\{1 - |c_{21}|^2\right\}^{-1/2}. \tag{A.5}$$

Next, we orthogonalize φ_3 to $\bar{\varphi}_2$ and $\bar{\varphi}_1$ ($\bar{\varphi}_1 \equiv \varphi_1$), and we put the normalized and orthogonal $\bar{\varphi}_3$ in the form

$$\bar{\varphi}_3 = A_3(\varphi_3 - c_{32}\bar{\varphi}_2 - c_{31}\bar{\varphi}_1), \tag{A.6}$$

where

$$A_3 = \left\{1 - |c_{32}|^2 - |c_{31}|^2\right\}^{-1/2}. \tag{A.7}$$

Proceeding this way, we put

$$\bar{\varphi}_k = A_k\left(\varphi_k - \sum_{s=1}^{k-1} c_{ks}\bar{\varphi}_s\right), \tag{A.8}$$

where

$$A_k = \left\{1 - \sum_{s=1}^{k-1} |c_{ks}|^2\right\}^{-1/2}. \tag{A.9}$$

The orbitals $\bar{\varphi}_1, \bar{\varphi}_2, \cdots, \bar{\varphi}_N$ are now orthogonal to each other and normalized.

Let us now form the determinantal wave function from the φ_k's. We get

$$\Psi'_T = \begin{vmatrix} \bar{\varphi}_1(1) & \bar{\varphi}_1(2) & \cdots & \bar{\varphi}_1(N) \\ \bar{\varphi}_2(1) & \bar{\varphi}_2(2) & \cdots & \bar{\varphi}_2(N) \\ \vdots & \vdots & & \vdots \\ \bar{\varphi}_N(1) & \bar{\varphi}_N(2) & \cdots & \bar{\varphi}_N(N) \end{vmatrix}, \tag{A.10}$$

where, as in Eq. (A.1), the determinant is not normalized. Substituting the expressions of Eqs. (A.8) and (A.9), we get

$$\Psi'_T = \begin{vmatrix} \bar{\varphi}_1(1) & \cdots & \bar{\varphi}_1(N) \\ A_2(\varphi_2(1) - c_{21}\bar{\varphi}_1(1)) & \cdots & A_2(\varphi_2(N) - c_{21}\bar{\varphi}_1(N)) \\ \vdots & & \vdots \\ A_N\left(\varphi_N(1) - \sum_{s=1}^{N-1} c_{Ns}\bar{\varphi}_s(1)\right) & \cdots & A_N\left(\varphi_N(N) - \sum_{s=1}^{N-1} c_{Ns}\bar{\varphi}_s(N)\right) \end{vmatrix}. \tag{A.11}$$

In the second row, we have $A_2\varphi_2$ from which the elements of the first row, multiplied by $A_2 c_{21}$, are substracted. Likewise, in the kth row, we have $A_k\varphi_k$ from which the elements of the first $k - 1$ rows, multiplied by constants, are substracted. We recall, however, that a determinant does not change if we substract from the kth row elements of the first $k - 1$ rows multiplied by constants. We recall also that a determinant, in which a row is multiplied by a constant, is equal to the original determinant multiplied by the same constant. Using these two properties of determinants, we obtain

$$\Psi_T' = A' \begin{vmatrix} \varphi_1(1) & \varphi_1(2) & \cdots & \varphi_1(N) \\ \varphi_2(1) & \varphi_2(2) & \cdots & \varphi_2(N) \\ \vdots & \vdots & & \vdots \\ \varphi_N(1) & \varphi_N(2) & \cdots & \varphi_N(N) \end{vmatrix}. \tag{A.12}$$

Here we have restored $\varphi_1 \equiv \bar{\varphi}_1$ and have put $A_1 = 1$, and we have

$$A' = A_1 A_2 \cdots A_N. \tag{A.13}$$

Comparing Eqs. (A.1) and (A.12), we see that

$$\Psi_T' = A' \Psi_T. \tag{A.14}$$

Therefore, if we replace the determinantal wave function in Eq. (A.1), in which the orbitals are normalized but not orthogonal, by the determinantal wave function of Eq. (A.10) in which the orbitals are orthogonalized to each other and normalized, we obtain, according to Eq. (A.14), the original determinant multiplied by the constant A'. Because a wave function is always determined only up to a multiplicative constant, this means that the determinantal wave function does not change by Schmidt orthogonalization, Q.E.D.

We turn now to the normalization of the determinantal wave function. We have, according to Eq. (2.37),

$$\Psi_T = A \det [\varphi_1\varphi_2 \cdots \varphi_N], \tag{A.15}$$

where now we can assume that

$$\langle \varphi_i | \varphi_j \rangle = \delta_{ij} \quad (i, j = 1, \cdots, N), \tag{A.16}$$

where δ_{ij} is the Kronecker symbol. We put Ψ_T in the form

$$\Psi_T = A \sum_P P[(-1)^p \varphi_1(k_1)\varphi_2(k_2) \cdots \varphi_N(k_N)], \tag{A.17}$$

where P is the permutation of the coordinates indicated by $k_1, k_2, \cdots, k_N$, p is the parity of the permutation, and the summation is for all permutations.

Using this notation, we get

$$\langle \Psi_T | \Psi_T \rangle = A^2 \sum_P \sum_{P'} (-1)^{p+p'}$$

$$\times \int [P\varphi_1^*(k_1) \cdots \varphi_N^*(k_N)][P'\varphi_1(k_1') \cdots \varphi_N(k_N')] \, dq. \quad \text{(A.18)}$$

Because of the orthonormality of the one-electron orbitals, it is clear that this integral will be zero whenever P' is a different permutation from P. When P' is the same permutation as P, we get 1 for the integral and thus

$$\langle \Psi_T | \Psi_T \rangle = A^2 N! = 1, \quad \text{(A.19)}$$

from which

$$A = (N!)^{-1/2}, \quad \text{(A.20)}$$

Q.E.D.

We now turn to the derivation of the formulas for the HF energy expression. Let Ψ_T be the determinantal wave function defined by Eq. (A.15) with the normalization given by Eq. (A.20). The Hamiltonian is given by Eq. (2.34). By putting

$$f \equiv t + g = -\frac{1}{2}\Delta - \frac{Z}{r}, \quad \text{(A.21)}$$

we can write the Hamiltonian in two terms:

$$H = \sum_{i=1}^{N} f_i + \frac{1}{2} \sum_{i,j=1}^{N} \frac{1}{r_{ij}}, \quad \text{(A.22)}$$

where the first term is a sum of one-electron operators and the second is a sum of two-electron potentials.

Let us first consider the diagonal matrix component of a typical one-electron term, f_1. The derivation will be done by using the Laplace expansions of the determinantal wave functions. We put

$$\Psi_T = (N!)^{-1/2} \sum_{k=1}^{N} (-1)^{1+k} \varphi_k(1) D^{(N-1)}(\varphi_k|1). \quad \text{(A.23)}$$

This is a Laplace expansion in terms of the first column of the determinant, which contains the coordinate q_1. The symbol $D^{(N-1)}(\varphi_k|1)$ indicates the complementary minor to $\varphi_k(1)$, that is, the $(N-1) \times (N-1)$ determinant that is obtained from Ψ_T by omitting the first column and the kth row.

Likewise, we put

$$\Psi_T^* = (N!)^{-1/2} \sum_{l=1}^{N} (-1)^{1+l} \varphi_l^*(1) D^{(N-1)*}(\varphi_l|1). \qquad (A.24)$$

Thus, we get

$$\langle \Psi_T|f_1|\Psi_T \rangle = (N!)^{-1} \sum_{k=1}^{N} \sum_{l=1}^{N} (-1)^{l+k} \int \varphi_k^*(1) f_1 \varphi_l(1) \, dq_1$$

$$\times \int D^{(N-1)*}(\varphi_k|1) D^{(N-1)}(\varphi_l|1) \, dq_2 \cdots dq_N. \quad (A.25)$$

It is clear, because of the orthonormality of the one-electron orbitals, that the integral over the two $(N-1) \times (N-1)$ determinants is zero unless the two determinants contain the same orbitals, that is, unless $k = l$. Thus, we get

$$\langle \Psi_T|f_1|\Psi_T \rangle = (N!)^{-1} \sum_{k=1}^{N} \langle \varphi_k|f_1|\varphi_k \rangle (N-1)!, \qquad (A.26)$$

where the factor $(N-1)!$ is the value of the integral over the two determinants when $k = l$. Using this result, we obtain

$$\sum_{i=1}^{N} \langle \Psi_T|f_i|\Psi_T \rangle = \sum_{i=1}^{N} \frac{1}{N} \sum_{k=1}^{N} \langle \varphi_k|f_i|\varphi_k \rangle$$

$$= \sum_{k=1}^{N} \langle \varphi_k|f|\varphi_k \rangle, \qquad (A.27)$$

because the matrix component $\langle \varphi_k|f_i|\varphi_k \rangle$ is the same for all i, this index being just a dummy integration variable.

Let us now consider the diagonal matrix component of a typical two-electron term, $1/r_{12}$. Here we construct Laplace expansions in terms of the first two columns of the HF determinants:

$$\Psi_T = (N!)^{-1/2} \sum_{k<l} (-1)^{k+l+1+2} \begin{vmatrix} \varphi_k(1) & \varphi_k(2) \\ \varphi_l(1) & \varphi_l(2) \end{vmatrix}$$

$$\times D^{(N-2)}(kl|12), \qquad (A.28)$$

and likewise

$$\Psi_T^* = (N!)^{-1/2} \sum_{s<t} (-1)^{s+t+1+2} \begin{vmatrix} \varphi_s^*(1) & \varphi_s^*(2) \\ \varphi_t^*(1) & \varphi_t^*(2) \end{vmatrix}$$

$$\times D^{(N-2)*}(s, t|12). \qquad (A.29)$$

In these Laplace expansions, $D^{(N-2)}(\alpha\beta|12)$ is the complementary minor to the 2×2 determinant $[\varphi_\alpha(1)\varphi_\beta(2) - \varphi_\alpha(2)\varphi_\beta(1)]$, that is, $D^{(N-2)}(\alpha\beta|12)$ is an $(N-2) \times (N-2)$ determinant, which is obtained from Ψ_T by crossing out the rows with the one-electron orbitals φ_α and φ_β and the columns containing the coordinates q_1 and q_2.

Using Equations (A.28) and (A.29), we obtain

$$\langle \Psi_T | \frac{1}{r_{12}} | \Psi_T \rangle = (N!)^{-1} \int \sum_{k<l} \sum_{s<t} (-1)^{k+l+s+t} \begin{vmatrix} \varphi_s^*(1) & \varphi_s^*(2) \\ \varphi_t^*(1) & \varphi_t^*(2) \end{vmatrix}$$

$$\times \frac{1}{r_{12}} \begin{vmatrix} \varphi_k(1) & \varphi_k(2) \\ \varphi_l(1) & \varphi_l(2) \end{vmatrix} dq_1\, dq_2$$

$$\times \int D^{(N-2)*}(st|12) D^{(N-2)}(kl|12)\, dq_3 \cdots dq_N. \quad \text{(A.30)}$$

The integral over the complementary minors will be zero unless $s = k$ and $t = l$. Thus, we obtain

$$\langle \Psi_T | \frac{1}{r_{12}} | \Psi_T \rangle$$

$$= (N!)^{-1} \sum_{k<l} \int \frac{1}{r_{12}}$$

$$\times [\varphi_k^*(1)\varphi_l^*(2) - \varphi_k^*(2)\varphi_l^*(1)][\varphi_k(1)\varphi_l(2) - \varphi_k(2)\varphi_l(1)]\, dq_1\, dq_2$$

$$\times (N-2)! = \frac{1}{N(N-1)} \sum_{k<l} \int \frac{1}{r_{12}} \big[|\varphi_k(1)|^2|\varphi_l(2)|^2 + |\varphi_k(2)|^2|\varphi_l(1)|^2$$

$$- \varphi_k^*(1)\varphi_l^*(2)\varphi_k(2)\varphi_l(1) - \varphi_k^*(2)\varphi_l^*(1)\varphi_k(1)\varphi_l(2) \big]\, dq_1\, dq_2$$

$$= \frac{2}{N(N-1)} \sum_{k<l} \int \frac{1}{r_{12}} \big[|\varphi_k(1)|^2|\varphi_l(2)|^2 - \varphi_k^*(1)\varphi_l^*(2)\varphi_k(2)\varphi_l(1) \big]\, dq_1\, dq_2$$

$$= \frac{2}{N(N-1)} \sum_{k<l} \left\{ \langle \varphi_k\varphi_l | \frac{1}{r_{12}} | \varphi_k\varphi_l \rangle - \langle \varphi_k\varphi_l | \frac{1}{r_{12}} | \varphi_l\varphi_k \rangle \right\}, \quad \text{(A.31)}$$

where we have used the notation introduced in the main text by Eq. (2.43). Using the result, we obtain

$$\langle \Psi_T | \sum_{i<j} \frac{1}{r_{ij}} | \Psi_T \rangle = \sum_{i<j} \langle \Psi_T | \frac{1}{r_{ij}} | \Psi_T \rangle$$

$$= \sum_{i<j} \frac{2}{N(N-1)} \sum_{k<l} \left\{ \langle \varphi_k\varphi_l | \frac{1}{r_{ij}} | \varphi_k\varphi_l \rangle - \langle \varphi_k\varphi_l | \frac{1}{r_{ij}} | \varphi_l\varphi_k \rangle \right\}$$

$$= \sum_{k<l} \left\{ \langle \varphi_k\varphi_l | \frac{1}{r_{12}} | \varphi_k\varphi_l \rangle - \langle \varphi_k\varphi_l | \frac{1}{r_{12}} | \varphi_l\varphi_k \rangle \right\}, \quad \text{(A.32)}$$

where the last step follows because i and j are dummy variables and

$$\sum_{i<j} = \binom{N}{2} = \frac{N(N-1)}{2}.$$ (A.33)

Combining Eqs. (A.27) and (A.32), we obtain for the diagonal matrix component of the Hamiltonian, which is also the HF energy expression,

$$
\begin{aligned}
E_F &= \langle \Psi_T | H | \Psi_T \rangle \\
&= \sum_{i=1}^{N} \langle \Psi_T | f_i | \Psi_T \rangle + \sum_{i<j} \langle \Psi_T | \frac{1}{r_{ij}} | \Psi_T \rangle \\
&= \sum_{i=1}^{N} \langle \varphi_i | f | \varphi_i \rangle + \sum_{i<j} \left\{ \langle \varphi_i \varphi_j | \frac{1}{r_{12}} | \varphi_i \varphi_j \rangle - \langle \varphi_i \varphi_j | \frac{1}{r_{12}} | \varphi_j \varphi_i \rangle \right\}.
\end{aligned}
$$ (A.34)

This is the formula quoted in the main text in Eq. (2.42).

Finally, we present the derivation of Eq. (2.76). The derivation of the first term in Eq. (2.68), the diagonal matrix components of the one-electron operators, is trivial. Let us consider the interaction terms. Our starting point is the second term in the HF energy expression, Eq. (A.34). Using the relationships given by Eqs. (2.59), (2.60), and (2.61), we obtain for the electrostatic interaction integral

$$
\begin{aligned}
\frac{1}{2} \sum_{ij} \langle \varphi_i \varphi_j | \frac{1}{r_{12}} | \varphi_i \varphi_j \rangle \\
&= \frac{1}{2} \sum_{ij} \int \frac{1}{r_{12}} | \varphi_i(1) |^2 | \varphi_j(2) |^2 \, dq_1 \, dq_2 \\
&= \frac{1}{2} \int \frac{1}{r_{12}} \left\{ \sum_{i=1}^{p} | \psi_i(1) |^2 | \alpha(1) |^2 + \sum_{i=p+1}^{N} | \psi_i(1) |^2 | \beta(1) |^2 \right\} \\
&\quad \times \left\{ \sum_{j=1}^{p} | \psi_j(2) |^2 | \alpha(2) |^2 + \sum_{j=p+1}^{N} | \psi_j(2) |^2 | \beta(2) |^2 \right\} \, dq_1 \, dq_2;
\end{aligned}
$$ (A.35)

$$
\begin{aligned}
\frac{1}{2} \sum_{ij} \langle \varphi_i \varphi_j | \frac{1}{r_{12}} | \varphi_i \varphi_j \rangle &= \frac{1}{2} \int \frac{1}{r_{12}} \rho_+(1) \rho_+(2) \, dv_1 \, dv_2 \\
&\quad + \frac{1}{2} \int \frac{1}{r_{12}} \rho_-(1) \rho_-(2) \, dv_1 \, dv_2 \\
&\quad + \int \frac{1}{r_{12}} \rho_+(1) \rho_-(2) \, dv_1 \, dv_2.
\end{aligned}
$$ (A.36)

We have obtained the electrostatic interaction terms in Eq. (2.68). Using Eq. (2.59), we obtain for the exchange integral in Eq. (A.34)

$$
\frac{1}{2}\sum_{ij}\langle\varphi_i\varphi_j|\frac{1}{r_{12}}|\varphi_j\varphi_i\rangle = \frac{1}{2}\sum_{ij}\int\frac{1}{r_{12}}\varphi_i^*(1)\varphi_j^*(2)\varphi_i(2)\varphi_j(1)\,dq_1\,dq_2
$$

$$
= \frac{1}{2}\sum_{ij}\int\frac{1}{r_{12}}\psi_i^*(1)\psi_j^*(2)\psi_i(2)\psi_j(1)\eta_i^*(1)
$$

$$
\times\,\eta_j(1)\eta_j^*(2)\eta_i(2)\,dv_1\,dv_2\,d\sigma_1\,d\sigma_2. \quad (A.37)
$$

We see from this expression that, in view of the spin integration, the integral will be different from zero only if i and j refer to the same spin. Thus, we obtain, using Eq. (2.60),

$$
\frac{1}{2}\sum_{ij}\langle\varphi_i\varphi_j|\frac{1}{r_{12}}|\varphi_j\varphi_i\rangle
$$

$$
= \frac{1}{2}\sum_{i=1}^{P}\sum_{j=1}^{P}\int\frac{1}{r_{12}}\psi_i^*(1)\psi_i(2)\psi_j^*(2)\psi_j(1)
$$

$$
\times\,|\alpha(1)|^2|\alpha(2)|^2\,d\sigma_1\,d\sigma_2\,dv_1\,dv_2
$$

$$
+ \frac{1}{2}\sum_{i=p+1}^{N}\sum_{j=p+1}^{N}\int\frac{1}{r_{12}}\psi_i^*(1)\psi_i(2)\psi_j^*(2)\psi_j(1)
$$

$$
\times\,|\beta(1)|^2|\beta(2)|^2\,d\sigma_1\,d\sigma_2\,dv_1\,dv_2
$$

$$
= \frac{1}{2}\int\frac{1}{r_{12}}[\rho_+(2,1)\rho_+(1,2)
$$

$$
+\rho_-(2,1)\rho_-(1,2)]\,dv_1\,dv_2. \quad (A.38)
$$

We have obtained the exchange terms in Eqs. (2.68a) and (2.68b). Thus, it is demonstrated that the energy expression given by Eq. (2.67) follows from the general HF energy expression, Eq. (2.42), for the case in which we have p electrons with up spins and $N - p$ with down spins.

APPENDIX B

THE HARTREE – FOCK EQUATIONS FOR A SINGLE-DETERMINANTAL WAVE FUNCTION

In this appendix, we show how the HF equations are derived from the variation principle and we discuss the behavior of the equations under unitary transformations, the diagonalization of the matrix of the Lagrangian multipliers, and the non-Hermitian character of the HF equations.

First, we derive Eq. (2.46) quoted in the main text. We vary the HF energy expression given by Eq. (2.42):

$$
\begin{aligned}
E_F &= \langle \Psi_T | H | \Psi_T \rangle \\
&= \sum_{i=1}^{N} \langle \varphi_i | t + g | \varphi_i \rangle + \frac{1}{2} \sum_{i,j=1}^{N} \left\{ \langle \varphi_i \varphi_j | \frac{1}{r_{12}} | \varphi_i \varphi_j \rangle \right. \\
&\qquad\qquad\qquad\qquad \left. - \langle \varphi_i \varphi_j | \frac{1}{r_{ij}} | \varphi_j \varphi_i \rangle \right\}.
\end{aligned} \tag{B.1}
$$

The subsidiary conditions are given by Eq. (2.41). Let λ_{ij} be a Lagrangian multiplier and let

$$
E_0 = -\frac{1}{2} \sum_{i,j=1}^{N} \lambda_{ij} \langle \varphi_i | \varphi_j \rangle. \tag{B.2}
$$

As is known, the variation of the energy expression with respect to φ_k^* and φ_k leads to the same results; thus, it is sufficient to vary the energy expression with respect to φ_k^*. Let δ_k be the variation with respect to φ_k^*. Our variation principle is

$$\delta_k(E_F + E_0) = 0 \qquad (k = 1, \cdots, N). \tag{B.3}$$

Let us consider the variation of the terms occurring in E_F and E_0 one by one. We get

$$\delta_k \sum_{i=1}^{N} \langle \varphi_i | t + g | \varphi_i \rangle = \langle \delta \varphi_k | t + g | \varphi_k \rangle, \tag{B.4}$$

$$\delta_k \frac{1}{2} \sum_{i,j=1}^{N} \langle \varphi_i \varphi_j | \frac{1}{r_{12}} | \varphi_i \varphi_j \rangle$$

$$= \frac{1}{2} \sum_j \langle \delta \varphi_k \varphi_j | \frac{1}{r_{12}} | \varphi_k \varphi_j \rangle + \frac{1}{2} \sum_i \langle \varphi_i \delta \varphi_k | \frac{1}{r_{12}} | \varphi_i \varphi_k \rangle$$

$$= \sum_j \langle \delta \varphi_k \varphi_j | \frac{1}{r_{12}} | \varphi_k \varphi_j \rangle, \tag{B.5}$$

where we have used the relationship

$$\langle \alpha \beta | \frac{1}{r_{12}} | \gamma \delta \rangle = \langle \beta \alpha | \frac{1}{r_{12}} | \delta \gamma \rangle, \tag{B.6}$$

which follows from Eq. (2.43). Also, we get

$$\delta_k \frac{1}{2} \sum_{i,j=1}^{N} \langle \varphi_i \varphi_j | \frac{1}{r_{12}} | \varphi_j \varphi_i \rangle$$

$$= \frac{1}{2} \sum_{j=1}^{N} \langle \delta \varphi_k \varphi_j | \frac{1}{r_{12}} | \varphi_j \varphi_k \rangle + \frac{1}{2} \sum_{i=1}^{N} \langle \varphi_i \delta \varphi_k | \frac{1}{r_{12}} | \varphi_k \varphi_i \rangle$$

$$= \sum_{j=1}^{N} \langle \delta \varphi_k \varphi_j | \frac{1}{r_{12}} | \varphi_j \varphi_k \rangle. \tag{B.7}$$

Finally,

$$\delta_k \frac{1}{2} \sum_{i,j=1}^{N} \lambda_{ij} \langle \varphi_i | \varphi_j \rangle = \sum_{j=1}^{N} \bar{\lambda}_{kj} \langle \delta \varphi_k | \varphi_j \rangle, \tag{B.8}$$

where $\bar{\lambda}_{kj} = \lambda_{kj}/2$. Substituting these expressions into Eq. (B.3), we obtain

$$\delta_k(E_F + E_0) = \langle \delta\varphi_k | t + g | \varphi_k \rangle + \sum_{j=1}^{N} \langle \delta\varphi_k \varphi_j | \frac{1}{r_{12}} \varphi_k \varphi_j \rangle$$

$$- \sum_{j=1}^{N} \langle \delta\varphi_k \varphi_j | \frac{1}{r_{12}} \varphi_j \varphi_k \rangle - \sum_{j=1}^{N} \bar{\lambda}_{kj} \langle \delta\varphi_k | \varphi_j \rangle = 0, \quad \text{(B.9)}$$

Let us consider the HF potential (operator) for which the definition was given by Eq. (2.49):

$$U_j(1)f(1) = \int \frac{1}{r_{12}} \Big[|\varphi_j(2)|^2 f(1) - \varphi_j(1)\varphi_j^*(2)f(2) \Big] dq_2. \quad \text{(B.10)}$$

Using this notation, we obtain from Eq. (B.9) that for an arbitrary variation δ_k, the expression will be zero if

$$\left\{ t + g + \sum_{j=1}^{N} U_j \right\} \varphi_k = \sum_{j=1}^{N} \bar{\lambda}_{kj} \varphi_j, \quad \text{(B.11)}$$

By writing

$$\varepsilon_k \equiv \bar{\lambda}_{kk}, \quad \text{(B.11a)}$$

we obtain

$$H_F \varphi_k = \varepsilon_k \varphi_k + \sum_{j=1}^{N} \bar{\lambda}_{kj} \varphi_j \quad (k = 1, 2, \cdots, N), \quad \text{(B.11b)}$$

where H_F was defined by Eqs. (2.47), (2.48), and (2.49). We have obtained Eq. (2.46) quoted in the main text.

Next, we discuss the behavior of the HF equations under a unitary transformation. Let

$$\hat{\varphi}_j = \sum_{k=1}^{N} C_{jk} \varphi_k \quad (j = 1, \cdots, N). \quad \text{(B.12)}$$

We assume that the operator of this transformation is unitary, that is, denoting the Hermitian adjoint by C^+ and the inverse by C^{-1}, we have

$$C^+ = C^{-1}. \quad \text{(B.13)}$$

Thus, the inverse transformation is

$$\varphi_k = \sum_{l=1}^{N} (C^{-1})_{kl} \hat{\varphi}_l = \sum_{l=1}^{N} C_{lk}^* \hat{\varphi}_l, \quad \text{(B.14)}$$

where C_{lk}^* is the complex conjugate of C_{lk}. Because the φ_k's are orthonormal and the transformation is unitary, the $\hat{\varphi}_j$'s will also be orthonormal.

We now show that the single-determinantal wave function does not change if the original orthonormal orbitals are replaced by the transformed orbitals. Using the properties of determinants, we obtain for the transformed determinant

$$
\Psi_T' = \begin{vmatrix}
\hat{\varphi}_1(1) & \hat{\varphi}_1(2) & \cdots & \hat{\varphi}_1(N) \\
\hat{\varphi}_2(1) & \hat{\varphi}_2(2) & \cdots & \hat{\varphi}_2(N) \\
\vdots & \vdots & & \vdots \\
\hat{\varphi}_N(1) & \hat{\varphi}_N(2) & \cdots & \hat{\varphi}_N(N)
\end{vmatrix}
$$

$$
= \begin{vmatrix}
C_{11} & C_{12} & \cdots & C_{1n} \\
C_{21} & C_{22} & \cdots & C_{2N} \\
\vdots & \vdots & & \vdots \\
C_{N1} & C_{N2} & \cdots & C_{NN}
\end{vmatrix} \cdot \begin{vmatrix}
\varphi_1(1) & \varphi_1(2) & \cdots & \varphi_1(N) \\
\varphi_2(1) & \varphi_2(2) & \cdots & \varphi_2(N) \\
\vdots & \vdots & & \vdots \\
\varphi_N(1) & \varphi_N(2) & \cdots & \varphi_N(N)
\end{vmatrix}. \quad \text{(B.15)}
$$

As we see, a unitary transformation of all orbitals yields a determinantal wave function that is, apart from a multiplicative constant, identical with the original single-determinantal wave function. Since a wave function is always determined only up to a multiplicative constant, this means that the single-determinantal wave function does not change under a unitary transformation of the orbitals. Indeed, when we calculate the expectation value of the Hamiltonian, we always use the formula

$$
E = \langle \Psi_T | H | \Psi_T \rangle / \langle \Psi_T | \Psi_T \rangle. \quad \text{(B.16)}
$$

It is evident that using the original determinant and the transformed determinant would lead to the same result because the multiplicative constant of Eq. (B.15) would drop out of Eq. (B.16).

Now, let us consider the HF equations, (B.11). We want to see what happens to these equations if the orbitals are subjected to the unitary transformation. The HF potential depends only on the first-order density matrix and on the local electron density. We show that these quantities are invariant under a unitary transformation. First, consider the density matrix formed from the transformed orbitals. We get

$$
\sum_i \hat{\varphi}_i(1) \hat{\varphi}_i^*(2) = \sum_{i,k,l} C_{ik} \varphi_k(1) C_{il}^* \varphi_l^*(2)
$$

$$
= \sum_{k,l} \sum_i C_{il}^* C_{ik} \varphi_k(1) \varphi_l^*(2)
$$

$$
= \sum_{k,l} \delta_{lk} \varphi_k(1) \varphi_l^*(2) = \sum_k \varphi_k(1) \varphi_k^*(2). \quad \text{(B.17)}
$$

In this argument, we have used the relationship

$$\sum_i C_{il}^* C_{ik} = \delta_{lk}, \tag{B.18}$$

which follows from the unitary property.

Equation (B.17) shows that the density matrix, and with the density matrix also the local density, is invariant under the transformation. Thus, we obtain by denoting the transformed HF Hamiltonian by $\hat{H}_F$:

$$\hat{H}_F = H_F. \tag{B.19}$$

Let us now write down the HF equations given by Eq. (B.11):

$$H_F \varphi_k = \sum_{j=1}^{N} \bar{\lambda}_{kj} \varphi_j. \tag{B.20}$$

We multiply the equation for φ_k from the left by C_{lk} and sum over k. Then we obtain

$$\sum_k C_{lk} H_F \varphi_k = H_F \hat{\varphi}_l = \sum_{\partial, k} C_{lk} \bar{\lambda}_{kj} \varphi_j. \tag{B.21}$$

Using Eq. (B.14), we get

$$H_F \hat{\varphi}_l = \sum_{j, k, i} C_{lk} \bar{\lambda}_{kj} C_{ij}^* \hat{\varphi}_i. \tag{B.22}$$

Let us introduce the notation

$$\hat{\lambda}_{li} = \sum_{j, k} C_{lk} \bar{\lambda}_{kj} C_{ij}^*. \tag{B.23}$$

Clearly, this is the matrix component of the transformed matrix

$$\hat{\lambda}_{li} = \sum_{j, k} C_{lk} \bar{\lambda}_{kj} C_{ji}^+ = (C \bar{\lambda} C^+)_{li}. \tag{B.24}$$

Thus, we can write Eq. (B.22) in the form

$$H_F \hat{\varphi}_l = \sum_{i=1}^{N} \hat{\lambda}_{li} \hat{\varphi}_i. \tag{B.25}$$

Finally, using Eq. (B.19), we get

$$\hat{H}_F \hat{\varphi}_l = \sum_{i=1}^{N} \hat{\lambda}_{li} \hat{\varphi}_i. \tag{B.26}$$

Thus, our results concerning the behavior of the HF equations under a unitary transformation are as follows. The HF Hamiltonian is invariant under the transformation. The transformed equations are formally the same as the originals. There will be an infinite number of equations for an infinite number of orbital sets, each belonging to one of the Lagrangian matrices $\bar{\lambda}_{li}$. The Lagrangian matrices will be connected by unitary transformations like in Eq. (B.24). Each of the infinite number of sets will yield the same density and the same total energy.

This situation looks surprising at first, but in fact it becomes plausible if we recall that we have applied the variation principle to the total energy, that is, we have varied the total energy. We have seen that the total energy depends on the total wave function and that this total wave function is invariant under a unitary transformation. Thus, by deriving the HF equations, we obtained the equations defining the energy minimum for any of the orbital sets that are connected by unitary transformations. The energy minimization cannot distinguish between the sets that are connected by unitary transformations because the total energy that was varied is itself invariant under such transformations.

If the Lagrangian matrix is Hermitian, then we can use the unitary transformation to diagonalize the matrix, that is, to transform it to the form

$$\hat{\lambda}_{li} = \varepsilon_l \delta_{li}. \tag{B.27}$$

Then, we obtain the HF equations in the form

$$\hat{H}_F \hat{\varphi}_l = \varepsilon_l \hat{\varphi}_l, \tag{B.28}$$

which is the form we quoted in the main text, Eq. (2.50). This equation is equivalent to all the other equations that are given by Eq. (B.26), except that due to the diagonalization, this equation has a very simple form.

The condition for the diagonalization is that the Lagrangian matrix be Hermitian. From Eq. (B.20), we obtain

$$\bar{\lambda}_{kj} = \langle \varphi_j | H_F | \varphi_k \rangle. \tag{B.29}$$

It is easy to see that the HF Hamiltonian is Hermitian from which follows the hermiticity of $\bar{\lambda}_{kj}$.

It is interesting to note that using Eq. (B.29), we can rewrite the HF equations as follows:

$$H_F \varphi_k = \varepsilon_k \varphi_k + \sum_{j=1}^{N} \varphi_j \langle \varphi_j | H_F | \varphi_k \rangle \qquad (j \neq k), \tag{B.30}$$

Let us introduce the projection operators Ω_k and Π_k with the definitions

$$\Omega_k \equiv \sum_{j=1}^{N} |\varphi_j\rangle\langle\varphi_j| \qquad (j \neq k), \tag{B.31}$$

and

$$\Pi_k \equiv 1 - \Omega_k. \tag{B.32}$$

Then the equations become

$$H_F \varphi_k = \varepsilon_k \varphi_k + \Omega_k H_F \varphi_k, \tag{B.33}$$

or

$$(1 - \Omega_k) H_F \varphi_k = \varepsilon_k \varphi_k, \tag{B.34}$$

and

$$(\Pi_k H_F) \varphi_k = \varepsilon_k \varphi_k. \tag{B.35}$$

Thus, we see that, in general, the HF orbitals are not the eigenfunctions of H_F, but of $\Pi_k H_F$. Only for the diagonalized equations (B.28) is it true that the orbitals are the eigenfunctions of the HF Hamiltonian H_F. Also, we note that the operator $\Pi_k H_F$ is not Hermitian since

$$\Pi_k H_F \neq H_F \Pi_k. \tag{B.36}$$

We must register the interesting fact that the HF orbitals, in general, are not the eigenfunctions of Hermitian operators. The exception to this rule is when we diagonalize the Lagrangian matrix and obtain Eq. (B.28). In that case, *and only in that case*, the orbitals are the eigenfunctions of the Hermitian operator H_F.

APPENDIX C

HARTREE–FOCK MATRIX COMPONENTS FOR CENTRAL-FIELD FUNCTIONS

In this appendix, we derive the formulas given by Eq. (2.70) in the main text. Our goal is to obtain the HF matrix components:

$$\langle i|t + g|i\rangle \equiv \langle \varphi_i|t + g|\varphi_i\rangle, \tag{C.1}$$

$$\langle ij|\frac{1}{r_{12}}|ij\rangle \equiv \langle \varphi_i\varphi_j|\frac{1}{r_{12}}|\varphi_i\varphi_j\rangle, \tag{C.2}$$

$$\langle ij|\frac{1}{r_{12}}|ji\rangle \equiv \langle \varphi_i\varphi_j|\frac{1}{r_{12}}|\varphi_j\varphi_i\rangle. \tag{C.3}$$

It will be useful to consider first the following general expressions

$$\langle i|t + g|j\rangle \tag{C.4}$$

and

$$\langle ij|\frac{1}{r_{12}}|st\rangle \tag{C.5}$$

where i, j, s, and t are arbitrary central-field orbitals.

470

Let us consider first $\langle i|t + g|j\rangle$ and let us put the orbitals i and j into the central-field form given by Eq. (2.69). Then, we get

$$\langle \varphi_i|t + g|\varphi_j\rangle = \int \varphi_i^*(t + g)\varphi_j \, dq$$

$$= \int \psi_i^*(t + g)\psi_j \, dv \int \eta_i^* \eta_j \, d\sigma$$

$$= \delta(m_{si}, m_{sj}) \int \psi_i^*(t + g)\psi_j \, dv, \qquad (C.6)$$

where δ is the Kronecker symbol. Turning to the spatial part of the integral, we get

$$\int \psi_i^*(t + g)\psi_j \, dv = \int \frac{P_{n_i l_i}}{r} Y_{l_i m_{l_i}}(\vartheta, \varphi)\left[-\frac{1}{2}\Delta - \frac{Z}{r}\right]$$

$$\times \frac{P_{n_j l_j}}{r} Y_{l_j m_{l_j}}(\vartheta, \varphi)r^2 \, dr \sin \vartheta \, d\vartheta \, d\varphi. \qquad (C.7)$$

We put the Laplacian into spherical polar coordinates:

$$t = -\frac{1}{2}\Delta = -\frac{1}{2r^2}\frac{\partial}{\partial r}\left(r^2 \frac{\partial}{\partial r}\right) - \frac{\nabla_{\vartheta\varphi}^2}{2r^2}, \qquad (C.8)$$

where $\nabla_{\vartheta\varphi}^2$ is the angular part that operates on the spherical harmonics in the following way:

$$-\nabla_{\vartheta\varphi}^2 Y_{lm} = l(l + 1)Y_{lm}. \qquad (C.9)$$

Using this line, we get

$$-\frac{1}{2}\Delta\left(\frac{P_{n_j l_j}}{r}Y_{l_j m_{l_j}}\right)$$

$$= -\frac{1}{2r^2}\frac{\partial}{\partial r}\left(r^2 \frac{\partial}{\partial r}\right)\frac{P_{n_j l_j}}{r}Y_{l_j m_{l_j}} - \frac{\nabla_{\vartheta\varphi}^2}{2r^2}\left(\frac{P_{n_j l_j}}{r}Y_{l_j m_{l_j}}\right)$$

$$= -\frac{1}{2r}\left(\frac{d^2}{dr^2}P_{n_j l_j}\right)Y_{l_j m_{l_j}} + \frac{l_j(l_j + 1)}{2r^2}\frac{P_{n_j l_j}}{r}Y_{l_j m_{l_j}}. \qquad (C.10)$$

Putting this line into Eq. (C.7) and taking into account the orthonormality of the spherical harmonics, we obtain

$$\int \psi_i^*(t+g)\psi_j \, dv = \delta(l_i, l_j)\,\delta(m_{li}, m_{lj})$$

$$\times \int P_{n_i l_i}\left[-\frac{1}{2}\frac{d^2}{dr^2} + \frac{l_j(l_j+1)}{2r^2} - \frac{Z}{r} \right] P_{n_j l_j} \, dr, \quad \text{(C.11)}$$

and from this, we obtain

$$\langle \varphi_i | t+g | \varphi_j \rangle = \delta(l_i, l_j)\,\delta(m_{li}, m_{lj})\,\delta(m_{si}, m_{sj})$$

$$\times \int P_{n_i l_i}\left[-\frac{1}{2}\frac{d^2}{dr^2} + \frac{l_j(l_j+1)}{2r^2} - \frac{Z}{r} \right] P_{n_j l_j} \, dr. \quad \text{(C.12)}$$

By putting $i = j$, we obtain from this formula Eq. (2.70a).

Let us now consider the general matrix component given by Eq. (C.5). We put the orbitals in the form given by Eq. (2.69) and we get

$$\langle ij | \frac{1}{r_{12}} | s, t \rangle = \int \varphi_i^*(1)\varphi_j^*(2)\frac{1}{r_{12}}\varphi_s(1)\varphi_t(2) \, dq_1 \, dq_2$$

$$= \int \psi_i^*(1)\psi_j^*(2)\frac{1}{r_{12}}\psi_s(1)\psi_t(2) \, dv_1 \, dv_2 \int \eta_i^* \eta_s \, d\sigma_1 \int \eta_j^* \eta_t \, d\sigma_2$$

$$= \delta(m_{si}, m_{ss})\,\delta(m_{sj}, m_{st}) \int \psi_i^*(1)\psi_j^*(2)\frac{1}{r_{12}}\psi_s(1)\psi_t(2) \, dv_1 \, dv_2.$$

$$\text{(C.13)}$$

Turning to the spatial part of the integral, we observe that because the one-electron orbitals are of the central-field type, it will be convenient to expand $1/r_{12}$ in terms of spherical harmonics. Thus, we put

$$\frac{1}{r_{12}} = \sum_{k=0}^{\infty} L_k(r_1, r_2) P_k(\cos \gamma_{12})$$

$$= \sum_{k=0}^{\infty} L_k(r_1, r_2)\frac{4\pi}{2k+1}\sum_{m=-k}^{+k} Y_{km}^*(\vartheta_1 \varphi_1) Y_{km}(\vartheta_2 \varphi_2), \quad \text{(C.14)}$$

where L_k is given by Eq. (2.72). Writing the one-electron orbitals in central-field form and using this expression, we get

$$\int \psi_i^*(1)\psi_j^*(2)\frac{1}{r_{12}}\psi_s(1)\psi_t(2)\,dv_1\,dv_2$$

$$= \int P_{n_il_i}(r_1)Y_{l_im_{li}}^*(\vartheta_1\varphi_1)P_{n_jl_j}(r_2)Y_{l_jm_{lj}}(\vartheta_2\varphi_2)\frac{1}{r_1r_2}$$

$$\times \sum_{k=0}^{\infty} L_k(r_1,r_2)\frac{4\pi}{2k+1}\sum_{m=-k}^{+k}Y_{km}^*(\vartheta_1\varphi_1)Y_{km}(\vartheta_2\varphi_2)$$

$$\times P_{n_sl_s}(r_1)Y_{l_sm_{ls}}(\vartheta_1\varphi_1)P_{n_tl_t}(r_2)Y_{l_tm_{lt}}(\vartheta_2\varphi_2)\frac{1}{r_1r_2}$$

$$\times r_1^2\,dr_1\,r_2^2\,dr_2\,d\omega_1\,d\omega_2, \tag{C.15}$$

where $d\omega_1$ and $d\omega_2$ are the angular integrations. We can integrate separately over the angles and over r_1 and r_2. Separating these integrations, we can write

$$\int \psi_i^*(1)\psi_j^*(2)\frac{1}{r_{12}}\psi_s(1)\psi_t(2)\,dv_1\,dv_2$$

$$= \sum_{k=0}^{\infty}\sum_{m=-k}^{+k}\frac{4\pi}{2k+1}$$

$$\times \int Y_{l_im_{li}}^*(1)Y_{km}^*(1)Y_{l_sm_{ls}}(1)\,d\omega_1$$

$$\times \int Y_{l_jm_{lj}}^*(2)Y_{km}(2)Y_{l_tm_{lt}}(2)\,d\omega_2$$

$$\times \int P_{n_il_i}(1)P_{n_jl_j}(2)P_{n_sl_s}(1)P_{n_tl_t}(2)L_k(1,2)\,dr_1\,dr_2. \tag{C.16}$$

From the angular integrations, we obtain the following relationships:

$$\begin{aligned}-m_{li}-m+m_{ls}&=0, & m&=m_{ls}-m_{li};\\-m_{lj}+m+m_{lt}&=0, & m&=m_{lj}-m_{lt};\end{aligned} \tag{C.17}$$

thus

$$\begin{aligned}m_{ls}-m_{li}&=m_{lj}-m_{lt},\\m_{ls}+m_{lt}&=m_{lj}+m_{li}.\end{aligned} \tag{C.18}$$

We see from Eq. (C.17) that for a given set of i, j, s, and t, the index m is restricted to one value. Now let us introduce the notations

$$c^k(lm, l'm') \equiv \sqrt{\frac{4\pi}{2k+1}} \int Y_{lm}^*(1) Y_{l'm'}(1) Y_{k(m-m')}(1) \, d\omega_1, \quad (C.19)$$

and

$$R_k(ij, st) \equiv \int P_{n_i l_i}(1) P_{n_j l_j}(2) P_{n_s l_s}(1) P_{n_t l_t}(2) L_k(1,2) \, dr_1 \, dr_2. \quad (C.20)$$

Using these notations and the relationship given by Eq. (C.18), we obtain

$$\int \psi_i^*(1) \psi_j^*(2) \frac{1}{r_{12}} \psi_s(1) \psi_t(2) \, dv_1 \, dv_2$$

$$= \delta(m_{li} + m_{lj}, m_{ls} + m_{lt})$$

$$\times \sum_{k=0}^{\infty} c^k(l_j m_{lj}, l_t m_{lt}) c^k(l_s m_{ls}, l_i m_{li}) R_k(ij, st). \quad (C.21)$$

The c^k coefficients were tabulated by Slater [Slater, II, 281] and we also give an explicit formula for them in App. E. It is easy to show that

$$c^k(lm, l'm') = 0, \quad (C.22)$$

unless

$$k \leq l + l'. \quad (C.23)$$

Thus, the upper limit of the k summation in Eq. (C.21) will be the lesser of $l_j + l_t$ and $l_s + l_i$.

We now easily obtain the desired matrix components given by Eqs. (C.2) and (C.3). First, we put $s = i$ and $t = j$ and obtain

$$\langle ij| \frac{1}{r_{12}} |ij\rangle = \sum_{k=0}^{l_m} a_k(l_i m_{li}, l_j m_{lj}) R_k(ij, ij), \quad (C.24)$$

where

$$a_k(lm, l'm') = c^k(lm, lm) c^k(l'm', l'm'), \quad (C.25)$$

and l_m is the lesser of $2l_i$ and $2l_j$. By putting $s = j$ and $t = i$, we get

$$\langle ij| \frac{1}{r_{12}} |ji\rangle = \sum_{k=0}^{l_i+l_j} b_k(l_i m_{li}, l_j m_{lj}) R_k(ij, ji), \quad (C.26)$$

where

$$b_k(lm, l'm') = \left[c^k(lm, l'm')\right]^2. \tag{C.27}$$

From the definition of R_k, Eq. (C.20), we see that

$$R_k(ij, ij) = F_k(n_i l_i n_j l_j), \tag{C.28}$$

and

$$R_k(ij, ji) = G_k(n_i l_i n_j l_j). \tag{C.29}$$

Using these relationships for Eqs. (C.24) and (C.26), we obtain directly the formulas given in Eqs. (2.70b) and (2.70c).

APPENDIX D

THE ELECTROSTATIC AND EXCHANGE POTENTIALS FOR ATOMS WITH COMPLETE GROUPS

In this appendix, we evaluate the electrostatic and exchange potentials that occur in the HF equations for complete groups. We want to evaluate first the exchange operator given by Eq. (2.91), that is, our first task is to get a formula for

$$A(1)\psi_i(1) = \sum_{j=1}^{N/2} \int \frac{1}{r_{12}} \psi_j(1)\psi_j^*(2)\psi_i(2)\, dv_2. \qquad (D.1)$$

Let us consider the kernel of this operator. We put ψ_j into central-field form and get

$$K_A(1,2) = \sum_{j=1}^{N/2} \frac{1}{r_{12}} \psi_j(1)\psi_j^*(2)$$

$$= \frac{1}{r_{12}} \sum_{n_j l_j m_{l_j}} \frac{P_{n_j l_j}(r_1)}{r_1} Y_{l_j m_{l_j}}(\vartheta_1 \varphi_1) \frac{P_{n_j l_j}(r_2)}{r_2} Y_{l_j m_{l_j}}^*(\vartheta_2 \varphi_2). \qquad (D.2)$$

Using the expansion theorem for the Legendre polynomials that we have already used in Eq. (C.14), we get

$$P_k(\cos \gamma_{12}) = \frac{4\pi}{2k+1} \sum_{m=-k}^{+k} Y_{km}(1)Y_{km}^*(2). \qquad (D.3)$$

In Eq. (D.2), the summation over m_{l_j} goes from $-l_j$ to $+l_j$ for any l_j. Thus, we get

$$K_A(1,2) = \frac{1}{r_{12}} \sum_{n_j l_j} \frac{1}{r_1 r_2} P_{n_j l_j}(1) P_{n_j l_j}(2) \frac{2l_j + 1}{4\pi} P_{l_j}(\cos \gamma_{12}). \quad (D.4)$$

where γ_{12} is the angle between vectors $\mathbf{r}_1$ and $\mathbf{r}_2$. Thus, it is advantageous to use Eq. (C.14) again and expand $1/r_{12}$ in terms of Legendre polynomials:

$$\frac{1}{r_{12}} = \sum_{k=0}^{\infty} L_k(1,2) P_k(\cos \gamma_{12}), \quad (D.5)$$

where the L_k is the function defined by Eq. (2.72).

If we substitute Eq. (D.5) into Eq. (D.4), we obtain products of two Legendre polynomials. Let us make a digression here and show how this problem can be handled. Consider one of these products of Legendre polynomials and let us express it as an infinite series:

$$P_{l_j}(\cos \gamma_{12}) P_k(\cos \gamma_{12}) = \sum_{l=0}^{\infty} A(k,l_j,l) P_l(\cos \gamma_{12}). \quad (D.6)$$

Such an expansion certainly exists because the polynomials form a complete system. Let us denote γ_{12} by ϑ. Multiply the equation from the left by P_{l_i} and integrate over the angular space. Denoting this integration by $d\omega$, we get

$$\int P_{l_i} P_{l_j} P_k \, d\omega = \sum_{l=0}^{\infty} A(k,l_j,l) \int P_{l_i} P_l \, d\omega. \quad (D.7)$$

The formula for the normalized spherical harmonics is

$$Y_{lm} = \sqrt{\frac{2l+1}{4\pi} \frac{(l-m)!}{(l+m)!}} P_l^m(\cos \vartheta) e^{im\varphi}, \quad (D.8)$$

where P_l^m is the associated Legendre polynomial. From this formula, we obtain, for $m = 0$,

$$Y_{l0} = \sqrt{\frac{2l+1}{4\pi}} P_l(\cos \vartheta), \quad (D.9)$$

and so

$$P_l(\cos \vartheta) = \sqrt{\frac{4\pi}{2l+1}} Y_{l0}. \quad (D.10)$$

Using this relationship, we obtain

$$\int P_{l_i} P_l \, d\omega = \sqrt{\frac{(4\pi)^2}{(2l_i + 1)(2l + 1)}} \int Y_{l_i,0}^* Y_{l0} \, d\omega$$

$$= \sqrt{\cdots} \; \delta(l_i, l), \tag{D.11}$$

and on putting this into Eq. (D.7), we get

$$\int P_{l_i} P_{l_j} P_k \, d\omega = \sum_{l=0}^{\infty} A(k, l_j, l) \sqrt{\frac{(4\pi)^2}{(2l_i + 1)(2l + 1)}} \; \delta(l_i, l)$$

$$= \frac{4\pi}{2l_i + 1} A(k, l_j, l_i). \tag{D.12}$$

Using the relationship given by Eq. (C.19), we obtain

$$A(k, l_j, l_i) = \frac{2l_i + 1}{4\pi} \int P_{l_i} P_{l_j} P_k \, d\omega$$

$$= \frac{2l_i + 1}{4\pi} \sqrt{\frac{(4\pi)^3}{(2l_i + 1)(2l_j + 1)(2k + 1)}} \int Y_{l_i,0}^* Y_{l_j,0} Y_{k0} \, d\omega$$

$$= \sqrt{\frac{2l_i + 1}{2l_j + 1}} \sqrt{\frac{4\pi}{2k + 1}} \int Y_{l_i,0}^* Y_{l_j,0} Y_{k0} \, d\omega$$

$$= \sqrt{\frac{2l_i + 1}{2l_j + 1}} \; c^k(l_i 0, l_j 0), \tag{D.13}$$

and on putting this into Eq. (D.6), we get

$$P_{l_j}(\cos \gamma_{12}) P_k(\cos \gamma_{12}) = \sum_{l=0}^{\infty} A(k, l_j, l) P_l(\cos \gamma_{12})$$

$$= \sum_{l=0}^{\infty} \sqrt{\frac{2l + 1}{2l_j + 1}} \; c^k(l0, l_j 0) P_l(\cos \gamma_{12}). \tag{D.14}$$

Now put the expansion of $1/r_{12}$ into the formula for the kernel, Eq. (D.4), and use, for the products of two Legendre polynomials, Eq. (D.14). Then, we get

$$K_A(1,2) = \sum_{n_j l_j} \frac{1}{r_1 r_2} P_{n_j l_j}(1) P_{n_j l_j}(2) \frac{2l_j + 1}{4\pi}$$

$$\times P_{l_j}(\cos \gamma_{12}) \sum_{k=0}^{\infty} L_k(1,2) P_k(\cos \gamma_{12})$$

$$= \sum_{n_j l_j} \sum_{l=0}^{\infty} \sum_{k=0}^{\infty} \sqrt{\frac{(2l+1)(2l_j+1)}{(4\pi)^2}} \, c^k(l0,l_j 0) \frac{1}{r_1 r_2}$$

$$\times P_{n_j l_j}(1) P_{n_j l_j}(2) L_k(1,2) P_l(\cos \gamma_{12}). \tag{D.15}$$

Our result is that the kernel of the exchange operator depends on the angle γ_{12} between the two vectors $\mathbf{r}_1$ and $\mathbf{r}_2$ but does not depend on the angles $(\vartheta_1 \varphi_1)$ and $(\vartheta_2 \varphi_2)$. This result was first obtained by Fock[5] and the derivation here was given by Slater [Slater I, 318].

By obtaining the kernel of the operator, it is easy to evaluate the expression given by Eq. (D.1). We put ψ_i in central-field form and integrate over dv_2. We obtain

$$A(1)\psi_i(1) = \sum_{j=1}^{N/2} \int \frac{1}{r_{12}} \psi_j(1) \psi_j^*(2) \psi_i(2) \, dv_2$$

$$= \int K_A(1,2) \psi_i(2) \, dv_2$$

$$= \sum_{n_j l_j} \sum_{l=0}^{\infty} \sum_{k=0}^{\infty} \int \sqrt{\frac{(2l+1)(2l_j+1)}{(4\pi)^2}} \, c^k(l0,l_j 0) \frac{1}{r_1 r_2} P_{n_j l_j}(1) P_{n_j l_j}(2)$$

$$\times L_k(1,2) P_l(\cos \gamma_{12}) \frac{P_{n_i l_i}(2)}{r_2} Y_{l_i m_{l_i}}(2)$$

$$\times r_2^2 \, dr_2 \sin \vartheta_2 \, d\vartheta_2 \, d\varphi_2. \tag{D.16}$$

For $P_l(\cos \gamma_{12})$, we use again the expression given by Eq. (D.3) and obtain for the angular integral:

$$\int P_l(\cos \gamma_{12}) Y_{l_i m_{l_i}}(2) \sin \vartheta_2 \, d\vartheta_2 \, d\varphi_2$$

$$= \frac{4\pi}{2l+1} \sum_{m=-l}^{+l} Y_{lm}(1) \int Y_{lm}^*(2) Y_{l_i m_{l_i}}(2) \, d\omega_2$$

$$= \frac{4\pi}{2l+1} \sum_{m=-l}^{+l} Y_{lm}(1) \, \delta(l,l_i) \, \delta(m,m_{l_i})$$

$$= \frac{4\pi}{2l+1} Y_{lm_{l_i}}(1) \, \delta(l,l_i), \tag{D.17}$$

and substituting this into Eq. (D.16), we get

$$
A(1)\psi_i(1) = \sum_{n_j l_j} \sum_{k=0}^{\infty} \sqrt{\frac{2l_j + 1}{2l_i + 1}}\, c^k(l_i 0, l_j 0)
$$

$$
\times \frac{1}{r_1} Y_k\!\left(n_j l_j n_i l_i | 1\right) \frac{P_{n_j l_j}(1)}{r_1} Y_{l_i m_{l_i}}(1). \tag{D.18}
$$

This is Eq. (2.92), which we have quoted in the main text. We note that due to the properties of the c^k coefficients, the upper limit of the summation over k is $l_i + l_j$.

Our next task is to derive the relationship given in Eq. (2.94). That is the formula for the electrostatic potential, and in order to obtain that, we first derive a formula for the total density, which we get easily from Eq. (D.4). If we omit $1/r_{12}$, then Eq. (D.4) is the density matrix, and by putting $\mathbf{r}_1 \equiv \mathbf{r}_2$, we obtain

$$
\sum_{j=1}^{N/2} \left| \psi_j(\mathbf{r}_2) \right|^2 = \sum_{n_j l_j} \frac{1}{r_2^2} P_{n_j l_j}^2(r_2) \frac{2l_j + 1}{4\pi} P_l(1), \tag{D.19}
$$

where we have put $\gamma_{12} = 0$. Because $P_l(1) = 1$, we get

$$
\rho(2) = \sum_{j=1}^{N/2} \left| \psi_j(2) \right|^2 = \sum_{n_j l_j} \frac{1}{r_2^2} P_{n_j l_j}(2) \frac{2l_j + 1}{4\pi}, \tag{D.20}
$$

and now again using Eq. (D.5), we obtain

$$
V(1) = 2 \sum_{j=1}^{N/2} \int \frac{1}{r_{12}} \left| \psi_j(2) \right|^2 dv_2
$$

$$
= \sum_{n_j l_j} 2(2l_j + 1) \frac{1}{r_1} Y_0\!\left(n_j l_j n_j l_j | r_1\right), \tag{D.21}
$$

which is Eq. (2.94), which we have quoted in the main text.

APPENDIX E

FORMULAS FOR THE AVERAGE ENERGY OF A CONFIGURATION IN THE RELATIVISTIC AND NONRELATIVISTIC HARTREE–FOCK MODELS

This appendix is a small formula collection. First, we list the formulas that define the angular/spin coefficients in the nonrelativistic and in the relativistic HF models. Then, we write down the joint formula for the average energy of a configuration for the relativistic and for the nonrelativistic models. Next, we show that the angular/spin coefficients that occur in the average energy formulas can be generated from the Clebsch–Gordan constants. For the latter, we present a closed explicit expression. The appendix closes with formulas for the magnetic interaction. It is shown that the constants in this interaction can also be generated from the Clebsch–Gordan coefficients.

We start by writing down the formulas defining the angular coefficients in the nonrelativistic HF model. We have, according to Eqs. (C.24), (C.25), (C.26), (C.27), (C.28), and (C.29),

$$\langle \varphi_i \varphi_j | \frac{1}{r_{12}} | \varphi_i \varphi_j \rangle = \sum_{k=0}^{l_m} a_k(l_i m_{li}, l_j m_{lj}) F_k(n_i l_i n_j l_j), \qquad (E.1)$$

$$\langle \varphi_i \varphi_j | \frac{1}{r_{12}} | \varphi_j \varphi_i \rangle = \sum_{k=0}^{l_i + l_j} b_k(l_i m_{li}, l_j m_{lj}) G_k(n_i l_i n_j l_j). \qquad (E.2)$$

In these formulas, φ_i is a central-field spin-orbital, defined by Eq. (2.69). The radial integrals were defined by Eqs. (C.20), (C.28), and (C.29). In Eq. (E.1), l_m is the lesser of $2l_i$ and $2l_j$. The a_k and b_k coefficients are defined as follows:

$$a_k(lm, l'm') = c^k(lm, lm)c^k(l'm', l'm'),$$ (E.3)

and

$$b_k(lm, l'm') = [c_k(lm, l'm')]^2.$$ (E.4)

Now we write down the formulas from the relativistic HFD theory. According to Eq. (4.108), we have

$$\langle \varphi_A \varphi_B | \frac{1}{r_{12}} | \varphi_A \varphi_B \rangle = \sum_k a^k(j_A m_{jA}, j_B m_{jB}) F_k^E(A, B),$$ (E.5)

and from Eq. (4.113), we obtain

$$\langle \varphi_A \varphi_B | \frac{1}{r_{12}} | \varphi_B \varphi_A \rangle = \sum_k b^k(j_A m_{jA}, j_B m_{jB}) G_k^E(A, B).$$ (E.6)

In these formulas, φ_A is a four-component Dirac-type function, given by Eq. (4.102). The radial integrals are defined by Eqs. (4.112) and (4.116). The range of k was given by Eq. (4.109) for the direct integral and by Eq. (4.115) for the exchange integral. The coefficients a^k and b^k are defined in terms of the d^k coefficients:

$$a^k(j_A m_{jA}, j_B m_{jB}) = d^k(j_A m_{jA}, j_A m_{jA})d^k(j_B m_{jB}, j_B m_{jB}),$$ (E.7)

and

$$b^k(j_A m_{jA}, j_B m_{jB}) = [d^k(j_A m_{jA}, j_B m_{jB})]^2.$$ (E.8)

Both the c^k and d^k coefficients can be generated from the Clebsch–Gordan constants. We write the formulas:

$$(-1)^k c^k(lm, l'm') = (-1)^{m'} \frac{[(2l + 1)(2l' + 1)]^{1/2}}{2k + 1}$$
$$\times C(ll'k; 0, 0)C(ll'k; -m, m'),$$ (E.9)

where $C(abc; d, e)$ is the Clebsch–Gordan constant. Then,

$$d^k(jm_j, j'm'_j) = (-1)^{m_j + \frac{1}{2}} \frac{[(2j + 1)(2j' + 1)]^{1/2}}{2k + 1}$$

$$\times C(jj'k; \tfrac{1}{2}, -\tfrac{1}{2}) C(jj'k; -m_j, m'_j). \qquad \text{(E.10)}$$

These two formulas, which are of conspicuously similar structure, were quoted in Eqs. (4.122) and (4.111).

Next, we write down the joint formula for the average energy of a configuration. In Sec. 4.3/C, we have seen how to construct a joint formula for the energies of complete groups. The formula that reproduced both the relativistic and nonrelativistic energy expressions was given by Eq. (4.181). For the average energies, it is equally simple to construct such a formula. We now show that the formula reproducing the average energy for the relativistic as well as for the nonrelativistic model is as follows:

$$
E_{\text{av}} = \sum_A \left\{ q(A) I(A) + \frac{q(A)}{2} [q(A) - 1] \right.
$$

$$
\times \left[F_0^E(A, A) - \sum_k \frac{\{q(A)\}}{2[\{q(A)\} - 1]} \Lambda(A, A, k) F_k^E(A, A) \right] \right\}
$$

$$
+ \sum_A \sum_{\substack{B \\ (A < B)}} q(A) q(B) \left\{ F_0^E(A, B) - \sum_k \frac{1}{2} \Lambda(A, B, k) G_k^E(A, B) \right\}.
$$

$$\text{(E.11)}$$

In this formula, $q(A)$ is the occupation number of the complete or incomplete group A. The radial integrals F_k^E and G_k^E were defined before. The symbol $\Lambda(A, B, k)$ is different from a similar symbol in Eq. (4.181). Now, we define, for the relativistic case,

$$\Lambda(A, B, k) = \Gamma^k_{j_A j_B}, \qquad \text{(E.12)}$$

where $\Gamma^k_{jj'}$ is the constant that we introduced in Eq. (4.130). The symbol $\{q(A)\}$ has the following meaning:

$$\{q(A)\} \equiv q(A)|_{\text{comp}} = 2j_A + 1. \qquad \text{(E.13)}$$

We now show that Eq. (E.11) reproduces our average energy formula in Eq. (4.196). We need to prove this only for the third term of Eq. (E.11) because the other terms are clearly identical with the corresponding terms in Eq.

(4.196). For the third term, we get

$$
\frac{\{q(A)\}}{2[\{q(A)\} - 1]} \Lambda(A, A, k) = \frac{2j_A + 1}{2[2j_A + 1 - 1]} \Gamma^k_{j_A j_A}
$$

$$
= \frac{2j_A + 1}{4j_A} \Gamma^k_{j_A j_A}, \tag{E.14}
$$

which is the same coefficient as in the third term of Eq. (4.196). Thus, our statement, regarding the reproduction of Eq. (4.196), is proved.

Next, we show that the formula in Eq. (E.11) also reproduces the nonrelativistic average energy formula given by Eq. (4.191). We show this by putting $A = (nl)$, $B = (n'l')$, and taking the nonrelativistic limit in the radial integrals, that is, putting $Q_A = Q_B = 0$. For the nonrelativistic case, we define Λ as follows:

$$
\Lambda(A, B, k) = \Lambda(nl, n'l', k)
$$

$$
= \frac{c^k(l0, l'0)}{\sqrt{(2l' + 1)(2l + 1)}}. \tag{E.15}
$$

With this definition, we get, from the third term of Eq. (E.11),

$$
\frac{1}{2} \frac{\{q(A)\}}{[\{q(A)\} - 1]} \Lambda(A, A, k)
$$

$$
= \frac{1}{2} \frac{\{q(nl)\}}{[\{q(nl)\} - 1]} \Lambda(nl, nl, k)
$$

$$
= \frac{2(2l + 1)}{2[2(2l + 1) - 1]} \frac{c^k(l0, l0)}{\sqrt{(2l + 1)(2l + 1)}}
$$

$$
= \frac{c^k(l0, l0)}{4l + 1}, \tag{E.16}
$$

which is the expression that we have in the third term of Eq. (4.191). For the fifth term of Eq. (E.11), we get, using the definition of Eq. (E.15),

$$
\frac{1}{2} \Lambda(A, B, k) = \frac{1}{2} \Lambda(nl, n'l', k)
$$

$$
= \frac{c^k(l0, l'0)}{2\sqrt{(2l + 1)(2l' + 1)}}, \tag{E.17}
$$

which is identical with the coefficient of the fifth term of Eq. (4.191). Therefore, it is proved that the formula in Eq. (E.11) also reproduces the nonrelativistic average energy formula.

Now, let us take a look at the formula in Eq. (E.11). The angular coefficients are contained in the symbol $\Lambda(A, B, k)$. For the relativistic case, this symbol is defined by Eq. (E.12) and for the nonrelativistic case by Eq. (E.15). We want to show that Λ can always be generated from the Clebsch–Gordan constants. For the c^k coefficients, this is evident from Eq. (E.9). For the $\Gamma_{jj'}^k$, we have, according to Eq. (4.131), the relationships

$$
\begin{aligned}
\Gamma_{jj'}^k &= \frac{2}{2k+1}\left[C\left(jj'k;\frac{1}{2},-\frac{1}{2}\right)\right]^2 \\
&= \frac{2}{2j'+1}\left[C\left(jkj';\frac{1}{2},0\right)\right]^2 \\
&= \frac{2}{2j+1}\left[C\left(j'kj;\frac{1}{2},0\right)\right]^2.
\end{aligned}
\tag{E.18}
$$

Therefore, it is proved that the coefficients in the joint formula given by Eq. (E.11) can always be computed from the Clebsch–Gordan constants.

Next, we present an explicit formula for the calculation of the Clebsch–Gordan coefficients. This formula is[†]

$$
C(j_1 j_2 j_3; m_1 m_2 m_3)
$$

$$
= \delta(m_3, m_1 + m_2)
$$

$$
\times \left[(2j_3+1)\frac{(j_3+j_1-j_2)!(j_3-j_1+j_2)!(j_1+j_2-j_3)!(j_3+m_3)!(j_3-m_3)!}{(j_1+j_2+j_3+1)!(j_1-m_1)!(j_1+m_1)!(j_2-m_2)!(j_2+m_2)!}\right]^{1/2}
$$

$$
\times \sum_\nu \frac{(-1)^{\nu+j_2+m_2}(j_2+j_3-m_1-\nu)!(j_1-m_1+\nu)!}{\nu!(j_3-j_1+j_2-\nu)!(j_3+m_3-\nu)!(\nu+j_1-j_2-m_3)!}.
\tag{E.19}
$$

In this formula, the summation over ν is for all integral values with which none of the factorial arguments are negative. All the arguments of the factorials are integers. The notation here is somewhat different from our previous notation. Previously, we had five numbers in the Clebsch–Gordan coefficients; here we have six. The connection is

$$
C(j_1 j_2 j_3; m_1 m_2 m_3) \equiv C(j_1 j_2 j_3; m_1, m_3 - m_1).
\tag{E.20}
$$

[†] E. Wigner, *Group Theory*, Academic Press, New York (1959). Another formula was derived by G. Racah, *Phys. Rev.* **62**, 438 (1942). This last is discussed by Slater [Slater II, 88].

We finish the appendix by showing that the formula for the magnetic energy can also be constructed from the Clebsch–Gordan coefficients. The formula for the magnetic interaction is given by Eq. (4.205). In view of the relationship given by Eq. (4.206), we actually have only one kind of constant here, $\hat{\Gamma}^k_{j_A j_B}(a_A a_B; \gamma)$. We recall that the γ parameter takes only the values 0, $+1$, and -1. For these values of γ, we have the formulas

$$\hat{\Gamma}^k_{j_A j_B}(a_A a_B; 0) = 2\Gamma^k_{j_A j_B}, \tag{E.21}$$

$$\hat{\Gamma}^k_{j_A j_B}(a_A a_B; a_A) = \frac{2j_A - 1}{j_A} \Gamma^k_{j_A - 1, j_B} + \frac{j_A + 1}{j_A} \Gamma^k_{j_A j_B}, \tag{E.22}$$

$$\hat{\Gamma}^k_{j_A j_B}(a_A a_B; -a_A) = [j_A/(j_A + 1)]\Gamma^k_{j_A j_B} + [(2j_A + 3)/(j_A + 1)]\Gamma^k_{j_A + 1, j_B}. \tag{E.23}$$

In these formulas, the $\Gamma^k_{jj'}$ coefficients without a caret are those for which we have the formulas in Eq. (E.18). Thus, it is proved that the formula for the magnetic interaction, Eq. (4.205), can be constructed by using the Clebsch–Gordan coefficients.

Finally, we note that some of the c^k coefficients were tabulated by Slater [Slater II, 281] and the d^k and $\Gamma^k_{jj'}$ parameters by Grant (Ref. 26).

DARWIN'S CLASSICAL RELATIVISTIC HAMILTONIAN FOR AN N-ELECTRON SYSTEM

In this appendix, we present the derivation by Darwin in which he formulated the classical relativistic Hamiltonian for an interacting system of N electrons. Let us consider first a single electron that is moving in the trajectory characterized by the vector $r_2(t)$ with velocity $\dot{r}_2(t)$. These functions are assumed to be known. The moving electron is creating an electromagnetic field that is described by the scalar potential φ and the vector potential A. We want to describe the electromagnetic field at point P, which is characterized by vector r_1. Figure F.1 shows all the quantities that occur in the derivation. At the instant t, the potentials are given, according to the theory of special relativity,

$$\varphi(P) = -\left.\frac{e}{r + \dfrac{1}{c}(\dot{r}_2 \cdot r_{21})}\right|_{t-\tau}, \qquad (F.1)$$

$$A(P) = -\left.\frac{-e\dot{r}_2}{c\left(r + \dfrac{1}{c}(\dot{r}_2 \cdot r_{21})\right)}\right|_{t-\tau}. \qquad (F.2)$$

In these formulas, we have c, which is the speed of light, and we use the notation

$$r_{21} \equiv r_2 - r_1, \qquad (F.3)$$

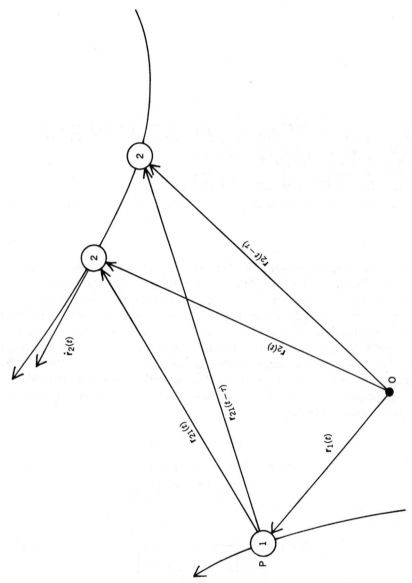

Fig. F.1. The electromagnetic interaction of two moving electrons.

and

$$r \equiv r_{21} \equiv |r_{21}| = |r_2 - r_1|. \tag{F.4}$$

The dot over the vector means, as usual, the time derivative.

The previous potentials are retarded expressions, that is, the quantities in them have to be evaluated not at the instant t, but at $t - \tau$, where τ is the time needed for the electromagnetic effects to move with the finite speed of the light from the location of the electron to point P. We evaluate the retarded potentials approximately, that is, with an accuracy that includes terms up to $1/c^2$.

We are interested in the potentials at time t. At that time, the position of the moving electron is r_2. Prior to that, at time $t - \tau$, the position of the electron was $r_2(t - \tau)$. This is the type of quantity that is needed in the potentials. We use a standard Taylor expansion around point a:

$$f(x) = f(a) + f'(a)(x - a) + \frac{1}{2!}f''(a)(x - a)^2 + \cdots \tag{F.5}$$

where we put $a = t$, $x = t - \tau$, and obtain

$$f(t - \tau) = f(t) - f'(t)\tau + \frac{1}{2!}f''(t)\tau^2 + \cdots \tag{F.6}$$

This formula gives us the desired quantities at $t - \tau$ in terms of quantities given at t. Thus, we get, for example,

$$r_2(t - \tau) = r_2(t) - \dot{r}_2(t)\tau + \tfrac{1}{2}\ddot{r}_2(t)\tau^2 + \cdots \tag{F.7}$$

For the potentials, we need $\dot{r}_2$, r_{21}, and r at time $t - \tau$. Let us evaluate these quantities on the basis of the previous formula. For $\dot{r}_2$, we obtain directly from Eq. (F.7)

$$\dot{r}_2(t - \tau) = \dot{r}_2(t) - \ddot{r}_2(t)\tau + \tfrac{1}{2}\dddot{r}_2(t)\tau^2 + \cdots \tag{F.8}$$

For r_{21}, we obtain by using Eq. (F.7)

$$
\begin{aligned}
r_{21}(t - \tau) &= r_2(t - \tau) - r_1(t - \tau) \\
&= r_2(t - \tau) - r_1(t) \\
&= r_2(t) - r_1(t) - \dot{r}_2(t)\tau + \tfrac{1}{2}\ddot{r}_2(t)\tau^2 + \cdots \\
&= r_{21}(t) - \dot{r}_2(t)\tau + \tfrac{1}{2}\ddot{r}_2(t)\tau^2 + \cdots ,
\end{aligned}
\tag{F.9}
$$

where we have used the relationship $r_1(t - \tau) = r_1(t)$ because P is a fixed point that is the same at $t - \tau$ as it is at t. In order to obtain r, we use the formula just derived and obtain

$$\left[r_{21}(t - \tau) \right]^2 = \left[r_{21} - \dot{r}_2 \tau + \tfrac{1}{2}\ddot{r}_2 \tau^2 \right]^2$$

$$= r_{21}^2 + \dot{r}_2^2 \tau^2 - 2 r_{21} \cdot \dot{r}_2 \tau + 2 r_{21} \cdot \tfrac{1}{2}\ddot{r}_2 \tau^2, \quad (F.10)$$

where we have omitted terms that contain higher than second powers of τ. As we will see presently, τ is proportional to $1/c$, thus the omission of higher than second powers of τ means the omission of higher than second powers of $1/c$. Beginning with Eq. (F.10), we do not indicate the functional dependence of quantities that are taken at time t. From the last formula, we get

$$r_{21}^2(t - \tau) = r_{21}^2 - 2(r_{21} \cdot \dot{r}_2)\tau + \left[(r_{21} \cdot \ddot{r}_2) + \dot{r}_2^2 \right]\tau^2$$

$$= r^2\left\{ 1 - \frac{2(r_{21} \cdot \dot{r}_2)}{r^2}\tau + \frac{(r_{21} \cdot \ddot{r}_2) + \dot{r}_2^2}{r^2}\tau^2 \right\}, \quad (F.11)$$

where we have put $r_{21} \equiv r$. From the last equation, we get the desired formula for $r(t - \tau)$:

$$r(t - \tau) = r\left\{ 1 - \frac{2(r_{21} \cdot \dot{r}_2)}{r^2}\tau + \frac{(r_{21} \cdot \ddot{r}_2) + \dot{r}_2^2}{r^2}\tau^2 \right\}^{1/2}$$

$$= r\left\{ 1 - \frac{r_{21} \cdot \dot{r}_2}{r^2}\tau + \frac{(\dot{r}_{21} \cdot \ddot{r}_2) + \dot{r}_2^2}{2r^2}\tau^2 \right.$$

$$\left. - \frac{1}{8}\left(\frac{2(r_{21} \cdot \dot{r}_2)}{r^2}\tau \right)^2 \right\}, \quad (F.12)$$

where the last expression on the right side comes from the third quadratic term of the binomial expansion.

We have obtained now the expressions for $\dot{r}_2$, r_{21}, and r, all taken at time $t - \tau$. The expressions give these quantities in terms of functions taken at point t and in terms of the retardation time τ. In order to get workable expressions that we can substitute into the potentials, we must determine the retardation time τ. In order to determine τ, we point out that

$$\tau = \frac{r_{21}(t - \tau)}{c} \quad (F.13)$$

because we have defined τ as the time needed by the electromagnetic effects to move from the position of the electron to point P, which is defined by vector r_1. The distance between the electron and point P at time $t - \tau$ is given by $r_{21}(t - \tau)$ because

$$\left| r_{21}(t - \tau) \right| = \left| r_2(t - \tau) - r_1 \right|, \tag{F.14}$$

and from this, Eq. (F.13) follows immediately. Thus, we get from Eq. (F.12)

$$
\begin{aligned}
\tau &= \frac{r_{21}(t - \tau)}{c} \\[2mm]
&= \frac{r}{c} - \frac{r_{21} \cdot \dot{r}_2}{cr}\tau + \left[\frac{(r_{21} \cdot \ddot{r}_2) + \dot{r}_2^2}{2cr} - \frac{(r_{21} \cdot \dot{r}_2)^2}{2cr^3} \right]\tau^2.
\end{aligned}
\tag{F.15}
$$

We solve this equation for τ by successive approximations. We get the zeroth approximation by omitting the terms containing τ on the right side:

$$\tau_0 = \frac{r}{c}. \tag{F.16}$$

This corresponds to replacing $r(t - \tau)$ by r in Eq. (F.13). For the next approximation, we omit the τ^2 terms in Eq. (F.15), and for τ, we substitute the zeroth approximation given by Eq. (F.16). We obtain

$$\tau_1 = \frac{r}{c} - \frac{r_{21} \cdot \dot{r}_2}{cr}\frac{r}{c} = \frac{r}{c} - \frac{1}{c^2}(r_{21} \cdot \dot{r}_2). \tag{F.17}$$

Finally, we take this formula and substitute it into the full expression given by Eq. (F.15). We get

$$
\tau = \frac{r}{c} - \frac{r_{21} \cdot \dot{r}_2}{cr}\left[\frac{r}{c} - \frac{1}{c^2}(r_{21} \cdot \dot{r}_2) \right] + \left[\frac{(r_{21} \cdot \ddot{r}_2) + \dot{r}_2^2}{2cr} - \frac{(r_{21} \cdot \dot{r}_2)^2}{2cr^3} \right]
$$
$$
\times \left[\frac{r}{c} - \frac{1}{c^2}(r_{21} \cdot \dot{r}_2) \right]^2, \tag{F.18}
$$

and from this, we get

$$
\begin{aligned}
\tau &= \frac{r}{c} - \frac{\boldsymbol{r}_{21} \cdot \dot{\boldsymbol{r}}_2}{cr}\frac{r}{c} + \frac{1}{rc^3}(\boldsymbol{r}_{21} \cdot \dot{\boldsymbol{r}}_2)^2 \\
&\quad + \left[\frac{(\boldsymbol{r}_{21} \cdot \ddot{\boldsymbol{r}}_2) + \dot{r}_2^2}{2cr} - \frac{(\boldsymbol{r}_{21} \cdot \dot{\boldsymbol{r}}_2)^2}{2cr^3} \right] \\
&\qquad \times \left[\frac{r^2}{c^2} - 2\frac{r}{c^3}(\boldsymbol{r}_{21} \cdot \dot{\boldsymbol{r}}_2) + O\!\left(\frac{1}{c^4}\right) \right] \\
&= \frac{r}{c} - \frac{\boldsymbol{r}_{21} \cdot \dot{\boldsymbol{r}}_2}{c^2} + \frac{1}{rc^3}(\boldsymbol{r}_{21} \cdot \dot{\boldsymbol{r}}_2)^2 \\
&\quad + \frac{(\boldsymbol{r}_{21} \cdot \ddot{\boldsymbol{r}}_2) + \dot{r}_2^2}{2cr}\cdot\frac{r^2}{c^2} - \frac{(\boldsymbol{r}_{21} \cdot \dot{\boldsymbol{r}}_2)^2}{2cr^3}\frac{r^2}{c^2} + O\!\left(\frac{1}{c^4}\right) \\
&= \frac{r}{c} - \frac{1}{c^2}(\boldsymbol{r}_{21} \cdot \dot{\boldsymbol{r}}_2) \\
&\quad + \frac{r}{2c^3}\left[\dot{r}_2^2 + \boldsymbol{r}_{21} \cdot \ddot{\boldsymbol{r}}_2 + \frac{(\boldsymbol{r}_{21} \cdot \dot{\boldsymbol{r}}_2)^2}{r^2} \right].
\end{aligned} \tag{F.19}
$$

This is our final expression for τ. The reason why we keep the $1/c^3$ term is that, as we see from the equations defining the potentials, Eqs. (F.1) and (F.2), for these expressions, we will need $r(t - \tau)$, which is proportional to $c\tau$. Thus, if we want to have $c\tau$ up to the $1/c^2$ terms, we must keep the $1/c^3$ term in the expression for τ.

Now we are ready to calculate the retarded potentials. Let us consider first the denominator of both expressions given by Eqs. (F.1) and (F.2). In order to calculate this denominator, we need r and $(1/c)(\boldsymbol{r}_{21} \cdot \dot{\boldsymbol{r}}_2)$ taken at time $t - \tau$. For r, we use the relationships given by Eqs. (F.13) and (F.19):

$$
\begin{aligned}
r(t - \tau) = c\tau &= r - \frac{1}{c}(\boldsymbol{r}_{21} \cdot \dot{\boldsymbol{r}}_2) \\
&\quad + \frac{r}{2c^2}\left[\dot{r}_2^2 + (\boldsymbol{r}_{21} \cdot \ddot{\boldsymbol{r}}_2) + \frac{(\boldsymbol{r}_{21} \cdot \dot{\boldsymbol{r}}_2)^2}{r^2} \right]. \tag{F.20}
\end{aligned}
$$

Using Eqs. (F.8) and (F.9), we get

$$
\begin{aligned}
\frac{1}{c}(\dot{\boldsymbol{r}}_2 \cdot \boldsymbol{r}_{21})\big|_{t-\tau} &= \frac{1}{c}\left\{ \dot{\boldsymbol{r}}_2 - \ddot{\boldsymbol{r}}_2\tau + \tfrac{1}{2}\dddot{\boldsymbol{r}}_2\tau^2 \right\} \\
&\quad \times \left\{ \boldsymbol{r}_{21} - \dot{\boldsymbol{r}}_2\tau + \tfrac{1}{2}\ddot{\boldsymbol{r}}_2\tau^2 \right\}. \tag{F.21}
\end{aligned}
$$

For τ, we use again Eq. (F.19), but we take into account that we need the expression only up to $1/c^2$ terms. Thus, we get

$$\frac{1}{c}(\dot{r}_2 \cdot r_{21})\big|_{t-\tau} = \frac{1}{c}\{(\dot{r}_2 \cdot r_{21}) - (r_{21} \cdot \ddot{r}_2)\tau - \dot{r}_2^2\tau\}$$

$$= \frac{1}{c}(\dot{r}_2 \cdot r_{21}) - \frac{1}{c}[(r_{21} \cdot \ddot{r}_2) + \dot{r}_2^2]\left(\frac{r}{c}\right)$$

$$= \frac{1}{c}(\dot{r}_2 \cdot r_{21}) - \frac{r}{c^2}[(r_{21} \cdot \ddot{r}_2) + \dot{r}_2^2]. \qquad \text{(F.22)}$$

Using Eqs. (F.20) and (F.22), we obtain

$$\left[r + \frac{1}{c}(\dot{r}_2 \cdot r_{21})\right]_{t-\tau} = r - \frac{1}{c}(r_{21} \cdot \dot{r}_2)$$

$$+ \frac{r}{2c^2}\left[\dot{r}_2^2 + (r_{21} \cdot \ddot{r}_2) + \frac{(r_{21} \cdot \dot{r}_2)^2}{r^2}\right]$$

$$+ \frac{1}{c}(\dot{r}_2 \cdot r_{21}) - \frac{r}{c^2}[(r_{21} \cdot \ddot{r}_2) + \dot{r}_2^2]$$

$$= r\left[1 - \frac{1}{2c^2}\left[\dot{r}_2^2 + (r_{21} \cdot \ddot{r}_2) - \frac{(r_{21} \cdot \dot{r}_2)^2}{r^2}\right]\right]. \qquad \text{(F.23)}$$

With the aid of the binomial expansion, we obtain

$$\left[\left[\left(r + \frac{1}{c}(\dot{r}_2 \cdot r_{21})\right)_{t-\tau}\right]^{-1} = r^{-1}\left[1 - \frac{1}{2c^2}\left[\dot{r}_2^2 + (r_{21} \cdot \ddot{r}_2) - \frac{(r_{21} \cdot \dot{r}_2)^2}{r^2}\right]\right]^{-1} \right.$$

$$= r^{-1}\left\{1 + \frac{1}{2c^2}\left[\dot{r}_2^2 + (r_{21} \cdot \ddot{r}_2) - \frac{(r_{21} \cdot \dot{r}_2)^2}{r^2}\right] \right.$$

$$\left. + O\left(\frac{1}{c^4}\right)\right\}. \qquad \text{(F.24)}$$

Keeping only the $1/c^2$ terms, we get the result

$$\left[\frac{1}{r + \frac{1}{c}(\dot{r}_2 \cdot r_{21})}\right]_{t-\tau}$$

$$= \frac{1}{r}\left\{1 + \frac{1}{2c^2}\left[\dot{r}_2^2 + (r_{21} \cdot \ddot{r}_2) - \frac{(r_{21} \cdot \dot{r}_2)^2}{r^2}\right]\right\}. \qquad \text{(F.25)}$$

From this expression, we get $\varphi(P)$ immediately by multiplying by $-e$:

$$\varphi(P) = -\frac{e}{r}\left\{1 + \frac{1}{2c^2}\left[\dot{r}_2^2 + (r_{21} \cdot \ddot{r}_2) - \frac{(r_{21} \cdot \dot{r}_2)^2}{r^2}\right]\right\}. \quad (F.26a)$$

For $A(P)$, we need also $\dot{r}_2$ again, which is given by Eq. (F.8), where we must substitute τ from Eq. (F.19). For reasons that will become clear presently, we need the vector potential only up to the $1/c$ terms. Now, as we see from Eq. (F.2), vector $\dot{r}_2$ in the numerator is multiplied by $1/c$. Thus, by substituting $\dot{r}_2$ from Eq. (F.8) and the denominator from Eq. (F.25), we need to keep only the zeroth-order terms in $1/c$, that is, the terms not containing $1/c$. We get

$$A(P) = -\frac{e}{c}\left(\frac{\dot{r}_2}{r}\right). \quad (F.26b)$$

The last two formulas giving us the electromagnetic field created by the electron that at time t is located at r_2 and is moving with velocity $\dot{r}_2$. The field is given at point P, which is located at r_1; the distance between point P and the location of the electron is given by vector r_{21} whose absolute value is $r = |r_{21}|$ (see Fig. F.1).

Now, let us assume that at time t, the momentary position of another electron will be r_1, that is, this electron will be at P at time t. Let us assume that $r_1(t)$ is the trajectory of this electron, which is moving in the electromagnetic field created by the other electron. We call the electron that is creating the electromagnetic field "electron two"; the electron that is moving in the field is called "electron one." For electron one, the equation of motion is

$$\frac{d}{dt}\frac{m\dot{r}_1}{\beta_1} = F \quad (F.27)$$

where F is the force acting on the electron and

$$\beta_1 = \sqrt{1 - \frac{V_1^2}{c^2}}, \quad (F.28)$$

and m is the rest mass. Using the equation for the Lorentz force, we get

$$\frac{d}{dt}\frac{m\dot{v}_1}{\beta_1} = -e\mathscr{E} - \frac{e}{c}(v_1 \times H), \quad (F.29)$$

where $\mathscr{E}$ is the electric field strength, and H is the magnetic field strength.

These are related to the scalar and vector potentials by the relationships

$$\mathscr{E} = -\operatorname{grad} \varphi - \frac{1}{c} \frac{\partial A}{\partial t}, \tag{F.30}$$

$$H = \operatorname{curl} A = \nabla \times A. \tag{F.31}$$

Using these equations, we get for the equation of motion

$$\frac{d}{dt} \frac{m v_1}{\beta_1} = e \left\{ \nabla \varphi + \frac{1}{c} \frac{\partial A}{\partial t} \right\} - \frac{e}{c} \{ v_1 \times (\nabla \times A) \}. \tag{F.32}$$

According to the special theory of relativity, this equation can be derived from the Lagrangian:

$$L = -mc^2 \beta_1 - \frac{e}{c} \Phi_k \dot{x}_k. \tag{F.33}$$

Here the four-vector $\dot{x}_k$ is defined as follows:

$$\dot{x}_1 = v_{1x}, \qquad \dot{x}_2 = v_{1y}, \qquad \dot{x}_3 = v_{1z}, \qquad \dot{x}_4 = ic, \tag{F.34}$$

and the four-vector Φ_k is given by

$$\Phi_1 = A_x, \qquad \Phi_2 = A_y, \qquad \Phi_3 = A_z, \qquad \Phi_4 = i\varphi. \tag{F.35}$$

Thus, we obtain for the Lagrangian the formula

$$L = -mc^2 \beta_1 - \frac{e}{c} (A \cdot v_1) + e\varphi. \tag{F.36}$$

By putting the approximate expressions for φ and A given by Eqs. (F.26a) and (F.26b) into the Lagrangian, we obtain

$$L = -mc^2 \beta_1 - \frac{e^2}{r}$$
$$- \frac{e^2}{2c^2} \left\{ \frac{\dot{r}_2^2 + (\ddot{r}_2 \cdot r_{21}) - 2(\dot{r}_1 \cdot \dot{r}_2)}{r} - \frac{(\dot{r}_2 \cdot r_{21})^2}{r^3} \right\}. \tag{F.37}$$

We note that the approximate expression for the vector potential, which is given by Eq. (F.26b), contains terms only up to $1/c$. The reason for this is that the vector potential term in the Lagrangian, Eq. (F.36), contains the factor $1/c$; thus, the substitution of the expression for the vector potential will yield terms up to $1/c^2$.

Next, we symmetrize the Lagrangian given by Eq. (F.37). This is accomplished by adding the terms

$$-mc^2\beta_2 + \frac{d}{dt} \frac{e^2}{2c^2} \frac{\dot{r}_2 \cdot r_{21}}{r}. \tag{F.38}$$

The first term does not depend on r_1 or $\dot{r}_1$, thus, it will yield zero when placed into the Euler–Lagrange equations, that is, it will yield zero when operated on by the operator

$$D_1 \equiv \frac{d}{dt} \frac{\partial}{\partial \dot{q}_1} - \frac{\partial}{\partial q_1}, \tag{F.39}$$

where q_1 and $\dot{q}_1$ are the coordinates and momenta of electron one, respectively. The second term of the expression given by Eq. (F.38) is a function of r_1 but not of $\dot{r}_1$. The operator D_1, when it is applied to this expression, yields zero. This is easy to see; let us write the second term of Eq. (F.38) in the form

$$\frac{d}{dt} \frac{e^2}{2c^2} \frac{\dot{r}_2 \cdot r_{21}}{r} \equiv \frac{d}{dt} f(x_1 y_1 z_1). \tag{F.40}$$

Then we get

$$D_1 \frac{df}{dt} = \left(\frac{d}{dt} \frac{\partial}{\partial \dot{q}_1} - \frac{\partial}{\partial q_1} \right) \frac{df}{dt}, \tag{F.41}$$

where q_1 is one of $(x_1 y_1 z_1)$. The first term yields

$$\begin{aligned}
\frac{d}{dt} \frac{\partial}{\partial \dot{q}_1} \frac{df}{dt} &= \frac{d}{dt} \frac{\partial}{\partial \dot{q}_1} \left(\frac{\partial f}{\partial t} + \sum_{j=1}^{3} \frac{\partial f}{\partial q_j} \dot{q}_j \right) \\
&= \frac{d}{dt} \frac{\partial}{\partial \dot{q}_1} \left(\frac{\partial f}{\partial q_1} \dot{q}_1 \right) \\
&= \frac{d}{dt} \frac{\partial f}{\partial q_1},
\end{aligned} \tag{F.42}$$

and by putting this into Eq. (F.41), we obtain

$$D_1 \frac{df}{dt} = \frac{d}{dt} \frac{\partial f}{\partial q_1} - \frac{\partial}{\partial q_1} \frac{df}{dt} = 0. \tag{F.43}$$

Therefore, it is demonstrated that the addition of the two terms of Eq. (F.38) does not have any effect on the equation of motion of electron one, that is, these terms yield zero when the Lagrangian is operated on by operator D_1.

Thus, the Lagrangian becomes (writing $r_{12} \equiv r$)

$$
L = -mc^2\beta_1 - mc^2\beta_2 - \frac{e^2}{r_{12}}
$$

$$
+ \frac{e^2}{2c^2} \left\{ \frac{\dot{r}_1 \cdot \dot{r}_2}{r_{12}} + \frac{(\dot{r}_1 \cdot r_{21})(\dot{r}_2 \cdot r_{21})}{r_{12}^3} \right\}. \tag{F.44}
$$

This is a completely symmetric expression, that is, it does not change when the two electrons are interchanged. Darwin concluded that this is the correct Lagrangian for two moving and interacting electrons, that is, it can be considered to describe the motion of electron one in the electromagnetic field of electron two and vice versa.

As a final step, we expand β_1 and β_2 in terms of $1/c$, so as to make them consistent with the last term of the expression in Eq. (F.44), which is accurate only to $1/c^2$. We get

$$
-mc^2\beta_1 = -mc^2\left(1 - \frac{v_1^2}{c^2}\right)^{1/2}
$$

$$
\approx -mc^2\left(1 - \frac{1}{2}\frac{v_1^2}{c^2} - \frac{1}{8}\frac{v_1^4}{c^4} + \cdots\right)
$$

$$
= -mc^2 + \frac{1}{2}mv_1^2 + \frac{1}{8}m\frac{v_1^4}{c^2} + \cdots \tag{F.45}
$$

We are now in the position to write down the Lagrangian for an *N*-electron system. We substitute the expression of Eq. (F.45) and a similar expression for electron two into the Lagrangian and sum up for all pairs in the system. We recall also that the Lagrangian given by Eq. (F.44) takes care of the electromagnetic interaction between the electrons but does not contain the effects of external electric and magnetic fields on the system. Let V and A be the scalar and vector potentials of the external electric and magnetic fields, respectively. Then these potentials will appear in the Lagrangian in the form

$$
+eV - \frac{e}{c}(A \cdot v_1), \tag{F.46}
$$

which can be seen from Eq. (F.36). Thus, we get

$$
\begin{aligned}
L = {} & \sum_{k=1}^{N} \frac{1}{2} m v_k^2 + \sum_{k=1}^{N} \frac{m}{8c^2} v_k^4 + \sum_{k=1}^{N} eV(k) \\
& - \sum_{k=1}^{N} \frac{e}{c} (A(k) \cdot v_k) - \frac{1}{2} \sum_{i,k=1}^{N} \frac{e^2}{r_{ik}} \\
& + \frac{1}{2} \sum_{i,k=1}^{N} \frac{e^2}{2c^2} \left\{ \frac{\dot{r}_i \cdot \dot{r}_k}{r_{ik}} + \frac{(\dot{r}_i \cdot r_{ki})(\dot{r}_k \cdot r_{ki})}{r_{ik}^3} \right\}.
\end{aligned}
\qquad \text{(F.47)}
$$

Having obtained the Lagrangian, we are able to write down the Hamiltonian immediately. According to the rules of Lagrangian mechanics, we obtain

$$
\begin{aligned}
H = {} & \sum_{k=1}^{N} \left\{ \frac{p_k^2}{2m} - \frac{p_k^4}{8c^2 m^3} - eV(k) + \frac{e}{cm}(A \cdot p_k) \right\} + \frac{1}{2} \sum_{i,k=1}^{N} \frac{e^2}{r_{ik}} \\
& - \frac{1}{2} \sum_{i,k=1}^{N} \frac{e^2}{2c^2 m^2} \left\{ \frac{p_i \cdot p_k}{r_{ik}} + \frac{(p_i \cdot r_{ki})(p_k \cdot r_{ki})}{r_{ik}^3} \right\}.
\end{aligned}
\qquad \text{(F.48)}
$$

This Hamiltonian is the classical relativistic expression for an interacting *n*-electron system. Each term has a definite physical meaning. The first term is the classical kinetic energy of the electrons. The second term is the first relativistic correction to the kinetic energy brought about by the increase of the electron mass at relativistic velocities. The third and fourth terms are the interaction energies of the electrons with external electric and magnetic fields, respectively. The fifth term is the electrostatic interaction energy of the electrons themselves. The sixth and last term represents the magnetic and retardation effects.

We want to identify these effects separately. The magnetic effects are obtained by substituting Eq. (F.26b) into the Lagrangian given by Eq. (F.36). The result is

$$
\frac{e^2}{c^2} \frac{v_1 \cdot v_2}{r_{12}}.
\qquad \text{(F.49)}
$$

This is the correct expression for the magnetic interaction. On the other hand, the final expression for the Lagrangian Eq. (F.44), contains the term

$$
\frac{1}{2} \frac{e^2}{c^2} \frac{v_1 \cdot v_2}{r_{12}}.
\qquad \text{(F.50)}
$$

Therefore, because Eq. (F.49) gives the correct magnetic interaction, but the final Lagrangian contains the term given by Eq. (F.50), there must be a contribution from the retardation amounting to

$$
- \frac{1}{2} \frac{e^2}{c^2} \frac{v_1 \cdot v_2}{r_{12}}.
\tag{F.51}
$$

When turning to the Hamiltonian, we must recall that the Hamiltonian contains the Lagrangian with a negative sign. Thus, we obtain for the electrostatic and magnetic interactions of the two electrons

$$
g_0(1,2) = \frac{e^2}{r_{12}} \left[1 - \frac{1}{c^2} (v_1 \cdot v_2) \right],
\tag{F.52}
$$

and for the retardation effects, we get

$$
g_1(1,2) = \frac{e^2}{2c^2 r_{12}} \left[v_1 \cdot v_2 - \frac{(v_1 \cdot r_{21})(v_2 \cdot r_{21})}{r_{12}^2} \right].
\tag{F.53}
$$

The final expression for the Hamiltonian, Eq. (F.48), contains the sum of these two terms summed over all pairs.

Let us summarize the approximations made in the course of the derivations. The retarded potentials given by Eqs. (F.1) and (F.2) are the exact relativistic expressions. Also, the Lagrangian, given by Eq. (F.36), reproduces exactly the equations of motion, given by Eq. (F.32). The only approximations that were made in the course of the derivations were the omission of terms containing higher than second powers of $1/c$ in the retarded potentials and in the final expression for the Lagrangian. Thus, the final Hamiltonian, given by Eq. (F.48), is an expression that is accurate up to terms containing $1/c^2$.

APPENDIX G

DIRAC'S RELATIVISTIC THEORY FOR ONE ELECTRON; THE EXACT FOUR-COMPONENT AND THE APPROXIMATE TWO-COMPONENT EQUATIONS

The exact relativistic quantum theory for one electron was formulated by Dirac.[16] Here we summarize those results of the theory that are needed for the buildup of the relativistic Hartree–Fock model.

The Hamiltonian operator is given by the formula

$$H = -c\boldsymbol{\alpha} \cdot \boldsymbol{p} - \beta mc^2 - eV. \tag{G.1}$$

Here c is the speed of light, and m and $-e$ are the rest mass and charge of the electron, respectively. V is the potential in which the electron is moving, and $\boldsymbol{p}$ is the momentum operator. The three components of $\boldsymbol{\alpha}$ and the quantity β are Dirac's 4×4 matrices whose properties we will discuss presently. The wave equation is

$$Hu = Eu, \tag{G.2}$$

where u is the wave function, and E is the relativistic energy of the electron. Equation (G.1) is valid in the absence of an external magnetic field. If, in addition to the electric field, which is described by the potential V, we also have a magnetic field characterized by the vector potential A, then the

500

momentum p must be replaced by the vector P, where

$$P = p + \frac{e}{c}A, \tag{G.3}$$

and the Hamiltonian becomes

$$H = -c\boldsymbol{\alpha} \cdot P - \beta mc^2 - eV. \tag{G.4}$$

The three components of $\boldsymbol{\alpha}$ and the quantity β are determined by comparing the Hamiltonian with the classical relativistic Hamiltonian function. This function is given by

$$H = -eV + c\{m^2c^2 + P^2\}^{1/2}. \tag{G.5}$$

Identifying the Hamiltonian function with the total energy of the system, we obtain from the last equation

$$(E + eV)^2 = c^2\{m^2c^2 + P^2\}. \tag{G.6}$$

By putting the Hamiltonian given by Eq. (G.4) into the wave equation, Eq. (G.2), we get

$$(E + eV)^2 = m^2c^4\beta^2 + c^2\sum_{k=1}^{3}\alpha_k^2 P_k^2, \tag{G.7}$$

if we assume that

$$\alpha_i\alpha_j + \alpha_j\alpha_i = 0 \quad (i, j = 1, 2, 3; \ i \neq j), \tag{G.8}$$

and

$$\alpha_k\beta + \beta\alpha_k = 0 \quad (k = 1, 2, 3). \tag{G.9}$$

If, in addition to these relationships, we also assume that

$$\alpha_k^2 = 1 \quad (k = 1, 2, 3), \tag{G.10}$$

and

$$\beta^2 = 1, \tag{G.11}$$

then the two equations, Eqs. (G.6) and (G.7), become identical. Thus, $\boldsymbol{\alpha}$ and β are determined by the relationships given in Eqs. (G.8), (G.9), (G.10), and (G.11). The three components of $\boldsymbol{\alpha}$ and the quantity β are called the Dirac matrices and have the form

$$
\alpha_1 = \left(\begin{array}{cc|cc} & & 0 & 1 \\ & & 1 & 0 \\ \hline 0 & 1 & & \\ 1 & 0 & & \end{array}\right), \qquad
\alpha_2 = \left(\begin{array}{cc|cc} & & 0 & -i \\ & & i & 0 \\ \hline 0 & -i & & \\ i & 0 & & \end{array}\right),
$$

$$
\alpha_3 = \left(\begin{array}{cc|cc} & & 1 & 0 \\ & & 0 & -1 \\ \hline 1 & 0 & & \\ 0 & -1 & & \end{array}\right), \qquad
\beta = \left(\begin{array}{cc|cc} 1 & 0 & & \\ 0 & 1 & & \\ \hline & & -1 & 0 \\ & & 0 & -1 \end{array}\right), \qquad \text{(G.12)}
$$

where the blank spaces mean zeros. Because the Hamiltonian is composed of 4×4 matrices, we must assume that the wave function u is a four-row one-column matrix

$$
u = \begin{pmatrix} u_1 \\ u_2 \\ u_3 \\ u_4 \end{pmatrix}. \tag{G.13}
$$

The Dirac matrices are connected to the Pauli 2×2 matrices. For the description of the spin of a single electron, Pauli introduced the spin operator

$$
s = \frac{\hbar}{2}\sigma, \tag{G.14}
$$

where the three components of the operator σ are 2×2 matrices given by

$$
\sigma_x = \begin{pmatrix} 0 & 1 \\ 1 & 0 \end{pmatrix}, \qquad \sigma_y = \begin{pmatrix} 0 & -i \\ i & 0 \end{pmatrix}, \qquad \sigma_z = \begin{pmatrix} 1 & 0 \\ 0 & -1 \end{pmatrix}. \tag{G.15}
$$

Thus, we can write the Dirac matrices in the form

$$
\alpha_1 = \left(\begin{array}{c|c} & \sigma_x \\ \hline \sigma_x & \end{array}\right), \qquad
\alpha_2 = \left(\begin{array}{c|c} & \sigma_y \\ \hline \sigma_y & \end{array}\right),
$$

$$
\alpha_3 = \left(\begin{array}{c|c} & \sigma_z \\ \hline \sigma_z & \end{array}\right), \qquad
\beta = \left(\begin{array}{c|c} I & \\ \hline & -I \end{array}\right), \tag{G.16}
$$

where I is the unit matrix:

$$
I = \begin{pmatrix} 1 & 0 \\ 0 & 1 \end{pmatrix}. \tag{G.17}
$$

Using Eq. (G.4) and (G.2), we obtain

$$
(E + eV + c\boldsymbol{\alpha} \cdot \boldsymbol{P} + \beta mc^2)u = 0. \tag{G.18}
$$

This is Dirac's relativistic wave equation in its exact form for an electron in external field. We write the equation out in its 4×4 form. Using the formulas for the Dirac matrices, we get

$$\alpha_1 u = \begin{pmatrix} u_4 \\ u_3 \\ u_2 \\ u_1 \end{pmatrix}, \qquad \alpha_2 u = i \begin{pmatrix} -u_4 \\ +u_3 \\ -u_2 \\ +u_1 \end{pmatrix},$$

$$\alpha_3 u = \begin{pmatrix} +u_3 \\ -u_4 \\ +u_1 \\ -u_2 \end{pmatrix}, \qquad \beta u = \begin{pmatrix} +u_1 \\ +u_2 \\ -u_3 \\ -u_4 \end{pmatrix}, \qquad (G.19)$$

and using these relationships, we obtain from Eq. (G.18)

$$(E + eV + mc^2)u_1 + cP_1 u_4 + cP_2(-iu_4) + cP_3 u_3 = 0, \quad (G.20a)$$

$$(E + eV + mc^2)u_1 + cP_1 u_3 + cP_2(iu_3) + cP_3(-u_4) = 0, \quad (G.20b)$$

$$(E + eV - mc^2)u_3 + cP_1 u_2 + cP_2(-iu_2) + cP_3 u_1 = 0, \quad (G.20c)$$

$$(E + eV - mc^2)u_4 + cP_1 u_1 + cP_2(iu_1) + cP_3(-u_2) = 0. \quad (G.20d)$$

The Iterated Dirac Equation. The physical interpretation of Eq. (G.18) is not easy. An equivalent equation, whose physical interpretation is plausible, can be derived as follows. Let us multiply the wave equation, Eq. (G.18), from the left by the operator

$$(E + eV - c\boldsymbol{\alpha} \cdot \boldsymbol{P} - \beta mc^2), \qquad (G.21)$$

that is, consider the iterated wave equation

$$\{(E + eV - c\boldsymbol{\alpha} \cdot \boldsymbol{P} - \beta mc^2)(E + eV + c\boldsymbol{\alpha} \cdot \boldsymbol{P} + \beta mc^2)\}u = 0. \quad (G.22)$$

The operator acting on u can be evaluated by using the commutation relationships obeyed by $\boldsymbol{\alpha}$, β, and $\boldsymbol{P}$. The result is the following equation:

$$\left\{ \frac{1}{c^2}(E + eV)^2 - m^2 c^2 + \hbar^2 \Delta + \frac{2ie\hbar}{c} \boldsymbol{A} \cdot \nabla - \frac{e^2}{c^2} A^2 \right.$$

$$\left. - \frac{ie\hbar}{c}(\boldsymbol{\mathscr{E}} \cdot \boldsymbol{\alpha}) + \frac{ie\hbar}{c} \sum_{k<l} \alpha_k \alpha_l \left(\frac{\partial A_l}{\partial x_k} - \frac{\partial A_k}{\partial x_l} \right) \right\} u = 0. \quad (G.23)$$

In this expression, the momentum operator p has been put into the standard form $p = -i\hbar\nabla$; $\mathscr{E}$ is the electric-field vector; in the last term of the equation, we have a summation over all (k,l) pairs, $A_1 = A_x$, $A_2 = A_y$,

$A_3 = A_z$ and likewise $x_1 = x$, $x_2 = y$, $x_3 = z$. In order to make the physical interpretation easier, let us introduce the following 4×4 matrices:

$$\sigma_1 = -i(\alpha_2 \alpha_3),$$
$$\sigma_2 = -i(\alpha_3 \alpha_1),$$
$$\sigma_3 = -i(\alpha_1 \alpha_2). \qquad (G.24)$$

These relationships can be written in the form

$$\sigma_j = -i(\alpha_k \alpha_l), \qquad (G.25)$$

where

$$(jkl) = \text{cyclic in } (123). \qquad (G.26)$$

Now, clearly, what we have in the last term of Eq. (G.23) is the curl of A, that is, the magnetic field strength. By using the definition of the curl, the definition of σ as given by (G.25) and the commutation relations for α, we obtain

$$\sum_{k<l} \alpha_k \alpha_l \left(\frac{\partial A_l}{\partial x_k} - \frac{\partial A_k}{\partial x_l} \right) = i\sigma \cdot (\text{curl } A) = i\sigma \cdot \mathcal{H}, \qquad (G.27)$$

and the last term of Eq. (G.23) becomes

$$\frac{ie\hbar}{c} i\sigma \cdot \mathcal{H} = -\frac{e\hbar}{c} \sigma \cdot \mathcal{H}. \qquad (G.28)$$

Let us define the nonrelativistic energy W by the relationship

$$W = E - mc^2. \qquad (G.29)$$

Then, we get for the first two terms of the Hamiltonian in Eq. (G.23)

$$\frac{1}{c^2}(E + eV)^2 - m^2 c^2 = \frac{1}{c^2}[W + mc^2 + eV] - m^2 c^2$$

$$= \frac{1}{c^2}\left[(W + eV)^2 + 2mc^2(W + eV) + m^2 c^4\right] - m^2 c^2$$

$$= \frac{1}{c^2}(W + eV)^2 + 2m(W + eV). \qquad (G.30)$$

Next, we take the expressions in Eqs. (G.30) and (G.28) and substitute them into the Hamiltonian of Eq. (G.23). Also, we divide the equation by $-2m$ and rearrange the result in such a way as to get the nonrelativistic kinetic energy and the potential energy up front. The result is

$$\left\{ -\frac{\hbar^2}{2m}\Delta - eV - \frac{1}{2mc^2}(W + eV)^2 - \frac{ie\hbar}{mc}(A \cdot \nabla) \right.$$

$$\left. + \frac{e^2A^2}{2mc^2} + \frac{ie\hbar}{2mc}(\mathscr{E} \cdot \boldsymbol{\alpha}) + \frac{e\hbar}{2mc}(\boldsymbol{\sigma} \cdot \mathscr{H}) \right\}u = Wu. \quad (G.31)$$

This is still the exact Dirac equation, but now we have it in the iterated second-order form. In this equation, we can give each term a plausible physical interpretation. The first two terms are the nonrelativistic kinetic energy and the potential energy of the electron in the external potential V. The meaning of the third term becomes clear if we put $W \approx p^2/2m - eV$ and get

$$-\frac{1}{2mc^2}(W + eV)^2 = -\frac{1}{2mc^2}\left(\frac{p^2}{2m} - eV + eV\right)^2 = -\frac{p^4}{8m^3c^2}. \quad (G.32)$$

Taking a look at Eq. (F.48) in which we have Darwin's classical relativistic Hamiltonian, we see that the term in Eq. (G.32) is the relativistic correction to the kinetic energy due to the increase of the electron mass at relativistic velocities. The fourth and fifth terms in Eq. (G.31) are the interaction energies of the electron with the external magnetic field. The fourth term is same that appeared already in Darwin's Hamiltonian. Indeed, if we restore the momentum by using $p = -i\hbar\nabla$, we get

$$-\frac{ie\hbar}{mc}(A \cdot \nabla) = \frac{e}{mc}(A \cdot p), \quad (G.33)$$

which is the fourth term in Darwin's expression. The fifth term does not appear in Darwin's expression, but it does in the nonrelativistic Hamiltonian, Eq. (4.13). The next to last term will be shown to be responsible for the spin–orbit interaction, that is, for the interaction of the intrinsic magnetic moment of the electron with the magnetic field generated by the orbital motion. Finally, the last term in Eq. (G.31) is the interaction energy of the intrinsic magnetic moment of the electron with the external magnetic field. This can be seen from the properties of the $\boldsymbol{\sigma}$ vector. From the definitions of the $\boldsymbol{\sigma}$ components, as they are given in Eq. (G.24), and from the commutation relations satisfied by the $\boldsymbol{\alpha}$ components, it is easy to show that $\boldsymbol{\sigma}$ satisfies the relationships

$$\sigma_1\sigma_2 - \sigma_2\sigma_1 = 2i\sigma_3, \quad (G.34)$$

that is,

$$\sigma_j\sigma_k - \sigma_k\sigma_j = 2i\sigma_l, \quad (G.35)$$

where

$$(jkl) = \text{cyclic in } (123).$$

These are the commutation relations defining the Pauli matrices. Indeed, we obtain from Eq. (G.24)

$$\sigma_1 = \left(\begin{array}{cc|cc} 0 & 1 & & \\ 1 & 0 & & \\ \hline & & 0 & 1 \\ & & 1 & 0 \end{array}\right) = \left(\begin{array}{c|c} \sigma_x & \\ \hline & \sigma_x \end{array}\right), \qquad \text{(G.36a)}$$

$$\sigma_2 = \left(\begin{array}{cc|cc} 0 & -i & & \\ i & 0 & & \\ \hline & & 0 & -i \\ & & i & 0 \end{array}\right) = \left(\begin{array}{c|c} \sigma_y & \\ \hline & \sigma_y \end{array}\right), \qquad \text{(G.36b)}$$

$$\sigma_3 = \left(\begin{array}{cc|cc} 1 & 0 & & \\ 0 & -1 & & \\ \hline & & 1 & 0 \\ & & 0 & -1 \end{array}\right) = \left(\begin{array}{c|c} \sigma_z & \\ \hline & \sigma_z \end{array}\right), \qquad \text{(G.36c)}$$

that is, these 4×4 matrices are equivalent to the Pauli matrices. From this, it follows that the expression in the last term of Eq. (G.31) is the intrinsic magnetic moment of the electron and the term itself is the interaction energy with the external magnetic field.

Reduction of the Dirac Equation to a Two-Component Form.[17, 18] Although the physical interpretation of the iterated Dirac equation, given by Eq. (G.31), is much easier than the original, Eq. (G.18), it is still an exact four-component wave equation. For some purposes, it will be convenient to approximate the exact four-component equation by a simpler two-component equation. This can be accomplished as follows.

First, we show that from the four components of the Dirac wave function u, two can be classified as "large" and two as "small" components. Symbolically, we can write

$$\begin{pmatrix} u_1 \\ u_2 \end{pmatrix} \ll \begin{pmatrix} u_3 \\ u_4 \end{pmatrix}, \qquad \text{(G.37)}$$

that is, we can say that u_1 and u_2 are the "small," and u_3 and u_4 are the "large" components.

In order to prove that such a statement indeed can be made, let us consider the exact equations, Eqs. (G.20a) to (G.20d). Let us put $A = 0$ and let us assume that in Eq. (G.20a), we can put

$$E + eV + mc^2 \approx 2mc^2. \qquad \text{(G.38)}$$

That this is a meaningful approximation can be seen from the classical relativistic energy formula given by Eq. (G.6). From that equation, we get

$$(E + eV)^2 = m^2c^4 + c^2p^2, \tag{G.39}$$

and so

$$E + eV = mc^2 \left[1 + \frac{p^2}{m^2c^2}\right]^{1/2}$$

$$\approx mc^2 \left[1 + \frac{1}{2}\frac{p^2}{m^2c^2} + O\left(\frac{1}{c^4}\right)\right]$$

$$= mc^2 + \frac{p^2}{2m} + O\left(\frac{1}{c^2}\right). \tag{G.40}$$

Thus,

$$E \approx -eV + \frac{p^2}{2m} + mc^2 = W + mc^2. \tag{G.41}$$

For an electron in an atom, the nonrelativistic energy, which we denoted by W, is at most a few thousand eV's. This can be seen from any Hartree–Fock energy-level diagram. On the other hand, the rest energy of the electron, mc^2, is about 0.5 MeV. Thus, the second term in the expression for E will be much larger than the first and we can put

$$E \approx mc^2, \tag{G.42}$$

from which we get Eq. (G.38) by a similar argument.

Using the approximation given by Eq. (G.38) in Eq. (G.20a), we obtain

$$2mc^2u_1 + cp_1u_4 - icp_2u_4 + cp_3u_3 = 0, \tag{G.43}$$

or

$$u_1 = -\frac{1}{2mc^2}\left[(cp_1 - icp_2)u_4 + cp_3u_3\right]. \tag{G.44}$$

On the right side of this equation, each term will contain a factor of the form

$$\frac{p_k}{2mc} = \frac{mv_k}{2mc} = \frac{v_k}{2c}, \tag{G.45}$$

where $k = 1, 2,$ and 3. Thus, each term will contain a factor v/c, where v is one of the speed components. Except in case of extreme relativistic velocities, these factors will be small and thus u_1 will be small compared to u_3 and u_4. A similar argument can be formulated for u_2 by making use of the second Dirac equation, (G.20b). From that equation, we obtain

$$u_2 = -\frac{1}{2mc^2}\left[(cp_1 + icp_2)u_3 - cp_3u_4\right], \tag{G.46}$$

which shows that u_2 will indeed be "small" compared to u_3 and u_4. Thus, it is shown that the symbolic relationship, Eq. (G.37), is valid.

The reduction of Eq. (G.31) to a two-component equation can be done as follows. Looking at the equation, we see that there is only one term, the sixth, that depends on the matrix-vector $\boldsymbol{\alpha}$. It is only this term that compels us to classify the equation as a four-component equation, and the reason that $\boldsymbol{\alpha}$ is a four-component matrix-vector is that the commutation relations, given by Eqs. (G.8) and (G.9), cannot be satisfied by matrices of less than fourth order. If the $\boldsymbol{\alpha}$ would not occur in Eq. (G.31), that is, if only the vector $\boldsymbol{\sigma}$ would occur, then we could classify the equation as a two-component equation, because the relations satisfied by $\boldsymbol{\sigma}$, Eq. (G.25), can be satisfied by the Pauli matrices as well as by the 4×4 matrices given by Eq. (G.36).

Our task is, therefore, to eliminate $\boldsymbol{\alpha}$ from the Dirac equation. This can be done by introducing the approximate relationships given by Eqs. (G.44) and (G.46). In these relationships, the small components of the wave function are expressed in terms of the large components. We use these equations to eliminate $\boldsymbol{\alpha}$, that is, to replace the term containing $\boldsymbol{\alpha}$ by terms that contain only $\boldsymbol{\sigma}$, which, as we have seen, can be classified as a two-component matrix-vector.

We now show that in the approximation in which Eqs. (G.44) and (G.46) are valid, we can put

$$\boldsymbol{\alpha}u = -\frac{1}{2mc}(\boldsymbol{p} + i[\boldsymbol{p} \times \boldsymbol{\sigma}])u + \boldsymbol{\varphi}, \tag{G.47}$$

where $\boldsymbol{\alpha}$ and $\boldsymbol{\sigma}$ are the four-component matrix-vectors, $\boldsymbol{p}$ is the momentum operator, and $\boldsymbol{\varphi}$ is a one-column matrix-vector in which the third and fourth rows are zero.

First, let us write down the x-component of the quantity on the left side of Eq. (G.47):

$$\alpha_1 u = \begin{pmatrix} u_4 \\ u_3 \\ u_2 \\ u_1 \end{pmatrix}, \tag{G.48}$$

which follows from Eq. (G.19). Using Eqs. (G.44) and (G.46), we obtain

$$
\alpha_1 u = -\frac{1}{2mc} \begin{pmatrix} \cdots \\ \cdots \\ (p_1 + ip_2)u_3 - p_3 u_4 \\ (p_1 - ip_2)u_4 + p_3 u_3 \end{pmatrix}. \tag{G.49}
$$

Next, we write down the x-component of the first term on the right side of Eq. (G.47):

$$
-\frac{1}{2mc} p_1 u = -\frac{1}{2mc} \begin{pmatrix} p_1 u_1 \\ p_1 u_2 \\ p_1 u_3 \\ p_1 u_4 \end{pmatrix}. \tag{G.50}
$$

Next, let us consider the vector product $(p \times \sigma)$. We have

$$
p \times \sigma = \begin{vmatrix} \hat{i} & \hat{j} & \hat{k} \\ p_1 & p_2 & p_3 \\ \sigma_1 & \sigma_2 & \sigma_3 \end{vmatrix}. \tag{G.51}
$$

From this, we obtain

$$
-\frac{i}{2mc}[p \times \sigma]_1 = -\frac{i}{2mc}(p_2 \sigma_3 - p_3 \sigma_2), \tag{G.52}
$$

and so

$$
\begin{aligned}
-\frac{i}{2mc}[p \times \sigma]_1 u &= -\frac{i}{2mc}(p_2 \sigma_3 - p_3 \sigma_2)u \\
&= -\frac{1}{2mc}(ip_2 \sigma_3 - ip_3 \sigma_2)u \\
&= \frac{-1}{2mc}\left[ip_2 \begin{pmatrix} u_1 \\ -u_2 \\ u_3 \\ -u_4 \end{pmatrix} + p_3 \begin{pmatrix} -u_2 \\ u_1 \\ -u_4 \\ u_3 \end{pmatrix} \right],
\end{aligned} \tag{G.53}
$$

where we have used the formulas for the σ vector given by Eq. (G.36). Combining Eqs. (G.50) and (G.53), we obtain

$$
-\frac{1}{2mc}(p + i[p \times \sigma])_1 u = -\frac{1}{2mc} \begin{pmatrix} \cdots \\ \cdots \\ (p_1 + ip_2)u_3 - p_3 u_4 \\ (p_1 - ip_2)u_4 + p_3 u_3 \end{pmatrix}. \tag{G.54}
$$

Next, let us compare Eqs. (G.49) and (G.54). We see that the third and fourth rows are identical. The first two rows are indicated by blanks and in order to establish the identity between the two matrices we have introduced, in Eq. (G.47), the quantity φ, which is a one-column matrix-vector. The first two rows of φ are used to establish the identity between the first two rows of the matrices given by Eqs. (G.49) and (G.54). The third and fourth rows of φ contain zeros. With this definition of φ, the relationship given by Eq. (G.47) is proved, provided that the approximations given by Eqs. (G.44) and (G.46) are valid.

We could easily write down the first two rows of φ, but they are irrelevant, as we shall see presently. The elimination of $\boldsymbol{\alpha}$ from the Dirac equation is accomplished by introducing the approximation given by Eq. (G.47). Using this relationship in Eq. (G.31), we obtain an equation in which only the $\boldsymbol{\sigma}$ matrix-vector occurs. We have seen that $\boldsymbol{\sigma}$ satisfies commutation rules that are also satisfied by the 2×2 Pauli matrices. Thus, we can reinterpret $\boldsymbol{\sigma}$ as a 2×2 vector-matrix identical with the Pauli matrix-vector given by Eq. (G.15), and we can reinterpret Eq. (G.31) as a two-component equation, that is, an equation for the large components of the original wave equation, u_3 and u_4. Since φ does not have third and fourth components, it can be safely omitted altogether from the equation, and what it contains in the first and second rows is irrelevant.

We should also note that the reinterpretation of $\boldsymbol{\sigma}$ as a 2×2 matrix is made possible because the original 4×4 $\boldsymbol{\sigma}$ does not interchange the small and large components. As we see from the relationships given by Eq. (G.19), the Dirac matrices α_1, α_2, and α_3 interchange the small and large components when they operate on the wave function u; on the other hand, it is easy to establish that $\boldsymbol{\sigma}$ operates on u as follows:

$$\sigma_1 u = \begin{pmatrix} u_1 \\ -u_2 \\ u_3 \\ -u_4 \end{pmatrix}, \tag{G.55a}$$

$$\sigma_2 u = i \begin{pmatrix} -u_2 \\ u_1 \\ -u_4 \\ u_3 \end{pmatrix}, \tag{G.55b}$$

$$\sigma_3 u = \begin{pmatrix} u_2 \\ u_1 \\ u_4 \\ u_3 \end{pmatrix}, \tag{G.55c}$$

and we see that the small and large components are not interchanged. This property of $\boldsymbol{\sigma}$ is crucial for the reinterpretation of the Dirac equation as a two-component relationship.

Substituting Eq. (G.47) into Eq. (G.31) and omitting φ, we get the approximate two-component form of the Dirac equation:

$$
\begin{aligned}
\Bigg\{ -\frac{\hbar^2}{2m}\Delta - eV - \frac{p^4}{8m^3c^2} - \frac{ie\hbar}{mc}(A \cdot \nabla) \\
+ \frac{e^2A^2}{2mc^2} - \frac{ie\hbar}{4m^2c^2}[(\mathscr{E}\cdot p) - i\mathscr{E}(\sigma \times p)] \\
+ \frac{e\hbar}{2mc}(\sigma \cdot \mathscr{H}) \Bigg\} u = Wu.
\end{aligned}
\tag{G.56}
$$

We have clarified before the physical meaning of each term, except that we postponed the discussion of the sixth term by saying that it is responsible for the spin–orbit interaction. We now analyze this term and show that it does indeed contain the spin–orbit interaction energy.

Let us assume that the external magnetic field is zero and the electric field can be generated by the central potential $V(r)$. Then, we have

$$
\mathscr{E} = -\nabla V = -\frac{dV}{dr}\frac{r}{r},
\tag{G.57}
$$

Using this formula, we obtain

$$
[(\mathscr{E}\cdot p) - i\mathscr{E}\cdot(\sigma \times p)] = -\frac{1}{r}\frac{dV}{dr}[(r \cdot p) - ir \cdot (\sigma \times p)].
\tag{G.58}
$$

We quote the standard vector relationship

$$
r \cdot (\sigma \times p) = -\sigma \cdot (r \times p),
\tag{G.59}
$$

with which we get

$$
[(\mathscr{E}\cdot p) - i\mathscr{E}\cdot(\sigma \times p)] = -\frac{1}{r}\frac{dV}{dr}[(r \cdot p) + i\sigma \cdot (r \times p)].
\tag{G.60}
$$

Putting this relationship into the sixth term of Eq. (G.56) and denoting the resultant operator by H_s, we obtain

$$
H_s = \frac{ie\hbar}{4m^2c^2}\frac{1}{r}\frac{dV}{dr}[(r \cdot p) + i\sigma \cdot (r \times p)].
\tag{G.61}
$$

This formula can be put in a form that is more familiar from the literature. Let L and s be the operators of the orbital and spin angular momenta. The definitions are

$$L = r \times p, \tag{G.62}$$

and

$$s = \frac{\hbar}{2}\sigma, \tag{G.63}$$

where p is the momentum operator. Substituting these definitions into Eq. (G.61), we obtain

$$H_s = \frac{ie\hbar}{4m^2c^2}\frac{1}{r}\frac{dV}{dr}(r \cdot p) - \frac{e}{2m^2c^2}\frac{1}{r}\frac{dV}{dr}(s \cdot L). \tag{G.64}$$

The second term is the familiar expression for the spin–orbit interaction energy, first derived by Thomas,[91] who used a semiclassical derivation based on the Bohr model. The usual form quoted is

$$H = \frac{1}{2m^2c^2}\frac{1}{r}\frac{dV}{dr}(s \cdot L). \tag{G.65}$$

We obtained $-e$ in our formula because the meaning of V is different in the two formulas. In our formula, V is the potential that is related to the electric field by the equation

$$\mathscr{E} = -\nabla V. \tag{G.66}$$

On the other hand, the V in Eq. (G.65) is related to the force by the relationship

$$F = -\nabla V^T, \tag{G.67}$$

where V^T is the potential in Thomas's formula. Now, because

$$F = -e\mathscr{E}, \tag{G.68}$$

we get

$$\mathscr{E} = -\nabla V = \frac{F}{-e} = \frac{-\nabla V^T}{-e}, \tag{G.69}$$

and so

$$-eV = V^T. \tag{G.70}$$

Thus, our expression is correct in Eq. (G.64) if V is the potential; if V is the potential energy, that is, the potential times $-e$, then the correct formula is Thomas's expression, Eq. (G.65). The difference between potential and potential energy should also be remembered in the evaluation of the first term of Eq. (G.64).

Our conclusion is that the sixth term of Eq. (G.56) is indeed responsible for the spin–orbit interaction. From Eq. (G.64), we see also that besides the spin–orbit energy, this term contains a term (the first in Eq. (G.64)) that does not contain the spin, but depends on the external potential. This term is called the Darwin term because Darwin has shown that this term arises from the Dirac equation when it is approximated by a two-component formula.[17] Thus, the correct statement is that the sixth term of Eq. (G.56) consists of the Darwin term and the spin–orbit interaction energy.

APPENDIX H

ANGULAR MOMENTUM AND SPIN PROPERTIES OF THE DIRAC EIGENFUNCTIONS

In this appendix, we summarize the main properties of the angular and spin parts of the Dirac eigenfunctions; these eigenfunctions are needed for the relativistic Hartree–Fock model. It will be convenient to investigate first the angular and spin parts of the solutions of the two-component Dirac equation. Let us consider the equation in the absence of an external magnetic field, that is, let us put $A = 0$, $\mathscr{H} = 0$, and let us assume that the external scalar potential is central. Then, we get

$$
\left\{ -\frac{\hbar^2}{2m}\Delta - eV - \frac{(-i\hbar\nabla)^4}{8m^3c^2} + \frac{e\hbar^2}{4m^2c^2}\frac{1}{r}\frac{dV}{dr}(\boldsymbol{r}\cdot\nabla) \right.
$$

$$
\left. - \frac{e}{2m^2c^2}\frac{1}{r}\frac{dV}{dr}(\boldsymbol{s}\cdot\boldsymbol{L}) \right\} u = Wu. \quad \text{(H.1)}
$$

Let us introduce the operators $\boldsymbol{L}$, $\boldsymbol{S}$, and $\boldsymbol{J}$ where $\boldsymbol{L}$ is the orbital angular momentum, $\boldsymbol{S}$ is the spin, and $\boldsymbol{J}$ is the total angular momentum defined as

$$
\boldsymbol{J} = \boldsymbol{L} + \boldsymbol{S}, \quad \text{(H.2a)}
$$

with

$$
J_z = L_z + S_z. \quad \text{(H.2b)}
$$

The Hamiltonian of Eq. (H.1) commutes with L^2, S^2, J^2, and J_z, but not with L_z and S_z. As we know, the nonrelativistic Hamiltonian, which consists

514

of the first two terms of Eq. (H.1), commutes with L^2, S^2, L_z, and S_z. We must construct the solutions of Eq. (H.1) in such a way that they will be eigenfunctions of L^2, S^2, J^2, and J_z; we accomplish this by the coupling of angular-momentum eigenfunctions that are eigenfunctions of L^2, S^2, L_z, and S_z.

Let u_{nljm_j} be the desired eigenfunction of Eq. (H.1). According to the quantum theory of angular momentum, this function can be written in the form

$$u_{nljm_j}(r, \vartheta, \varphi, \sigma)$$

$$= R_{nlj}(r) \sum_{m_s} C\left(l\tfrac{1}{2}j; (m_j - m_s), m_s\right) Y_{l(m_j - m_s)} \eta_{m_s}. \qquad (H.3)$$

In this formula, j and m_j are the quantum numbers associated with the operators J^2 and J_z. C is a Clebsch–Gordan coefficient, $Y_{lm}(\vartheta, \varphi)$ is the normalized spherical harmonics, and $\eta_{m_s}(\sigma)$ is the spin function with σ being the spin variable. $R_{nlj}(r)$ is the radial part of the eigenfunction. The summation is over the two possible values of m_s, that is, over the values $\tfrac{1}{2}$ and $-\tfrac{1}{2}$.

In Eq. (H.3), the uncoupled products, the products of the spherical harmonics and of the spin function, are eigenfunctions of L^2, S^2, L_z, and S_z. Indeed, we have

$$L^2 Y_{lm_l} = l(l + 1)\hbar^2 Y_{lm_l}, \qquad (H.4a)$$

$$L_z Y_{lm_l} = m_l \hbar Y_{lm_l}, \qquad (H.4b)$$

$$S^2 \eta_{m_s} = \tfrac{1}{2}\left(\tfrac{1}{2} + 1\right)\hbar^2 \eta_{m_s}, \qquad (H.4c)$$

and

$$S_z \eta_{m_s} = m_s \hbar \eta_{m_s}. \qquad (H.4d)$$

As we see from Eq. (H.3), the Clebsch–Gordan expansion generates a wave function that is an eigenfunction of L^2, S^2, J^2, and J_z from the uncoupled representation. Among the indices of the Clebsch–Gordan coefficients, the first three, l, $\tfrac{1}{2}$, and j, indicate that eigenfunctions of L^2 and S^2 are coupled into eigenfunctions of J^2; on the right side of the semicolon, the indices indicate that the summation is over m_s and the result will be an eigenfunction of J_z with m_j.

We want to examine the function u_{nljm_j}. In order to do that, we recall now the properties of the spin functions. We have introduced the spin function in connection with the introduction of spin-orbitals, Eqs. (2.35) and (2.36). The spin function $\eta_{m_s}(\sigma)$ has the properties described by the equations

$$\eta_{1/2}\left(\tfrac{1}{2}\right) = 1, \qquad \eta_{1/2}\left(-\tfrac{1}{2}\right) = 0,$$
$$\eta_{-1/2}\left(\tfrac{1}{2}\right) = 0, \qquad \eta_{-1/2}\left(-\tfrac{1}{2}\right) = 1. \tag{H.5}$$

From this, we see that both the index m_s and the argument σ can take only the values $\tfrac{1}{2}$ and $-\tfrac{1}{2}$. The functions are orthonormal, that is, they satisfy the relationship

$$\int \eta_{m_s}^*(\sigma)\eta_{m'_s}(\sigma)\,d\sigma = \sum_\sigma \eta_{m_s}^*(\sigma)\eta_{m'_s}(\sigma)$$
$$= \eta_{m_s}^*\left(\tfrac{1}{2}\right)\eta_{m'_s}\left(\tfrac{1}{2}\right) + \eta_{m_s}^*\left(-\tfrac{1}{2}\right)\eta_{m'_s}\left(-\tfrac{1}{2}\right)$$
$$= \delta_{m_s m'_s}, \tag{H.6}$$

and we see that the integration over σ is replaced by summation over the two possible values.

The spin functions can be written in matrix form. The matrices are the eigenfunctions of the spin operators that were defined by Eqs. (G.15). The matrix form is

$$\eta_{1/2} = \begin{pmatrix} 1 \\ 0 \end{pmatrix}, \qquad \eta_{-1/2} = \begin{pmatrix} 0 \\ 1 \end{pmatrix}. \tag{H.7}$$

Using Eq. (G.15), we obtain

$$S^2 = \frac{\hbar^2}{4}\boldsymbol{\sigma}^2 = \frac{\hbar^2}{4}\left(\sigma_x^2 + \sigma_y^2 + \sigma_z^2\right)$$
$$= \frac{1}{2}\left(\frac{1}{2} + 1\right)\hbar^2 \begin{pmatrix} 1 & 0 \\ 0 & 1 \end{pmatrix}, \tag{H.8}$$

and

$$S^2\eta_{m_s} = \frac{1}{2}\left(\frac{1}{2} + 1\right)\hbar^2\eta_{m_s}, \tag{H.9}$$

for both values of m_s. Thus, the square of the spin is always $\tfrac{3}{4}\hbar^2$. Also, we get

$$S_z\eta_{m_s} = \frac{\hbar}{2}\sigma_z\eta_{m_s} = \pm\frac{1}{2}\eta_{m_s} = m_s\hbar\eta_{m_s} \qquad \left(m_s = \pm\frac{1}{2}\right). \tag{H.10}$$

We now turn back to Eq. (H.3). Writing it out in detail, we get

$$
\frac{u_{nljm_j}(r,\vartheta,\varphi,\sigma)}{R_{nlj}(r)}
$$

$$
= C\left(l\frac{1}{2}j;\left(m_j-\frac{1}{2}\right),\frac{1}{2}\right)Y_{l,m_j-1/2}(\vartheta,\varphi)\eta_{1/2}(\sigma)
$$

$$
+ C\left(l\frac{1}{2}j;\left(m_j+\frac{1}{2}\right),-\frac{1}{2}\right)Y_{l,m_j+1/2}(\vartheta,\varphi)\eta_{-1/2}(\sigma). \quad \text{(H.11)}
$$

According to the quantum theory of angular momentum, for each value of l, we get two values of j:

$$
j = \left(l \pm \tfrac{1}{2}\right) \qquad (l = 1, 2, \dots), \quad \text{(H.12)}
$$

and for $l = 0$, $j = l + \tfrac{1}{2} = \tfrac{1}{2}$, that is,

$$
j = \frac{1}{2} \qquad (l = 0). \quad \text{(H.13)}
$$

Thus, for each value of l, we will have two j values, and for each (lj) combination, there will be two values of m_s, that is, we can build up the Clebsch–Gordan series from a total of four types of coefficients. We have analytic formulas for these coefficients that we write down here (Slater II, 92):

$j = l + \tfrac{1}{2}$:

$$
C\left(l\frac{1}{2}j;\left(m_j-\frac{1}{2}\right),\frac{1}{2}\right) = \left[\frac{l+m_j+1/2}{2l+1}\right]^{1/2}; \quad \text{(H.14)}
$$

$$
C\left(l\frac{1}{2}j;\left(m_j+\frac{1}{2}\right),-\frac{1}{2}\right) = \left[\frac{l-m_j+1/2}{2l+1}\right]^{1/2}. \quad \text{(H.15)}
$$

$j = l - 1/2$:

$$
C\left(l\frac{1}{2}j;\left(m_j-\frac{1}{2}\right),\frac{1}{2}\right) = -\left[\frac{l-m_j+1/2}{2l+1}\right]^{1/2}; \quad \text{(H.16)}
$$

$$
C\left(l\frac{1}{2}j;\left(m_j+\frac{1}{2}\right),-\frac{1}{2}\right) = \left[\frac{l+m_j+1/2}{2l+1}\right]^{1/2}. \quad \text{(H.17)}
$$

The range of m_j is always

$$
-j \leq m_j \leq j. \quad \text{(H.18)}
$$

We have seen from its derivation that Eq. (H.1) is a two-component equation. We have stated that the function given in Eq. (H.3) is the solution of this equation; however, this function does not have the appearance of a two-component function. That it is indeed a two-component function can be demonstrated if we put the spin functions into matrix form. Let us introduce the notation given by Eq. (H.7) into the expansion given by Eq. (H.11). Then, we get

$$u_{nljm_j}(r, \vartheta, \varphi, \sigma) = aR_{nlj}(r)Y_{l,(m_j-1/2)}\begin{pmatrix}1\\0\end{pmatrix} + bR_{nlj}(r)Y_{l,(m_j+1/2)}\begin{pmatrix}0\\1\end{pmatrix}$$

$$= \begin{pmatrix} aR_{nlj}(r)Y_{l(m_j-1/2)}(\vartheta, \varphi) \\ bR_{nlj}(r)Y_{l(m_j+1/2)}(\vartheta, \varphi) \end{pmatrix}, \tag{H.19}$$

where we have denoted the Clebsch–Gordan coefficients by a and b. It is clear now that our eigenfunction is a two-component function as it must be in order to be an eigenfunction of the two-component equation (H.1). We note also that although the angular part of the two components are different, the radial part is the same. Thus having a two-component function does not mean that we have two different radial functions.

Finally, let us consider the orthogonality integral

$$I = \int u^*_{nljm_j} u_{n'l'j'm'_j}\, dq, \tag{H.20}$$

where dq is the integration over the space functions and summation over the spin part. Function u is constructed in such a way that it is the eigenfunction of L^2, S^2, J^2, and J_z. The quantum numbers associated with these operators are $l, \frac{1}{2}, j$, and m_j. Thus, there will be orthogonality with respect to these quantum numbers. In addition, because the function will also be an eigenfunction of the Hamiltonian of Eq. (H.1), we can assume orthogonality with respect to n also. Thus, we get

$$\int u^*_{nljm_j} u_{n'l'j'm'_j}\, dq = \delta(n, n')\,\delta(l, l')\,\delta(j, j')\,\delta(m_j, m'_j). \tag{H.21}$$

If we condense the four indices into a single index, we can write

$$\int u^*_k u_j\, dq = \delta_{kj}. \tag{H.22}$$

We turn now to the exact four-component Dirac equation. The Hamiltonian was given by Eq. (G.1):

$$H = -c\boldsymbol{\alpha} \cdot \boldsymbol{p} - \beta mc^2 - eV. \tag{H.23}$$

The first thing that we observe is that the constants of the motion associated with this Hamiltonian are different from the constants of the motion of the two-component equation. The Dirac Hamiltonian commutes with the operators J^2, J_z, and S^2, but not with L^2. As a result of this, we must seek eigenfunctions that, besides being eigenfunctions of the Dirac Hamiltonian, will also satisfy the relationships

$$J^2\psi = j(j+1)\hbar^2\psi, \tag{H.24a}$$

$$S^2\psi = \tfrac{1}{2}(\tfrac{1}{2}+1)\hbar^2\psi, \tag{H.24b}$$

and

$$J_z\psi = m_j\hbar\psi. \tag{H.24c}$$

First, let us consider Eq. (H.3). Using the quantum number κ, which we introduced in the main text, we can write

$$u_{nljm_j} = u_{n\kappa m_j}, \tag{H.25}$$

where κ replaces l and j. If we omit the radial part of the function in Eq. (H.3), then we can write

$$u_{\kappa m_j} = \sum_{m_s} C\big(l\tfrac{1}{2}j; (m_j - m_s), m_s\big) Y_{l,(m_j-m_s)} \eta_{m_s}. \tag{H.26}$$

The eigenfunctions of the Dirac Hamiltonian, Eq. (H.23), which also satisfy the relationships given by Eqs. (H.24), will have the form

$$\psi = \begin{pmatrix} iu_{-\kappa m_j} f(r) \\ u_{\kappa m_j} g(r) \end{pmatrix}, \tag{H.27}$$

where the radial function $f(r)$ and $g(r)$ are introduced in place of the single radial function $R_{nlj}(r)$ that we had in Eq. (H.3). This is clearly a four-component function in which $(u_{\kappa m_j} g)$ is the "large" component and $(iu_{-\kappa m_j} f)$ is the "small."

Let us substitute ψ into the Dirac equation. We write the equation in the form

$$(H - E)\psi = 0, \tag{H.28}$$

where H is given by Eq. (H.23). Then it is easy to show that

$$(H - E)\psi = (H - E) \begin{pmatrix} iu_{-\kappa m_j} f(r) \\ u_{\kappa m_j} g(r) \end{pmatrix}$$

$$= \begin{pmatrix} iu_{-\kappa m_j} \left[-c\hbar \left(\dfrac{dg}{dr} + \dfrac{\kappa + 1}{r} g \right) - (eV + mc^2 + E)f \right] \\ u_{\kappa m_j} \left[c\hbar \left(\dfrac{df}{dr} - \dfrac{\kappa - 1}{r} f \right) - (eV - mc^2 + E)g \right] \end{pmatrix}$$

$$= 0. \tag{H.29}$$

Thus, the condition for ψ to be an eigenfunction of H is that f and g satisfy the equations

$$-c\hbar \left(\frac{dg}{dr} + \frac{\kappa + 1}{r} g \right) - (eV + mc^2 + E)f = 0,$$

$$c\hbar \left(\frac{df}{dr} - \frac{\kappa - 1}{r} f \right) - (eV - mc^2 + E)g = 0. \tag{H.30}$$

It can also be shown that the function ψ satisfies the eigenvalue equations (H.24). Thus, ψ, given by Eq. (H.27), is a joint eigenfunction of the Dirac Hamiltonian and of the operators J^2, S^2, and J_z.

The Dirac equations given by Eq. (H.30) can easily be put into a form that resembles more closely the nonrelativistic formulation. Let

$$g = \frac{P}{r}, \qquad f = \frac{Q}{r}. \tag{H.31}$$

We obtain by simple manipulations the two equations for P and Q:

$$\frac{dP}{dr} + \frac{\kappa}{r} P + \frac{1}{c\hbar} (eV + mc^2 + E)Q = 0, \tag{H.32a}$$

$$\frac{dQ}{dr} - \frac{\kappa}{r} Q - \frac{1}{c\hbar} (eV - mc^2 + E)P = 0. \tag{H.32b}$$

Introducing the nonrelativistic energy W with the definition

$$W = E - mc^2, \tag{H.33}$$

we obtain

$$\frac{dP}{dr} + \frac{\kappa}{r} P + \frac{1}{c\hbar} (eV + W + 2mc^2)Q = 0, \tag{H.34a}$$

$$\frac{dQ}{dr} - \kappa \frac{Q}{r} - \frac{1}{c\hbar} (eV + W)P = 0. \tag{H.34b}$$

Now in the nonrelativistic approximation, we can put

$$W \approx \frac{p^2}{2m} - eV, \tag{H.35}$$

and

$$eV + W \approx \frac{p^2}{2m}, \tag{H.36}$$

that is, we get

$$eV + W + 2mc^2 \approx \frac{p^2}{2m} + 2mc^2. \tag{H.37}$$

As we have argued following Eq. (G.41), the kinetic energy of the electron is at most a few thousand eV's, and $2mc^2$ is about 1 MeV. Thus, in the previous expression, we can omit $p^2/2m$ beside the $2mc^2$. We obtain in this approximation from Eq. (H.34a)

$$\frac{dP}{dr} + \frac{\kappa}{r}P + \frac{1}{c\hbar}2mc^2Q = 0. \tag{H.38}$$

Differentiating with respect to r, we get

$$\frac{d^2P}{dr^2} - \frac{\kappa}{r^2}P + \frac{\kappa}{r}\frac{dP}{dr} + \frac{2mc}{\hbar}\frac{dQ}{dr} = 0. \tag{H.39}$$

We substitute dQ/dr from Eq. (H.34b) and dP/dr from Eq. (H.38). In this way, we obtain

$$\frac{d^2P}{dr^2} - \frac{\kappa^2 + \kappa}{r^2}P + \frac{2m}{\hbar^2}(eV + W)P = 0. \tag{H.40}$$

It is easy to see that $\kappa^2 + \kappa$ is always equal to $l(l + 1)$. Indeed, from Eq. (4.95), we get

$$j = l + \tfrac{1}{2}, \qquad a = 1, \qquad \kappa = -(l + 1),$$

$$\kappa^2 = (l + 1)^2, \qquad \kappa^2 + \kappa = l(l + 1); \tag{H.41}$$

$$j = \bar{l} - \tfrac{1}{2}, \qquad a = -1, \qquad \kappa = \bar{l},$$

$$\kappa^2 = \bar{l}^2, \qquad \kappa^2 + \kappa = \bar{l}(\bar{l} + 1). \tag{H.42}$$

We substitute these results into Eq. (H.40), which becomes, after multiplying by $-\hbar^2/2m$:

$$-\frac{\hbar^2}{2m}\frac{d^2P}{dr^2} + \frac{\hbar^2}{2m}\frac{l(l+1)}{r^2}P - eVP = WP. \qquad (H.43)$$

This is the nonrelativistic Schroedinger equation for a single electron in the potential field V. Thus, we see that, in the nonrelativistic limit, the Dirac equations given by Eqs. (H.34a) and (H.34b) become identical with the nonrelativistic Schroedinger equation.

APPENDIX I

A DERIVATION OF BREIT'S FORMULA FOR THE RELATIVISTIC MANY-ELECTRON HAMILTONIAN

As we have indicated in the main text, Breit derived the relativistic many-electron Hamiltonian by an application of the correspondence principle, that is, he formulated the Hamiltonian by demanding that in the classical limit, the Hamiltonian should become identical with the classical Hamiltonian function. In order to present this derivation, we start by reviewing the classical formulas for the equation of motion of an electron in an electromagnetic field. Having obtained the classical formulas, we will move to the Dirac formulation of relativistic electron theory and derive the equation of motion in that formulation. A comparison between the classical and the quantum mechanical equations of motion will provide us with the clue for the correct formulation of the relativistic many-electron Hamiltonian.

Let the cartesian coordinates of the electron be q_1, q_2, and q_3 and its velocities $\dot{q}_1$, $\dot{q}_2$, and $\dot{q}_3$. Let p_1, p_2, and p_3 be the canonically conjugate momenta and let H be the classical Hamiltonian function. Then, according to the Hamiltonian formulation of classical mechanics, we have the relationships

$$\dot{q}_i = \frac{\partial H}{\partial p_i}, \tag{I.1}$$

$$\dot{p}_i = -\frac{\partial H}{\partial q_i}. \tag{I.2}$$

The time derivative of a function F, which depends on coordinates q_i, momenta p_i, and time t, is given by

$$\frac{dF}{dt} = \frac{\partial F}{\partial t} + \{F, H\}, \tag{I.3}$$

where

$$\{F, H\} = \sum_i \left(\frac{\partial F}{\partial q_i} \frac{\partial H}{\partial p_i} - \frac{\partial H}{\partial q_i} \frac{\partial F}{\partial p_i} \right), \tag{I.4}$$

is the Poisson bracket.

If the electron is moving in an arbitrary electromagnetic field that is characterized by the scalar potential $V = V(q_i t)$ and by the vector potential $A = A(q_i t)$, then its Hamiltonian function is given by

$$H = \frac{P^2}{2m} - eV, \tag{I.5}$$

where

$$P \equiv p + \frac{e}{c} A, \tag{I.6}$$

and, accordingly,

$$P^2 = \left(p_1 + \frac{e}{c} A_1 \right)^2 + \left(p_2 + \frac{e}{c} A_2 \right)^2 + \left(p_3 + \frac{e}{c} A_3 \right)^2. \tag{I.7}$$

First, we derive the formula for $\dot{q}_i$. Using Eqs. (I.1) and (I.5), we get

$$\dot{q}_i = \frac{1}{m} \left(p_i + \frac{e}{c} A_i \right). \tag{I.8}$$

Next, we derive the equation of motion, that is, we construct the expression for $m(d\dot{q}_i/dt)$. From Eq. (I.8), we get

$$m \frac{d}{dt} \dot{q}_i = \frac{dp_i}{dt} + \frac{e}{c} \frac{dA_i}{dt}. \tag{I.9}$$

For the first term, we use the Hamiltonian equation (I.2) and for dA_i/dt, we use Eq. (A.3). We obtain

$$m\ddot{q}_i = -\frac{\partial H}{\partial q_i} + \frac{e}{c} \left[\frac{\partial A_i}{\partial t} + \sum_k \left(\frac{\partial A_i}{\partial q_k} \frac{\partial H}{\partial p_k} - \frac{\partial H}{\partial q_k} \frac{\partial A_i}{\partial p_k} \right) \right]. \tag{I.10}$$

The last term on the right side is zero because A_i does not depend on p_i. Using the expression for H, we get

$$-\frac{\partial H}{\partial q_i} = -\sum_k \left(p_k + \frac{e}{c} A_k \right) \frac{e}{mc} \frac{\partial A_k}{\partial q_i} + e \frac{\partial V}{\partial q_i}, \tag{I.11}$$

and

$$\frac{\partial H}{\partial p_i} = \frac{1}{m} \left(p_i + \frac{e}{c} A_i \right) = \dot{q}_i. \tag{I.12}$$

Using the notation of Eq. (I.6), we can write

$$-\frac{\partial H}{\partial q_i} = -\sum_k \frac{e}{mc} P_k \frac{\partial A_k}{\partial q_i} + e \frac{\partial V}{\partial q_i}, \tag{I.13}$$

and

$$\frac{\partial H}{\partial p_i} = \frac{1}{m} P_i. \tag{I.14}$$

Substituting the last two expressions into Eq. (I.10), we get

$$\begin{aligned} m\ddot{q}_i &= -\sum_k \frac{e}{mc} P_k \frac{\partial A_k}{\partial q_i} + e \frac{\partial V}{\partial q_i} \\ &\quad + \frac{e}{c} \frac{\partial A_i}{\partial t} + \frac{e}{mc} \sum_k P_k \frac{\partial A_i}{\partial q_k} \\ &= e \frac{\partial V}{\partial q_i} + \frac{e}{c} \frac{\partial A_i}{\partial t} + \frac{e}{mc} \sum_k P_k \left(\frac{\partial A_i}{\partial q_k} - \frac{\partial A_k}{\partial q_i} \right). \end{aligned} \tag{I.15}$$

In order to see the meaning of this expression, let us consider

$$\begin{aligned} \boldsymbol{P} \times (\mathrm{curl}\ \boldsymbol{A}) &= \boldsymbol{P} \times \mathscr{H} \\ &= \begin{vmatrix} \hat{i} & \hat{j} & \hat{k} \\ P_1 & P_2 & P_3 \\ (\mathrm{curl}\ \boldsymbol{A})_1 & (\mathrm{curl}\ \boldsymbol{A})_2 & (\mathrm{curl}\ \boldsymbol{A})_3 \end{vmatrix}, \end{aligned} \tag{I.16}$$

where $\mathscr{H}$ is the magnetic-field vector. We will need the formulas for curl $\boldsymbol{A}$, thus it will be useful to write them down here:

$$(\mathrm{curl}\ \boldsymbol{A})_1 = \frac{\partial A_3}{\partial q_2} - \frac{\partial A_2}{\partial q_3}, \tag{I.16a}$$

$$(\mathrm{curl}\ \boldsymbol{A})_2 = \frac{\partial A_1}{\partial q_3} - \frac{\partial A_3}{\partial q_1}, \tag{I.16b}$$

and

$$(\text{curl } A)_3 = \frac{\partial A_2}{\partial q_1} - \frac{\partial A_1}{\partial q_2}. \tag{I.16c}$$

Using these relationships, we obtain

$$
\begin{aligned}
P \times (\text{curl } A) \\
= P \times \mathscr{H} \\
= \hat{\imath}[P_2(\text{curl } A)_3 - P_3(\text{curl } A)_2] \\
+ \hat{\jmath}[P_3(\text{curl } A)_1 - P_1(\text{curl } A)_3] \\
+ \hat{k}[P_1(\text{curl } A)_2 - P_2(\text{curl } A)_1] \\
= \hat{\imath}\left[P_2\left(\frac{\partial A_2}{\partial q_1} - \frac{\partial A_1}{\partial q_2} \right) - P_3\left(\frac{\partial A_1}{\partial q_3} - \frac{\partial A_3}{\partial q_1} \right) \right] \\
+ \hat{\jmath}\left[P_3\left(\frac{\partial A_3}{\partial q_2} - \frac{\partial A_2}{\partial q_3} \right) - P_1\left(\frac{\partial A_2}{\partial q_1} - \frac{\partial A_1}{\partial q_2} \right) \right] \\
+ \hat{k}\left[P_1\left(\frac{\partial A_1}{\partial q_3} - \frac{\partial A_3}{\partial q_1} \right) - P_2\left(\frac{\partial A_3}{\partial q_2} - \frac{\partial A_2}{\partial q_3} \right) \right]. \tag{I.17}
\end{aligned}
$$

In the last term of the expression given by Eq. (I.15), index i can be 1, 2, or 3 while we sum up over index k. It is clear that the term with $k = i$ is zero. Let $i = 1$; then we get

$$
\begin{aligned}
\sum_k P_k\left(\frac{\partial A_1}{\partial q_k} - \frac{\partial A_k}{\partial q_1} \right) &= P_2\left(\frac{\partial A_1}{\partial q_2} - \frac{\partial A_2}{\partial q_1} \right) + P_3\left(\frac{\partial A_1}{\partial q_3} - \frac{\partial A_3}{\partial q_1} \right) \\
&= -(P \times \text{curl } A)_1. \tag{I.18}
\end{aligned}
$$

A similar result is obtained by choosing $i = 2$ or 3. Thus, we get for Eq. (I.15), by returning to the conventional notation ($q_i \equiv x_i$),

$$m\ddot{x}_i = e(\nabla V)_i + \frac{e}{c}\frac{\partial A_i}{\partial t} - \frac{e}{mc}(P \times \text{curl } A)_i, \tag{I.19}$$

or, in vector notation,

$$m\ddot{r} = e(\nabla V) + \frac{e}{c}\frac{\partial A}{\partial t} - \frac{e}{mc}[P \times (\nabla \times A)]. \tag{I.20}$$

This is the equation of motion of the electron. We recall that, according to Eq. (I.8), we have

$$\dot{r} = \frac{dr}{dt} = \frac{1}{m}\left(p + \frac{e}{c}A \right) = \frac{1}{m}P, \tag{I.21}$$

and so Eq. (I.20) can be written also in the form

$$m\ddot{\mathbf{r}} = m\frac{d}{dt}\dot{\mathbf{r}} = \frac{d\mathbf{P}}{dt}$$

$$= e(\nabla V) + \frac{e}{c}\frac{\partial \mathbf{A}}{\partial t} - \frac{e}{mc}[\mathbf{P} \times (\nabla \times \mathbf{A})]. \tag{I.22}$$

This expression can be written in a more familiar form if we recall that the connection between the scalar potential V, the vector potential A, and the electric- and magnetic-field vectors $\mathscr{E}$ and $\mathscr{H}$, respectively, is given by Eqs. (F.30) and (F.31):

$$\mathscr{E} = -\nabla V - \frac{1}{c}\frac{\partial \mathbf{A}}{\partial t}, \tag{I.23}$$

and

$$\mathscr{H} = \operatorname{curl} A = \nabla \times A. \tag{I.24}$$

Using these relationships, we can write

$$m\ddot{\mathbf{r}} = \frac{d\mathbf{P}}{dt} = -e\mathscr{E} - \frac{e}{mc}[\mathbf{P} \times \mathscr{H}]. \tag{I.25}$$

We recover the conventional form of the equation of motion if we use again Eq. (I.21). According to that equation, the velocity of the electron is given by

$$\mathbf{v} = \frac{\mathbf{P}}{m}, \tag{I.26}$$

and so we get

$$m\frac{d\mathbf{v}}{dt} = -e\mathscr{E} - \frac{e}{c}[\mathbf{v} \times \mathscr{H}]. \tag{I.27}$$

where the left side is the mass times acceleration and the right side is the formula for the Lorentz force, that is, the force that is acting on an electron that is moving in an electromagnetic field.

We now turn to Dirac's relativistic electron theory. Our goal is to derive the equation of motion. We turn to the matrix formulation of quantum mechanics (see, for example, Ref. 20, p. 148). The essence of this formulation is that we work with operators and derive operator equations; these equations must be understood in such a way that the resulting equations are relationships between the matrix representations of operators. The great advantage of this formulation is that the formalism is closely analogous to the classical Hamiltonian formulation.

Let $(x_1 x_2 x_3)$ be the cartesian coordinates of the electron and $(p_1 p_2 p_3)$ the canonically conjugate momenta. According to the prescription of quantum mechanics, we replace these quantities by operators. The coordinate operator is identical with itself, while for the momentum vector, we put

$$p = -i\hbar \nabla. \tag{I.28}$$

The coordinates and the momenta commute with themselves, whereas between each other, they have to satisfy the commutation relations

$$f(x_1 x_2 x_3) p_k - p_k f(x_1 x_2 x_3) = i\hbar \frac{\partial f}{\partial x_k}. \tag{I.29}$$

For the equation of motion, we use the Heisenberg picture. We define the derivative of the operator Ω as

$$\frac{d\Omega}{dt} = \frac{\partial \Omega}{\partial t} + \frac{1}{i\hbar}[\Omega H - H\Omega]. \tag{I.30}$$

For the Hamiltonian, we substitute Dirac's operator, which was given in Eqs. (G.3) and (G.4):

$$H = -c\boldsymbol{\alpha} \cdot \boldsymbol{P} - \beta mc^2 - eV, \tag{I.31}$$

with

$$\boldsymbol{P} = \boldsymbol{p} + \frac{e}{c}\boldsymbol{A}. \tag{I.32}$$

First, let us consider the derivative of the coordinate x_1. We obtain from Eqs. (I.30), (I.31), and (I.32)

$$
\begin{aligned}
\frac{dx_1}{dt} &= \frac{1}{i\hbar}[x_1 H - H x_1] \\
&= \frac{1}{i\hbar}\{-x_1(c\boldsymbol{\alpha} \cdot \boldsymbol{P} + \beta mc^2 + eV) \\
&\qquad + (c\boldsymbol{\alpha} \cdot \boldsymbol{P} + \beta mc^2 + eV)x_1\}.
\end{aligned}
\tag{I.33}
$$

the x_1 commutes with β and also with V because V depends only on the coordinates. Thus,

$$\frac{dx_1}{dt} = \frac{c}{i\hbar}\left\{\boldsymbol{\alpha} \cdot \left(\boldsymbol{p} + \frac{e}{c}\boldsymbol{A}\right)x_1 - x_1\boldsymbol{\alpha} \cdot \left(\boldsymbol{p} + \frac{e}{c}\boldsymbol{A}\right)\right\}. \tag{I.34}$$

Here we observe again that $\boldsymbol{A}$ depends only on the coordinates; therefore, it commutes with x_1. We get

$$\frac{dx_1}{dt} = \frac{c}{i\hbar}\{\boldsymbol{\alpha}\cdot\boldsymbol{p}x_1 - x_1\boldsymbol{\alpha}\cdot\boldsymbol{p}\}$$
$$= \frac{c}{i\hbar}\{(\alpha_1 p_1 + \alpha_2 p_2 + \alpha_3 p_3)x_1$$
$$- x_1(\alpha_1 p_1 + \alpha_2 p_2 + \alpha_3 p_3)\}. \tag{I.35}$$

From Eq. (I.28), we see that x_1 commutes with p_2 and p_3. Thus, we get

$$\frac{dx_1}{dt} = \frac{c}{i\hbar}\{\alpha_1 p_1 x_1 - x_1\alpha_1 p_1\}$$
$$= \frac{c}{i\hbar}\{\alpha_1(p_1 x_1 - x_1 p_1)\}$$
$$= \frac{c}{i\hbar}\alpha_1(-i\hbar) = -c\alpha_1, \tag{I.36}$$

where we have used Eq. (I.29). Thus, our result is that for any coordinate x_k,

$$\frac{dx_k}{dt} = -c\alpha_k. \tag{I.37}$$

Next, we derive the equation of motion. In the classical case, we started with $d\dot{r}/dt$ and then we made the substitution

$$m\frac{d}{dt}\dot{r} = \frac{d\boldsymbol{P}}{dt}, \tag{I.38}$$

according to Eq. (I.8). Here we start again with the relationship given by Eq. (I.38), but now we turn to Eq. (I.30), which gives the derivative of an operator. Thus, we evaluate the expression

$$\frac{d\boldsymbol{P}}{dt} = \frac{\partial\boldsymbol{P}}{\partial t} + \frac{1}{i\hbar}[\boldsymbol{P}H - H\boldsymbol{P}]. \tag{I.39}$$

Using Eqs. (I.31) and (I.32), we obtain

$$\frac{dP_1}{dt} = \frac{\partial}{\partial t}\left[p_1 + \frac{e}{c}A_1\right]$$
$$+ \frac{1}{i\hbar}\left[-P_1(c\boldsymbol{\alpha}\cdot\boldsymbol{P} + \beta mc^2 + eV) + (c\boldsymbol{\alpha}\cdot\boldsymbol{P} + \beta mc^2 + eV)P_1\right]$$
$$= \frac{e}{c}\frac{\partial A_1}{\partial t} + \frac{1}{i\hbar}\left[(c\boldsymbol{\alpha}\cdot\boldsymbol{P})P_1 - P_1(c\boldsymbol{\alpha}\cdot\boldsymbol{P})\right.$$
$$\left. + eV\left(p_1 + \frac{e}{c}A_1\right) - \left(p_1 + \frac{e}{c}A_1\right)eV\right]. \tag{I.40}$$

Here we have taken into account that P_1 commutes with β and that $\partial p_1/\partial t = 0$.

Let us consider the first two terms in the square bracket. We have

$$(c\boldsymbol{\alpha} \cdot \boldsymbol{P})P_1 - P_1(c\boldsymbol{\alpha} \cdot \boldsymbol{P})$$

$$= c\{(\alpha_1 P_1 + \alpha_2 P_2 + \alpha_3 P_3)P_1$$

$$- P_1(\alpha_1 P_1 + \alpha_2 P_2 + \alpha_3 P_3)\}$$

$$= c\{\alpha_2(P_2 P_1 - P_1 P_2) + \alpha_3(P_3 P_1 - P_1 P_3)\}. \qquad (I.41)$$

We evaluate the first term:

$$P_2 P_1 - P_1 P_2 = \left(p_2 + \frac{e}{c}A_2\right)\left(p_1 + \frac{e}{c}A_1\right) - \left(p_1 + \frac{e}{c}A_1\right)\left(p_2 + \frac{e}{c}A_2\right)$$

$$= \frac{e}{c}(A_2 p_1 + p_2 A_1 - A_1 p_2 - p_1 A_2)$$

$$= \frac{e}{c}(A_2 p_1 - p_1 A_2) - \frac{e}{c}(A_1 p_2 - p_2 A_1). \qquad (I.42)$$

For these expressions, we use Eq. (I.29). We obtain

$$P_2 P_1 - P_1 P_2 = \frac{e}{c}i\hbar\frac{\partial A_2}{\partial x_1} - \frac{e}{c}i\hbar\frac{\partial A_1}{\partial x_2}$$

$$= \frac{i\hbar e}{c}\left(\frac{\partial A_2}{\partial x_1} - \frac{\partial A_1}{\partial x_2}\right). \qquad (I.43)$$

Using Eq. (I.16c), we obtain

$$P_2 P_1 - P_1 P_2 = \frac{i\hbar e}{c}(\text{curl } \boldsymbol{A})_3 = \frac{i\hbar e}{c}\mathscr{H}_3, \qquad (I.44)$$

where $\mathscr{H}$ is the magnetic field vector. Evaluating the second term of Eq. (I.41) similarly, we get

$$(c\boldsymbol{\alpha} \cdot \boldsymbol{P})P_1 - P_1(c\boldsymbol{\alpha} \cdot \boldsymbol{P}) = c\frac{i\hbar e}{c}[\alpha_2 \mathscr{H}_3 - \alpha_3 \mathscr{H}_2]. \qquad (I.45)$$

This result gives us the first two terms in the square brackets of Eq. (I.40). Let us consider now the third and fourth terms. The vector-potential component, A_1, commutes with the scalar potential because both depend on the coordinates only. Thus, we get

$$eV\left(p_1 + \frac{e}{c}A_1\right) - \left(p_1 + \frac{e}{c}A_1\right)eV$$
$$= e(Vp_1 - p_1V) = ei\hbar \frac{\partial V}{\partial x_1}, \tag{I.46}$$

where we have used again Eq. (I.29). Substituting the results contained in Eqs. (I.45) and (I.46) into Eq. (I.40), we obtain

$$\frac{dP_1}{dt} = \frac{e}{c}\frac{\partial A_1}{\partial t} + e\frac{\partial V}{\partial x_1} + e[\alpha_2 \mathscr{H}_3 - \alpha_3 \mathscr{H}_2]. \tag{I.47}$$

Putting this into vector form, we obtain

$$\frac{d\mathbf{P}}{dt} = e(\nabla V) + \frac{e}{c}\frac{\partial \mathbf{A}}{\partial t} + e[\boldsymbol{\alpha} \times \mathscr{H}], \tag{I.48}$$

and recalling the relationship given by Eq. (I.23), we get

$$\frac{d\mathbf{P}}{dt} = -e\mathscr{E} + e[\boldsymbol{\alpha} \times \mathscr{H}]. \tag{I.49}$$

This is the equation of motion in the Dirac formulation.

We summarize the results. For an electron that is moving in an electromagnetic field characterized by the scalar potential V and vector potential $\mathbf{A}$, we obtained from the classical theory, the relationships given by Eqs. (I.21) and (I.27):

$$\mathbf{v} = \frac{1}{m}\mathbf{P} = \frac{1}{m}\left(\mathbf{p} + \frac{e}{c}\mathbf{A}\right), \tag{I.50}$$

and

$$m\frac{d\mathbf{v}}{dt} = \frac{d\mathbf{P}}{dt} = -e\mathscr{E} - \frac{e}{c}[\mathbf{v} \times \mathscr{H}]. \tag{I.51}$$

For the same problem, we have obtained, from Dirac's relativistic formulation of the quantum mechanics,[†] Eqs. (I.37) and (I.49):

$$\mathbf{v} = \frac{d\mathbf{r}}{dt} = -c\boldsymbol{\alpha}, \tag{I.52}$$

[†]If the reader finds that these relationships look rather strange in view of the form of the $\boldsymbol{\alpha}$ operator, remember that all equations are in the matrix formulation of quantum mechanics, that is, an equation like Eq. (I.55) means that the matrix components of $\mathbf{v}$ are identical with the matrix components of $-c\boldsymbol{\alpha}$.

and

$$\frac{d\boldsymbol{P}}{dt} = -e\mathscr{E} + e[\boldsymbol{\alpha} \times \mathscr{H}]. \tag{I.53}$$

It is evident that the equation of motion, Eq. (I.53), becomes identical with the classical equation, Eq. (I.51), if we use the relationship given by Eq. (I.52). Substituting Eq. (I.52) into (I.53), we obtain

$$\frac{d\boldsymbol{P}}{dt} = -e\mathscr{E} - \frac{e}{c}[\boldsymbol{v} \times \mathscr{H}], \tag{I.54}$$

which is identical with the classical equation, Eq. (I.51). Thus, the correspondence principle tells us that the transition from classical theory into the relativistic quantum mechanical formulation is accomplished by the substitution

$$\boldsymbol{v} \rightarrow -c\boldsymbol{\alpha}, \tag{I.55}$$

where $\boldsymbol{\alpha}$ is the Dirac matrix-vector.

After the derivation of this correspondence relationship, it is easy to set up the many-electron Hamiltonian. The classical Hamiltonian function is given by Darwin's expression, Eq. (F.48). The last two terms of that expression are the electron–electron interaction terms. As we have shown in App. F, those two terms can be written as the sum of two distinctly defined functions, the first of which contains the electrostatic and magnetic interactions and the second represents the retardation effects. According to Eq. (F.52), the electrostatic and magnetic interaction is given by

$$g_0(1,2) = \frac{e^2}{r_{12}}\left[1 - \frac{1}{c^2}\boldsymbol{v}_1 \cdot \boldsymbol{v}_2\right], \tag{I.56}$$

and according to Eq. (F.53), the retardation effect is

$$g_1(1,2) = \frac{e^2}{2c^2 r_{12}}\left[(\boldsymbol{v}_1 \cdot \boldsymbol{v}_2) - \frac{(\boldsymbol{v}_1 \cdot \boldsymbol{r}_{21})(\boldsymbol{v}_2 \cdot \boldsymbol{r}_{21})}{r_{12}^2}\right]. \tag{I.57}$$

The last two terms of Eq. (F.48) consist of the sum of these two added for all electron pairs.

Now we obtain the electron–electron interaction in Breit's relativistic Hamiltonian by making the substitution of Eq. (I.55) in the previous expressions. The result is

$$g_0(1,2) = \frac{e^2}{r_{12}}[1 - \boldsymbol{\alpha}(1) \cdot \boldsymbol{\alpha}(2)], \tag{I.58}$$

and

$$g_1(1,2) = \frac{e^2}{2r_{12}} \left[\boldsymbol{\alpha}(1) \cdot \boldsymbol{\alpha}(2) - \frac{(\boldsymbol{\alpha}(1) \cdot \boldsymbol{r}_{21})(\boldsymbol{\alpha}(2) \cdot \boldsymbol{r}_{21})}{r_{12}^2} \right]. \quad (I.59)$$

In these expressions, the Dirac matrix-vector $\boldsymbol{\alpha}(i)$ is operating on the wave functions depending on the coordinates of the ith electron.

Having obtained the interaction operator, it is easy to set up the many-electron Hamiltonian. We replace the one-electron terms in Darwin's classical Hamiltonian, Eq. (F.48), by the Dirac Hamiltonian, Eq. (G.4). Let $H_D(k)$ be the Dirac Hamiltonian operating on the coordinates of the kth electron. Then the many-electron relativistic Hamiltonian becomes

$$H = \sum_{k=1}^{N} H_D(k) + \sum_{i<k} \left[g_0(i,k) + g_1(i,k) \right], \quad (I.60)$$

where g_0 and g_1 are the operators given by Eqs. (I.58) and (I.59).

Breit presented further arguments for the validity of Eq. (I.60). He has shown for the case of a two-electron system that using this Hamiltonian one obtains the relationships

$$\frac{dr_1}{dt} = -c\boldsymbol{\alpha}(1), \qquad \frac{dr_2}{dt} = -c\boldsymbol{\alpha}(2), \quad (I.61)$$

where r_1 and r_2 are the positions of the two electrons. These relationships are the generalization of Eq. (I.37). Next, Breit derived the equation of motion for one of the electrons in the two-electron system. He performed this calculation using the classical Hamiltonian of Darwin, Eq. (F.48), and also using the quantum-mechanical Hamiltonian, Eq. (I.60). He has shown that the two equations of motions become identical if, in the classical equation, the substitution implied by Eq. (I.61) is made. Thus, he was able to show that, according to the correspondence principle, the classical analog of the operator given by Eq. (I.60) is the Hamiltonian function derived by Darwin, which is given by Eq. (F.48).

APPENDIX J

THE FORMULA FOR THE EINSTEIN TRANSITION PROBABILITIES

As we have formulated in Sec. 6.5, our goal is to derive an approximate formula for the Einstein transition coefficient $B_{n \to m}$. We start with the time-dependent Schroedinger equation:

$$H\Phi = i\hbar \frac{\partial \Phi}{\partial t}, \tag{J.1}$$

where

$$\Phi = \Phi(q, t). \tag{J.2}$$

The letter q stands for all spatial and spin coordinates of the electrons. We write H in the form given by Eq. (6.131), that is, we put

$$H = H_0 + H', \tag{J.3}$$

where, according to (6.137), we have

$$H' = \sum_{j=1}^{N} \frac{e}{mc} (A_j \cdot p_j) \tag{J.4}$$

The unperturbed Hamiltonian, H_0, is a time-independent operator; therefore, we have the equations

$$H_0 \Phi_n^0 = i\hbar \frac{\partial \Phi_n^0}{\partial t}, \tag{J.5}$$

where Φ_n^0 has the form

$$\Phi_n^0(q, t) = \psi_n^0(q) e^{-i\omega_n t}, \tag{J.6}$$

with

$$\omega_n = E_n/\hbar. \tag{J.7}$$

We expand the solution of Eq. (J.1) in terms of the complete set of unperturbed solutions:

$$\Phi(q,t) = \sum_{n=1}^{\infty} c_n(t)\Phi_n^0(q,t). \tag{J.8}$$

Substitution into Eq. (J.1) gives

$$H\Phi(q,t) = \sum_n c_n(t)(H_0 + H')\Phi_n^0(q,t)$$

$$= \sum_n c_n(t)H_0\Phi_n^0(q,t) + \sum_n c_n(t)H'\Phi_n^0(q,t)$$

$$= i\hbar\frac{\partial}{\partial t}\sum_n c_n(t)\Phi_n^0(q,t)$$

$$= i\hbar\sum_n \frac{\partial c_n(t)}{\partial t}\Phi_n^0(q,t) + i\hbar\sum_n c_n(t)\frac{\partial\Phi_n^0(q,t)}{\partial t}. \tag{J.9}$$

In view of Eq. (J.5), we get

$$\sum_n c_n(t)H'\Phi_n^0(q,t) = i\hbar\sum_n \frac{\partial c_n(t)}{\partial t}\Phi_n^0(q,t). \tag{J.10}$$

Multiply from the left by Φ_m^{0*} and integrate over q. Then we get

$$\sum_n c_n(t)\int \Phi_m^{0*}H'\Phi_n^0\,dq = i\hbar\sum_n \frac{\partial c_n}{\partial t}\delta_{mn} = i\hbar\frac{\partial c_m}{\partial t}, \tag{J.11}$$

where we have used the orthogonality of the solutions of Eq. (J.5). Thus, we get

$$\frac{dc_m}{dt} = \frac{1}{i\hbar}\sum_n c_n(t)\langle\Phi_m^0|H'|\Phi_n^0\rangle. \tag{J.12}$$

This equation is still exactly equivalent to Eq. (J.1).

Now, let us consider a plane-polarized light wave for which the vector potential has the form

$$A_x = A_x^0 \cos 2\pi\nu\left(t - \frac{z}{c}\right),$$

$$A_y = 0,$$

$$A_z = 0. \tag{J.13}$$

This represents a wave moving in the z direction with velocity c. We will assume that the wavelength of this light is in the visible region; thus, it will be about 1000 times larger than the size of atoms. Thus, we will assume that $A =$ constant over the atom. Then, we get for the matrix component needed for Eq. (J.12)

$$\langle \Phi_m^0 | H' | \Phi_n^0 \rangle = \langle \Phi_m^0 | \sum_{j=1}^{N} \frac{e}{mc} (A \cdot p)_j | \Phi_n^0 \rangle$$

$$= \sum_{j=1}^{N} \frac{e}{mc} A \cdot \langle \Phi_m^0 | p_j | \Phi_n^0 \rangle. \tag{J.14}$$

In this equation, p_j is the time-independent momentum operator. Using Eq. (J.6), we get

$$\langle \Phi_m^0 | p_j | \Phi_n^0 \rangle = \langle \psi_m^0 | p_j | \psi_n^0 \rangle e^{i(\omega_m - \omega_n)t}. \tag{J.15}$$

Writing p_j into operator form, we get

$$\langle \Phi_m^0 | p_j | \Phi_n^0 \rangle = -i\hbar \langle \psi_m^0 | \nabla_j | \psi_n^0 \rangle e^{i(\omega_m - \omega_n)t}. \tag{J.16}$$

We want to transform the matrix component of ∇_j into a matrix component of r_j. We demonstrate with a one-dimensional case how this can be done. Let the one-dimensional wave functions ψ_m^{0*} and ψ_n^0 be solutions of the Schroedinger equations:

$$\frac{d^2\psi_m^{0*}}{dx^2} + \frac{2m}{\hbar^2}[E_m - V(x)]\psi_m^{0*} = 0,$$

$$\frac{d^2\psi_n^0}{dx^2} + \frac{2m}{\hbar^2}[E_n - V(x)]\psi_n^0 = 0. \tag{J.17}$$

Multiply the first equation from the left by $\psi_n^0 x$ and the second by $\psi_m^{0*} x$, integrate over x, and subtract the second equation from the first. Then, we get

$$\int \left(\psi_n^0 x \frac{d^2 \psi_m^{0*}}{dx^2} - \psi_m^{0*} x \frac{d^2 \psi_n^0}{dx^2} \right) dx$$

$$= \frac{2m}{\hbar^2} (E_m - E_n) \int \psi_m^{0*} x \psi_n^0 \, dx. \tag{J.18}$$

We integrate the left side in parts:

$$\int \left(\psi_n^0 x \frac{d^2 \psi_m^{0*}}{dx^2} - \psi_m^{0*} x \frac{d^2 \psi_n^0}{dx^2} \right) dx$$

$$= (\psi_n^0 x) \frac{d\psi_m^{0*}}{dx} \bigg|_{-\infty}^{+\infty} - \int \frac{d}{dx} (\psi_n^0 x) \frac{d\psi_m^{0*}}{dx} \, dx$$

$$- (\psi_m^{0*} x) \frac{d\psi_n^0}{dx} \bigg|_{-\infty}^{+\infty} + \int \frac{d}{dx} (\psi_m^{0*} x) \frac{d\psi_n^0}{dx} \, dx. \tag{J.19}$$

The integrated terms will be zero because the eigenfunctions go to zero at infinity stronger than any power of x. Thus, we get

$$\int \left(\psi_n^0 x \frac{d^2 \psi_m^{0*}}{dx^2} - \psi_m^{0*} x \frac{d^2 \psi_n^0}{dx^2} \right) dx$$

$$= \int \frac{d}{dx} (\psi_m^{0*} x) \frac{d\psi_n^0}{dx} \, dx - \int \frac{d}{dx} (\psi_n^0 x) \frac{d\psi_m^{0*}}{dx} \, dx$$

$$= \int \psi_m^{0*} \frac{d\psi_n^0}{dx} \, dx - \int \psi_n^0 \frac{d\psi_m^{0*}}{dx} \, dx. \tag{J.20}$$

Integrating the second term on the right side again in parts, we get

$$\int \left(\psi_n^0 x \frac{d^2 \psi_m^{0*}}{dx^2} - \psi_m^{0*} x \frac{d^2 \psi_n^0}{dx^2} \right) dx = 2 \int \psi_m^{0*} \frac{d\psi_n^0}{dx} \, dx. \tag{J.21}$$

Substituting this into Eq. (J.20), we get

$$2 \int \psi_m^{0*} \frac{d\psi_n^0}{dx} \, dx = \frac{2m}{\hbar^2} (E_m - E_n) \int \psi_m^{0*} x \psi_n^0 \, dx. \tag{J.22}$$

From this, we obtain

$$\langle \psi_m^0 | \frac{d}{dx} | \psi_n^0 \rangle = \frac{m}{\hbar} (E_m - E_n) \langle \psi_m^0 | x | \psi_n^0 \rangle. \tag{J.23}$$

Generalizing this result to three dimensions and to N electrons, we get

$$\langle \psi_m^0 | \nabla_j | \psi_n^0 \rangle = \frac{m}{\hbar^2} (E_m - E_n) \langle \psi_m^0 | r_j | \psi_n^0 \rangle. \tag{J.24}$$

Using this relationship in Eq. (J.16), we obtain

$$\langle \Phi_m^0 | \boldsymbol{p}_j | \Phi_n^0 \rangle = -i\hbar \langle \psi_m^0 | \nabla_j | \psi_n^0 \rangle e^{i(\omega_m - \omega_n)t}$$

$$= -\frac{i}{\hbar} m (E_m - E_n) \langle \psi_m^0 | r_j | \psi_n^0 \rangle e^{i(\omega_m - \omega_n)t}, \tag{J.25}$$

and putting this into Eq. (J.14), the matrix component of H' becomes

$$\langle \Phi_m^0 | H' | \Phi_n^0 \rangle = \sum_{j=1}^{N} \frac{e}{mc} A \cdot \left\{ -\frac{im}{\hbar} (E_m - E_n) \langle \psi_m^0 | r_j | \psi_n^0 \rangle e^{i(\omega_m - \omega_n)t} \right\}$$

$$= \frac{i}{c\hbar} (E_m - E_n) A \cdot \boldsymbol{R}_{mn} e^{i(\omega_m - \omega_n)t}, \tag{J.26}$$

where

$$\boldsymbol{R}_{mn} = \langle \psi_m^0 | \sum_{j=1}^{N} (-er_j) | \psi_n^0 \rangle. \tag{J.27}$$

In our case A_y and $A_z = 0$; therefore,

$$\langle \Phi_m^0 | H' | \Phi_n^0 \rangle = \frac{i}{c\hbar} (E_m - E_n) A_x X_{mn} e^{i(\omega_m - \omega_n)t}, \tag{J.28}$$

where

$$X_{mn} = \langle \psi_m^0 | \sum_{j=1}^{N} (-ex_j) | \psi_n^0 \rangle. \tag{J.29}$$

We now turn back to Eq. (J.12) and solve that equation approximately by iteration. Let us assume that when the light is turned on, that is, at $t = 0$, the atom is in state ψ_n, which means that at this time, $c_n = 1$ and all other c's are zero. What we do is to put all c's equal to zero on the right side of Eq. (J.12) with the exception of c_n for which we put $c_n = 1$. Then, we get

$$
\begin{aligned}
\frac{dc_m}{dt} &= \frac{1}{i\hbar} \langle \Phi_m^0 | H' | \Phi_n^0 \rangle \\
&= \frac{1}{i\hbar} \frac{i}{c\hbar} (E_m - E_n) A_x X_{mn} e^{i(\omega_m - \omega_n)t}.
\end{aligned}
\tag{J.30}
$$

By putting A_x into complex exponential form, we get

$$
A_x = \frac{A_x^0}{2} (e^{2\pi i \nu t} + e^{-2\pi i \nu t}),
\tag{J.31}
$$

and with this, we obtain

$$
\begin{aligned}
\frac{dc_m}{dt} &= \frac{1}{c\hbar^2} (E_m - E_n) A_x^0 X_{mn} \\
&\quad \times \left\{ e^{i \frac{E_m - E_n + h\nu}{\hbar} t} + e^{i \frac{E_m - E_n - h\nu}{\hbar} t} \right\}.
\end{aligned}
\tag{J.32}
$$

We integrate and choose the constant of integration in such a way that $c_m = 0$ when $t = 0$. Then

$$
\begin{aligned}
c_m &= -\frac{i}{2c\hbar} A_x^0 X_{mn} (E_m - E_n) \\
&\quad \times \left\{ \frac{e^{i \left(\frac{E_m - E_n + h\nu}{\hbar} t \right)} - 1}{E_m - E_n + h\nu} + \frac{e^{i \left(\frac{E_m - E_n - h\nu}{\hbar} t \right)} - 1}{E_m - E_n - h\nu} \right\}.
\end{aligned}
\tag{J.33}
$$

We consider the case when $E_n < E_m$. We obtain a large c_m when

$$
E_m - E_n \approx h\nu,
\tag{J.34}
$$

that is, when the second term in braces, is large. Neglecting the first term in braces, we obtain for the probability of transition

$$
\begin{aligned}
c_m^* c_m &\approx \frac{t^2}{4c^2 \hbar^4} |A_x^0|^2 |X_{mn}|^2 (E_m - E_n)^2 \\
&\quad \times \frac{\sin^2 \left\{ \frac{E_m - E_n - h\nu}{2\hbar} t \right\}}{\left\{ \frac{E_m - E_n - h\nu}{2\hbar} t \right\}^2}.
\end{aligned}
\tag{J.35}
$$

We have written the vector potential in Eq. (J.13) so that it contains only one frequency, ν. If we have light, which is a mixture of all frequencies, we have to integrate our result for all ν. But we have seen that our result is large only for $h\nu = E_m - E_n$. We take this into account in such a way that we integrate our result for all ν, but we put

$$A_x^0 \approx A_x^0(\nu_{mn}), \tag{J.36}$$

that is, we assume that only the amplitude belonging to ν_{mn} is significant. In this way, we obtain

$$|c_m|^2 = \frac{\pi^2 \nu_{mn}^2}{c^2 \hbar^2} |A_x^0(\nu_{mn})|^2 |X_{mn}|^2 t, \tag{J.37}$$

where $\nu_{mn} = (E_m - E_n)/h$. Now, if we have, instead of the vector potential given by Eq. (J.13), a vector potential that has all three components, then our result will become

$$|c_m|^2 = \frac{\pi^2 \nu_{mn}^2}{c^2 \hbar^2} \left\{ |A_x^0|^2 |X_{mn}|^2 + |A_y^0|^2 |Y_{mn}|^2 + |A_z^0|^2 |Z_{mn}|^2 \right\} t, \tag{J.38}$$

where all three A's are functions of ν_{mn} only. If the radiation is isotropic, we get

$$|A_x^0|^2 = |A_y^0|^2 = |A_z^0|^2 = \tfrac{1}{3}|A^0|^2. \tag{J.39}$$

According to classical electromagnetic theory, we can write for the energy density of the electromagnetic field

$$\rho(\nu_{mn}) = \frac{\overline{\mathscr{E}^2(\nu_{mn})}}{4\pi}, \tag{J.40}$$

where $\mathscr{E}$ is the electric field associated with ν_{mn} and $\overline{\mathscr{E}^2}$ means the time average. According to Eq. (4.1), we have

$$\mathscr{E} = -\frac{1}{c}\frac{\partial A}{\partial t}, \tag{J.41}$$

and so

$$\mathscr{E}(\nu_{mn}) = -\frac{1}{c}\frac{\partial}{\partial t}A^0(\nu_{mn})\cos(2\pi\nu t)$$

$$= \frac{2\pi\nu_{mn}}{c}A^0(\nu_{mn})\sin(2\pi\nu t). \tag{J.42}$$

For the time average, we get

$$\overline{\mathscr{E}^2} = \frac{4\pi^2 \nu_{mn}^2}{c^2} |A^0(\nu_{mn})|^2 \overline{\sin(2\pi\nu t)}$$

$$= \frac{4\pi^2 \nu_{mn}^2}{c^2} |A^0(\nu_{mn})|^2 \tfrac{1}{2}. \tag{J.43}$$

Putting this result into Eq. (J.40), we get

$$\rho(\nu_{mn}) = \frac{\overline{\mathscr{E}^2}}{4\pi} = \frac{4\pi^2 \nu_{mn}^2}{4\pi c^2} |A^0(\nu_{mn})|^2 \tfrac{1}{2}$$

$$= \frac{\pi \nu_{mn}^2}{2c^2} |A^0(\nu_{mn})|^2. \tag{J.44}$$

Using Eqs. (J.39) and (J.44), we obtain for the probability of transition

$$|c_m|^2 = \frac{\pi^2 \nu_{mn}^2}{3c^2\hbar^2} |A^0|^2 \{|X_{mn}|^2 + |Y_{mn}|^2 + |Z_{mn}|^2\} t$$

$$= \frac{\pi^2 \nu_{mn}^2}{c^2\hbar^2} \frac{2c^2\rho(\nu_{mn})}{3\pi\nu_{mn}^2} \{|X_{mn}|^2 + \cdots\} t$$

$$= \frac{2\pi}{3\hbar^2} \rho(\nu_{mn}) \{|X_{mn}|^2 + \cdots\} t. \tag{J.45}$$

The probability of $n \to m$ transition per unit time becomes

$$|c_m|^2 = \frac{2\pi}{3\hbar^2} \rho(\nu_{mn}) |R_{mn}|^2, \tag{J.46}$$

where

$$|R_{mn}|^2 = |X_{mn}|^2 + |Y_{mn}|^2 + |Z_{mn}|^2. \tag{J.47}$$

Equating this with the expression in Eq. (6.125), we get

$$B_{n \to m} \rho(\nu_{mn}) = \frac{2\pi}{3\hbar^2} |R_{mn}|^2 \rho(\nu_{mn}), \tag{J.48}$$

from which we see that

$$B_{n \to m} = \frac{2\pi}{3\hbar^2} |R_{mn}|^2, \tag{J.49}$$

where $|R_{mn}|^2$ is given by Eq. (J.47). These are the formulas quoted in Eqs. (6.138) and (6.139).

APPENDIX K

NON-HERMITIAN OPERATORS IN ATOMIC STRUCTURE THEORY

Throughout the discussions of this book, we have encountered situations in which the eigenfunctions under discussion were eigenfunctions of non-Hermitian operators. The non-Hermiticity is arising, in every case, from the presence of Lagrangian multipliers or from the presence of pseudopotentials that are replacing the Lagrangian multipliers. In fact, the only cases in which this is *not* happening are the case of single-determinantal HF approximation and the HF equations for the core orbitals of an atom with complete groups in the core and an arbitrary valence-electron configuration. In the single-determinantal HF approximation, the Lagrangian multipliers can be eliminated from the HF equations with a unitary transformation. In the case of an atom with complete groups in its core, the Lagrangian multipliers can again be eliminated by a unitary transformation from the HF equations of the core orbitals (but not from the equations of the valence orbitals). In all other cases, including the conventional effective Hamiltonians for n valence electrons, we have Lagrangian multipliers in the equations as a result of which the final equations will contain non-Hermitian operators.

The purpose of this appendix is to summarize the main properties of these operators and to show that they do not represent a break in the formalism of quantum mechanics. In particular, we show that the eigenvalues of these operators are always real, which means that there is no violation of basic quantum mechanical principles.

First, let us consider the HF equation for one of the valence electrons of an atom with n valence electrons. Special cases of this situation were given in Eq. (3.47) for an l^q configuration and in Eq. (6.75) for an $l^q l'$ configuration. The HF equations for a general valence configuration were written in Eq.

(8.72). It is easy to see that both Eqs. (3.47) and (6.75) are special cases of Eq. (8.72). Our discussion will be generally valid if we consider Eq. (8.72).

Let $\varphi_1 \cdots \varphi_N$ be the core orbitals and φ_v $(v = 1, 2, \ldots, n)$ be the valence orbitals. The HF equation for φ_v is given in the form

$$H_v \varphi_v = \varepsilon_v \varphi_v + \sum_{\lambda=1}^{N} \lambda_{iv} \varphi_i, \tag{K.1}$$

where H_v is the Hamiltonian:

$$H_v = t + g + U_C + U_v, \tag{K.2}$$

where U_C is the HF potential of the core, and U_v the potential of the other valence electrons. The subscript on H_v indicates that we may have different operators for different valence orbitals. Introducing the projection operators Ω and Π with the usual definitions

$$\Omega = \sum_{j=1}^{N} |\varphi_j\rangle\langle\varphi_j|, \tag{K.3}$$

and

$$\Pi = 1 - \Omega, \tag{K.4}$$

we obtain for Eq. (K.1)

$$\hat{H}_v \varphi_v = \varepsilon_v \varphi_v, \tag{K.5}$$

where

$$\hat{H}_v = \Pi H_v = H_v - \Omega H_v. \tag{K.6}$$

$\hat{H}_v$ is not Hermitian because, in general, H_v and Π do not commute. Thus, the presence of Lagrangian multipliers changes the character of the HF equations into eigenvalue equations with non-Hermitian operators. A special case of Eq. (K.5) was given in Eq. (3.58).

It is interesting to note that we could have arrived at Eq. (K.5) without introducing the multipliers. Let us consider the Schroedinger equation

$$H_v \varphi_v' = \varepsilon_v' \varphi_v', \tag{K.7}$$

where the eigenfunctions are *not* orthogonal to the core orbitals. If we want to change the Hamiltonian H_v into another Hamiltonian with orthogonalized eigenfunctions, we have to make the substitution

$$H_v \to \Pi H_v = \hat{H}_v. \tag{K.8}$$

This is easy to verify. Consider

$$\hat{H}_v \varphi_v = \varepsilon_v \varphi_v. \tag{K.9}$$

Multiply from the left by φ_k^* and integrate (φ_k is any core orbital). We get

$$\langle \varphi_k | \hat{H}_v | \varphi_v \rangle = \varepsilon_v \langle \varphi_k | \varphi_v \rangle. \tag{K.10}$$

Because Π is Hermitian, we obtain

$$\begin{aligned}
\langle \varphi_k | \hat{H}_v | \varphi_v \rangle &= \langle \varphi_k | \Pi H_v | \varphi_v \rangle \\
&= \langle \Pi \varphi_k | H_v | \varphi_v \rangle \\
&= \langle (1 - \Omega) \varphi_k | H_v | \varphi_v \rangle = 0, \tag{K.11}
\end{aligned}$$

because

$$\Pi \varphi_k = (1 - \Omega) \varphi_k = \varphi_k - \varphi_k = 0. \tag{K.12}$$

Inserting Eq. (K.11) into Eq. (K.10), we get

$$\langle \varphi_k | \varphi_v \rangle = 0 \qquad (k = 1, 2, \ldots, N). \tag{K.13}$$

Thus, the eigenfunctions of Eq. (K.9) are indeed orthogonal to the core orbitals, Q.E.D.

A by-product of this consideration is that we realize clearly the origin of pseudopotentials. Instead of making the substitution of Eq. (K.8), we can orthogonalize the valence orbitals by making the substitution

$$H_v \to H_v + V_P, \tag{K.14}$$

where V_P is the pseudopotential. The eigenfunctions of this operator, ψ_v, will not be orthogonal to the core orbitals, but $(1 - \Omega)\psi_v$ will be orthogonal and *identical with* φ_v. Thus, the pseudopotential in the substitution given by Eq. (K.14) accomplishes *exactly* the same as the projection operator Π in Eq. (K.8). *Thus, the pseudopotential is not an ad hoc device, but a uniquely defined alternative to the Lagrangian multipliers.*

It is easy to show that the eigenvalues of Eq. (K.9) are *always* real. Let us write the conjugate complex equation

$$\hat{H}_v^* \varphi_v^* = \varepsilon_v^* \varphi_v^*. \tag{K.15}$$

Multiply Eq. (K.9) by φ_v^*, Eq. (K.15) by φ_v, integrate, and subtract Eq. (K.15) from Eq. (K.9). We get

$$\langle \varphi_v | \hat{H}_v | \varphi_v \rangle - \langle \varphi_v | \hat{H}_v | \varphi_v \rangle^*$$
$$= (\varepsilon_v - \varepsilon_v^*) \langle \varphi_v | \varphi_v \rangle = \varepsilon_v - \varepsilon_v^*. \tag{K.16}$$

Because Π is Hermitian, we obtain

$$\langle \varphi_v | \hat{H}_v | \varphi_v \rangle - \langle \varphi_v | \hat{H}_v | \varphi_v \rangle^*$$
$$= \langle \Pi \varphi_v | H_v | \varphi_v \rangle - \langle \Pi \varphi_v | H_v | \varphi_v \rangle^*$$
$$= \langle \varphi_v | H_v | \varphi_v \rangle - \langle \varphi_v | H_v | \varphi_v \rangle^* = 0, \tag{K.17}$$

because H_v is Hermitian. Here we used the relationship

$$\Pi \varphi_v = (1 - \Omega) \varphi_v = \varphi_v, \tag{K.18}$$

which is valid for any function that is orthogonal to the core orbitals and the solutions of Eq. (K.9) certainly have that property. Inserting Eq. (K.17) into Eq. (K.16), we obtain

$$\varepsilon_v = \varepsilon_v^*, \tag{K.19}$$

Q.E.D.

It is easy to see that the physically important properties of Eq. (K.9) are essentially the same as the physically important properties of an equation with a Hermitian operator.

Next, we consider the HF equations for the average of a configuration. That equation is given in Eq. (3.117) and the non-Hermitian character of the Hamiltonian is demonstrated in Eq. (3.118). It is clear that the discussion presented for Eq. (K.9) is equally valid for Eq. (3.118), so we are dealing with the same kind of equation.

Next, we consider the conventional effective Hamiltonians for n valence electrons given by Eqs. (8.60) and (8.67). Both Hamiltonians have the form

$$H_{\text{eff}} = \Pi H, \tag{K.20}$$

where H is a Hermitian n-electron operator, and Π is an n-electron projection operator defined as

$$\Pi = \Pi_1 \Pi_2 \cdots \Pi_n, \tag{K.21}$$

where

$$\Pi_k = 1 - \Omega_k, \tag{K.22}$$

and

$$\Omega_k = \sum_{j=1}^{N} \langle k|\varphi_j\rangle\langle\varphi_j|. \tag{K.23}$$

As we see, H_{eff} has the same form as $\hat{H}_v$, that is, both have the form = (projection operator) × (Hermitian Hamiltonian). We have seen in the text, Sec. 8.2, that the n-electron eigenfunctions of H_{eff} will be strong-orthogonal to the core orbitals. Thus, the rule for orthogonalization is the same in the many-electron case as it is in the one-electron case.

The main property of operators of the type of Eq. (K.20) is that they are *partially Hermitian* [Szasz, 129]. By this, we mean the property that these operators are Hermitian with respect to n-electron functions that are strong-orthogonal to the core orbitals, that is, satisfy the relationship

$$f(1,2,\ldots,n) = \Pi f(1,2,\ldots,n). \tag{K.24}$$

Let us prove this statement. Let f and g be n-electron functions satisfying Eq. (K.24). Consider

$$\langle f| H_{\text{eff}}|g\rangle = \langle f|\Pi H|g\rangle = \langle \Pi f|H|g\rangle$$
$$= \langle f|H|g\rangle = \langle g|H|f\rangle^*. \tag{K.25}$$

Here we used the Hermiticity of Π and H. Using Eq. (K.24) again, we get

$$\langle g|H|f\rangle = \langle \Pi g|H|f\rangle = \langle g|H_{\text{eff}}|f\rangle. \tag{K.26}$$

Inserting Eq. (K.26) into Eq. (K.25), we obtain

$$\langle f|H_{\text{eff}}|g\rangle = \langle g|H_{\text{eff}}|f\rangle^*, \tag{K.27}$$

which is the statement we wanted to prove.

From the statement of partial Hermiticity, it follows that the effective Schroedinger equation that contains H_{eff} will have all the properties expected from a regular Schroedinger equation. In order to see this, it is enough to recall that only strong-orthogonal wave functions are meaningful in connection with the operator H_{eff}. We have seen this in both the FVP and BD theories in Secs. 8.2 and 8.3. All eigenfunctions of H_{eff} will be strong-orthogonal; even in a variational procedure, only strong-orthogonal wave functions can be admitted as trial functions. Thus, H_{eff} will be Hermitian with respect to all wave functions that are physically meaningful for this operator.

This being the case, we do not need to prove, for example, that the eigenvalues of H_{eff} are real. That follows from the partial Hermiticity because from that it follows that H_{eff} is Hermitian with respect to its eigenfunctions and from that it follows that the eigenvalues will be real. Likewise, the orthogonality of eigenfunctions belonging to different eigenvalues follows from the partial Hermiticity. The approximate energies computed with strong-orthogonal trial functions will be real and upper limits to the exact eigenvalues.

This short discussion demonstrates that the special kind of non-Hermiticity that is brought about by the presence of Lagrangian multipliers, that is, by the Pauli exclusion principle, does not disturb the basic postulates of quantum mechanics. For equations with pseudopotentials, the situation is not so straightforward because, for example, the effective pseudopotential Hamiltonian in Eq. (8.101) is not Hermitian and not partially Hermitian. It is easy to see, however, that the eigenvalues are always real in this case also [Szasz, 291]. If the exact Hamiltonian of Eq. (8.101) is replaced by the approximate operators in Eq. (8.120) or (8.123) and if the modified potentials V_M are replaced by local or semilocal potentials, then, of course, the operators will be Hermitian. This is the form that is used in most calculations.

APPENDIX L

THE ELIMINATION OF THE PROJECTION OPERATORS FROM THE EFFECTIVE PSEUDOPOTENTIAL HAMILTONIAN

In this appendix, we show the approximate validity of the relationship given by Eqs. (8.109) and (8.114). Our starting point is Eq. (8.100), where H_{eff} is given by Eq. (8.108). We have

$$H_{\text{eff}}\Psi = E\Psi, \tag{L.1}$$

where

$$H_{\text{eff}} = \sum_{i=1}^{n} \left(H_i + V_P(i) \right) + Q - S\Omega. \tag{L.2}$$

The operator S is given by Eq. (8.104):

$$S = Q - W, \tag{L.3}$$

where Q is the electrostatic interaction of the valence electrons:

$$Q = \frac{1}{2} \sum_{i,j=1}^{n} \frac{1}{r_{ij}}, \tag{L.4}$$

and W is the operator

$$W = \sum_{j=1}^{n} U_v(j), \tag{L.5}$$

where U_v is the HF potential of the valence electrons in the Hamiltonian given by Eq. (8.71). For the purposes of this derivation, we approximate U_v by the HF potential of a complete group, which is given by

$$U_v = \sum_{j=1}^{n} U_j, \tag{L.6}$$

where U_j is the HF potential, electrostatic and exchange, generated by the valence orbital φ_j. Thus, we obtain for W

$$W = \sum_{i=1}^{n} \sum_{j=1}^{n} U_i(j). \tag{L.7}$$

The many-electron part of the Pauli exclusion principle is represented by the operator Ω in the last term of Eq. (L.2). As the first task, we estimate the ratio

$$\eta = \langle S\Omega \rangle / \langle S \rangle, \tag{L.8}$$

which represents the ratio between the expectation values of the core part of S and the whole S.

In order to handle S efficiently, let us write it in the form

$$S = \sum_{\text{(all pairs, } i \neq j)} \left(1/r_{ij} - U_i(j) - U_j(i) \right) - \sum_{i=1}^{n} U_i(i). \tag{L.9}$$

Let

$$A_{ij} \equiv \frac{1}{r_{ij}} - U_i(j) - U_j(i). \tag{L.10}$$

First, we consider the matrix components of A_{ij}. Let

$$I_{ij} \equiv \langle A_{ij} \rangle = \int \Psi^* A_{ij} \Psi \, dq, \tag{L.11}$$

and

$$K_{ij} \equiv \langle A_{ij}\Omega \rangle = \int \Psi^* A_{ij}\Omega\Psi \, dq, \qquad (L.12)$$

where Ψ is the eigenfunction of Eq. (L.1). Let us approximate Ψ by the product

$$\Psi = \psi_1(1)\psi_2(2) \cdots \psi_n(n), \qquad (L.13)$$

where $\psi_1 \cdots \psi_n$ are normalized pseudoorbitals for the valence electrons. These orbitals are the solutions of Eq. (8.74), that is, they satisfy

$$(H_v + V_P)\psi_k = \varepsilon_k \psi_k \qquad (k = 1, 2, \ldots, n), \qquad (L.14)$$

It is a plausible choice to approximate the exact wave function by a product formed from those one-electron pseudo-orbitals that are the exact pseudoorbitals for the valence electrons in the HF approximation. It will be shown in what follows that the results of the derivation are not appreciably affected by the omission of the exchange and correlation effects from the trial function, Eq. (L.13). Using Eq. (L.13), we obtain

$$I_{ij} = \int \psi_1^*(1) \cdots \psi_n^*(n)$$

$$\times \left[\frac{1}{r_{ij}} - U_i(j) - U_j(i) \right] \psi_1(1) \cdots \psi_n(n) \, dq$$

$$= \int \psi_i^*(1)\psi_j^*(2) \frac{1}{r_{12}} \psi_i(1)\psi_j(2) \, dq$$

$$- \int \psi_j^*(1) U_i(1)\psi_j(1) \, dq - \int \psi_i^*(1) U_j(1)\psi_i(1) \, dq. \qquad (L.15)$$

In order to calculate K_{ij}, we have to consider first $\Omega\Psi$. We get

$$\Omega\Psi = \Psi - \Pi\Psi = \Psi - (\Pi_1\Pi_2 \cdots \Pi_n)\Psi, \qquad (L.16)$$

and by putting this expression into Eq. (L.12), we obtain

$$K_{ij} = \int \Psi^* A_{ij}\Psi \, dq - \int \Psi^* A_{ij}(\Pi_1\Pi_2 \cdots \Pi_n)\Psi \, dq$$

$$= I_{ij} - \int \psi_1^*(1) \cdots \psi_n^*(n) A_{ij}\hat{\psi}_1(1) \cdots \hat{\psi}_n(n) \, dq, \qquad (L.17)$$

where

$$\hat{\psi}_k = \Pi_k \psi_k. \tag{L.18}$$

We obtain further

$$K_{ij} = I_{ij} - \int \psi_i^*(i)\psi_j^*(j) A_{ij} \hat{\psi}_i(i)\hat{\psi}_j(j)\, dq_{12}$$

$$\times \int \psi_1^* \hat{\psi}_1\, dq_1 \cdots (ij) \cdots \int \psi_n^* \hat{\psi}_n\, dq_n, \tag{L.19}$$

where the (ij) indicates that these indices are missing from the product. Taking into account Eq. (L.18), we obtain

$$\int \psi_k^* \hat{\psi}_k\, dq = \int \psi_k^*(1 - \Omega)\psi_k\, dq = 1 - \int \psi_k^* \Omega \psi_k\, dq = 1 - \langle \Omega \rangle_k, \tag{L.20}$$

where $\langle \Omega \rangle_k$ is the expectation value of the one-electron projection operator Ω with respect to the pseudoorbital ψ_k. Using Eq. (L.20), we get

$$K_{ij} = I_{ij} - \langle \psi_i \psi_j | A_{ij} | \hat{\psi}_i \hat{\psi}_j \rangle B(ij), \tag{L.21}$$

where

$$B(ij) = (1 - \langle \Omega \rangle_1) \cdots (ij) \cdots (1 - \langle \Omega \rangle_n). \tag{L.22}$$

Now, let us consider the integral in Eq. (L.21):

$$J_{ij} \equiv \langle \psi_i \psi_j | A_{ij} | \hat{\psi}_i \hat{\psi}_j \rangle. \tag{L.23}$$

Here we have the expression

$$\hat{\psi}_i(1)\hat{\psi}_j(2) = \Pi_1 \Pi_2 \psi_i(1)\psi_j(2) = (1 - \Omega_1)(1 - \Omega_2)\psi_i(1)\psi_j(2)$$

$$= (1 - \Omega_1 - \Omega_2 + \Omega_1 \Omega_2)\psi_i(1)\psi_j(2). \tag{L.24}$$

By putting this back into Eq. (L.23), we obtain

$$J_{ij} = \langle \psi_i \psi_j | A_{ij} | \psi_i \psi_j \rangle - \langle \psi_i \psi_j | A_{ij} | (\Omega_1 + \Omega_2)\psi_i \psi_j \rangle$$

$$+ \langle \psi_i \psi_j | A_{ij} | \Omega_1 \Omega_2 \psi_i \psi_j \rangle. \tag{L.25}$$

The first term is again I_{ij}. Considering the second and the third expressions, we obtain

$$-\langle\psi_i\psi_j|A_{ij}|(\Omega_1+\Omega_2)\psi_i\psi_j\rangle + \langle\psi_i\psi_j|A_{ij}|\Omega_1\Omega_2\psi_i\psi_j\rangle$$

$$= -\sum_{s=1}^{N}\langle\varphi_s|\psi_i\rangle\langle\psi_i\psi_j|A_{ij}|\varphi_s\psi_j\rangle$$

$$-\sum_{s=1}^{N}\langle\varphi_s|\psi_j\rangle\langle\psi_i\psi_j|A_{ij}|\psi_i\varphi_s\rangle$$

$$+\sum_{s=1}^{N}\sum_{t=1}^{N}\langle\varphi_s|\psi_i\rangle\langle\varphi_t|\psi_j\rangle\langle\psi_i\psi_j|A_{ij}|\varphi_s\varphi_t\rangle. \qquad (L.26)$$

Inserting this into Eq. (L.25), we get for K_{ij}

$$K_{ij} = I_{ij} - J_{ij}B(ij)$$

$$= I_{ij} - B(ij)\left\{ I_{ij} - \sum_{s=1}^{N}\langle\varphi_s|\psi_i\rangle\langle\psi_i\psi_j|A_{ij}|\varphi_s\psi_j\rangle \right.$$

$$-\sum_{s=1}^{N}\langle\varphi_s|\psi_j\rangle\langle\psi_i\psi_j|A_{ij}|\psi_i\varphi_s\rangle$$

$$\left. +\sum_{s=1}^{N}\sum_{t=1}^{N}\langle\varphi_s|\psi_i\rangle\langle\varphi_t|\psi_j\rangle\langle\psi_i\psi_j|A_{ij}|\varphi_s\varphi_t\rangle \right\}. \qquad (L.27)$$

This expression can be simplified further by introducing the following approximation. Consider

$$\langle\psi_i\psi_j|A_{ij}|\varphi_s\psi_j\rangle = \int\psi_i^*(1)\psi_j^*(2)\,A_{ij}(1,2)\,\varphi_s(1)\psi_j(2)\,dq_1\,dq_2. \qquad (L.28)$$

In this double integral, we can first integrate over q_2 and then the integration over q_1 will be essentially only over the core area because of the presence of $\varphi_s(1)$, which is a core orbital. Thus, we can approximate this integral by replacing φ_s with $|\psi_i\rangle\langle\psi_i|\varphi_s\rangle$, which is the projection of φ_s in the direction of ψ_i, that is, by making the substitution

$$|\varphi_s\rangle \rightarrow |\psi_i\rangle\langle\psi_i|\varphi_s\rangle. \qquad (L.29)$$

Replacing φ_s by this expression means that the integration over q_1 will now be extended over all space, but the whole integral will be reduced by the factor $\langle\psi_i|\varphi_s\rangle$, which is that fraction of ψ_i that is inside of the core. Carrying out this approximation in the last three terms of Eq. (L.27), we get

$$
\begin{aligned}
K_{ij} = I_{ij} - B(ij) \bigg\{ I_{ij} &- \sum_{s=1}^{N} \langle \psi_i | \varphi_s \rangle \langle \varphi_s | \psi_i \rangle \langle \psi_i \psi_j | A_{ij} | \psi_i \psi_j \rangle \\
&- \sum_{s=1}^{N} \langle \psi_j | \varphi_s \rangle \langle \varphi_s | \psi_j \rangle \langle \psi_i \psi_j | A_{ij} | \psi_i \psi_j \rangle \\
&+ \sum_{s,t=1}^{N} \langle \psi_i | \varphi_s \rangle \langle \varphi_s | \psi_i \rangle \langle \psi_j | \varphi_t \rangle \langle \varphi_t | \psi_j \rangle \langle \psi_i \psi_j | A_{ij} | \psi_i \psi_j \rangle \bigg\} \\
= I(ij) \big\{ 1 &- B(ij) \big[(1 - \langle \Omega \rangle_i)(1 - \langle \Omega \rangle_j) \big] \big\} \\
= I(ij) \big\{ 1 &- B \big\},
\end{aligned}
\tag{L.30}
$$

where now

$$
B = (1 - \langle \Omega \rangle_1)(1 - \langle \Omega \rangle_2) \cdots (1 - \langle \Omega \rangle_n). \tag{L.31}
$$

Let us consider now the diagonal terms in Eq. (L.9). We need the expressions

$$
I_i = \int \Psi^* U_i(i) \dot{\Psi} \, dq, \tag{L.32}
$$

and

$$
K_i = \int \Psi^* U_i(i) \Omega \Psi \, dq. \tag{L.33}
$$

Using Eq. (L.13), we get

$$
I_i = \int \psi_i^*(1) U_i(1) \psi_i(1) \, dq. \tag{L.34}
$$

Also,

$$
\begin{aligned}
K_i &= \int \Psi^* U_i(i) [\Psi - \Pi \Psi] \, dq \\
&= I_i - B(i) \int \psi_i^*(1) U_i(1) \hat{\psi}_i(1) \, dq,
\end{aligned}
\tag{L.35}
$$

where

$$
B(i) = (1 - \langle \Omega \rangle_1)(1 - \langle \Omega \rangle_2) \cdots (i) \cdots (1 - \langle \Omega \rangle_n). \tag{L.36}
$$

Now, we carry out the approximation in Eq. (L.35) that we have used in Eq. (L.27). The sequence is as follows (the symbol $\approx$ indicates the step where the approximation is introduced):

$$\int \psi_i^*(1)U_i(1)\hat{\psi}_i(1)\,dq = \int \psi_i^*(1)U_i(1)\psi_i(1)\,dq$$

$$- \int \psi_i^*(1)U_i(1)\Omega_1\psi_i(1)\,dq$$

$$= I_i - \sum_{s=1}^{N} \langle \varphi_s|\psi_i \rangle \int \psi_i^*(1)U_i(1)\varphi_s(1)\,dq$$

$$\approx I_i - \sum_{s=1}^{N} \langle \varphi_s|\psi_i \rangle\langle \psi_i|\varphi_s \rangle \int \psi_i^*(1)U_i(1)\psi_i(1)\,dq$$

$$= I_i\{1 - \langle \Omega \rangle_i\}, \tag{L.37}$$

and putting this expression into Eq. (L.35), we obtain

$$K_i = I_i\{1 - B\}, \tag{L.38}$$

where B is given by Eq. (L.31).

Using Eq. (L.9) we obtain

$$\langle S \rangle = \sum_{\substack{\text{(all pairs, } i \neq j)}} I_{ij} - \sum_{i=1}^{n} I_i, \tag{L.39}$$

and

$$\langle S\Omega \rangle = \sum_{\substack{\text{(all pairs, } i \neq j)}} K_{ij} - \sum_{i=1}^{n} K_i. \tag{L.40}$$

Taking into account Eqs. (L.30) and (L.38), we get

$$\langle S\Omega \rangle = (1 - B)\left\{ \sum_{\text{(all pairs)}} I_{ij} - \sum_{i=1}^{n} I_i \right\}, \tag{L.41}$$

which, together with Eq. (L.39), gives

$$\eta = \frac{\langle S\Omega \rangle}{\langle S \rangle} = 1 - B$$

$$= 1 - [(1 - \langle \Omega \rangle_1)(1 - \langle \Omega \rangle_2) \cdots (1 - \langle \Omega \rangle_n)], \tag{L.42}$$

where we have taken into account Eq. (L.31).

Next, we discuss the omission of exchange and correlation effects in the derivation of Eq. (L.42). In order to take into account these effects, we would have to replace the product in Eq. (L.13) by a fully antisymmetric correlated n-electron function. Thus, in our formulas, we would have to make the substitution

$$\psi_1(1)\psi_2(2) \cdots \psi_n(n) \rightarrow \Psi(1,2,\ldots,n). \qquad (L.43)$$

The derivation of Eq. (L.42) rested on the approximation

$$\varphi_s \rightarrow \langle \psi_i | \varphi_s \rangle \psi_i,$$

which we have used in all integrals and for all core orbitals φ_s. Using this approximation, we obtained the formulas

$$\int \psi_1^*(1) \cdots \psi_n^*(n) A_{ij}\Omega(\psi_1(1) \cdots \psi_n(n)) \, dq$$

$$= \int \psi_1^*(1) \cdots \psi_n^*(n) A_{ij}\psi_1(1) \cdots \psi_n(n) \, dq$$

$$\times \{1 - [(1 - \langle\Omega\rangle_1)(1 - \langle\Omega\rangle_2) \cdots (1 - \langle\Omega\rangle_n)]\}, \quad (L.44)$$

and

$$\int \psi_1^*(1) \cdots \psi_n^*(n) U_i\Omega(\psi_1(1) \cdots \psi_n(n)) \, dq$$

$$= \int \psi_1^*(1) \cdots \psi_n^*(n) U_i\psi_1(1) \cdots \psi_n(n) \, dq$$

$$\times \{1 - [(1 - \langle\Omega\rangle_1)(1 - \langle\Omega\rangle_2) \cdots (1 - \langle\Omega\rangle_n)]\}. \quad (L.45)$$

Let us consider the physical meaning of Eq. (L.45). This approximate relationship means that instead of multiplying Ψ by the operator Ω, which projects Ψ onto the core functions and thereby effectively eliminates the valence part of the integral, we can take the integral itself and reduce it by the factor shown in braces. The factor in braces is nothing else but

$$\langle\Psi|\Omega|\Psi\rangle = 1 - \langle\Psi|\Pi|\Psi\rangle, \qquad (L.46)$$

where Ψ is given by Eq. (L.13). Therefore, it is very plausible that after making the substitution of Eq. (L.43), we would obtain, instead of Eq. (L.45), the relationship

$$\int \Psi^* U_i\Omega\Psi \, dq = \int \Psi^* U_i\Psi \, dq \times \langle\Psi|\Omega|\Psi\rangle, \qquad (L.47)$$

where Ψ is now the exact n-electron function. Similarly, we would obtain, instead of Eq. (L.44), the relationship

$$\langle \Psi | A_{ij} \Omega | \Psi \rangle = \langle \Psi | A_{ij} | \Psi \rangle \langle \Psi | \Omega | \Psi \rangle. \tag{L.48}$$

Inspecting the derivation that led to Eq. (L.42), we recognize that if, instead of starting with the product of Eq. (L.13), we would start with the exact antisymmetric correlated wave function Ψ, then, by going through the same steps as before, we would arrive at the result

$$\eta = \langle \Psi | \Omega | \Psi \rangle = 1 - \langle \Psi | \Pi | \Psi \rangle, \tag{L.49}$$

where Ψ is now the exact n-electron function. We obtain Eq. (L.42) from Eq. (L.49) by substituting the product of Eq. (L.13) for the exact Ψ. Thus, we see that Eq. (L.42) is a reasonable approximation even though the exchange and correlation effects were not taken into account.

The relationship of Eq. (L.42) expresses the connection between the operators $S\Omega$ and S. Taking a look at Eq. (L.2), however, we see that expressing $\langle S\Omega \rangle$ in terms of $\langle S \rangle$ is not convenient because the quantity that we have in Eq. (L.2) is not S but Q. Fortunately, it is easy to show that $\langle S\Omega \rangle$ can be expressed in terms of $\langle Q \rangle$, thereby establishing a connection between the operators $(S\Omega)$ and Q.

Inspecting the derivation that led to Eq. (L.42), we realize that by using the approximations embodied in Eqs. (L.47) and (L.48), we would obtain the same result for $\langle Q\Omega \rangle / \langle Q \rangle$ and for $\langle W\Omega \rangle / \langle W \rangle$ as for $\langle S\Omega \rangle / \langle S \rangle$. Thus, we can write

$$\langle Q\Omega \rangle / \langle Q \rangle = \eta, \tag{L.50}$$

and, similarly,

$$\langle W\Omega \rangle / \langle W \rangle = \eta. \tag{L.51}$$

From the last equations, it follows that

$$\langle W\Omega \rangle = \langle W \rangle \langle Q\Omega \rangle / \langle Q \rangle. \tag{L.52}$$

Using this result, we obtain, by taking into account the definition of S,

$$\begin{aligned}
\langle S\Omega \rangle &= \langle Q\Omega \rangle - \langle W\Omega \rangle \\
&= \langle Q\Omega \rangle - \langle W \rangle \frac{\langle Q\Omega \rangle}{\langle Q \rangle} \\
&= \langle Q\Omega \rangle \left(1 - \frac{\langle W \rangle}{\langle Q \rangle} \right).
\end{aligned} \tag{L.53}$$

Dividing the equation by $\langle Q \rangle$, we get

$$\frac{\langle S\Omega \rangle}{\langle Q \rangle} = \frac{\langle Q\Omega \rangle}{\langle Q \rangle}\left(1 - \frac{\langle W \rangle}{\langle Q \rangle}\right)$$

$$= \eta\left(1 - \frac{\langle W \rangle}{\langle Q \rangle}\right), \tag{L.54}$$

where we have taken into account Eq. (L.50).

In order to get a simple formula for $\langle S\Omega \rangle / \langle Q \rangle$, we have to estimate $\langle W \rangle / \langle Q \rangle$. Let us consider first $\langle Q \rangle$. Using Eq. (L.13) for Ψ, we obtain

$$\langle Q \rangle = \int \Psi^* Q \Psi \, dq$$

$$= \sum_{\substack{\text{(all pairs)}}} \langle \psi_i \psi_j | \frac{1}{r_{12}} | \psi_i \psi_j \rangle$$

$$= \frac{1}{2} \sum_{\substack{i,j=1 \\ (i \neq j)}}^{n} \langle \psi_i | V_j | \psi_i \rangle, \tag{L.55}$$

where we have used the notation

$$V_j(1) = \int \frac{\left|\psi_j(2)\right|^2 dq_2}{r_{12}}. \tag{L.56}$$

Similarly, we obtain

$$\langle W \rangle = \sum_{i=1}^{n} \sum_{j=1}^{n} \langle \psi_i | U_j | \psi_i \rangle. \tag{L.57}$$

We now make the assumption that it is a meaningful approximation to put

$$U_j \approx V_j, \tag{L.58}$$

that is, we can replace the HF potential formed with the HF orbital φ_j by the functionally identical HF potential formed with the pseudoorbital ψ_j. More precisely, the two potentials would be functionally identical if V_j would contain the exchange potential like U_j. This would be the case if, instead of using Eq. (L.13), we would have used an antisymmetrized product. Assuming that Eq. (L.58) is valid, we obtain

$$U_i \psi_i = V_i \psi_i = 0, \tag{L.59}$$

because the HF potential annihilates its own orbital. Thus, the diagonal terms drop out from Eq. (L.57) and we obtain, by comparing that equation with Eq. (L.55),

$$\langle W \rangle / \langle Q \rangle \approx 2. \tag{L.60}$$

The reader can easily verify that, by using a Slater determinant for the wave function Ψ, formed from the valence-electron HF orbitals $\varphi_1 \cdots \varphi_n$, one obtains the result of Eq. (L.60) exactly.

Returning now to Eq. (L.54), we obtain, by using Eq. (L.60),

$$\langle S\Omega \rangle / \langle Q \rangle = \eta(1 - 2) = -\eta. \tag{L.61}$$

From this, we obtain that

$$\langle Q \rangle - \langle S\Omega \rangle = (1 + \eta)\langle Q \rangle. \tag{L.62}$$

Thus, the combination occurring in Eq. (L.2) can be written in the form

$$Q - S\Omega = \hat{\eta}Q, \tag{L.63}$$

where

$$\hat{\eta} = 1 + \eta, \tag{L.64}$$

and η is given by Eq. (L.42). This completes our derivation; the relationship quoted in Eq. (8.109) is given by Eq. (L.63), and Eq. (8.114) follows from Eq. (L.50). In the derivation, we have assumed everywhere that the relationships between the expectation values can be transferred to the operators themselves.

The last detail to be cleared up is the question of complete versus incomplete groups in the valence shell. It is easy to see, by inspecting the derivation, that the results would be the same if we had used the general expression given by Eq. (L.5) for the valence-electron potential instead of the expression given by Eq. (L.6), which is valid only for complete groups. In fact, an interesting feature of this derivation is that the results are not sensitive to the exact form of the potentials in operator S; they are certainly the same for complete and incomplete groups of valence electrons.

APPENDIX M

THE FORMULA FOR THE SPIN–ORBIT INTERACTION IN EFFECTIVE HAMILTONIAN THEORY

In this appendix, we derive Eq. (9.55). Let us consider an atom with N electrons in complete groups and n valence electrons. In the effective Hamiltonian theory, the wave function used for this situation is given by Eq. (8.54):

$$\Psi = \left[(N + n)!\right]^{-1/2} \tilde{A}\{\Phi(1, 2, \ldots, n)$$

$$\times \det[\varphi_1(n + 1)\varphi_2(n + 2) \cdots \varphi_N(n + N)]\}.$$

$$(M.1)$$

Instead of using this, we will consider the somewhat simpler wave function

$$\Psi = \left\{(N!)^{-1/2} \det[\varphi_1(1) \cdots \varphi_N(N)]\Phi(N + 1, \ldots, N + n)\right\}. \quad (M.2)$$

The main difference between the two functions is that in Eq. (M.1), we have the partial antisymmetrizer operator $\tilde{A}$, which makes the whole function antisymmetric. The function in Eq. (M.2) does not have this operator, therefore, this function is not antisymmetrized between the core and valence wave functions. Because Φ is antisymmetric and the determinant is antisymmetric by definition, the wave function is antisymmetric between the core and valence electrons themselves. We assume that Φ is strong-orthogonal to the core orbitals and is normalized.

We want to compute the expectation value of the operator in Eq. (9.54), that is, we want an expression for

$$I_{so} = \langle \Psi | V_{so} | \Psi \rangle / \langle \Psi | \Psi \rangle, \tag{M.3}$$

where Ψ is given by Eq. (M.2) and

$$V_{so} = -\frac{\alpha^2}{2} \sum_{i=N+1}^{N+n} \sum_{j=1}^{N} \frac{1}{r_{ij}^3} (r_{ij} \times p_i) \cdot s_i. \tag{M.4}$$

We obtain first

$$\langle \Psi | V_{so} | \Psi \rangle$$

$$= \frac{1}{N!} \int \det[\varphi_1^*(1) \cdots \varphi_N^*(N)] \Phi^*(N+1, \ldots, N+n)$$

$$\times \left(-\frac{\alpha^2}{2} \right) \sum_{i=N+1}^{N+n} \sum_{j=1}^{N} \frac{1}{r_{ij}^3} (r_{ij} \times p_i) \cdot s_i$$

$$\times \det[\varphi_1(1) \cdots \varphi_N(N)] \Phi(N+1, \ldots, N+n) \, dq, \tag{M.5}$$

where dq is the integration over all coordinates.

Let i be one of the valence-electron coordinates, $i = N+1, \ldots, (N+n)$, and let us consider

$$I_i = -\frac{\alpha^2}{2} \frac{1}{N!} \int \det[\varphi_1^*(1) \cdots \varphi_N^*(N)] \Phi^*(N+1, \ldots, N+n)$$

$$\times \sum_{j=1}^{N} \frac{1}{r_{ij}^3} (r_{ij} \times p_i) \cdot s_i$$

$$\times \det[\varphi_1(1) \cdots \varphi_N(N)] \Phi(N+1, \ldots, N+n) \, dq_C, \tag{M.6}$$

where dq_C means integration over the core coordinates.

I_i can be written as

$$I_i = -\frac{\alpha^2}{2} \frac{1}{N!} \Phi^*(N+1, \ldots, N+n)$$

$$\times \int \det[\varphi_1^*(1) \cdots \varphi_N^*(N)] \sum_{j=1}^{N} \left\{ \frac{1}{r_{ij}^3} (r_{ij} \times p_i) \cdot s_i \right\}$$

$$\times \det[\varphi_1(1) \cdots \varphi_1(N)] \, dq_C \, \Phi(N+1, \ldots, N+n). \tag{M.7}$$

Next, we introduce the transformation

$$\frac{\mathbf{r}_{ij}}{r_{ij}^3} = -\nabla_i\left(\frac{1}{r_{ij}}\right). \tag{M.8}$$

The quantity in braces in Eq. (M.7) depends only on one core coordinate, q_j. Thus, the summation over j can be written in the form

$$\sum_{j=1}^{N}\left\{\frac{1}{r_{ij}^3}(\mathbf{r}_{ij}\times\mathbf{p}_i)\cdot\mathbf{s}_i\right\} = \sum_{j=1}^{N}\left\{\left[-\nabla_i\left(\frac{1}{r_{ij}}\right)\times\mathbf{p}_i\right]\cdot\mathbf{s}_i\right\} = \sum_{j=1}^{N} f_j, \tag{M.9}$$

where we used Eq. (M.8). In Eq. (M.7), the sum of Eq. (M.9) is averaged with respect to the determinantal core functions. According to Eq. (A.27), we obtain

$$\frac{1}{N!}\int \det[\varphi_1^*(1)\cdots\varphi_N^*(N)]\sum_{j=1}^{N}f_j\det[\varphi_1(1)\cdots\varphi_N(N)]\,dq_C$$

$$= \sum_{k=1}^{N}\langle\varphi_k|f|\varphi_k\rangle$$

$$= \sum_{k=1}^{N}\int\varphi_k^*(2)\left[-\nabla_i\left(\frac{1}{r_{i2}}\right)\times\mathbf{p}_i\right]\cdot\mathbf{s}_i\varphi_k(2)\,dq_2$$

$$= [-\nabla_i U_C(i)\times\mathbf{p}_i]\cdot\mathbf{s}_i, \tag{M.10}$$

where U_C is the electrostatic potential of the core:

$$U_C(1) = \sum_{k=1}^{N}\int\frac{1}{r_{12}}\varphi_k^*(2)\varphi_k(2)\,dq_2. \tag{M.11}$$

By putting Eq. (M.10) into Eq. (M.7), we obtain

$$I_i = -\frac{\alpha^2}{2}\Phi(N+1,\ldots,N+n)$$
$$\times[-\nabla_i U_C(i)\times\mathbf{p}_i]\cdot\mathbf{s}_i\Phi(N+1,\ldots,N+n). \tag{M.12}$$

Using this expression, we get

$$\langle\Psi|V_{so}|\Psi\rangle$$
$$= \sum_{i=N+1}^{N+n}\left(-\frac{\alpha^2}{2}\right)\int\Phi^*(N+1,\ldots,N+n)$$
$$\times[-\nabla_i U_C(i)\times\mathbf{p}_i]\cdot\mathbf{s}_i\Phi(N+1,\ldots,N+n)\,dq_v, \tag{M.13}$$

where dq_v is integration over the valence coordinates. This is the expression we are seeking, except that we must divide by the normalization constant in order to get I_{so} in Eq. (M.3). We have

$$\langle \Psi | \Psi \rangle = \frac{1}{N!} \int \det[\varphi_1^*(1) \cdots \varphi_N^*(N)]\Phi(N+1,\ldots,N+n)$$

$$\times \det[\varphi_1(1) \cdots \varphi_1(N)]\Phi(N+1,\ldots,N+n)\, dq, \quad (M.14)$$

where dq is the integration over all coordinates. Using the results of App. A, we obtain

$$\frac{1}{N!}\int |\det[\varphi_1(1) \cdots \varphi_1(N)]|^2 \, dq_C = 1, \quad (M.15)$$

and using this, we get

$$\langle \Psi | \Psi \rangle = \int |\Phi(1,2,\ldots,n)|^2 \, dq_v, \quad (M.16)$$

where dq_v again means integration with respect to the valence coordinates. If Φ is normalized, then

$$\langle \Phi | \Phi \rangle = 1. \quad (M.17)$$

Using this equation and Eq. (M.13), we obtain for the expectation value of the operator

$$I_{so} = \frac{\langle \Psi | V_{so} | \Psi \rangle}{\langle \Psi | \Psi \rangle}$$

$$= \int \Phi^*(1,2,\ldots,n)\left\{ \frac{\alpha^2}{2} \sum_{i=1}^{n} [\nabla_i U_C(i) \times p_i] \cdot s_i \right.$$

$$\left. \times \Phi(1,2,\ldots,n) \right\} dq_v. \quad (M.18)$$

In the main text, we stated that V_{so} can be combined with the first term of Eq. (9.53) to give Eq. (9.55). The first term of Eq. (9.53) reads

$$\frac{\alpha^2}{2} \sum_{\substack{i=1 \\ (\text{valence})}}^{n} \frac{Z}{r_i^3}(r_i \times p_i) \cdot s_i. \quad (M.19)$$

Using the relationship

$$\frac{Z}{r_i^3} \boldsymbol{r}_i = \nabla_i \left(-\frac{Z}{r_i} \right), \tag{M.20}$$

we obtain for Eq. (M.19)

$$\frac{\alpha^2}{2} \sum_{\substack{i=1 \\ \text{(valence)}}}^{n} \left[\nabla_i \left(-\frac{Z}{r_i} \right) \times \boldsymbol{p}_i \right] \cdot \boldsymbol{s}_i, \tag{M.21}$$

and adding this expression to the operator in Eq. (M.18), we get

$$\frac{\alpha^2}{2} \sum_{i=1}^{n} \left[\nabla_i \left(-\frac{Z}{r_i} + U_C(i) \right) \times \boldsymbol{p}_i \right] \cdot \boldsymbol{s}_i, \tag{M.22}$$

which is the expression in Eq. (9.55) that we wanted to derive.

From the derivation, we see that this result is an approximation in the sense that the core–valence exchange is neglected in the derivation of Eq. (M.18). We see this clearly because the wave function of Eq. (M.2) is not antisymmetric with respect to the interchange of core and valence coordinates. If this deficiency is removed, then we get an additional term in Eq. (M.22). We do not derive that term here, but discuss its physical meaning in the main text.

REFERENCES

A. BOOK REFERENCES[†]

Hartree	D. R. Hartree, *The Calculation of Atomic Structures*, John Wiley, New York (1957).
Slater I	J. C. Slater, *Quantum Theory of Atomic Structure*, Vol. 1, McGraw-Hill, New York (1960).
Slater II	J. C. Slater, *Quantum Theory of Atomic Structure*, Vol. 2, McGraw-Hill, New York (1960).
Fraga, Karkowski, Saxena (*FKS*)	S. Fraga, J. Karkowski, K. M. Saxena, *Handbook of Atomic Data*, Elsevier, Amsterdam (1976).
Atomic Energy Level Tables AET I, II, III, IV	I, II, III: C. E. Moore, *Atomic Energy Levels*, National Bureau of Standards, No. 467 (1949/1970). IV: W. C. Martin, R. Zalubas, L. Hagan, *Atomic Energy Levels*: *The Rare Earth Elements*, National Bureau of Standards, Washington, D.C. (1978).
Szasz	L. Szasz, *Pseudopotential Theory of Atoms and Molecules*, Wiley-Interscience, New York (1985).

[†]Cited by the name(s) of the author(s) or by abbreviation of title.

B. REFERENCES†

1. D. R. Hartree, *Proc. Camb. Phil. Soc.* **24**, 89, 111 (1928).
2. D. R. Hartree, W. Hartree, *Proc. Roy. Soc.* **A149**, 210 (1935).
3. V. Fock, *Z. Physik* **61**, 126 (1930); **62**, 795 (1930).
4. M. Delbrück, *Proc. Roy. Soc.* **A129**, 686 (1930).
5. W. Fock, *Z. Physik* **81**, 195 (1933).
6. J. C. Slater, *Phys. Rev.* **81**, 385 (1951); **91**, 528 (1953).
7. E. Wigner and F. Seitz, *Phys. Rev.* **43**, 84 (1933); **46**, 509 (1934).
8. L. C. Allen, *Phys. Rev.* **118**, 167 (1960).
9. H. D. Cohen and C. C. J. Roothaan, *J. Chem. Phys.* **43**, 34 (1965).
10. T. A. Koopmans, *Physica*, **1**, 104 (1933).
11. P. A. M. Dirac, *Proc. Camb. Phil. Soc.* **26**, 376 (1930).
12. J. C. Slater, *Phys. Rev.* **34**, 1293 (1929).
13. G. H. Shortley, *Phys. Rev.* **50**, 1072 (1936).
14. C. W. Nielson and G. F. Koster, *Spectroscopic Coefficients for the p^q, d^q and f^q Configurations*, the MIT Press, Cambridge, MA (1963).
15. C. G. Darwin, *Phil. Mag.* **39**, 537 (1920).
16. P. A. M. Dirac, *Proc. Roy. Soc.* **A117**, 610 (1928); **A118**, 351 (1928).
17. C. G. Darwin, *Proc. Roy. Soc.* **A118**, 654 (1928).
18. H. Bethe, *Handbuch der Physik*, Vol. 24/1, p. 304, Springer, Berlin (1933).
19. G. Breit, *Phys. Rev.* **34**, 553 (1929); **36**, 383 (1930); **39**, 616 (1932).
20. L. Schiff, *Quantum Mechanics*, McGraw-Hill, New York (1968). See p. 176.
21. H. Bethe and E. Salpeter, *Quantum Mechanics of One- and Two-Electron Atoms*, p. 170, Springer, Berlin (1957).
22. L. G. Malli, Ed., *Relativistic Effects in Atoms, Molecules and Solids*, Plenum Press, New York (1983). See article by J. Sucher, p. 1.
23. *Ibid*. See article by J. H. Detrich and C. C. J. Roothaan, p. 169.
24. H. Narumi and I. Shimamura, Eds., *Atomic Physics* 10, *Proceedings of the Tenth International Conference on Atomic Physics*, North-Holland, Amsterdam (1987). See article by J. Sapirstein, p. 77.
25. B. Swirles, *Proc. Roy. Soc.* **A152**, 625 (1935).
26. I. P. Grant, *Proc. Roy. Soc.* **A262**, 555 (1961); *Proc. Phil. Soc.* **86**, 526 (1965); *Advances in Physics* **19**, 747 (1970).
27. G. Racah, *Phys. Rev.* **62**, 438 (1942).
28. Y. K. Kim, *Phys. Rev.* **154**, 17 (1967).
29. P. A. M. Dirac, *The Principles of Quantum Mechanics*, p. 23, Oxford University Press, London (1958).
30. E. Wigner, *Phys. Rev.* **46**, 1002 (1934).
31. C. Froese-Fischer, *The Hartree–Fock Method for Atoms*, Wiley-Interscience, New York (1977).

†Cited by number.

32. See Ref. 22, article by J. P. Desclaux, p. 115.

33. J. P. Desclaux, *Atomic Data and Nuclear Data Tables*, **12**, 311 (1973).

34. P. Gombas, *The Statistical Theory of Atoms*, p. 96, Springer, Vienna (1949).

35. L. Szasz, I. Berrios-Pagan, and G. McGinn, *Z. Naturforschung*, **30a**, 1516 (1975).

36. C. E. Moore, *Ionization Potentials*, Reference Data Series, No. 34, National Bureau of Standards, Washington, D.C. (1970).

37. J. B. Mann and J. T. Waber, *Atomic Data* **5**, 201 (1973).

38. A. C. Larson and J. T. Waber, *Self-Consistent Field Hartree Calculations for Atoms and Ions*, Preprint LA-DC-8508, Los Alamos Scientific Laboratory, Los Alamos, NM (1967).

39. B. Fricke, W. Greiner, and J. T. Waber, *Theoret. Chim. Acta* (*Berlin*) **21**, 235 (1971).

40. J. Mali and M. Hussonois, *Theoret. Chim. Acta* (*Berlin*) **28**, 363 (1973).

41. C. Froese-Fischer, *Atomic Data* **4**, 301 (1972).

42. E. Clementi, *IBM J. Res. Dev.* **9**, 2 (1965).

43. K. M. S. Saxena, *J. Phys.* **B5**, 766 (1972).

44. V. M. Burke and I. P. Grant, *Proc. Roy. Soc.* **90**, 297 (1967).

45. L. R. Kahn, P. J. Hay, and R. D. Cowan, *J. Chem. Phys.* **68**, 2386 (1978).

46. J. C. Slater, *Phys. Rev.* **98**, 1039 (1955).

47. J. A. Beardeen, *Rev. Mod. Phys.* **39**, 78 (1967).

48. J. A. Beardeen and A. F. Burr, *Rev. Mod. Phys.* **39**, 125 (1967).

49. W. Lotz, *J. Optical Soc. Am.* **60**, 206 (1970).

50. V. Fock and M. J. Petrashen, *Phys. Z. Sowjetunion* **6**, 368 (1934); **8**, 359 (1935).

51. P. M. Morse, L. A. Young, and E. S. Haurwitz, *Phys. Rev.* **48**, 948 (1935).

52. J. C. Slater, *Phys. Rev.* **36**, 57 (1930).

53. L. Szasz, *J. Chem. Phys.* **73**, 5212 (1980).

54. C. C. J. Roothaan, *Rev. Mod. Phys.* **23**, 69 (1951); **32**, 179 (1960).

55. S. P. Goldman, *Phys. Rev.* **A37**, 16 (1988).

56. E. Clementi and C. Roetti, *Atomic Data and Nuclear Data Tables*, **14**, 177 (1974); *J. Chem. Phys.* **60**, 4725 (1975).

57. G. Racah, *Phys. Rev.* **61**, 186 (1942); **62**, 438 (1942); **63**, 367 (1943); **76**, 1352 (1949).

58. R. D. Cowan, *The Theory of Atomic Structure*, University of California Press, Los Angeles (1981).

59. M. Cohen and P. S. Kelly, *Can. J. Phys.* **43**, 1867 (1965); **44**, 3227 (1966); **45**, 1661 (1967).

60. A. Einstein, *Phys. Z.* **18**, 121 (1917).

61. W. Kuhn, *Z. Physik*, **33**, 408 (1925); W. Thomas, *Naturwiss.* **13**, 627 (1925).

62. E. Wigner, *Phys. Z.* **11**, 450 (1931).

63. P. A. M. Dirac, *Proc. Roy. Soc.* **A111**, 281 (1926).

64. P. Güttinger and W. Pauli, *Z. Physik*, **67**, 743 (1931).

65. D. H. Menzel and L. Goldberg, *Astrophys. J.* **84**, 1 (1936).

66. E. U. Condon and H. Odabasi, *Atomic Structure*, Cambridge University Press, London (1980).

67. H. Hellmann, *J. Chem. Phys.* **3**, 61 (1935); *Einfuerung in die Quantenchemie*, Fritz Deutike, Leipzig and Vienna (1937).

68. P. Szepfalusy, *Acta Phys. Hung.* **5**, 325 (1955).

69. J. C. Phillips and L. Kleinman, *Phys. Rev.* **116**, 287 (1959); **118**, 1153 (1960).

70. J. D. Weeks and S. Rice, *J. Chem. Phys.* **49**, 2741 (1968).

71. M. H. Cohen and V. Heine, *Phys. Rev.* **122**, 1821 (1961).

72. L. Szasz and G. McGinn, *J. Chem. Phys.* **47**, 3495 (1967).

73. G. McGinn, *J. Chem. Phys.* **51**, 5090 (1969); **52**, 3358 (1970).

74. J. Callaway, *Phys. Rev.* **106**, 868 (1957).

75. C. Bottcher and A. Dolgarno, *Proc. Roy. Soc.* **A340**, 187 (1974).

76. H. Bethe, Ref. 18, p. 339.

77. R. M. Sternheimer, *Phys. Rev.* **96**, 951 (1954); **107**, 1565 (1957); **115**, 1198 (1959).

78. J. N. Bardsley, *Chem. Phys. Lett.* **7**, 517 (1970).

79. A. Dolgarno, C. Bottcher, and G. A. Victor, *Chem. Phys. Lett.* **7**, 265 (1970).

80. W. Prokofjew, *Z. Physik* **58**, 255 (1929).

81. Y. Sugiura, *Phil. Mag.* **4**, 495 (1927); W. Thomas, *Z. Phys.* **24**, 169 (1924).

82. C. Bottcher, *J. Phys.* **B**, 1140 (1971).

83. V. Fock, M. Veselov, and M. Petrashen, *J. Exp. Theoret. Phys.* (*USSR*) **10**, 723 (1940).

84. L. Szasz, *Z. Naturforschung* **14a**, 1014 (1959).

85. L. Szasz and L. Brown, *J. Chem. Phys.* **63**, 4560 (1975); **65**, 1393 (1976).

86. L. R. Kahn, P. Baybutt, and D. G. Truhlar, *J. Chem. Phys.* **65**, 3826 (1976).

87. G. McGinn, *J. Chem. Phys.* **51**, 5090 (1969); **52**, 3358 (1970).

88. J. N. Bardsley, *Case Studies in Atomic Physics*, Vol. 4, No. 5, p. 299, North-Holland, Amsterdam (1974).

89. R. D. Cowan and D. C. Griffin, *J. Op. Soc. Am.* **66**, 1010 (1976).

90. P. J. Hay and W. R. Wadt, *J. Chem. Phys.* **82**, 270 (1985); **82**, 284 (1985); **82**, 299 (1985).

91. L. H. Thomas, *Nature* **117**, 514 (1926).

92. M. Blume and R. E. Watson, *Proc. Roy. Soc.* **A270**, 127 (1962).

93. L. R. Kahn, *Int. J. Quant. Chem.* **25**, 149 (1984).

94. Y. S. Lee, W. C. Ermler, and K. S. Pitzer, *J. Chem. Phys.* **67**, 5861 (1977).

95. Y. Ishikawa and G. Malli, *J. Chem. Phys.* **75**, 5423 (1981).

INDEX